Cockroaches

Cockroaches

ECOLOGY, BEHAVIOR, AND NATURAL HISTORY

William J. Bell
Louis M. Roth
Christine A. Nalepa

Foreword by
Edward O. Wilson

The Johns Hopkins University Press
Baltimore

Printed in the United States of America on acid-free paper

Johns Hopkins Paperback edition, 2016

9 8 7 6 5 4 3 2 1

The Johns Hopkins University Press
2715 North Charles Street
Baltimore, Maryland 21218-4363
www.press.jhu.edu

The Library of Congress has cataloged the hardcover edition of this book as follows:

Bell, William J.
Cockroaches : ecology, behavior, and natural history / William J. Bell, Louis M. Roth, Christine A. Nalepa ; foreword by Edward O. Wilson.
p. cm.
Includes bibliographical references and index.
ISBN-13: 978-0-8018-8616-4 (hardcover : alk. paper)
ISBN-10: 0-8018-8616-3 (hardcover : alk. paper)
1. Cockroaches. I. Roth, Louis M. (Louis Marcus), 1918– II. Nalepa, Christine A. III. Title.
QL505.5.B43 2007
595.7′28 — dc22 2006033232

A catalog record for this book is available from the British Library.

ISBN-13: 978-1-4214-2114-8
ISBN-10: 1-4214-2114-3

To the families, friends, and colleagues of
William J. Bell and Louis M. Roth

Contents

Foreword

Let the lowly cockroach crawl up, or, better, fly up, to its rightful place in human esteem! Most of us, even the entomologists in whose ranks I belong, have a stereotype of revolting little creatures that scatter from leftover food when you turn on the kitchen light and instantly disappear into inaccessible crevices. These particular cockroaches are a *problem*, and the only solution is *blatticide*, with spray, poison, or trap.

I developed a better understanding when I came to realize that the house pests and feces-consuming sewer dwellers are only the least pleasant tip of a great blattarian biodiversity. My aesthetic appreciation of these insects began during one of my first excursions to the Suriname rainforest, where I encountered a delicate cockroach perched on the leaf of a shrub in the sunshine, gazing at me with large uncockroach-like eyes. When I came too close, it fluttered away on gaily colored wings like a butterfly.

My general blattarian education was advanced when I traveled with Lou Roth to Costa Rica in 1959, and further over the decades we shared at Harvard's Museum of Comparative Zoology, as he worked as a taxonomist through the great evolutionary radiation of the blattarian world fauna.

This volume lays out, in detail suitable for specialists but also in language easily understood by naturalists, the amazing panorama of adaptations achieved by one important group of insects during hundreds of millions of years of evolution. Abundant in most terrestrial habitats of the world, cockroaches are among the principal detritivores (their role, for example, in our kitchens), but some species are plant eaters as well. The species vary enormously in size, anatomy, and behavior. They range in habitat preference from old-growth forests to deserts to caves. They form intricate symbioses with microorganisms. The full processes of their ecology, physiology, and other aspects of their biology have only begun to be explored. This book will provide a valuable framework for the research to come.

Edward O. Wilson

Preface

The study of roaches may lack the aesthetic values of bird-watching and the glamour of space flight, but nonetheless it would seem to be one of the more worthwhile of human activities.

—H.E. Evans, *Life on a Little Known Planet*

Most available literature on cockroaches deals with domestic pests and the half dozen or so other species that are easily and commonly kept in laboratories and museums. It reflects the extensive efforts undertaken to find chinks in the armor of problematic cockroaches, and the fact that certain species are ideal for physiological and behavioral investigations under controlled conditions. These studies have been summarized in some excellent books, including those by Guthrie and Tindall (1968), Cornwell (1968), Huber et al. (1990), Bell and Adiyodi (1982a), and Rust et al. (1995). The last two were devoted to single species, the American and the German cockroaches, respectively. As a result of this emphasis on Blattaria amenable to culture, cockroaches are often discussed as though they are a homogeneous grouping, typified by species such as *Periplaneta americana* and *Blattella germanica*. In reality the taxon is amazingly diverse. Cockroaches can resemble, among other things, beetles, wasps, flies, pillbugs, and limpets. Some are hairy, several snorkel, some chirp, many are devoted parents, and males of several species, surprisingly, light up.

The publication most responsible for alerting the scientific community to the diversity exhibited by the 99+% of cockroaches that have never set foot in a kitchen is *The Biotic Associations of Cockroaches,* by Louis M. Roth and Edwin R. Willis, published in 1960. Its encyclopedic treatment of cockroach ecology and natural history was an extraordinary achievement and is still, hands down, the best primary reference on the group in print. Now, nearly 50 years later, we feel that the subject matter is ripe for revisitation. The present volume was conceived as a grandchild of the Roth and Willis book, and relies heavily on the information contained in its progenitor. Our update, however, narrows the focus, includes recent studies, and when possible and appropriate, frames the information within an ecological and evolutionary context.

This book is intended primarily as a guided tour of non-domestic cockroach species, and we hope that it is an eye-opening experience for students and researchers in behavioral ecology and evolution. Even we were surprised at some recent findings, such as the estimate by Basset (2001) that cockroaches constitute approximately 24% of the arthropod biomass in tropical tree canopies worldwide, and hints from various studies suggesting that cockroaches may ecologically replace termites in some habitats (Chapter 10). We address previously unexplored aspects of their biology, such as the relationship with microbes that lies at the heart of their image as anathema to civilized households (Chapter 5). As our writing progressed, some chapters followed unpredicted paths, particularly evident in the one on mating strategies (Chapter 6). We became fascinated with drawings of male and female genitalia that are buried in the taxonomic literature and that suggest ongoing, internally waged battles to determine paternity of offspring. It is the accessibility of this kind of information that can have the most impact on students searching for a dissertation topic, and we cover it in detail at the expense of addressing more familiar aspects of cockroach mating biology. We planned the book so that each chapter can be mined for new ideas, new perspectives, and new directions for future work.

An interesting development since Roth and Willis (1960) was published is that the definition of a cockroach is somewhat less straightforward than it used to be. Cockroaches are popularly considered one of the oldest terrestrial arthropod groups, because insects with a body plan closely resembling that of extant Blattaria dominated the fossil record of the Carboniferous, "The Age of Cockroaches." The lineage that produced extant cockroaches, however, radiated sometime during the early to mid-Mesozoic (e.g., Labandeira, 1994; Vršanský, 1997; Grimaldi and Engel, 2005). Although the Carboniferous fossils probably include the group that gave rise to modern Blattaria, they also include basal forms of other taxa. Technically, then, they cannot be considered cockroaches, and the Paleozoic group has been dubbed "roachoids" (Grimaldi and Engel, 2005), among other things. Recent studies of extant species are also blurring our interpretation of what may be considered a cockroach. Best evidence currently supports the view that termites are nested within the cockroaches as a subgroup closely related to the cockroach genus *Cryptocercus*. We devote Chapter 9 to developing the argument that termites evolved as eusocial, juvenilized cockroaches.

Roth (2003c) recognized six families that place most cockroach species: Polyphagidae, Cryptocercidae, Nocti-

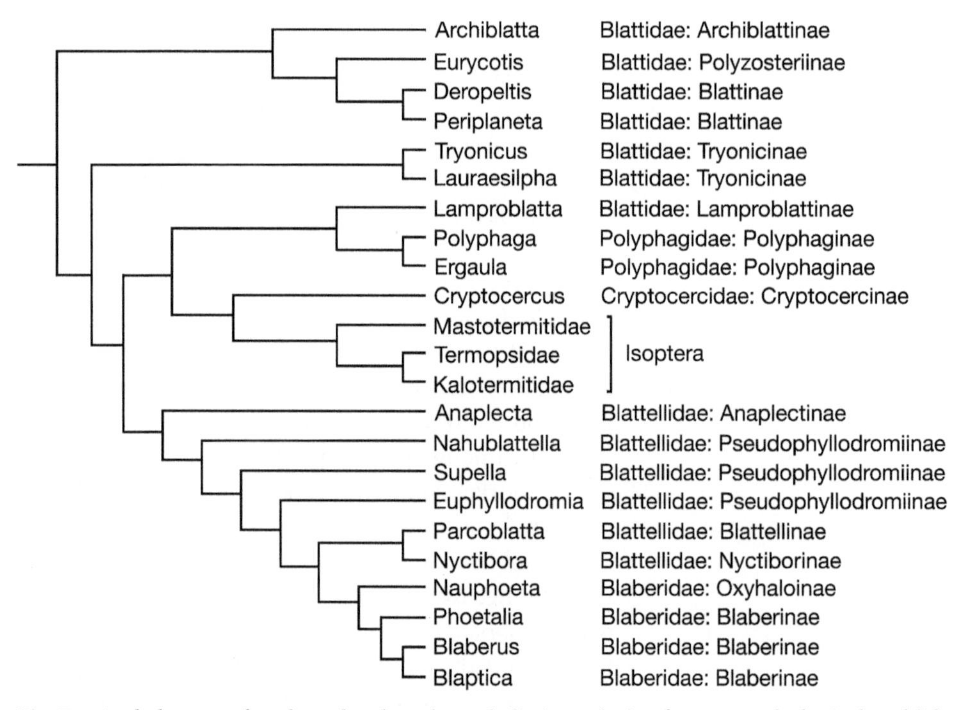

Fig. P.1 A phylogeny of cockroaches based on cladistic analysis of 175 morphological and life history characters; after Klass and Meier (2006), courtesy of Klaus Klass. Assignation of genera to subfamilies is after Roth (2003c) and differs somewhat from that of K & M, who place *Archiblatta* in the Blattinae and *Phoetalia* in the Epilamprinae. Pseudophyllodromiinae used here is Plecopterinae in K & M. Based on their results, K & M suggest that Lamproblattinae and Tryonicinae be elevated to family-level status. Mukha et al. (2002, Fig. 2) summarize additional hypotheses of higher-level relationships. Phylogenetic trees of Vršanský et al. (2002, Fig. 364) and Grimaldi and Engel (2005, Fig. 7.60) include fossil groups. Lo et al. (2000), Klass (2001, 2003), and Roth (2003c) discuss major issues.

colidae, Blattidae, Blattellidae, and Blaberidae; the majority of cockroaches fall into the latter three families. His paper was used as the basis for assigning the cockroach genera discussed in this book to superfamily, family, and subfamily, summarized in the Appendix. Despite recent morphological and molecular analyses, the relationships among cockroach lineages are still very much debated at many levels; Roth (2003c) summarizes current arguments. For general orientation, we offer a recent, strongly supported hypothesis by Klass and Meier (2006) (see fig. P.1). In it, there is a basal dichotomy between the family Blattidae and the remaining cockroaches, with the rest falling into two clades. The first consists of Cryptocercidae and the termites as sister groups, with these closely related to the Polyphagidae and to *Lamproblatta*. The other clade consists of the Blattellidae and Blaberidae, with the Anaplectinae as most basal and Blattellidae strongly paraphyletic with respect to Blaberidae. One consequence of the phylogenetic uncertainties that exist at so many taxonomic levels of the Blattaria is that mapping character states onto phylogenetic trees is in most cases premature. An analysis of the evolution of some wing characters in Panesthiinae (Blaberidae) based on the work of Maekawa et al. (2003) is offered in Chapter 2, a comparative phylogeny of cockroaches and their fat body endosymbionts (Lo et al., 2003a) is included in Chapter 5, and key symbiotic relationships are mapped onto a phylogenetic tree of major Dictyopteran groups in Chapter 9.

Since the inception of this book nearly 15 years ago, the world of entomology has lost two of its giants, William J. Bell and Louis M. Roth. It was an enormous responsibility to finish the work they initiated, and I missed their wise counsel in bringing it to completion. If just a fraction of their extraordinary knowledge of and affection for cockroaches shines through in the pages that follow, I will consider the book a success. This volume contains unpublished data, observations, and personal communications of both men, information that otherwise would have been lost to the scientific community at large. Bill Bell's observations of aquatic cockroaches are in Chapter 2, and his unpublished research on the diets of tropical species is summarized in Chapter 4. Lou Roth was the acknowledged world expert on all things cockroach, and was the "go to" man for anyone who needed a specimen identified or with a good cockroach story to share. The content of his conversations and personal observations color the text throughout the book. Bill's and Lou's notes and papers were kindly loaned to me by their colleagues at the University of Kansas and Harvard University, respectively. I found it revealing that on Lou's copy of a paper by Asahina (1960) entitled "Japanese cockroaches as household pest," the *s* in the last word was rather emphatically scratched out.

A large number of colleagues were exceedingly generous in offering their time and resources to this project, and without their help this volume never would have seen the light of day. For advice, information, encouragement, references, photographs, illustrations, permission to use material, or for supplying reprints or other written matter I am glad to thank Gary Alpert, Dave Alexander, David Alsop, L.N. Anisyutkin, Jimena Aracena, Kathie Atkinson, Calder Bell, David Bignell, Christian Bordereau, Michel Boulard, Michael Breed, John Breznak, Remy Brossut, Valerie Brown, Kevin Carpenter, Randy Cohen, Stefan Cover, J.A. Danoff-Burg, Mark Deyrup, R.M. Dobson, C. Durden, Betty Faber, Robert Full, César Gemeno, Fabian Haas, Johannes Hackstein, Bernard Hartman, Scott Hawkes, W.F. Humphreys, T. Itioka, Ursula Jander, Devon Jindrich, Susan Jones, Patrick Keeling, Larry Kipp, Phil Koehler, D. Kovach, Conrad Labandeira, Daniel Lebrun, S. Le Maitre, Tadao Matsumoto, Betty McMahan, John Moser, I. Nagamitsu, M.J. O'Donnell, George Poinar, Colette Rivault, Edna Roth, Douglas Rugg, Luciano Sacchi, Coby Schal, Doug Tallamy, Mike Turtellot, L. Vidlička, Robin Wootton, T. Yumoto, and Oliver Zompro.

I am particularly indebted to Horst Bohn, Donald Cochran, Jo Darlington, Thomas Eisner, Klaus Klass, Donald and June Mullins, Piotr Naskrecki, David Rentz, Harley Rose, and Ed Ross for their generosity in supplying multiple illustrations, and to George Byers, Jo Darlington, Lew Deitz, Jim Hunt, Klaus Klass, Nathan Lo, Kiyoto Maekawa, Donald Mullins, Patrick Rand, David Rentz, and Barbara Stay for reviewing sections or chapters of the book and for spirited and productive discussions. Anne Roth and the Interlibrary Loan and Document Delivery Services at NCSU were instrumental in obtaining obscure references. I thank Vince Burke and the Johns Hopkins University Press for their patience during the overlong gestation period of this book. I am sure that there are a great number of people whose kindness and contributions eased the workload on Bill Bell and Lou Roth during the early stages of this endeavor, and I thank you, whoever you are.

Christine A. Nalepa

Cockroaches

ONE Shape, Color, and Size

many a cockroach
believes himself as beautiful
as a butterfly
have a heart o have
a heart and
let them dream on

—archy, "archygrams"

The image that floats to consciousness at mention of the word *cockroach* is one based on experience. For most people, it is the insect encountered in the sink during a midnight foray into the kitchen, or the one that is pinned splay legged on a wax tray in entomology class. While these domestic pests and lab "rats" do possess a certain subtle beauty, they are rather pedestrian in appearance when compared to the exuberance of design and color that characterizes insects such as beetles and butterflies. Nonetheless, these dozen or so familiar cockroaches constitute a half percent or less of described species and can be rather poor ambassadors for the group as a whole. Our goal in this chapter, and indeed, the book, is not only to point out some rather extraordinary features of the cockroaches with which we are already acquainted but to expand the narrow image of the group. Here we address their outward appearance, the externally visible morphological features, and how their environment helps shape them.

GENERAL APPEARANCE AND ONTOGENY

The standard cockroach body is flattened and broadly oval, with a large, shield-like pronotum covering the head, ventrally deployed, chewing mouthparts, and long, highly segmented antennae. The forewings (tegmina) are typically leathery and the hindwings more delicate and hyaline. The coxae are flattened and modified to house the femur, so that when the legs are tucked in close to the body the combined thickness of the two segments is reduced. A comprehensive discussion of the morphological features of cockroaches, particularly those of importance in recognizing and describing species, is given in Roth (2003c).

Like other hemimetabolous insects, cockroach nymphs generally resemble adults except for the absence of tegmina and wings; these structures are, however, sometimes indicated by non-articulated, lobe-like extensions of the meso- and metanotum in later developmental stages. Early instars of both sexes have styles on the subgenital plate; these are usually lost in older female instars and are absent in adult females. Juveniles have undeveloped and poorly sclerotized genitalia and they often lack other characters useful in species identification. Nymphs of Australian soil-burrowing cockroaches, for example, are difficult to tell apart because the pronotal and tergal features that distinguish the various species are not fully developed (Walker et al., 1994). In some taxa, nymphal coloration and markings differ markedly from those of adults, making them scarcely recognizable as the same species (e.g., Australian *Polyzosteria* spp.—Tepper, 1893; Mackerras, 1965a). In general, the first few instars of a given species can be distinguished from each other on the basis of non-overlapping measurements of sclerotized morphological features such as head width or leg segments. In older stages, however, accumulated variation results in overlap of these measurements, making it difficult to determine the stage of a given nymph. This variation results from intermolt periods that differ greatly from individual to individual, not only in different stages, but also within a stage (Scharrer, 1946; Bodenstein, 1953; Takagi, 1978; Zervos, 1987). The difficulty in distinguishing different developmental stages within a species and the nymphs of different species from each other often makes young developmental stages intractable to study in the field. Consequently, the natural history of cockroach juveniles is virtually unknown.

Dimorphism

In addition to dimorphism in the presence of wings (Chapter 2) and overall body size (discussed below), male and female cockroaches may differ in the color and shape of the body or in the size, color, and shape of specific body parts. The general shape of the male, particularly the abdomen, is often more attenuated than that of the female. Several sex-specific morphological differences suggest that the demands of finding and winning a mate are highly influential in cockroach morphological evolution. Dimorphism is most pronounced in species where males are active, aerial insects, but the females have reduced wings or are apterous. These males may have large, bulging, nearly contiguous eyes while those of the more sedentary female are flattened and farther apart, for example, several species of *Laxta* and *Neolaxta* (Mackerras, 1968b; Roth, 1987a, 1992) and *Colapteroblatta compsa* (Roth, 1998a). Male morphology in the blattellid genera *Escala* and *Robshelfordia* is completely different from that of the opposite sex (Roth, 1991b). Such strong sexual dimorphism makes associating the sexes difficult, particularly when related species are sympatric (Roth, 1992); as a result, conspecific males and females are sometimes described as separate species. Additional sexual dimorphisms include the presence of tergal glands on males of many species, and the size and shape of the pronotum.

Asymmetry

Cockroaches tend to have an unusually high level of fluctuating asymmetry (Hanitsch, 1923), defined as small, random differences in bilateral characters. The cockroach tarsus is normally composed of five segments, but on one leg it may have just four. Spines on the femora also may vary in number between the right and left sides of the same individual. In both characters a reduction more often occurs on the left side of the body. Wing veins may be simple on one side and bifurcated on the other. This tendency often makes it difficult to interpret the fossil record, where so much of our information is based on wings. Asymmetries of this type are widely used as a measure of fitness because they result from developmental instability, the ability of an organism to withstand developmental perturbation. Of late, fluctuating asymmetry has become a major but controversial topic in evolutionary biology (e.g., Markow, 1995; Nosil, 2001), but is unstudied in the Blattaria. Less subtle bilateral asymmetries also occur in cockroaches; gynandromorphs are reported in *Periplaneta americana, Byrsotria fumigata* (Willis and Roth, 1959), *Blattella germanica* (Ross and Cochran, 1967), and *Gromphadorhina portentosa* (Graves et al., 1986).

Pronotum

The large, shield-shaped pronotum is a defining characteristic of cockroaches and its size, shape, curvature, and protuberances have systematic value in certain groups (e.g., Perisphaeriinae, Panesthiinae). Some cockroaches are more strongly hooded than others, that is, the head ranges from completely covered by the pronotum to almost entirely exposed. In some species the pronotum is flat, in others it has varying degrees of declivity. At its extreme it may form a cowl, shaped like an upside down *U* in section. The border of the pronotum may be recurved to varying degrees, forming a gutter around the sides, which sometimes continues into the cephalic margin. The majority of species of *Colapteroblatta,* for example, have the lateral wings of the pronotum deflexed and the edges may be ridged or swollen (Hebard, 1920 [1919]; Roth, 1998a, Fig. 1-6). In a few cases the pronotum can resemble the headpiece of certain orders of nuns (Fig. 1.1A). Some species of *Cyrtotria* have pronota perforated with large, semilunar pores in both sexes; these may be the openings of glands (Fig. 1.1B) (Shelford, 1908). The shape of the pronotum can vary within a species, with distinct forms correlated with varying degrees of wing re-

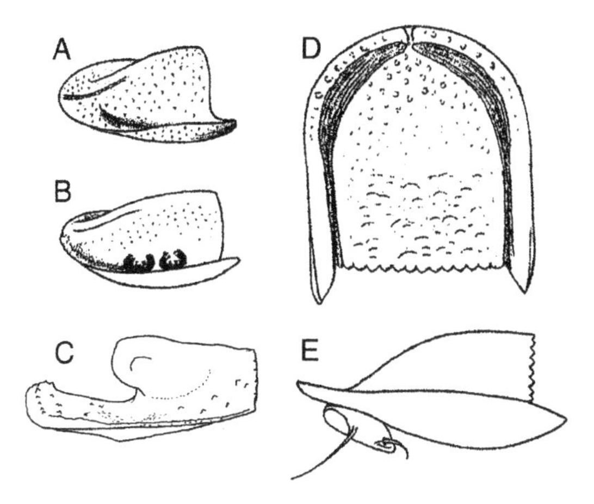

Fig. 1.1 Variations in pronotal morphology. (A) Female of *Cyrtotria marshalli*, three-quarter view. (B) Female of *Cyrtotria pallicornis*, three-quarter view; note large lateral pores. (C) Male of *Princisia vanwaerebeki*, lateral view. (D) Female of *Pilema mombasae*, dorsal view. (E) ditto, lateral view. After Shelford (1908) and Van Herrewege (1973). Not drawn to scale.

Fig. 1.2 Male *Microdina forceps* (Panesthiinae) from India. Photo by L.M. Roth.

duction (e.g., African *Ectobius*—Rehn, 1931). Both males and females of *Microdina forceps* have the anterior pronotal margins extended into a pair of curved spines, resembling the forceps of earwigs or the mandibles of staghorn beetles (Fig. 1.2) (Roth, 1979b). In females these are about 2 mm long, and in males they are slightly longer (2.5 mm). In *Bantua valida* the lateral margins of the pronotum in both sexes are curved upward, but only in the female are the caudad corners prolonged into "horns" (Kumar, 1975).

Functionally, the pronotum is a versatile tool that can serve as a shield, shovel, plug, wedge, crowbar, and battering ram. Those cockroaches described as "strongly hooded," with the head concealed under the extended anterior edge of the pronotum, are often burrowers. The large, flat pronotum of *Blaberus craniifer*, for example, serves as a wedge and protects the head when used in the oscillating digging motion described by Simpson et al. (1986). In museum specimens of *Pilema* spp. the channel between the pronotal disc and lateral bands is often chocked with dirt, leading Shelford (1908) to conclude that the pronotum (Fig. 1.1D,E) is used in digging the neat round holes in which these cockroaches are found. Adult *Cryptocercus* have been observed using the pronotum as a tool in two different contexts. When they are cleaning and maintaining their galleries, the insects use the pronotum as a shovel to move frass and feces from place to place and to tamp these materials against gallery walls (CAN, unpubl. obs.). During aggressive encounters the pronotum is used to block access to galleries and to push and butt intruders (Seelinger and Seelinger, 1983; Park and Choe, 2003b). In male *Nauphoeta cinerea*, combatants try to flip rivals onto their backs by engaging the edge of their pronotum under that of their opponents (Ewing, 1967). In species with strong sexual differences in pronotal morphology, dimorphism is likely related to sexual competition among males. In *Elliptorhina*, *Princisia*, and *Gromphadorhina*, males have heavy, well-developed knobs on their pronota and use them to battle rivals (Fig. 1.1C) (Van Herrewege, 1973; Beccaloni, 1989). When males charge, their knobbed pronotal shields come together with an audible sound (Barth, 1968c). In Geoscapheini (Blaberidae), males often have conspicuous pronotal tubercles that are absent in the female, and have the anterior edge thickened and prominently upturned (Walker et al., 1994); *Macropanesthia rhinoceros* is named for the blunt, horn-like processes projecting from the surface of the pronotum in males (Froggatt, 1906). Individuals of *M. rhinoceros* are most often observed above ground when they have "fallen on to their backs and are unable to right themselves" (Day, 1950). It is unknown if these are all males, and the result of nocturnal battles. The allometry of male combat weaponry has not been examined in cockroaches.

In some cockroach species the pronotum is used to both send and receive messages and thus serves as a tool in communication. In *N. cinerea* there are about 40 parallel striae on the ventral surface of the latero-posterior edges of the pronotum. The insects stridulate by rubbing these against the costal veins of the tegmina (Roth and Hartman, 1967). The pronotum is also very sensitive to tactile stimulation in this species. Patrolling dominant males of *N. cinerea* tap members of their social group on the pronotum with their antennae, evoking a submissive posture in lower-ranking members (Ewing, 1972). Similarly, reflex immobilization in *Blab. craniifer* can result from antennal tapping of the pronotal shield by another individual (Gautier, 1967).

COLOR

As in many other insect groups, the suborder Blattaria encompasses species with both cryptic and conspicuous coloration. The former decreases the risk of detection, and the latter is often used in combination with chemical defenses and specific behaviors that discourage predators. Color patterns can vary considerably within a species, contributing to taxonomic difficulties (Mackerras, 1967a), and in a few cockroaches color variation is correlated with geographic features, seasonal factors, or both. Two subspecies of *Ischnoptera rufa* collected at high elevations in Costa Rica and Mexico are darker than their counterparts collected near sea level (Hebard, 1916b). Adults of *Ectobius panzeri* in Great Britain are darker at higher latitudes, and females have a tendency to darken toward the end of the breeding season (Brown, 1952). *Parcoblatta divisa* individuals are typically dark in color, but a strikingly pale morph is found in Alachua County, Florida. No dark individuals were found in a series of several hundred specimens taken from this location, and the pale form has not been collected elsewhere (Hebard, 1943). Color variation among developmental stages within a species may be associated with changing requirements for crypsis, mimicry, or aposematicism. Adults of *Panchlora nivea,* for example, are pale green, while the juvenile stages are brown (Roth and Willis, 1958b).

Many cockroaches are dark, dull-colored insects, a guise well suited to both their cryptic, nocturnal habits and their association with decaying plant debris. Several species associated with bark have cuticular colors and patterns that harmonize with the backgrounds on which they rest. *Trichoblatta sericea* lives on *Acacia* trees, blending nicely with the bark of their host plant (Reuben, 1988). *Capucina rufa* lives on and under the mottled bark of fallen trees and seems to seek compatibly patterned substrates on which to rest (WJB, pers. obs.). A cloak of background substrate enhances crypsis in some species. Female *Laxta* spp. may be encrusted with soil or a parchment-like membrane (Roth, 1992), and *Monastria biguttata* nymphs are often covered with dust (Pellens and Grandcolas, 2003).

Not unexpectedly (Cott, 1940), there are dramatic differences in coloration between the cockroaches on the dayshift versus the nightshift. Day-active cockroaches tend to fall into three broad categories: first, the small, active, colorful, canopy cockroaches; second, the chemically defended, aposematically colored species; and third, those that are Batesian mimics of other taxa. Patterned, brightly colored insects active in the canopy in brilliant sunshine have a double advantage against predators. They are not only cryptic against colorful backgrounds, but they are obscured by rapidly changing contrast when moving in and out of sun flecks (Endler, 1978). A number of aerial cockroach species have translucent wing covers, tinted green or tan, that provide camouflage when they are sitting exposed on leaves (Perry, 1986).

Among the best examples of aposematic coloration are in the Australian Polyzosteriinae (Blattidae). Nocturnal species in the group are usually striped yellow and brown, but the majority are large, wingless, slow-moving, diurnal cockroaches fond of sunning themselves on stumps and shrubs. They are very attractive insects, often metallically colored, or spotted and barred with bright orange, red, or yellow markings (Rentz, 1996; Roach and Rentz, 1998). When disturbed, they may first display a warning signal before resorting to defensive measures. *Platyzosteria castanea* and *Pl. ruficeps* adults assume a characteristic stance with the head near the ground and the abdomen flexed upward at a sharp angle, revealing orange-yellow markings on the coxae and venter. Continued harassment results in the discharge of an evil-smelling liquid "so execrable and pungent that it drove us from the spot" (Shelford, 1912a). Elegant day-flying cockroaches in the genera *Ellipsidion* and *Balta* (Blattellidae) can be observed basking in the sun and exhibit bright orange colors suggestive of Müellerian mimicry rings (Rentz, 1996). Cockroaches in the genus *Eucorydia* (Polyphaginae) are usually metallic blue insects, often with orange or yellow markings on the wings (Asahina, 1971); little is known of their habits. The beautiful wing patterns of some fossil cockroaches are suggestive of warning coloration. Some Spiloblattinidae, for example, had opaque, black, glossy wings with red hyaline windows (Durden, 1972; Schneider and Werneburg, 1994).

Several tropical cockroaches mimic Coleoptera in size, color, and behavior. This is evident in their specific names, which include *lycoides, buprestoides, coccinelloides, dytiscoides,* and *silphoides.* Shelford (1912a) attributes beetle-mimicry in the Blattaria to the similar body types of the two taxa. Both have large pronota and membranous wings covered by thickened elytra or tegmina. "Only a slight modification of the cockroach form is required to produce a distinctly coleopterous appearance." Vršanský (2003) described beautifully preserved fossils of small, beetle-like cockroaches that were day active in Mesozoic forests (140 mya). Extant species of *Prosoplecta* (Pseudophyllodromiinae) (Fig. 1.3) have markedly convex oval or

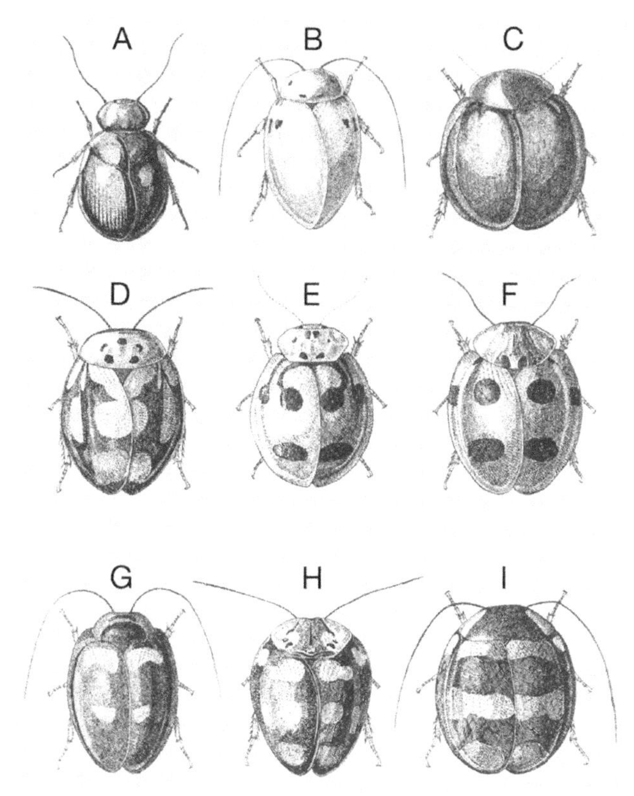

Fig. 1.3 Species of *Prosoplecta* that mimic beetles. (A) *Pr. bipunctata;* (B) Female *Pr. trifaria,* which resembles the light morph of the leaf beetle *Oides biplagiata;* (C) *Pr. nigra;* (D) *Pr. gutticolis;* (E) *Pr. nigroplagiata;* (F) *Pr. semperi,* which resembles the coccinellid *Leis dunlopi;* (G) *Pr. quadriplagiata;* (H) *Pr. mimas;* (I) *Pr. coelophoroides,* which resembles the coccinellid *Coelophora formosa.* After Shelford (1912a). Information on coleopteran models is from Wickler (1968).

circular bodies, smooth and shiny tegmina that do not exceed the tip of the abdomen, and short legs and antennae; they are colored in brilliant shades of orange, red, and black. These cockroaches are considered generalized mimics of coccinellids and chrysomelids, as in most cases their models are unknown. Wickler (1968), however, indicates that females of *Pr. trifaria* (Fig.1.3B) resemble the light morph of the leaf beetle *Oides biplagiata,* while males of this cockroach species resemble the dark morph of the same beetle. Both models and mimics can be collected at the same sites and at the same time of year in the Philippines. Members of the blattellid subfamily Anaplextinae in Australia are diurnal and resemble members of the chrysomelid genus *Monolepta* with which they occur (Rentz, 1996). *Schultesia lampyridiformis* resembles fireflies (Lampyridae) so closely that they cannot be distinguished without close examination (Belt, 1874); on his first encounter with them LMR took them into a darkened hold of the research vessel *Alpha Helix* to see if they would flash (they did not). Other cockroach species have the black and yellow coloration associated with stinging Hymenoptera, and *Cardacopsis shelfordi* (Nocticolidae) runs and sits like an ant, with the body held high off the ground (Karny, as cited by Roth, 1988). All these mimics are thought to be palatable. There is at least one suggested instance of a cockroach serving as a model: Conner and Conner (1992) indicate that a South American arctiid moth (*Cratoplastis* sp.) mimics chemically protected Blattaria.

Cockroaches may be devoid of pigmentation in three general situations. The most common includes new hatchlings and freshly molted individuals of any species (Fig. 1.4), often reported to extension agents as albinos. These typically gain or regain their normal coloration within a few hours. The second are the dependent young nymphs of cockroach species that display extensive parental care. The first few instars of *Cryptocercus, Salganea,* and some other subsocial cockroaches are altricial, with pale, fragile cuticles (Nalepa and Bell, 1997). In *Cryptocercus* pigmentation is acquired gradually over the course of their extended developmental period. Lastly, cockroaches adapted to the deep cave environment lack pigment as part of a correlated character loss typical of many taxa adapted to subterranean life. Color has no signal value for guiding behavior in aphotic environments; neither is there a need for melanin, which confers protection from ultraviolet radiation. Desiccation resistance afforded by a thick cuticle is superfluous in the consistently high humidity of deep caves, and mechanical strength is not demanded of insects that live on the cave walls and floor (Kalmus, 1941; Culver, 1982; Kayser, 1985).

Adults of burrowing cockroaches, on the other hand, typically possess dark, thick cuticles that are abrasion resistant, are able to withstand mechanical stress, and provide insertions of considerable rigidity for the attachment of muscles, particularly leg muscles (Kalmus, 1941; Day, 1950). This thick-skinned group includes the desert-burrowing *Arenivaga,* as well as the soil- and wood-burrowing Panesthiinae and Cryptocercidae. Adults of

Fig. 1.4 Freshly ecdysed *Blaberus* sp. in stump, Ecuador. Photo courtesy of Edward S. Ross.

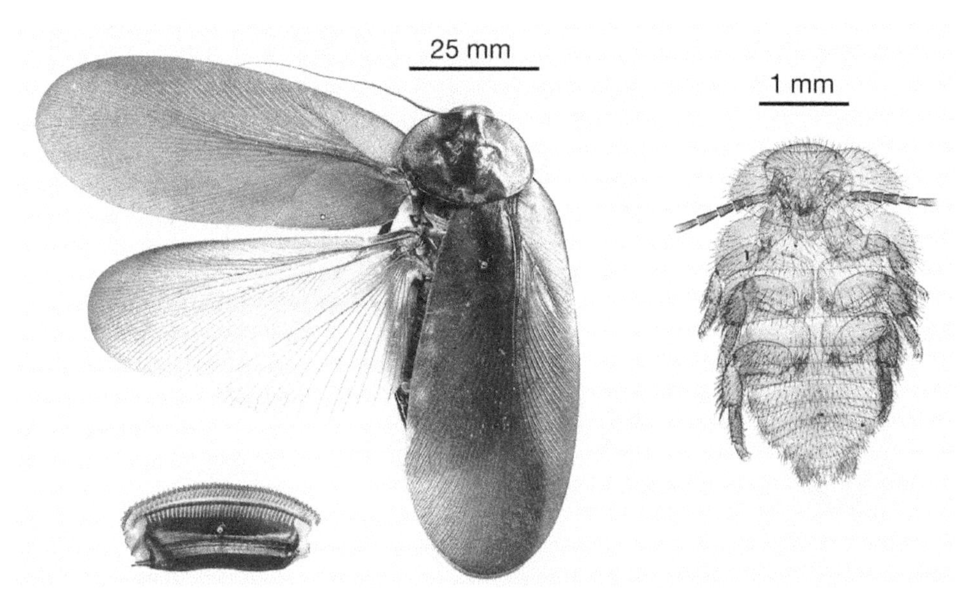

Fig. 1.5 One of the largest and one of the smallest known cockroaches. *Left,* adult female of *Megaloblatta blaberoides* from Costa Rica; the ootheca is that of *Megaloblatta regina* from Ecuador. *Right,* female nymph of *Attaphila fungicola;* ventral view of specimen cleared and mounted on a slide, courtesy of John Moser. Photos by L.M. Roth and E.R. Willis.

these taxa are long lived, requiring a sturdy body to weather the wear and tear of an extended adult life (Kalmus, 1941; Karlsson and Wickman, 1989). They also can be large-bodied insects, with allometric scaling of cuticle production resulting in disproportionately heavy integuments (Cloudsley-Thompson, 1988). The pronotum of *M. rhinoceros* is 100 μ thick, and the cuticle of the sternites is 80 μ, almost twice that of the tergites. The considerable bulk of the abdomen normally rests on the ground, thus requiring greater abrasion resistance (Day, 1950).

BODY SIZE

The general public has always been fascinated with "giant" cockroaches. Discoveries of large species, whether alive or in the fossil record, are thus guaranteed a certain amount of attention. The concept of body size, however, is qualitative and multivariate in nature (McKinney, 1990). Consider two cockroaches that weigh the same but differ in linear dimensions. Is a lanky, slender species bigger than one with a stocky morphotype? Neotropical *Megaloblatta blaberoides* (Nyctiborinae) triumphs for overall length (head to tip of folded wing) (Fig. 1.5). The body measures 66 mm, and when the tegmina are included in the measurement, its length tops out at 100 mm. This species has a wingspan of 185 mm (Gurney, 1959), about the length of a new pencil. Also in contention among the attenuated, lighter-bodied cockroaches are several in the oft-cultured genus *Blaberus. Blaberus giganteus* may measure 80 mm overall (60 mm body length) and female *Blab. craniifer* 62 mm. Pregnant females of the latter weigh about 5 g (Nutting, 1953a). A male *Archimandrita tessalata* measured by Gurney (1959) stretched to 85 mm, and one of the largest species in West Africa (more than 60 mm) is *Rhyparobia* (= *Leucophaea*) *grandis* (Kumar, 1975). Recently, a large cockroach in the genus *Miroblatta* was discovered in caves and rock shelters in limestone formations in East Kalimantan, the Indonesian section of Borneo.[1] The cockroach was widely reported as being 100 mm in length (e.g., BBCNews, 23 December 2004). Two males measured by Drs. Anne Bedos and Louis Deharveng were 60 mm, but they noted that some specimens, particularly females, may be larger. The cockroach is a streamlined, long-legged species that moves very slowly on tiptoe, with the body elevated up over the substrate. It is a beautiful reddish-brown, with lighter-colored legs and wings that are about half the length of the abdomen.

In the heavyweight division, the undisputed champs are the wingless, burrowing types. The Australian soil-burrowing behemoth *M. rhinoceros* weighs in at 30 g or more, and can measure 85 mm in length. *Macropanesthia rothi* is sized similarly to *M. rhinoceros,* but is more robust in the thorax and legs (Rugg and Rose, 1991; Walker et al.,

1. For information on the species, we thank Patricia Crane, Leonardo Salas, Scott Stanley, and Louisa Tuhatu of the Nature Conservancy, and Louis Deharveng, Anne Bedos, Yayuk Suhardjono, and Cahyo Rachmadi, the entomologists in the expedition that discovered the species. The cockroach was identified by P. Grandcolas.

1994). Males of *Macropanesthia* are frequently mistaken for small tortoises during periods of surface activity (Rentz, 1996). The Malagasian *G. portentosa* can reach 78 mm in length (Gurney, 1959), and *G. grandidieri,* with a body length of 85 mm, rivals *M. rhinoceros* in size (Walker et al., 1994).

The oft-repeated myth that the Carboniferous was the "Age of Giant Cockroaches" is based on the size of fossil and modern cockroaches that were known during the late 1800s. More recently described species of extant cockroaches raise the modern mean, and scores of recently collected small fossil species will no doubt lower the Paleozoic mean (Durden, 1988). The fossil record also may be biased in that large organisms have better preservation potential, are easier to find, and can better survive incarceration in fine- and coarse-grained sediments (Carpenter, 1947; Benton and Storrs, 1996). Small cockroaches, on the other hand, may be filtered from the fossil record because they are more likely to be swallowed whole by fish during transport in flowing water (Vishniakova, 1968). The largest fossil cockroach to date is an undescribed species from Columbiana County, Ohio, which has a tegmen length of at least 80 mm (Hansen, 1984 in Durden, 1988); a complete fossil from the same location has recently received media attention (e.g., Gordner, 2001). Nonetheless, the tenet that no fossil cockroach exceeds in size the largest living species (Scudder, 1886; F.M. Carpenter in Gurney, 1959) still applies. It would not be unreasonable to suggest that we are currently in the age of giant cockroaches (C. Durden, pers. comm. to CAN)!

At the other end of the scale, the smallest recorded cockroaches are mosquito sized species collected from the nests of social insects, where a minute body helps allow for integration into colony life. The myrmecophile *Attaphila fungicola* is a mere 2.7 mm long (Cornwell, 1968) (Fig. 1.5), and *Att. flava* from Central America is not much larger—2.8 mm (Gurney, 1937). Others include *Myrmecoblatta wheeleri* from Florida at less than 3 mm (Deyrup and Fisk, 1984), and *Pseudoanaplectinia yumotoi* (4 mm) from Sarawak (Roth, 1995c). Australian species of *Nocticola* measure as little as 3 mm and have been collected from both termite nests and caves (Rentz, 1996). Another category of cockroaches that can be quite small are those that mimic Coleoptera. *Plecoptera poeyi,* for example, lives on foliage of holly (*Ilex*) in Florida and is 5–6 mm long (Helfer, 1953). To put the sizes of these cockroaches into perspective, it is worthwhile to note that the fecal pellets of *M. rhinoceros* are 10 mm in length (Day, 1950).

As a group, blattellids are generally small in size, but several genera are known to include moderately large members (Rentz, 1996). A number of tiny aerial Blattellidae live in the canopy of tropical rainforests, where "their size is suited to hiding in the crease of a leaf or by a small bit of moss" (Perry, 1986). Small bodies may confer a survival advantage in graduate student lounges; Park (1990) noted that American cockroaches live for about 5 sec when placed in a microwave oven set on "high," but the more diminutive German cockroach lasts for twice that long. Small cockroaches usually mature more rapidly and have shorter lives than the larger species (Mackerras, 1970).

Intraspecific variation in cockroach body size can be considerable, with the difference between the largest and the smallest specimens so great that they appear to be different species (Roth, 1990b). Male length in *Laxta granulosa,* for example, ranges from 14.8 to 25.4 mm (Roth, 1992). In most cockroaches, the abdominal segments can telescope. Extension of the abdomen in live specimens and shrinkage in the dead ones, then, may contribute to noted variation when body length is the measurement of choice. Body size may vary within (e.g., *Platyzosteria melanaria*—Mackerras, 1967b), and between (e.g., Parcoblattini—Roth, 1990b), geographic locations, or be rather consistent over an extensive range (e.g., *Ectobius larus, E. involutus*—Rehn, 1931). No latitudinal clines in body size have been reported in cockroaches.

As in most invertebrates (Fairbairn, 1997; Teder and Tammaru, 2005), sexual dimorphism in body size of adult cockroaches is common. All patterns are exhibited, but a female size bias seems to predominate (Fig. 1.6). Examples include *Colapteroblatta surinama,* where females are 18.5–19.0 mm and males are 13.0–15.5 mm in length (Roth, 1998a), and the cave-adapted species *Trogloblattella nullarborensis,* with females measuring 34.5–38.5 mm and males 24–27.5 mm (Roth, 1980). Because of intraspecific variation and the multivariate nature of size, however, generalizations can be difficult to make. Males may measure longer than females, especially when wings are included in the measurement, but females are usually broader and bulkier, particularly in the abdomen. Both *P. americana* and *Supella longipalpa* fall into this category (Cornwell, 1968) (Fig. 1.7). Several burrowing cockroaches exhibit little, if any size dimorphism. There is no significant difference in the fresh weight or head capsule width of males and females of field-collected pairs of *Cryptocercus punctulatus,* but the dry weight of females is slightly higher (Nalepa and Mullins, 1992). In most Geoscapheini, males and females are of similar size (Fig. 1.8) (e.g., Walker et al., 1994), as are several species of *Salganea,* such as *Sal. amboinica* and *Sal. rugulata* (Roth, 1979b). In some *Salganea,* however, the male is distinctly smaller than the female. These include *Sal. rectangularis* (Roth, 1999a) and *Sal. morio,* where males average 41.9 mm in length and females 46.6 mm (Roth, 1979b). Species in which males outsize females include several

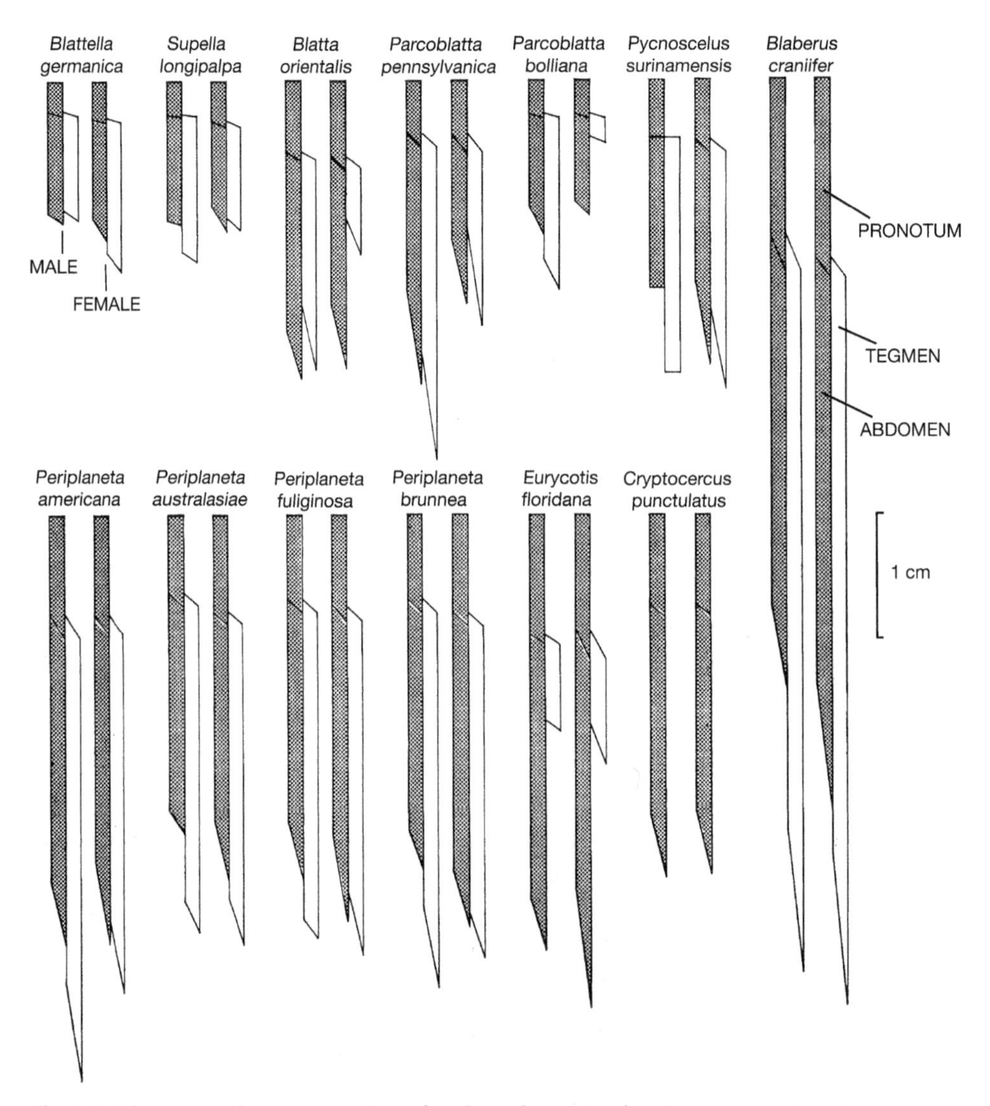

Fig. 1.6 Diagrammatic representation of cockroach species showing comparative size, comparison between males (*left*) and females (*right*), degree of size variation within a sex (minimum measurement on *left,* maximum measurement on *right*), and relationship between tegmen and body length. From Cornwell (1968), based on data from Hebard (1917). With permission of Rentokil Initial plc.

Parcoblatta species (Fig 1.6) *(Parc. lata, Parc. bolliana, Parc. divisa, Parc. pennsylvanica).* Males of the latter are 22–30 mm in length, while females measure 13–20 mm. In *Parc. fulvescens,* however, females outsize the males (Cornwell, 1968; Horn and Hanula, 2002).

Like other animals, the pattern of sexual size dimorphism within a cockroach species is related to the relative influence of body size on fecundity in females and mating success in males. In *G. portentosa,* males tend to be larger than females, and big males are the more frequent victors in male-male contests (Barth, 1968c; Clark and Moore, 1995). In species where males offer food items to the female as part of courtship and mating, nuptial gifts may reduce the value of large size in females and increase its value in males (Leimar et al., 1994; Fedorka and Mousseau, 2002). This hypothesis is unexplored in the cockroach species that employ such a mating strategy. One proximate cause of female-biased sexual size dimorphism in cockroaches is protandry. Males may mature faster than females because it gives them a mating advantage, but become smaller adults as a consequence. Males of *Diploptera punctata,* for example, usually undergo one fewer molt than do females, and require a shorter period of time to mature (Willis et al., 1958). Males of *Anisogamia tamerlana* mature in five instars, and females in six (Kaplin, 1995).

Physiological correlates of body size have been examined in some cockroaches; these include studies of metabolic rate and the ability to withstand extremes of temperature, desiccation, and starvation. Coelho and Moore

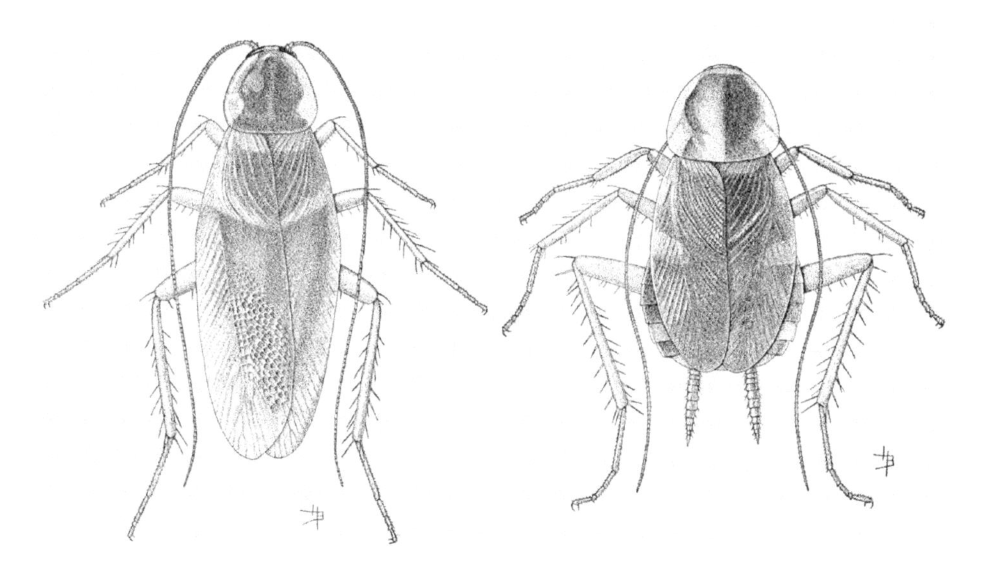

Fig. 1.7 Male (*left*) and female *Supella longipalpa,* showing dissimilarity in form between the sexes. The female is stouter, and the head is broader with a larger interocular space; the pronotum is also larger than that of the male. The tegmina of the female reach only to the end of the abdomen and are more chitinous than those of the male (Hebard, 1917). From Back (1937), with permission from the Entomological Society of Washington.

(1989) found that resting metabolic rate for 11 species scales allometrically ($VO_2 = 0.261\ M^{0.776}$) with mass. As in other animals, then, it is metabolically more expensive for a small cockroach to maintain a gram of tissue than it is for a large one. Relative brain size has been compared in two cockroach species. The brain (supra-esophageal ganglia) of *B. germanica* occupies about 10 times as much of the cranial cavity as does that of *M. rhinoceros,* a species that weighs 320 times more (Day, 1950) (Fig 1.9). There is, however, no marked difference in the size of individual nerve cell bodies. Day thought that the large size of *Macropanesthia* could be attributed to its burrowing habit, which "greatly reduces the effectiveness of gravity in limiting size." More likely factors include the ability to withstand predation, the power required to dig in indurate soils, and the lower rate of water loss associated with a small surface to volume ratio. The latter was suggested as being influential in *G. portentosa*'s ability to thrive in the long tropical dry season of Madagascar (Yoder and Grojean, 1997); in the laboratory adult females survived 0% humidity without food and free water for a month.

Fig. 1.8 Harley A. Rose, The University of Sydney, displaying male-female pairs of Australian soil-burrowing cockroaches (Geoscapheini). Photo by C.A. Nalepa.

The social environment experienced during development influences adult body size in cockroaches. Isolated cockroach nymphs mature into larger adults than nymphs that have been reared in groups, but a smaller adult body size occurs when nymphs are reared under crowded conditions (e.g., Willis et al., 1958; Woodhead and Paulson, 1983). Unlike laboratory studies, however, overpopulation in nature may be relatively rare, except perhaps in some cave populations. Crowded adults are likely to disperse or migrate when competition for food and space becomes fierce. In all known cases where biotic or abiotic factors affect cockroach adult size, these factors act by influencing the duration of juvenile growth. In *D. punc-*

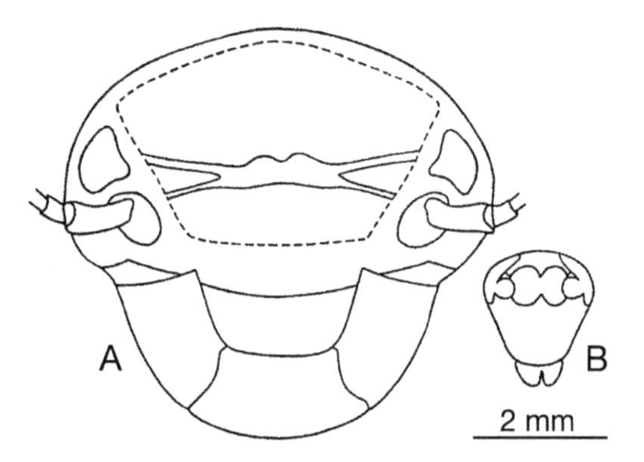

Fig. 1.9 Comparison of the relative size of the head and anterior nervous system in (A) *Macropanesthia,* and (B) *Blattella.* From Day (1950), with permission from CSIRO Publishing.

tata, the greater adult weight of isolated animals results from a longer nymphal development. Males normally have three or four instars, but isolation results in a higher proportion of the four-instar type (Woodhead and Paulson, 1983). A longer postembryonic development induced by suboptimal diet resulted in heavier adults in *Blaptica dubia* (Hintze-Podufal and Nierling, 1986). In three families of *Cryptocercus clevelandi* monitored under field conditions, some of each litter matured to adults a year before their siblings did. Those that matured in 6 yr had larger head widths than those that matured in 5 (Nalepa et al., 1997).

THE ECOLOGY OF MORPHOTYPE

The smooth, flattened body typical of many cockroaches is functionally related to their crevice-inhabiting lifestyle; it allows them to slip into narrow, horizontally extended spaces like those found in strata of matted, decayed leaves. There are, however, a number of variations on the basic body type that are exhibited by groups of often distantly related cockroaches occupying more or less the same ecological niche. These possess a complex of similar morphological characters reflecting the demands of their environment. Here we briefly profile seven distinct morphological groups. Two are defensive morphotypes, and two are forms specialized for penetrating solid substrates. Desert dwellers, those living in social insect nests, and cave cockroaches round out the gallery.

The Pancake Syndrome

The dorsoventrally compressed morphotype typical of the "classic" cockroach has been taken to extremes in several distantly related taxa. These extraordinarily flattened insects resemble limpets and live in deep, narrow clefts such as those found under loose bark, at the log-soil interface, under stones, or in the cracks of boulders and rocks. In most species, the borders of the tergites are extended, flattened, and held flush with the substrate so that a close seal is formed (Fig. 1.10). The proximal parts of the femora may be distinctively flattened as part of the overall pancake syndrome (Mackerras, 1967b; Roth, 1992). Included in this group are female West Indian *Homalopteryx laminata* (Epilamprinae) (Kevan, 1962) and several Australian taxa. A number of *Leptozosteria* and *Platyzosteria* spp. (Polyzosteriinae) live in deep, narrow clefts under rocks or bark (Mackerras, 1967b; Roach and Rentz, 1998). Members of the genus *Laxta* (Epilamprinae) live under eucalypt bark and are common under large slabs at the bases of trees (Roth, 1992; Rentz, 1996). Some Central and South American Zetoborinae (e.g., *Lanxoblatta emarginata, Capucina patula)* and Blaberinae (e.g., *Mon. biguttata* nymphs) have a comparable body type and habitat (Roth, 1992; Grandcolas and Deleporte, 1994; Pellens and Grandcolas, 2003; WJB, unpubl. obs.). Highly compressed morphotypes are associated

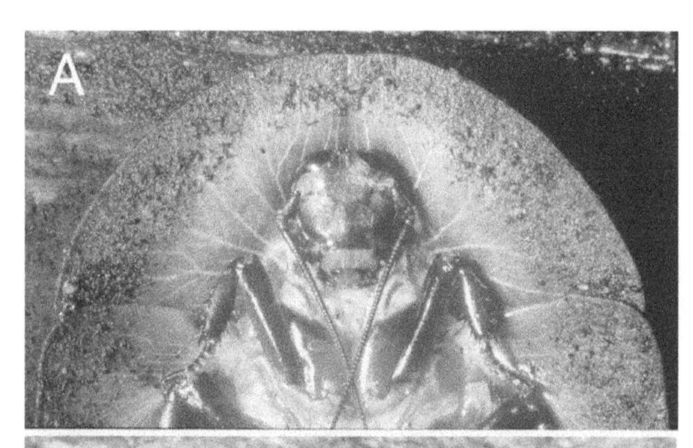

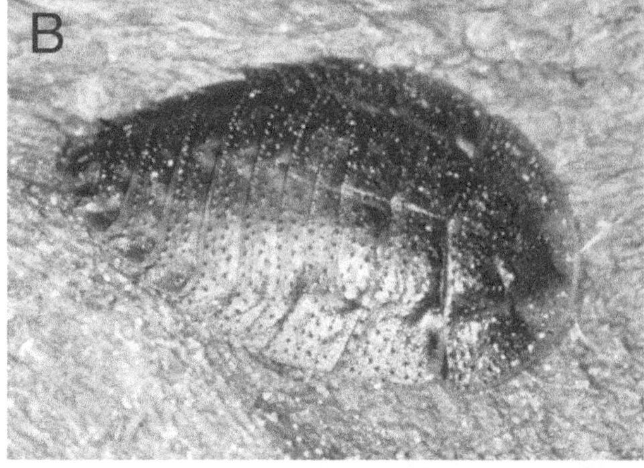

Fig. 1.10 (A) Ventral view of head and expanded pronotum and metanotum of an unidentified, dorsoventrally flattened cockroach collected under bark in Brazil; most likely a female or nymph of *Capucina patula* or *Phortioeca phoraspoides* (LMR, pers. obs.). Note debris attached to the pronotal edges, which were closely applied to the wood surface. Photo courtesy of Edward S. Ross. (B) Female of *Laxta friedmani* (named after LMR's urologist). Photo courtesy of David Rentz.

Fig. 1.11 Mechanisms of cockroach defense against ants. (A) Chemical defense by *Diploptera punctata. Pogonomyrmex badius* is attacking the cockroach on the *left,* whose defensive glands have been removed. The intact cockroach on the *right* was also attacked by the ants, but it discharged a spray of quinones and repelled the attackers. The spray pattern is shown by indicator paper on which the cockroach is standing. From Eisner (1958). (B) Defense by conglobulation. Adult female of *Perisphaerus semilunatus* from Thailand, protected from attack by rolling up into a ball. From Roth (1981b). (C) Defense by adhesion. A flattened *Capucina patula* nymph protected from attack by hugging the substrate. The body of the cockroach is clearly seen through the lateral extensions of the tergites. All photographs courtesy of Thomas Eisner.

with defense against both abiotic and biotic hazards. In the intensely arid climate of Australia, these cockroaches squeeze into deep, narrow clefts and cracks to avoid desiccation (Mackerras, 1967b). In the Neotropical species, it has been demonstrated that compressed bodies confer protection against ant attacks (Fig. 1.11C). The insects become immobile and cling so tightly to the substrate that their vulnerable undersurfaces cannot be harmed (Grandcolas and Deleporte, 1994; Pellens and Grandcolas, 2003; Roth, 2003a).

The Conglobulators

Another variation of defensive morphotype is exhibited by the wingless half-ellipsoids, those cockroaches that are rounded on top and flat on the bottom, like a watermelon cut on its long axis. Species of this shape in several genera of Perisphaeriinae (*Perisphaeria, Perisphaerus,* and *Pseudoglomeris*) are able to roll themselves into a ball, that is, conglobulate, when alarmed (Fig. 1.12) (Shelford, 1912a; Roth, 1981b). They are usually rather small, black species with a tough cuticle. When enrolled, the posterior abdomen fits tightly against the edge of the pronotum. All sense organs are covered; there are no gaps for an enemy to enter nor external projections for them to grab (Fig. 1.11B). In some species, the female encloses young nymphs that are attached to her venter when she rolls up (Chapter 8). Not only are small predators like ants thwarted, but the rounded form is very resistant to pressure and requires considerable force to crush (Lawrence, 1953). In other taxa exhibiting this behavior (e.g., isopods, myriapods), the rolled posture is maintained during long periods of quiescence, so that the animal is protected from desiccation as well as enemies (Lawrence,

Fig. 1.12 *Perisphaerus semilunatus* female: dorsal, ventral, lateral, and nearly conglobulated. Photos by L.M. Roth.

1953); it is unknown whether that is the case in these cockroaches.

The Burrowers

Cockroaches that burrow in wood or soil exhibit a remarkable convergence in overall body plan related to the ability to loosen, transport, and travel through the substrate, and to maneuver in confined spaces. These insects are often wingless, with a hard, rigid, pitted exoskeleton and a thick, scoop-shaped pronotum. The body is stocky and compact, and the legs are powerful and festooned with stout, articulated spines that provide anchorage within the tunnels and leverage during excavation (Fig. 1.13). The cerci are short, and can be withdrawn into the body in *Cryptocercus* (thus the name) and *Macropanesthia.* Long cerci make backward movement in enclosed spaces inconvenient (Lawrence, 1953).

The similarity in the external morphology of *Cryptocercus* and wood-feeding Panesthiinae is so striking that they were initially placed in the same family (Wheeler, 1904; Roth, 1977). McKittrick (1964, 1965), however, examined their genitalia and internal anatomy and demonstrated that the resemblance was superficial. Her studies resulted in placing the two taxa into distantly related families (Cryptocercidae and Blaberidae). They currently offer an opportunity to scientists interested in sorting the relative influences of phylogeny and ecology in structuring life history and behavior.

The Borers

Although little to nothing is known of their biology, several small cockroaches have a heavy pronotum and exhibit the elongated, cylindrical body form typical of many wood-boring beetles (Cymorek, 1968). Their appearance suggests that these cockroaches drill into solid wood or soil because the shape minimizes cross-sectional area, reducing the tunnel bore and the force required to advance a given body weight. This morphotype is exhibited by the genus *Colapteroblatta* (Epilamprinae) (Roth, 1998a), as well as some species of Perisphaeriinae in the genera *Compsagis, Cyrtotria, Bantua,* and *Pilema* (Shelford, 1908; Roth, 1973c). *Compsagis lesnei* typifies this type of cockroach (Fig. 1.14) and is a small (9.5 mm in length) African species found inside of tree branches (Chopard, 1952).

Fig. 1.13 Adult *Cryptocercus punctulatus.* Photo courtesy of Piotr Naskrecki.

Desert Dwellers

Cockroaches that live in the desert typically have morphological adaptations allowing for the conservation of water and for ease in negotiating their sandy environment. Adult females and nymphs are shaped like smooth, truncated ovals, with short, spined legs (e.g., *Arenivaga investigata*—Friauf and Edney, 1969). The head is strongly hooded by the pronotum, and cuticular extensions of the thoracic and abdominal tergites cover the

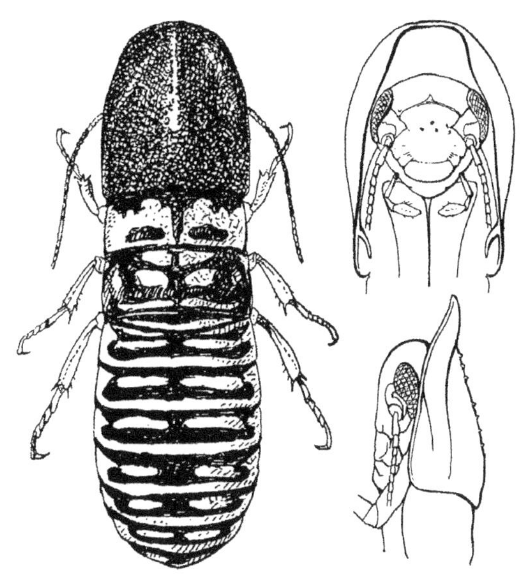

Fig. 1.14 Female of the wood-boring cockroach *Compsagis lesnei. Left,* whole body. *Right,* head and pronotum: ventral view (*top*), lateral view (*bottom*). From Chopard (1952), with permission of Société Entomologique de France.

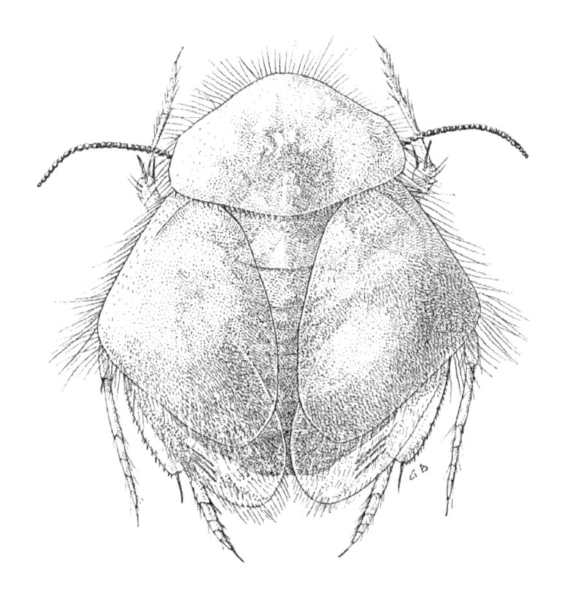

Fig. 1.15 Male of the desert-dwelling Iranian cockroach *Leiopteroblatta monodi,* exhibiting the long hairs that create an insulating boundary layer of air in many desert-dwelling cockroaches. From Chopard (1969), with permission of the Société Entomologique de France.

body and the legs. The periphery of the body is fringed by hairs that directly contact the substrate when the insect is on the desert surface, creating a boundary layer of air and trapping respiratory water (Fig. 1.15). A microclimate that is more favorable than the general desert atmosphere is thus maintained under the body (Vannier and Ghabbour, 1983). Most of these desert dwellers are in the Polyphagidae, but some *Polyzosteria* spp. (Blattidae) that inhabit dry areas of Australia are apterous, are broadly

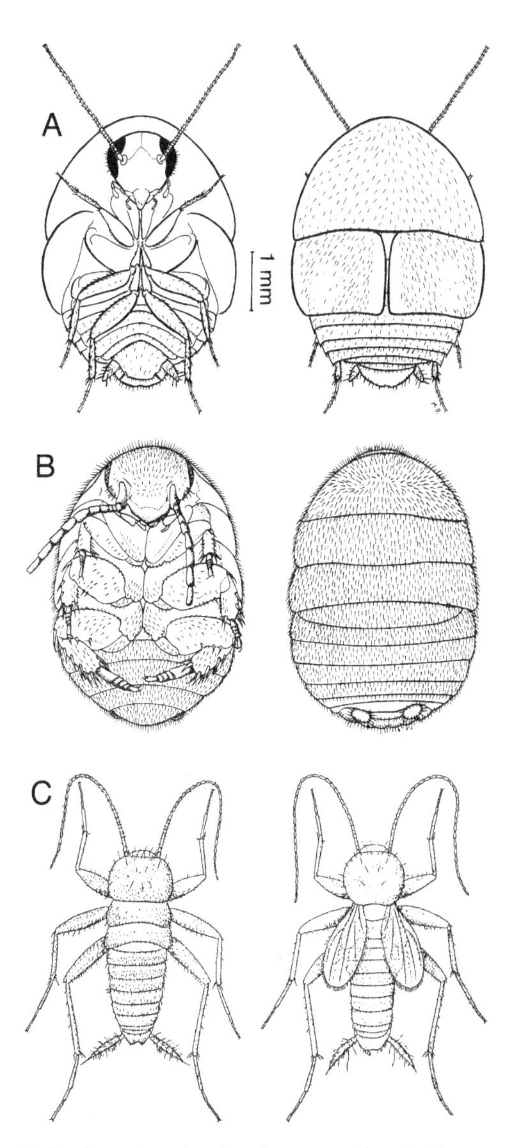

Fig. 1.16 Cockroaches that live in nests of social insects. (A) Male myrmecophile *Myrmecoblatta wheeleri; left,* ventral view; *right,* dorsal view. From Deyrup and Fisk (1984), with permission of M.A. Deyrup. (B) Female myrmecophile *Attaphila fungicola.* From Wheeler (1900). (C) Termitophile *Nocticola termitophila; left,* female; *right,* male. From Silvestri (1946). Not drawn to scale.

oval, and have a "remarkably hairy covering" (Mackerras, 1965a).

Myrmecophiles/Termitophiles

Myrmecophiles are just a few millimeters long, oval in shape, strongly convex, and rather uniformly covered with short, fine setae (Fig. 1.16A,B). They are typically apterous or brachypterous, the legs and antennae are short, and in some species the eyes are reduced. *Att. fungicola* (Blattellidae) have no more than 70 ommatidia per eye (Wheeler, 1900; Roth, 1995c). No glands are obvious that may function in appeasing their hosts. *Myrme-*

oblatta wheeleri (Polyphagidae) run rapidly, and when disturbed withdraw their appendages under the body and adhere tightly to the substrate (Deyrup and Fisk, 1984). This behavior is similar to the defensive behavior of flattened Neotropical species (Fig. 1.11C) and suggests that although they appear integrated into colony life, a wariness of their predator hosts remains of selective value. Wheeler (1900) suggested that *Att. fungicola* is a "truly cavernicolous form, living in caves constructed by its emmet hosts." It is the species of *Nocticola* taken from termite nests, however, that exhibit the delicate, elongate body, attenuated appendages, and pale cuticle typical of cave-adapted insects (and of most other Nocticolidae — Roth, 1988, 1991a; Fig. 1.16C).

Cave Dwellers

Cave-adapted cockroaches exhibit a suite of morphological characters common to cave-dwelling taxa around the world. These include depigmentation and thinning of cuticle, the reduction or loss of eyes, the reduction or loss of tegmina and wings, the elongation and attenuation of appendages, and a more slender body form (Howarth, 1983; Gilbert and Deharveng, 2002). A large nymph of the genus *Nelipophygus* collected in Chiapas, Mexico, for example, cannot survive outside of its cave and is colorless, slender, and 20 mm long; it has extremely long antennae and limbs, and has no trace of compound eyes or pigment (Fisk, 1977). Males of *Alluaudellina cavernicola* exhibit a remarkable parallel reduction of eyes and wings (Fig. 1.17) (Chopard, 1932). Eye size ranges from well developed to just three ommatidia, with intermediates between. Individuals of *Nocticola australiensis* from the Chillagoe region of Australia also show a consistent gradation of forms, from less troglomorphic in southern caves to more troglomorphic in the north (Stone, 1988). The pattern of variation is very regular, unlike the more complex variation seen in some other taxa. The Australian species *Paratemnopteryx howarthi,* for example, also demonstrates the entire range of morphological variation, but both the reduced-eye, brachypterous forms and the large-eyed, winged morphs can occur in the same cave (Chopard, 1932; Roth, 1990b).

One consequence of regressive evolution of visual structures in cave-adapted animals is that orientation and communication have to be mediated by non-visual systems. Thus, the loss of the visual modality is often complemented by the hypertrophy of other sensory organs (Nevo, 1999; Langecker, 2000). In cockroaches, this may include the elongation of the legs, antennae, and palps (Fig. 1.18). In *All. cavernicola* the antennae are three times the length of the body (Vandel, 1965), and both *Noc. australiensis* and *Neotrogloblattella chapmani* have very long,

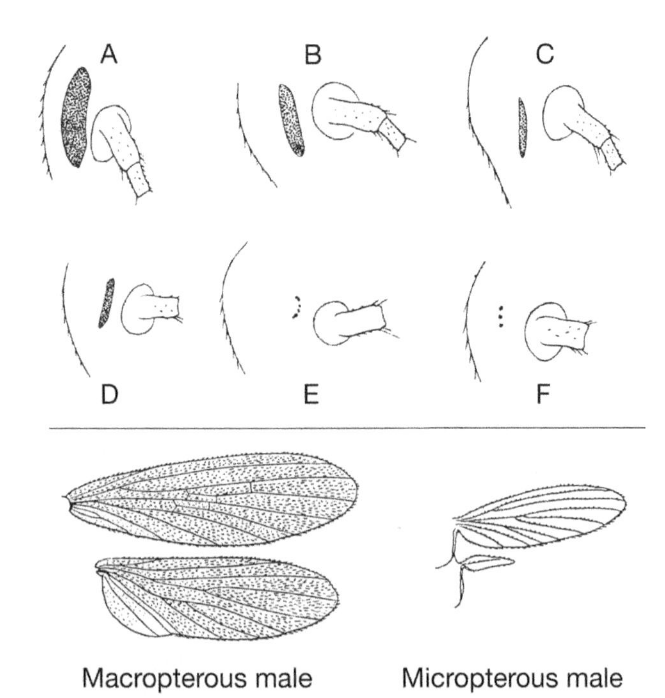

Fig. 1.17 Variation in eye and wing development in cave-dwelling *Alluaudellina cavernicola.* (A,B) Eye development in macropterous males; (C) eye development in a micropterous males; (D,E,F) eye development in wingless females. After Chopard (1938).

slender legs and elongated maxillary palps. Palps are long in *Ischnoptera peckorum* as well (Roth, 1980, 1988). In nymphs of some species of *Spelaeoblatta* from Thailand it is only the front pair of legs that is elongated, which together with their narrow, elongated pronotum confers a mantid-like appearance (Vidlička et al., 2003). Long legs are adaptive in reaching across gaps, negotiating irregular substrates, and covering larger areas per unit of expended energy (Howarth, 1983). Elongated antennae and palps function in extending the sensory organs, allowing the insects to detect food and mates faster and at a greater distance from their bodies. Consequently, less energy is required for resource finding (Hüppop, 2000), a decided advantage in a habitat where food may be scarce and population densities low. Cave-dwelling *Paratemnopteryx* exhibit subtle shifts in the number and type of antennal and mouthpart sensilla as compared to surface-dwelling relatives (Bland et al., 1998a, 1998b). There is a moderate reduction in the mechano – contact receptors and an increase in the number of olfactory sensilla in the cave dwellers when compared to similar sized epigean species. The elongation of appendages is typically correlated with a behavioral change. Troglomorphic cockroaches move with slow deliberation while probing with their long appendages. They "thereby avoid entering voids from which no escape is possible" (Howarth, 1983). Weinstein and Slaney (1995) found that highly troglomorphic species of

Fig. 1.18 Male of the Western Australian troglobitic cockroach *Nocticola flabella* from a cave in the Cape Range, Western Australia (Roth, 1991c). *Top,* dorsal view; *bottom,* grooming its metathoracic leg.; photo courtesy of the Western Australia Museum, via W.F. Humphreys.

Paratemnopteryx were able to avoid baited pitfall traps, but the slightly troglomorphic species readily entered them. Overall, cockroaches may experience less selection pressure for improved non-visual sensory organs than many other insects; cave colonizers that are already nocturnal may require little sensory improvement (Langecker, 2000).

Selection Pressures

Food limitation is most commonly suggested as the selective basis of the syndrome of characters associated with cave-dwelling organisms. First, many of the characters are directed toward improved food detection (e.g., elongation of appendages) and food utilization (e.g., lower metabolic and growth rate, starvation resistance, slow movement, fewer eggs) (Poulson and White, 1969; Hüppop, 2000; Gilbert and Deharveng, 2002). Second, troglomorphic species are more often found in caves that lack sources of vertebrate guano (Vandel, 1965; Culver, 1982). It is the combination of scarce food and the consistently dark, humid environment of deep caves, however, that best accounts for the reductions and losses that characterize troglomorphism. Eyes are complex organs, expensive to develop and maintain. Animals rarely have sophisticated visual systems unless there is substantial selection pressure to favor them (Prokopy, 1983). Optical sensors are useless in the inky blackness of deep caves and "compete" with non-visual systems for available metabolites and energy (Culver, 1982; Nevo, 1999). Photoreception is also related to a complex of behavioral and morphological traits that become functionless in the permanent darkness of a cave. These include visually guided flight and signaling behavior based on cuticular pigmentation (Langecker, 2000). Cave-dwelling cockroaches in north Queensland, Australia, display a remarkable degree of correlation between levels of troglomorphy and the cave zone in which they occur. In the genera *Nocticola* and *Paratemnopteryx,* the most modified species described by LMR are found only in the stagnant air zones of deep caves, while the slightly troglomorphic species of *Paratemnopteryx* are concentrated in twilight transition zones (Howarth, 1988; Stone, 1988). Because cockroaches live in a variety of stable, dark, humid, organic, living spaces, however, reductive evolutionary trends are not restricted to cavernicolous species (discussed in Chapter 3). *Nocticola* (= *Paraloboptera*) *rohini* from Sri Lanka, for example, lives under stones and fallen tree trunks. The female is apterous; the males have small, lateral tegminal lobes but lack wings, and the eyes are represented by just a few ommatidia (Fernando, 1957).

Many cave cockroaches diverge from the standard character suite associated with cave-adapted insects. They may exhibit no obvious troglomorphies, or display some characters, but not others. *Blattella cavernicola* is a habitual cave dweller but shows no structural modifications for a cave habitat (Roth, 1985). Neither does the premise that some cave organisms diverge from the morphological profile because they live in energy-rich environments such as guano piles (Culver et al., 1995) always hold true for cockroaches. *Paratemnopteryx kookabinnensis* and *Para. weinsteini* are associated with bats (Slaney, 2001), yet both show eye and wing reduction. Heterogeneity in these characters may occur for a variety of reasons. The surface-dwelling ancestor may have exhibited varying levels of morphological reduction or loss prior to becoming established in the cave (i.e., some losses are plesiomorphic traits) (Humphreys, 2000a). Such is likely the case for the two species of *Paratemnopteryx* mentioned above; most species in the genus have reduced eyes, lack pulvilli, and are apparently "pre-adapted" for cave dwelling (Roth, 1990b). Species also may be at different stages of adaptation to the underground environment (Peck, 1998). Generally, regression increases and variability decreases with phylogenetic age (Culver et al., 1995; Langecker, 2000). *Nocticola flabella* is probably the most troglobitic cockroach known (Fig. 1.18); the male is 4–5 mm long, eyeless, with reduced tegmina and no hindwings, has very long legs and antennae, and is colorless except for amber mouthparts and tegmina (Roth, 1991c). This high level of regressive evolution is also found in other species found in deep caves of the Cape Range in western Australia and is consistent with the apparent great age of this fauna (Humphreys, 2000b). Other sources of variation that may play a role include ecological differences within and among caves, continued gene flow between epigean and cave populations, the accumulation of neutral mutations, developmental constraints, or some combination of these (Culver, 1982; Slaney and Weinstein, 1997b; Hüppop, 2000; Langecker, 2000).

Retention of Sexually Selected Characters

In several cave-adapted cockroaches, male tergal glands, which serve as close-range enticements to potential mates, do not vary in concert with other morphological features. The glands can be large, or numerous and complex, despite the otherwise troglomorphic features displayed by the male. *Trogloblattella nullarborensis* is found deep within limestone caves in Australia, and is much larger than other blattellids. It lacks eyes, and has reduced wings and elongated appendages and antennae. Its color, however, has not been modified. Adults are medium to dark brown, and the male has huge tergal glands (Mackerras, 1967c; Richards, 1971; Rentz, 1996). Similarly, males in the genus *Spelaeoblatta* are pale in color and have reduced eyes, brachypterous wings, and long legs and antennae; however, they have large, elaborate tergal glands on two different tergites, and in *Sp. myugei,* large tubercles of unknown function on tergites 5 through 8 (Fig. 1.19) (Roth and McGavin, 1994; Vidlička et al., 2003). Tergal glands are rare in *Nocticola* spp., but *Noc. uenoi uenoi* living in the dark zone of caves on the Ryukyu Islands has a prominent one (Asahina, 1974). The genitalia of male cave cockroaches also can be very complex, despite the regressive evolution evident in other body parts, for example, *Nocticola brooksi* (Roth, 1995b) as well as other Nocticolidae (Roth, 1988). Mating behavior in cave-adapted cockroaches has not been described.

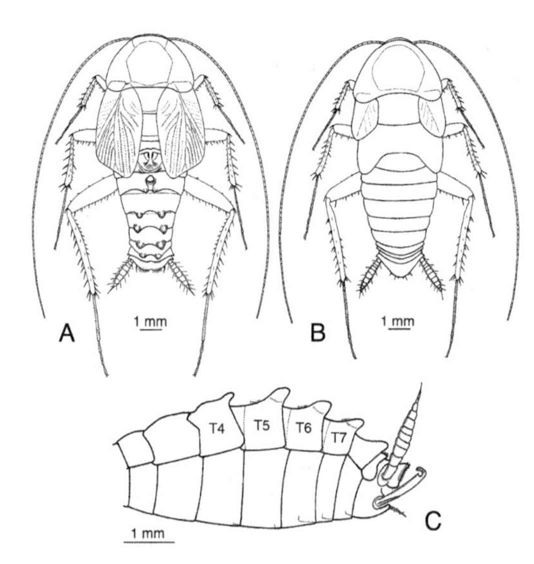

Fig. 1.19 The cave-adapted cockroach species *Spelaeoblatta myugei* from Thailand. (A) Dorsal view of male. Note large tergal glands on tergites 3 and 4, and paired tubercles on tergites 5–8. (B) Dorsal view of female. (C) Lateral view of male abdomen and its tubercles. From Vidlička et al. (2003), with permission from Peter Vršanský and the Taylor & Francis Group.

TWO

Locomotion: Ground, Water, and Air

i can walk on six feet
or i can walk on four feet
maybe if i tried hard enough
i could walk on two feet
but i cannot walk on five feet
or on three feet
or any odd number of feet
it slews me around
so that i go catercornered

—archy, "a wail from little archy"

Cockroaches were once placed in the suborder Cursoria (Blatchley, 1920) (Lat., runner) because the familiar ones, the domestic pests, are notorious for their ground speed on both horizontal and vertical surfaces. Indeed, the rapid footwork of these species has made cockroach racing a popular sport in a number of institutions of higher learning. Like most animal taxa, however, cockroaches exhibit a range of locomotor abilities, reflecting ease of movement in various habitats. On land, the limits of the range are mirrored in body designs that maximize either speed or power: the lightly built, long-legged runners, and the bulkier, more muscular burrowers. There is a large middle ground of moderately fast, moderately powerful species; however, research has focused primarily on the extremes, and it is on these that we center our discussion of ground locomotion. We touch on cockroach aquatics, then address the extreme variation in flight capability exhibited within the group. Finally, we discuss ecological and evolutionary factors associated with wing retention or loss.

GROUND LOCOMOTION: SPEED

Periplaneta americana typifies a cockroach built to cover ground quickly and is, relative to its mass, one of the fastest invertebrates studied. It has a lightly built, somewhat fragile body and elongated, gracile legs capable of lengthy strides. The musculature is typical of running insects, but the orientation of the appendages with respect to the body differs. The middle and hind pairs point obliquely backward, and the leg articulations are placed more ventrally than in most insects (Hughes, 1952; Full and Tu, 1991). *Periplaneta americana* has a smooth, efficient stride, and at most speeds, utilizes an alternating tripod gait, that is, three legs are always in contact with the ground. The insect can stop at any point in the walking pattern because its center of gravity is always within the support area provided by the legs. At a very slow walk the gait grades into a metachronal wave, moving from back to front, that is, left 3-2-1, then right 3-2-1 (Hughes, 1952; Del-

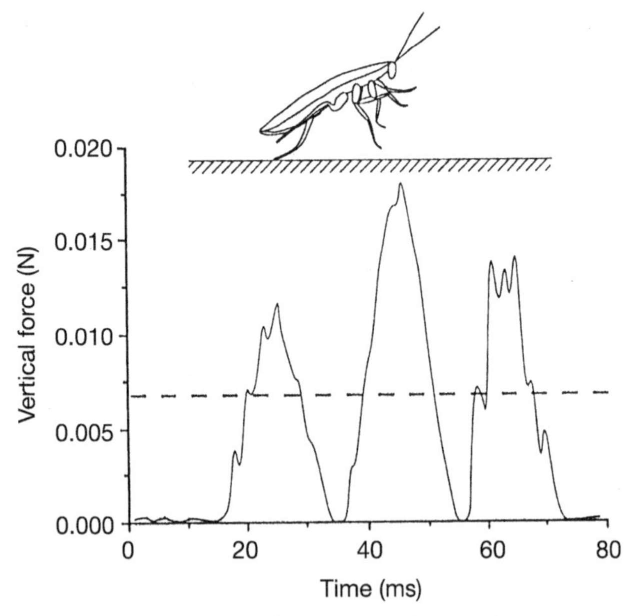

Fig. 2.1 Ground reaction force pattern for *Periplaneta americana* running bipedally, with the metathoracic legs propelling the body. Vertical forces periodically decrease to zero, indicating that all six legs are off the ground in an aerial phase. From Full and Tu (1991), with the permission of Robert J. Full and Company of Biologists Ltd.

comyn, 1971; Spirito and Mushrush, 1979). At its highest speed, *P. americana* shifts its body weight posteriorly and becomes bipedal by sprinting on its hind legs. The body is raised well off the ground and an aerial phase is incorporated into each step in a manner remarkably similar to bipedal lizards (Fig. 2.1). *Periplaneta* can cover 50 body lengths/sec in this manner (Full and Tu, 1991). As pointed out by Heinrich (2001), by that measure they can run four times faster than a cheetah. Other studied cockroaches are slower and less efficient. The maximum speed for *Blaberus discoidalis,* for example, is less than half of that of *P. americana.* The former is a more awkward runner, with a great deal of wasted motion (Full and Tu, 1991). Speed is known to vary with temperature (*Blab. discoidalis*), substrate type, sex, and developmental stage (*B. germanica*) (Wille, 1920; Full and Tullis, 1990). Hughes and Mill (1974) note that it is the ability to change direction very rapidly that often gives the impression of great speed. The ability to run swiftly and to fly effectively are not mutually exclusive. *Imblattella panamae,* a species that lives among the roots of epiphytic orchids, is fast moving both on wing and on foot (Rentz, 1987, pers. comm. to CAN). Hebard (1916a) noted that *Cariblatta,* a genus of diminutive insects, "ran about with great speed and took wing readily, though usually flying but short distances. When in flight, they appeared very much like small brownish moths." As a group, blattellids are generally very fast moving, especially when pursued. Most are long-legged with the ventral surfaces of the tarsi spined (Rentz, 1996).

Stability and Balance

Impressive locomotor performances are not limited to flat surfaces; cockroaches can scamper over uneven ground and small obstacles with agility and speed. Their vertically oriented joint axes act in concert with a sprawled posture to allow the legs to perform like damped springs during locomotion. As much as 50% of the energy used to displace a leg is stored as elastic strain energy, then returned (Spirito and Mushrush, 1979; Dudek and Full, 2000; Watson et al., 2002). In experiments on rough terrain, running *P. americana* maintained their speed and their alternating tripod gait while experiencing pitch, yaw, and roll nearly 10-fold greater than on flat surfaces (Full et al., 1998). *Blaberus discoidalis* scaled small objects (5.5 mm) with little change in running movements. Larger (11 mm) objects, however, required some changes in kinematics. The insects first assessed the obstacle, then reared up, placed their front tarsi on it, elevated their center of mass to the top of the object, then leveled off. The thorax was capable of substantial ventral flexion during these movements (Watson et al., 2002).

In a remarkable and no doubt entertaining series of experiments, Jindrich and Full (2002) studied self-stabilization in *Blab. discoidalis* by outfitting cockroaches with miniature cannons glued to the thorax. They then triggered a 10 ms lateral blast designed to knock the cockroach suddenly off balance in mid-run (Fig. 2.2). The insects successfully regained their footing in the course of a single step, never breaking stride. Stabilization occurred too quickly to be controlled by the nervous system; the mechanical properties of the muscles and exoskeleton were sufficient to account for the preservation of balance.

Fig. 2.2 *Blaberus discoidalis* with an exploding cannon backpack attempting to knock it off balance. Photo courtesy of Devin Jindrich.

There is some concern over gangs of these armed research cockroaches escaping and riddling the ankles of unsuspecting homeowners with small-bore cannon fire (Barry, 2002).

A healthy cockroach flipped onto its back is generally successful in regaining its footing. In most instances righting involves body torsion toward one side, flailing movements of the legs on the same side, and extension of the opposite hind leg against the substrate to form a strut. The turn may be made to either the right or left, but some individuals were markedly biased toward one side. In some cases a cockroach will right itself by employing a forward somersault, a circus technique particularly favored by *B. germanica* (Guthrie and Tindall, 1968; Full et al., 1995). If flipped onto its back on a smooth surface *Macropanesthia rhinoceros* is unable to right itself and will die (H. Rose, pers. comm. to CAN).

Aging cockroaches tend to dodder. There is a decrease in spontaneous locomotion, the gait is altered, slipping is more common, and there is a tendency for the prothoracic leg to "catch" on the metathoracic leg. The elderly insects develop a stumbling gait, and have difficulty climbing an incline and righting themselves (Ridgel et al., 2003).

The recent spate of sophisticated research on mechanisms of cockroach balance and control during locomotion is in part the result of collaborative efforts between robotic engineers and insect biologists to develop blattoid walking robots. The ultimate goal of this "army of biologically inspired robots" (Taubes, 2000) is to carry sensory and communication devices to and from areas that are difficult or dangerous for humans to enter, including buildings collapsed by earthquakes, bombs, or catastrophic weather events. In some cases living cockroaches have been outfitted with small sensory and communication backpacks ("biobots"), and their movement steered via electrodes inserted into the bases of the antennae (Moore et al., 1998). *Gromphadorhina portentosa* was the species selected for these experiments because they are large, strong enough to carry a reasonable communications payload, easy to maintain, and "no one would get too upset if we were mean to them" (T. E. Moore, pers. comm. to LMR). One limitation is that biobots could be employed only in the tropics or during the summer in temperate zones. Perhaps engineers should start thinking about making warm clothing for them, modeled after spacesuits (LMR, pers. obs.).

Orientation by Touch

Like many animals active in low-light conditions, cockroaches often use tactile cues to avoid obstacles and guide their locomotion. The long filiform antennae are positioned at an angle of approximately 30 degrees to the body's midline when the insect is walking or running in open spaces (*P. americana*). These serve as elongate probes that "cut a sensory swath" approximately 5.5 cm wide (Camhi and Johnson, 1999). The antennae are also used to maintain position relative to walls and other vertical surfaces. One antenna is dragged along the wall, and when it loses touch the cockroach veers in the direction of last contact. The faster they run the closer their position to the wall. Experimentally trimming the antennae also results in a path closer to the wall. The insects quickly compensate for projections or changes in wall direction, but depart from convex walls with diameters of less than 1 m (Creed and Miller, 1990; Camhi and Johnson, 1999). German cockroaches placed in a new environment tend to follow edges, but wander more freely in a familiar environment (Durier and Rivault, 2003).

GROUND LOCOMOTION: CLIMBING

The ability of a cockroach to walk on vertical and inverted horizontal surfaces (like ceilings) is predicated on specific features of the tarsi. The tarsus is comprised of five subsegments or tarsomeres. Each of the first four of these may bear on its ventral surface a single, colorless pad-like swelling called the euplanta, plantula, or tarsal pulvillus. At the apex of the fifth tarsal subsegment is a soft adhesive lobe called the arolium, which lies between two large articulated claws (Fig. 2.3). The surface of the arolium is sculptured and bears a number of different types of sensillae. Both arolia and euplantae deform elastically to assure maximum contact with a substrate and to conform to the microsculpture of its surface. Little cockroach footprints left behind on glass surfaces indicate that secretory material aids in forming a seal with the substrate. Generally, when a cockroach walks on a smooth or rough surface, some of the euplantae touch the substrate, but the arolia do not. The tarsal claws function only when the insect climbs rough surfaces, sometimes assisted by spines at the tip of the tibiae. The arolium is employed primarily when a cockroach climbs smooth vertical surfaces such as glass; the claws spread laterally and the aroliar pad presses down against the substrate (Roth and Willis, 1952b; Arnold, 1974; Brousse-Gaury, 1981; Beutel and Gorb, 2001). These structures can be quite effective; an individual of *Blattella asahinai* that landed on a car windshield was not dislodged until the vehicle reached a speed of 45 mph (= 72 kph) (Koehler and Patterson, 1987).

Cockroach species vary in the way they selectively employ their tarsal adhesive structures. *Diploptera punctata*, for example, stands and walks with the distal tarsomeres raised high above the others, and lowers them only when climbing, but in *Blaberus* the distal tarsomeres are always

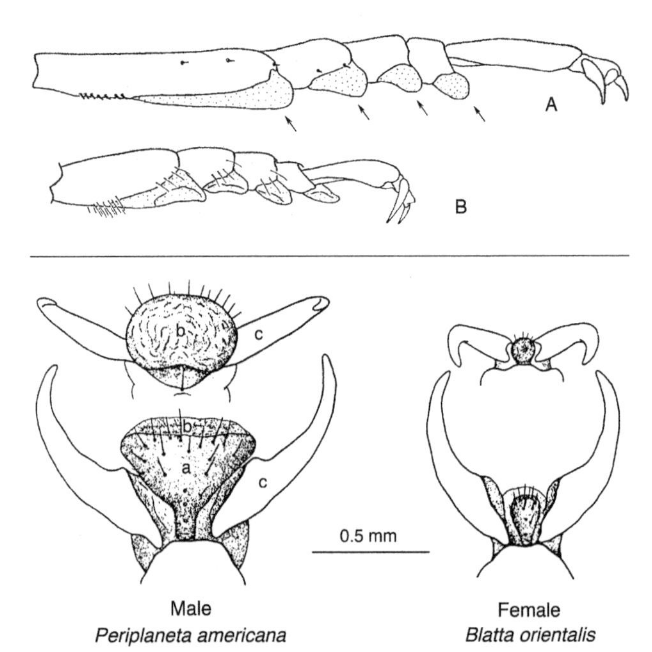

Fig. 2.3 Adhesive structures on the legs of cockroaches. *Top,* euplantae (arrows) on tarsal segments of two cockroach species. (A) Hind tarsus of male *Opisthoplatia orientalis;* (B) hind tarsus of male *Comptolampra liturata.* From Anisyutkin (1999), with permission of L.N. Anisyutkin. *Bottom,* apical and dorsal view of the pretarsi of the prothoracic legs in two cockroach species, showing the claws and arolia. *Left,* a cockroach able to walk up a vertical glass surface (male *Periplaneta americana*); *right,* one unable to do so (female *Blatta orientalis*). a = arolium; b = aroliar pad; c = tarsal claw. After Roth and Willis (1952b).

in contact with the substrate (Arnold, 1974). Within a species, there may be ontogenetic differences. Unlike adults, first instars of *B. germanica* are 50% faster on glass than they are on rough surfaces, probably because they use euplantae more than claws or spines during locomotion (Wille, 1920). Variation in employing adhesive structures is related to the need to balance substrate attachment with the need to avoid adhesion and consequent inability to move quickly on various surfaces. Both *Blatta orientalis* and *Periplaneta australasiae* walk readily on horizontal glass surfaces if they walk "on tiptoe" with the body held high off the substrate. If the euplantae of the mid and hind legs are allowed to touch the surface, they become attached so firmly that the cockroach can wrench itself free only by leaving the tarsi behind, clinging to the glass (Roth and Willis, 1952b).

Tarsal Morphology: Relation to Environment

Cockroaches vary in their ability to climb (i.e., escape) glass containers (Willis et al., 1958). This is due principally to the development of the arolium, which varies in size, form, and sculpturing and may be absent in some species (Arnold, 1974). *Blatta orientalis,* for example, has subobsolete, nonfunctional arolia and is incapable of climbing glass (Fig. 2.3). Euplantae may also differ in size and shape on the different tarsomeres, be absent from one or more, or be completely lacking. The presence or absence of these adhesive structures can be used as diagnostic characters in some genera (e.g., the genus *Allacta* has euplantae only on the fourth tarsomere of all legs), but are of minor taxonomic significance in others (e.g., the genera *Tivia, Tryonicus, Neostylopyga, Paratemnopteryx*) (Roth, 1988, 1990b, 1991d). Intraspecifically, variation may occur among populations, between the sexes, and among developmental stages (Roth and Willis, 1952b; Mackerras, 1968a). In *Paratemnopteryx* (= *Shawella*) *couloniana* and *Neotemnopteryx* (= *Gislenia*) *australica* euplantae are acquired at the last ecdysis (Roth, 1990b).

Although arolia and euplantae are considered adaptive characters related to functional requirements for climbing in different environments (Arnold, 1974), it is not currently obvious what habitat-related features influence their loss or retention in cockroaches. Adhesive structures are frequently reduced or lost in cave cockroaches, perhaps because clinging mud or the surface tension of water on moist walls reduces their effectiveness (Mackerras, 1967c; Roth, 1988, 1990b, 1991a). It would be instructive to determine if the variation in adhesive structures exhibited by different cave populations of species like *Paratemnopteryx stonei* can be correlated with variation among surfaces in inhabited caves. Arolia are absent in all Panesthiinae (Mackerras, 1970), and the two cockroaches listed by Arnold (1974) as having both arolia and euplantae absent or "only vaguely evident"—*Arenivaga investigata* and *Cryptocercus punctulatus*—are both burrowers. Nonetheless, the loss of arolia and euplantae is not restricted to cave and burrow habitats (Roth, 1988); many epigean species lack them. Arnold (1974) found it "surprising" that the tarsal features are so varied within cockroach families and among species that inhabit similar environments. A number of authors, however, have emphasized that it is the behavior of the animal within its habitat, rather than the habitat itself, that most influences locomotor adaptations (Manton, 1977; Evans and Forsythe, 1984; Evans, 1990). The presence and nature of appendage attachment devices is thought to be strongly associated with a necessity for negotiating smooth, often vertical plant surfaces (Gorb, 2001). Thus in a tropical forest, a cockroach that perches or forages on leaves during its active period may retain arolia and euplantae, but these structures may be reduced or lost in a species that never ventures from the leaf litter. Pulvilli and arolia are very well developed, for example, in *Nyctibora acaciana,* a species that oviposits on ant-acacias (Deans and Roth,

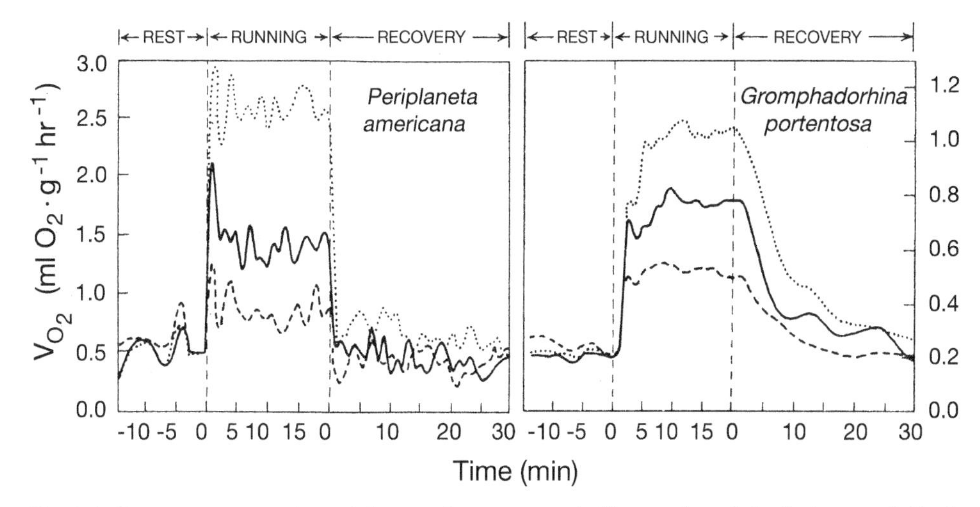

Fig. 2.4 Oxygen consumption while running on a treadmill: a cockroach built for speed (*Periplaneta americana*) versus one built for power (*Gromphadorhina portentosa*). Oxygen peaks rapidly in *P. americana*, and afterward the insect recovers rapidly. There is a lag time before oxygen peaks in *G. portentosa*, and a slow recovery time while the insect "catches its breath." Note difference in scale of y-axis. Reprinted from Herreid and Full (1984), with permission from Elsevier.

2003). In cockroaches that possess them, variation in sculpturing on the arolia may function in maximizing tenacity and agility on specific plant surface morphotypes (Bernays, 1991). Many species of tropical cockroach do not run when on leaves, but instead stilt-walk (WJB, pers. obs.). The slow leg movements produce little vibration in the substrate, and may allow them to ease past spiders without eliciting an attack, a phenomenon called "vibrocrypticity" (Barth et al., 1988).

GROUND LOCOMOTION: POWER

At the other end of the spectrum from sleek, fast-running cockroaches such as *P. americana* are the muscular, shorter-legged species that burrow into soil or wood. Their legs are usually ornamented with sturdy spines, particularly at the distal end of the tibiae; these function to brace the insect against the sides of the burrow, providing a stable platform for the transmission of force. Fossorial cockroaches are built for power, not speed. When forced to jog on a treadmill, all tested cockroach species exhibited a classic aerobic response to running; oxygen consumption (VO_2) rapidly rose to a steady state that persisted for the duration of the workout. When exercise was terminated, the recovery time of *P. americana* and *Blab. discoidalis* rivaled or exceeded the performance of the best vertebrate runners (Fig. 2.4). Among the slowest to recover was the heavy-bodied *G. portentosa*, which took 15–45 min, depending on the speed of the run (Herreid et al., 1981; Herreid and Full, 1984). Some individuals of *G. portentosa* exhibited obvious signs of fatigue. They stopped, carried their body closer to the substrate, and had a hard time catching their breath: respiratory movements were exaggerated and the insects maintained their spiracles in a wide-open position.

Burrowing

Digging behavior in cockroaches has not been studied, but the little, mostly anecdotal information we have indicates substantial variation, both in the behavior employed and in the body part used as a digging tool. There are at least two modes of creating tunnels in a hard substrate (soil, wood), both of which are accomplished by moving the substrate mechanically from in front of the insect and depositing it elsewhere. There are also two methods of digging into more friable material (guano, leaf litter, sand), achieved by insinuating the body into or through preexisting spaces. Cockroaches use refined excavation and building techniques in burying oothecae (Chapter 9).

Scratch-Digging (Geoscapheini)

All members of the uniquely Australian Geoscapheini excavate permanent underground living quarters in the compact, semi-arid soils of Queensland and New South Wales. The unbranched burrows of *M. rhinoceros* can reach a meter beneath the surface (Chapter 10); the tunnel widens near the bottom into a compartment that functions as a nursery and a storage chamber for the dried vegetation that serves as food. The distal protibiae are impressively expanded to act as clawed spades, driven by the

Fig. 2.5 *Macropanesthia rhinoceros,* initiating descent into sand; photo courtesy of David Rentz. *Inset:* Detail of mole-like tibial claw used for digging; photo courtesy of Kathie Atkinson.

large muscles of the bulky body (Fig. 2.5). The hard, stout spines flick the soil out behind the cockroach as it digs. When the insect is moving through an established burrow, the spines fold neatly out of the way against the shank of the tibia. The tarsi are small and dainty (Park, 1990). The large, scoop-like pronotum probably serves as a shovel. Tepper (1894) described the behavior of *Geoscapheus robustus* supplied with moist, compressed soil: "they employ not only head and forelegs, but also the other two pairs, appearing to sink into the soil without raising any considerable quantity above the surface, nor do they appear to form an unobstructed tunnel, as a part of the dislodged soil appears to be pressed against the sides, while the remainder fills up the space behind the insect. A few seconds suffice them to get out of sight." Soil texture and compaction no doubt determine the energetic costs of digging and whether burrows remain open or collapse behind the excavator.

Tooth-Digging (Cryptocercidae)

Cryptocercus spp. chew irregular tunnels in rotted logs, but the tunnels are clearly more than a by-product of feeding activities. Numerous small pieces of wood are obvious in the frass pushed to the outside of the gallery. When entering logs, the cockroaches often take advantage of naturally occurring crevices (knotholes, cracks), particularly at the log-soil interface. Burrows then generally follow the pattern of moisture and rot in individual logs. Rotted spring wood between successive annual layers is often favored. In well-rotted logs, the cockroaches will in part mold their living spaces from damp frass. In fairly sound logs, galleries are only slightly larger than the diameter of the burrower, and may be interspersed with larger chambers (Nalepa, 1984, unpubl. obs.).

Adult *Cryptocercus* have been observed manipulating feces and loosened substrate within galleries. The material is pushed to their rear via a metachronal wave of the legs. The insect then turns and uses the broad surface of the pronotum to tamp the material into place. The tarsi are relatively small, and stout spines on the tibiae serve to gain purchase during locomotion. The cockroach is often upside down within galleries, and like many insects living in confined spaces (Lawrence, 1953), frequently walks backward, allowing for a decrease in the number of turning movements. The body also has a remarkable degree of lateral flexion, which allows the insect to bend nearly double when reversing direction in galleries (CAN, unpubl. obs).

Sand-Swimming (Desert Polyphagidae)

During their active period, fossorial desert Polyphagidae form temporary subsurface trails as they "swim" through the superficial layers of the substrate. Their activities generate a low rise on the surface as the loosely packed sand collapses in their wake. The resultant serpentine ridges look like little mole runs (Fig. 2.6) (Hawke and Farley, 1973). During the heat of day, the cockroaches (*Arenivaga*) may burrow to a depth of 60 cm (Hawke and Farley, 1973). The bodies of adult females and nymphs are streamlined, with a convex thorax and sharp-edged pronotum. Tibial spines on the short, stout legs facilitate their pushing ability and serve as the principal digging tools. These spines are often flattened or serrated, with sharp tips. Anterior spines are sometimes united around the apex in a whorl, forming a powerful shovel (Chopard, 1929; Friauf and Edney, 1969). *Eremoblatta subdiaphana*, for example, has seven spines projecting from the front tibiae (Helfer, 1953). Also aiding subterranean move-

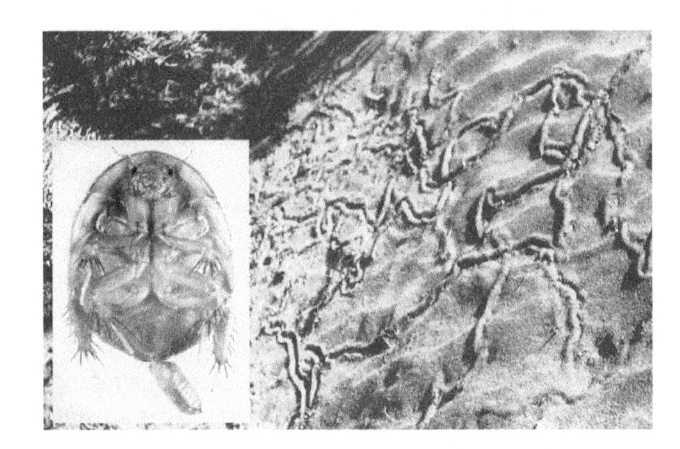

Fig. 2.6 Tracks (2–3 cm wide) of *Arenivaga* sp. at the base of a mesquite shrub near Indigo, California. Females and nymphs burrow just beneath the surface at night. From Hawke and Farley (1973), courtesy of Scott Hawke. *Inset:* Ventral view of female *Arenivaga cerverae* carrying an egg case. The orientation of the egg case is likely an adaptation for carrying it while the female "swims" through the sand. Note well-developed tibial spines. Photo by L.M. Roth and E.R. Willis.

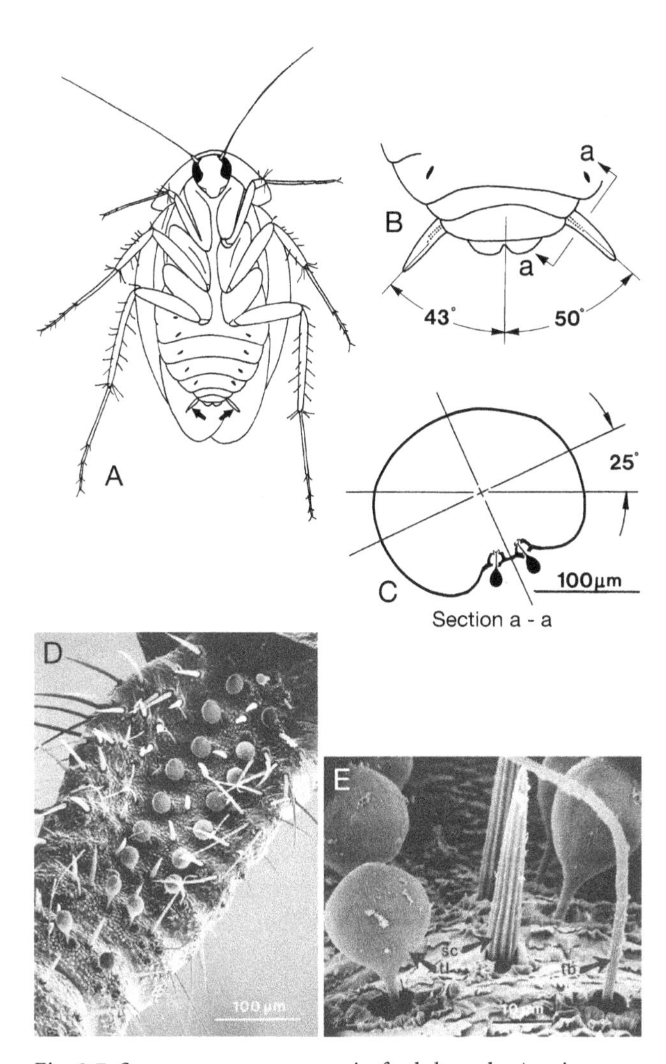

Fig. 2.7 Sensory organs on cerci of adult male *Arenivaga* sp. (A) Ventral view of insect, with the cerci indicated by arrows. (B) Posterior end of the abdomen showing the orthogonal position of the cerci and rows of tricholiths. (C) Cross section through the left cercus to illustrate that the cerci are rotated laterally from the horizontal plane. (D–E) Scanning electron micrographs showing details of tricoliths on the cerci. (D) Ventral view of left cercus; note two parallel rows of tricholiths. (E) View from the distal end of the tricholith (tl) rows showing sensilla chaetica (sc) and a trichobothrium (tb). Courtesy of H. Bernard Hartman. From Hartman et al. (1987), with permission from Springer Verlag.

ments are large spherical sense organs (tricholiths) on the ventral surface of the cerci in *Arenivaga* and other polyphagids (Roth and Slifer, 1973). These act like tiny plumb bobs in assisting orientation of the cockroaches while they move through their quasifluid environment (Walthall and Hartman, 1981; Hartman et al., 1987) (Fig. 2.7). First instars of *Arenivaga* have only one tricholith on each cercus; new ones are added at each molt. Adult females have six pairs and males have seven pairs (Hartman et al., 1987).

Head-Raising (*Blaberus craniifer*)

In studying the burrowing tendencies of *Blab. craniifer,* Simpson et al. (1986) supplied the cockroaches with a mixture of peat moss and topsoil, then filmed them as they dug into the substrate. The insects were able to bury themselves in just a few seconds using a rapid movement of the legs, combined with a stereotyped dorsal-ventral flexion of the head and pronotum. The combined head-raising, leg-pushing behavior seems well suited to digging in light, loose substrates (litter, dust, guano), but may also facilitate expanding existing crevices, like those in compacted leaf litter or under bark. This digging technique does not require the profound body modifications exhibited by cockroaches specialized for burrowing in hard substrates, and is therefore compatible with the ability to run rapidly. Indeed, the behavior seems well suited to the "standard" cockroach body type displayed by *Blab. craniifer:* an expanded, hard-edged pronotum, inflexed head, slick, flattened, rather light body, and moderately strong, spined legs.

SWIMMING

It seems logical that cockroaches are not easily drowned, as they are members of a taxon whose ancestors were associated with swamp habitats and "almost certainly able to swim" (North, 1929). As anyone who has tried to flush a cockroach down the toilet can verify, these insects have positive buoyancy and will bob to the surface of the water if forced under. A water-repellent cuticle aids surface tension in keeping them afloat (Baudoin, 1955). *Periplaneta americana* is a fine swimmer, and can move in a straight line at 10 cm/sec. The body is usually arched, with the antennae held clear of the water and moving in normal exploratory fashion. If the antennae touch a solid substrate, the insect turns toward the source of stimulation and swims faster. While swimming, the legs are coordinated in the same alternating tripod pattern seen while walking on land; this differs from the pattern of synchronous leg pairs seen in other terrestrial and aquatic insects in water. Articulated spines on the tibia of each leg are strongly stimulated by movement through water and may provide feedback in regulating swimming behavior. All developmental stages can swim, but the youngest instars are hampered by surface tension (Lawson, 1965; Cocatre-Zilgein and Delcomyn, 1990).

Most *P. americana* isolated on an artificial island will escape within 10 min, with escape more rapid in experienced insects (Lawson, 1965). Two strategies are employed, reminiscent of those seen in humans at any swimming pool. (1) Gradual immersion (the "wader"): the surface of the water is first explored with the forefeet (Fig.

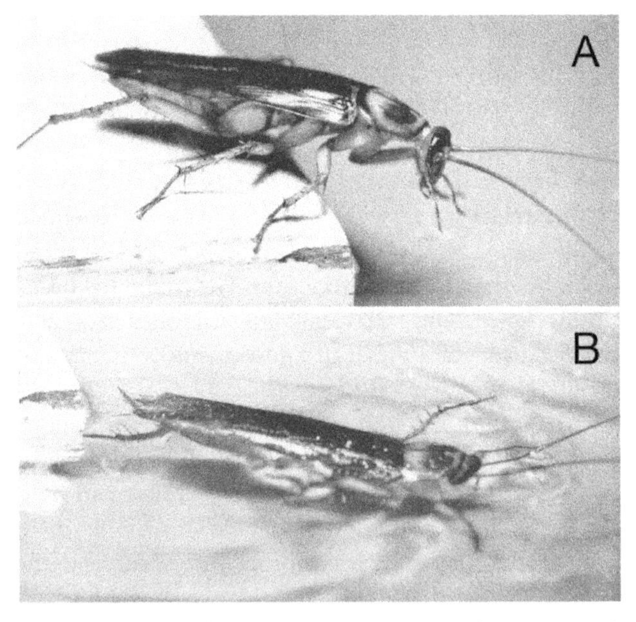

Fig. 2.8 (A) *Periplaneta americana* testing the water with forelegs before (B) taking the plunge. Courtesy of R.M. Dobson.

2.8). The middle legs then attempt to reach the bottom beneath the water, while clinging to the island with the rear legs and with the front of the body afloat. Finally, the cockroach releases the hind legs, enters completely, and swims away. (2) The "cannonball" strategy: after initial exploration, the insect retires slightly from the edge, crouches, then jumps in, often while fluttering the wings.

The legs of amphibious cockroaches do not exhibit any morphological adaptations for swimming and are no different from those of non-aquatic species (Shelford, 1909; Takahashi, 1926). Nymphs of many *Epilampra* spp. swim rapidly below the surface (Crowell, 1946; Wolcott, 1950); newborn nymphs as well as adults of *Ep. wheeleri* (= *Ep. abdomennigrum*) swim easily and remain under water a good deal of the time (Séin, 1923). Individuals of *Poeciloderrhis cribrosa verticalis* can swim against a current velocity of 0.15 m/sec (Rocha e Silva Albuquerque et al., 1976). *Opisthoplatia maculata,* on the other hand, rarely swims, but instead walks on submerged rocks along stream bottoms (Takahashi, 1926).

WINGS AND FLIGHT

Adult cockroaches with fully developed flight organs have two sets of wings that reach or surpass the end of the abdomen, completely covering the abdominal terga. The hindwings are membranous, but the forewings (tegmina) are somewhat sclerotized. In most species the tegmina cross each other, with the left tegmen covering a portion of the right, and with the covered portion of a different texture and color. There are also cases where the forewings are transparent and similar in size and texture to the hindwings (e.g., *Paratemnopteryx suffuscula, Pilema cribrosa, Nocticola adebratti, Cardacus* (= *Cardax*) *willeyi*), or hardened and elytra-like (e.g., *Diploptera* and other beetle mimics).

The entire wing apparatus of cockroaches shows clear adaptations for a concealed lifestyle (Brodsky, 1994). Dorsoventral flattening has altered the structure of the thoracic skeleton and musculature, and when at rest the wings are folded flat against the abdomen. One exception is *Cardacopsis shelfordi,* whose wings do not lie on the abdomen with the tips crossing distally, but diverge as in flies (Karny, 1924 in Roth, 1988). Elaborate mechanisms of radial and transverse folding allow the delicate hindwings to fit under the more robust tegmina. In repose, the anal lobe of the hindwing is always tucked under the anterior part of the wing (remigium). Polyphagids accomplish this with a single fold line (Fisk and Wolda, 1979), but in other cockroaches this area is folded along radial lines into a simple fan. There may be apical rolling (e.g., *Prosoplecta nigrovariegata, Pr. coccinella, Choristima* spp.) or folding (e.g., *Anaplecta*) of the remigium. In some species (e.g., *D. punctata*), this crease is in the middle of the wing, allowing for a folded wing with only half the length and a quarter of the area of the unfolded wing (Fig. 2.9). These more elaborate strategies of wingfolding are common in beetle mimics, as it allows for the protection of hindwings that exceed the length of the tegmina (Shelford, 1912a; Roth, 1994). Patterns of wingfolding, together with other wing characters, can be useful in cockroach classification (Rehn, 1951; Haas and Wootton, 1996; Haas and Kukalova-Peck, 2001). A number of generic names originate from wing characters, for example, *Plecoptera* (Gr., plaited + wing), *Chorisoneura* (Gr., separate + veins), *Symploce* (Gr., woven together), *Ischnoptera* (Gr., slender + wing) (Blatchley, 1920).

Cockroaches are "hindmotor" flyers. The hindwing is

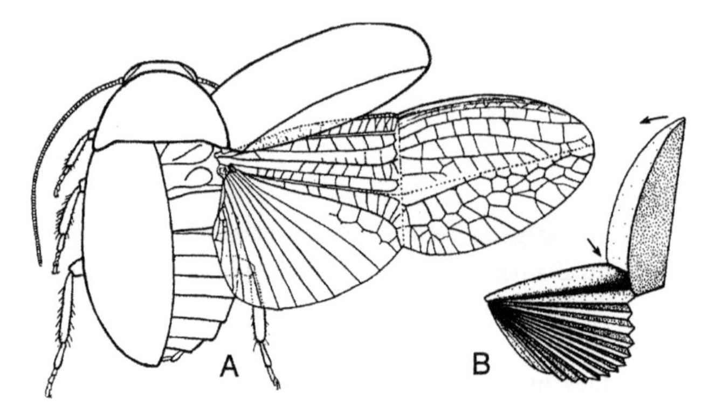

Fig. 2.9 Wing folding in *Diploptera punctata;* (A) dorsal view, right tegmen and wing expanded, longitudinal and transverse folds marked as dotted lines; from Tillyard (1926). (B) Posterodorsal view of a wing in the process of folding. Drawing by Robin Wootton, courtesy of Robin Wootton and Fabian Haas.

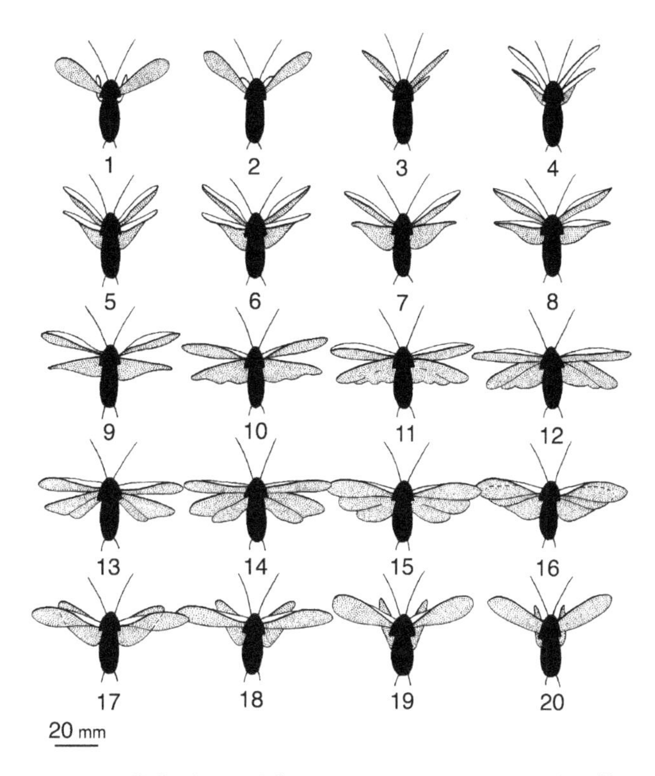

Fig. 2.10 Flight in *Periplaneta americana;* consecutive film tracings of a single wingbeat. The forewings reach the top of the stroke just as the hindwings pass the top of the stroke and begin to pronate (#3). As a result, both pairs pronate nearly simultaneously (#4), so that the hindwings, moving faster, are ahead of the forewings (#5), approach the bottom of the stroke, supinate, and go up (#12–20). From Brodsky (1994), by permission of Oxford University Press.

the main source of propulsion (Brodsky, 1994), and the two pairs of wings operate independently and slightly out of phase (Fig. 2.10). In basal cockroaches the tegmina seem to be an integral part of the flight mechanism, but in the more derived species their direct use in flight is less common (Rehn, 1951). During flight, aerodynamically induced bending of the cerci serves as a feedback in regulating wingbeat frequency (Lieberstat and Camhi, 1988). It is generally believed that the majority of winged cockroaches are rather inept fliers and lack the ability to sustain long-distance flight (Peck and Roth, 1992). Flight ability within the group varies, of course, and even weak fliers can be quite maneuverable in the air, with various strategies for evading predators. A number of small tropical species are known to be strong fliers, capable of sustained flights in a straight line or with slight lateral curves. They are able to increase altitude but cannot hover (Farnsworth, 1972).

Wing Reduction and Flightlessness

All taxonomic groups of cockroaches include species with variably reduced or absent tegmina and hindwings, exposing all or part of the dorsal surface of the abdomen. The exceptions are those groups in which the distal portion of the hindwing is set off by a transverse fold (e.g., Diplopterinae, Ectobiinae, Anaplectinae—Rehn, 1951). Wing reduction typically affects the hindwings more than the tegmina (Peck and Roth, 1992). Even when they are reduced, wings are always flexibly joined to the thorax. Adults with reduced wings can be distinguished from older nymphs, then, because the wing pads of the latter are nonflexible extensions of the posterior margins of the wing-bearing thoracic segments (Fisk and Wolda, 1979). Although in some cockroach groups apterous species are tiny and may be passed over by collectors because they resemble nymphs (Mackerras, 1968a), some of the largest known cockroaches (*Macropanesthia*) also lack wings.

Based on information in Rehn (1932b) and Roth and Willis (1960), Roff (1990, Table 8) estimated that more than 50% of all cockroaches and 50–60% of temperate species lack the ability to fly. Vastly different figures also have been published. Roff (1994) indicated that just 4% of cockroaches are flightless in both sexes, and 24% are sexually dimorphic, with males flying and females flightless (data from North America, French Guiana, Africa, and Malagasy). There are reasons to be cautious when assessing cockroach flight ability. First, only a fraction of the more than 4000 known cockroach species are included in these estimates; volant canopy species in particular may be underestimated. Second, flight capability in cockroaches is typically based on published descriptions of wing morphology in museum specimens. The possession of fully developed wings, however, does not necessarily mean that a cockroach can fly (Farnsworth, 1972; Peck and Roth, 1992).

A more accurate measure of cockroach flight capability may lie in the color of the thoracic musculature of freshly killed insects. Kramer (1956) found that the pterothoracic musculature of apterous, brachypterous, and flightless or feebly flying macropterous cockroaches appears hyaline white, while that of strong fliers is opaque and conspicuously pink (Table 2.1). These color differences are correlated with distinct metabolic differences, as reflected in enzymatic activity and oxygen uptake (Kramer, 1956). Consequently, cockroaches with white musculature may not be able to release energy rapidly enough to sustain wing beating (Farnsworth, 1972). In cockroaches with pink musculature, the muscles of the mesothorax and metathorax are equally pigmented. One exception is the "beetle" cockroach *D. punctata* (= *dytiscoides*), which derives its common name from the fact that the somewhat reduced, hardened tegmina resemble elytra and cover a pair of long hindwings (Fig. 2.9). In this species the mesothoracic muscles are hyaline white, but the metathorax bearing the elongated hindwings con-

Table 2.1. Wing development and its relationship to pigmentation of the thoracic musculature. Based on Kramer (1956) and Roth and Willis (1960).

Cockroach species	Color of pterothoracic musculature	
	Mesothorax (wing condition)[1]	Metathorax (wing condition)
Blaberus craniifer	Pink (M)	Pink (M)
Blaberus giganteus	Pink (M)	Pink (M)
Blatta orientalis	White (R)	White (R)
Blattella germanica	White (M)	White (M)
Blattella vaga	Pink (M)	Pink (M)
Cryptocercus punctulatus	White (A)	White (A)
Diploptera punctata	White (R)	Pink (M)
Eurycotis floridana	White (R)	White (R)
Nauphoeta cinerea	White (R)	White (R)
Neostylopyga rhombifolia	White (R)	White (R)
Parcoblatta pennsylvanica		
Male	Pink (M)	Pink (M)
Female	White (R)	White (R)
Parcoblatta virginica		
Male	Pink (M)	Pink (M)
Female	White (R)	White (R)
Periplaneta fuliginosa		
Male	Pink (M)	Pink (M)
Female	White (M)	White (M)
Periplaneta brunnea		
Male	Pink (M)	Pink (M)
Female	White (M)	White (M)
Periplaneta australasiae		
Male	Pink (M)	Pink (M)
Female	White (M)	White (M)
Pycnoscelus surinamensis[2]	Pink (M)	Pink (M)
Rhyparobia maderae	Pink (M)	Pink (M)
Supella longipalpa		
Male	Pink (M)	Pink (M)
Female	White (R)	White (R)

[1]M = macropterous, R = reduced, A = absent.
[2]Female morphs with reduced wings exist.

tains pigmented muscle (Kramer, 1956). Macropterous adults with white musculature include *Blattella germanica*, females of *Supella longipalpa* (= *supellectilium*), and three species of *Periplaneta*. Both sexes of *B. germanica* and *Blattella vaga* have fully developed wings (see Plate 5 of Roth and Willis, 1960), but *B. germanica* is incapable of sustained flight (Brenner et al., 1988).[2] The rosy flight muscles of *B. vaga* are an indication that it is volant, but its flight behavior is unknown. The Asian cockroach *Blattella asahinai* is morphologically very similar (Lawless, 1999) and very closely related (Pachamuthu et al., 2000) to *B. germanica*, but flies readily and strongly (Brenner et al., 1988); presumably, dissections would indicate that it has pigmented flight muscles. Males of *Su. longipalpa* are fleet runners and can take to the air for short distances, but females are unable to fly (Hafez and Afifi, 1956). Another example of a macropterous but flightless species is *Thorax porcellana* (Epilamprinae). Both sexes are fully winged, but only the male uses them for short flights and only rarely (Reuben, 1988).

The correlation between flight muscle pigmentation and the physiological ability to sustain flight has been examined most extensively in *P. americana*. In tests on laboratory strains tethered females (white flight muscles) could sustain no more than a 3–12 sec flight, compared to 5–15 min in males (pink flight muscles). Moreover, freshly ecdysed male *P. americana* have white pterothoracic muscles and flight behavior similar to that of adult females: they flutter weakly or plummet when tossed into the air. The flight behavior of these young males changes in conjunction with the postmetamorphic development of pink pigmentation in their musculature (Kramer, 1956; Farnsworth, 1972; Stokes et al., 1994). In the tropics *P. americana* is reportedly an excellent flyer, and is known in some locales as the "Bombay canary." It has been observed flying out of sewers and into buildings. It was also spotted in a German zoo flying distances of up to 30 m, in fairly straight lines or in flat arcs about 0.5 to 1.5 m above the ground (Roth and Willis, 1957). It is unclear, however, whether these volant *P. americana* are males only, or if both sexes in natural populations can fly. Rehn (1945) indicated that the flying ability of *Periplaneta* (species unspecified) is "often exercised and by both sexes." Female *P. americana* from laboratory cultures in two U.S. locations and one in Germany, however, remained earthbound during flight tests (Kramer, 1956). Appel and Smith (2002) report that *P. fuliginosa* females with fully formed oothecae are capable of sustained flight on warm, humid evenings in the southern United States, but laboratory-reared females of this species sank like rocks when tossed in the air (Kramer, 1956). Perhaps females lose the ability to fly when raised in culture. At least one study demonstrated that flight initiation in *P. americana* was significantly affected by the temperature at which they were reared (Diekman and Ritzman, 1987), and flight performance in other insects is known to quickly suffer under laboratory selection (Johnson, 1976).

A physiological change in flight musculature no doubt precedes or accompanies morphological wing reduction, but may be the only modification if the tegmina and wings have a functional significance other than flight. Full-sized wings may be retained in flightless species be-

2. It is, however, a frequent flier on airplanes (Roth and Willis, 1960).

cause they may act as parachutes, controlling the speed and direction of jumps and falls. German cockroaches, for example, will glide short distances when disturbed (Koehler and Patternson, 1987). Tegmina and wings may be used as tools in territorial or sexual signaling; males in several species flutter their wings during courtship. They also may serve as stabilizers during high-speed running, as physical protection for the abdomen and associated tergal glands, in visual defense from enemies (crypsis, mimicry, aposematicism), and, in rare cases, as shelter for first instars.

Ecological Correlates of Flight Condition

A number of papers have focused on the ecological determinants that may select for wing retention versus loss in various insect groups. Chopard (1925) was the first to examine the phenomenon in cockroaches, and divided cockroach genera into one of three wing categories: (1) tegmina and hindwings developed in both sexes; (2) wings short or absent in females only; and (3) wings short or absent in both sexes. He then arranged genera by collection locality and concluded that flightlessness was correlated with certain geographic locations. Rehn (1932b), however, demonstrated that each of the three listed conditions can be displayed by different species within the same genus, and refuted the idea that flightlessness was correlated with geography. Rehn could find no single factor that selected for wing reduction in the cockroaches he studied (New World continental and West Indian species), but thought that "altitude and possibly humidity or aridity under special conditions" might be involved. More recently, Roff (1990, Table 1) surveyed the literature and concluded that cockroaches as well as other insects that live in deserts, caves, and social insect nests have a higher than average incidence of flightlessness. He also found that a lack of flight ability was not exceptionally high on islands, in contrast to conventional thought.

Generalizations on the correlation between flight ability and habitat are difficult to make for cockroaches. With few exceptions, conclusions are based on wing length, and habitat type is inferred from daytime resting sites or baited traps. As discussed above, the possession of full-sized wings is not always a reliable index of flight ability, and the location of diurnal shelter is only a partial indication of cockroach habitat use. Although it is safe to assume that cockroaches attracted to light traps have some degree of flight ability, the traps collect only night-active species that are attracted to light, and the ecological associations of these remain a mystery. Males of *Neolaxta*, for example, are very rarely seen in the field, but can be collected in considerable numbers from light traps (Monteith, in Roth, 1987a). Given those caveats (there will be more later), we will here examine wing trends in some specific habitat categories.

Islands

Darwin (1859) first suggested that the isolation imposed by living on an island selects for flightless morphologies, because sedentary organisms are less likely to perish by being gusted out to sea. More recent authors, however, have questioned the hypothesis (e.g., Darlington, 1943). For one thing, scale is not taken into account. Conditions are different for a large insect on a small island versus a tiny insect on a substantial one (Dingle, 1996). Roff (1990) analyzed the wing condition of insects on oceanic islands versus mainland areas (corrected for latitude) and found no correlation between island life and a sedentary lifestyle. Denno et al.'s (2001a) work on planthoppers in the British Virgin Islands also supports this view.

The observation that a flightless cockroach lives on an island does not necessarily mean that the wingless condition evolved there. Cockroaches have greater over-water dispersal powers than is generally assumed, because they raft on or in floating debris and vegetation (Peck, 1990; Peck and Roth, 1992). Moreover, cockroaches that live under bark or burrow in wood or other dead vegetation may be the most likely sailors; this category includes a relatively high percentage of wing-reduced species (discussed below). Trewick (2000) recently analyzed DNA sequences in the blattid *Celatoblatta*, a flightless genus found in New Zealand and in the Chatham Islands, habitats separated by about 800 km of Pacific Ocean. The island populations were monophyletic, and probably dispersed from New Zealand to the islands by rafting sometime during the Pliocene (2–6 mya). Members of this genus are known to shelter in logs during the day.

When six small mangrove isles off the coast of Florida were experimentally sterilized, *Latiblattella rehni* and an undescribed species in the same genus were early re-invaders on several of them (Simberloff and Wilson, 1969). Males of *Lat. rehni* have fully developed, "very delicate" (Blatchley, 1920) wings; those of the female are slightly reduced, but it is unknown if they are functional. Colonization, then, could have been by active or passive flight, or by rafting. The Krakatau Islands offered a unique opportunity to study the reintroduction of cockroaches into a tropical ecosystem from a sterile baseline after a series of volcanic eruptions in 1883 stripped them of plant and animal life. A 1908 survey found a few cockroach species already present, with a subsequent steep colonization curve that flattened out after the 1930s (Thornton et al., 1990). The 17 species reported from the islands by 1990 include pantropical species (*P. americana*, *Blatta orientalis*) probably introduced by humans, fully winged species (e.g., *Balta notulata*, *Haanina major*), those with reduced wings (*Lobopterella dimidiatipes*), and species in which there is a great deal of variation in wing reduction in both sexes (e.g., *Hebardina concinna*). *Neostylopyga picea,* which has short tegminal pads and lacks wings, also is present on the islands and probably arrived by rafting. It is generally found in humus and decaying wood (Roth, 1990a).

Studies in the Galapagos offer the best evidence that the evolution of flightlessness may occur on islands. Eighteen species are reported on the Galapagos (Peck and Roth, 1992). Of these, the introduced or native (naturally occurring tropical American and Galapagos) cockroaches are fully winged as adults, except for female *Symploce pallens.* The five endemic species are all partially or wholly flightless. Peck and Roth (1992) suggest that three natural colonization events took place. First, an early colonization by *Ischnoptera* and loss of flight wings in three descendent species, a later colonization by *Chorisoneura* and partial reduction of flight wings in two descendent species, and lastly, a recent colonization by *Holocampsa nitidula* and perhaps another *Holocampsa* sp. These authors give a detailed analysis of the process of wing reduction in the studied cockroaches, and conclude that their data fit the generalization that loss of flight capability often accompanies speciation on islands. The authors do note, however, that the flightless condition "may not be a result of island life per se, but may be a specialization for life in more homogenous leaf litter or cave habitats at higher elevations on the islands."

Mountains

There are several indications that wing reduction or loss in cockroaches may be correlated with altitude. On Mt. Kilimanjaro in Africa, for example, fully alate *Ectobius africanus* females were collected only below 1000 m (Rehn, 1932b). In Australia, males in the genus *Laxta* may be macropterous, brachypterous, or apterous, but all known females lack wings. In the two cases where males are not fully winged, both were collected at altitude: *Lax. aptera* (male apterous) from the Brindabella Ranges and Snowy Mountains, and *Lax. fraucai* (male brachypterous) from northeastern Australia at 670–880 m (Mackerras, 1968b; Roach and Rentz, 1998; Roth, 1992). Although most *Ischnoptera* species are fully winged, the flightless *Ischnoptera rufa debilis* occurs at high altitude in Costa Rica (Fisk, 1982). The metabolic cost of flight may be substantial at the cold temperatures typical of high elevations (Wagner and Liebherr, 1992).

Deserts

Females of desert cockroach species are generally apterous or brachypterous, but males are fully alate (Rehn, 1932b). The high cost of desiccation during flight may account for many cases of wing reduction in desert insects (Dingle, 1996), but may be less of a problem for night-active insects like many Blattaria. Rehn (1932b) noted that the number of brachypterous and subapterous cockroaches in deserts was comparable to that of humid rainforest areas of tropical America. It has been suggested that the strong tendency for wing reduction among all families of Australian cockroaches (Mackerras, 1965a) is a response to desert conditions (Chopard, in Rehn, 1932b). Almost all of the large Australian group Polyzosteriinae are brachypterous or apterous, but not all live in the desert. *Scabina antipoda,* for example, is brachyterous and found under bark in the rainforests of eastern Australia (Roach and Rentz, 1998).

Insect Nests

Cockroaches adapted to living in the nests of social insects are always apterous or have wings reduced to varying degrees. *Pseudoanaplectinia yumotoi,* associated with *Crematogaster* sp. ants in canopy epiphytes in Sarawak, is among those with the longest wings. The tegmina and wings reach to about the sixth tergite in the female, and to about the supra-anal plate in the male (Roth, 1995c); it is unknown as to whether these allow for flight. Females of *Nocticola termitofila,* from nests of *Termes* sp. and *Odontotermes* sp. termites, are apterous (Fig. 1.16C). Males are brachypterous, with transparent wings about half the length of the abdomen (Silvestri, 1946); these are fringed around the edges (like thrips) and may allow for passive wind transport. *Attaphila* living in the fungus gardens of leaf-cutting ants have apterous females and brachypterous or apterous males (Gurney, 1937; Roth, 1991a). Both *Att. fungicola* and *Att. bergi* have evolved a unique solution for moving between nests—they are phoretic on ant alates leaving the nest on their mating flight (Fig. 2.11) (Wheeler, 1900; Bolívar, 1901; Moser, 1964; Waller and Moser, 1990). These myrmecophiles have large arolia (Gurney, 1937) that may assist them in clinging to their transport. Several questions arise concerning this phoretic relationship. Do both male and female cockroaches disperse with the alates, or only fertilized females? Since the nuptial flight of male ants is invariably fatal (Hölldobbler and Wilson, 1990), do the cockroaches choose the sex of their carrier? If cockroaches do choose male alates, perhaps they can transfer to female alates while the ants are copulating. The vast majority of the thousands of released virgin queens die within hours of leaving the nest (Hölldobbler and Wilson, 1990); do their associated cockroaches subsequently search for nests on foot? Because they disperse together, would molecular analysis reveal a co-evolutionary relationship between this myrmecophile and its host? A comparison of *Attaphila* to *Myrmecoblatta wheeleri* also

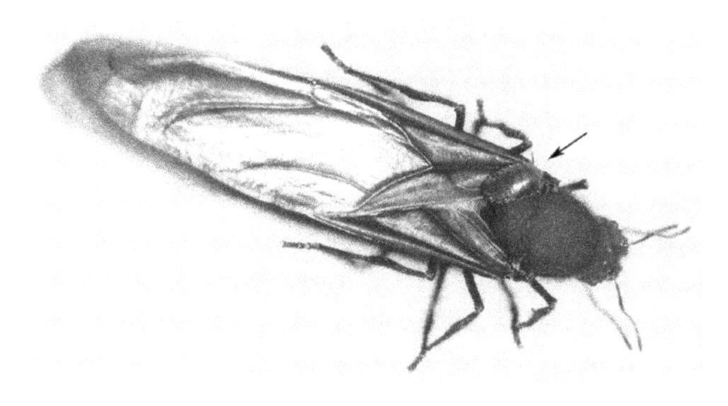

Fig. 2.11 Phoretic female of *Attaphila fungicola* attached to the wing base of *Atta* sp. host. The cockroach is about 2.7 mm in length. Courtesy of John Moser.

would be of interest. The latter lives in the nests of a variety of ant genera (*Campanotus, Formica, Solenopsis*), but have no arolia or pulvilli on the tarsi, and there are no records of host transport (Fisk et al., 1976).

Arboreal

Species that live in trees are generally expected to be good fliers, because the alternative is a long down-and-up surface trip when moving between limbs or trunks (Roff, 1990; Masaki and Shimizu, 1995). Fisk (1983) identified the cockroaches that fell during canopy fogging experiments conducted in rainforests in Panama and Costa Rica. Of the 25 species for which wing condition is known in both males and females, 23 (92%) are winged in both sexes, one (*Nesomylacris asteria*) has reduced tegmina and wings in both sexes, and one (*Compsodes deliculatus*) has winged males and apterous females (analyzed by LMR). Small blattellid species were the most abundant and diverse group collected during the study. These data support the notion that cockroaches that spend the day in trees are generally flight-capable. Further support comes from behavioral observations in Costa Rica. Flight between perches was noted in all winged species observed during their active period (Schal and Bell, 1986). Some cockroach species, however, spend their entire lives within specialized arboreal niches, are unlikely to be collected during canopy fogging, and are not necessarily volant. These include cockroaches that live under bark, in epiphytes, in arboreal litter, and in insect and bird nests. Of the 31 species of Brazilian cockroaches collected in bromeliads by Rocha e Silva Albuquerque and Lopes (1976), 55% were apterous or brachypterous.

Caves

As discussed in the following chapter, caves are at one end of a continuum of subterranean spaces frequented by cockroaches, with the border between caves and other such habitats often vague. Variation in wing reduction, as well as associated morphological changes, may reflect different degrees of adaptation to these specialized habitats. In Australian *Paratemnopteryx,* species found in caves usually exhibit some degree of wing reduction (Table 2.2). Several species in this genus are intraspecifically variable; both macropterous and reduced-wing morphs of *Para. howarthi* can even be found in the same cave (Roth, 1990b). Epigean species in the genus living under bark or in leaf litter are often macropterous, but also may exhibit wing reduction. The area of the cave inhabited (deep cave versus twilight zone), nutrient availability (is there a source of vertebrate excrement?), and length of time a population has been in residence all potentially influence the morphological profiles of the cave dwellers. Like other invertebrates, cockroaches that are obligate cavernicoles (troglobites) typically exhibit wing reduction or loss.

Table 2.2. Wing development in cavernicolous and epigean species of the Australian genus *Paratemnopteryx,* based on Roth (1990b), Roach and Rentz (1998), and Slaney (2001). Those species described as epigean were found under bark and in litter.

Species	Habitat	Wing condition
Para. atra	Cavernicolous, in mines	Slightly reduced
Para. australis	Epigean, one record from termite nest	Reduced
Para. broomehillensis	Epigean	Macropterous
Para. centralis	Epigean	Macropterous
Para. couloniana	Epigean, in houses	Variably reduced, some males macropterous
Para. glauerti	Epigean	Male macropterus, female reduced
Para. howarthi[1]	Cavernicolous and epigean	Macropterous and reduced males, females reduced
Para. kookabinnensis	Cavernicolous	Reduced
Para. rosensis	Epigean	Male macropterous, female reduced
Para. rufa	Cavernicolous and epigean	Reduced
Para. stonei	Cavernicolous and epigean	Variably reduced[2]
Para. suffuscula	Epigean	Macropterous
Para. weinsteini	Cavernicolous	Reduced, female more so

[1]Brachypterous and macropterous morphs can be found in same cave.
[2]Female wings slightly longer than male's.

Wing Variation within Closely Related Groups

A number of closely related cockroach taxa unassociated with caves can show as much variation as *Paratemnopteryx*. Wing condition is therefore of little value as a diagnostic generic character unless it occurs in conjunction with one or more stable and distinctive characters (Hebard, 1929; Rehn, 1932b). The three native species of the genus *Ectobius* in Great Britain clearly depict an evolutionary trend in female wing reduction. Males are macropterous in all three species. Females of *E. pallidus* also have fully developed wings, but in *E. lapponicus* the tegmina of the female are about two-thirds the length of the abdomen and the wings are reduced. In *E. panzeri* the tegmina of the female are just a little longer than wide and the wings are micropterous (Kramer, 1956). The subfamily Tryonicinae illustrates the degree of wing variation that can occur at higher taxonomic levels. Table 2.3 displays the genera of these blattids arranged to exhibit a detailed gradient of wing development from one extreme (macropterous) to the other (apterous).

Case Study: Panesthiinae

Those members of the Panesthiinae for which we have ecological information are known to burrow in soil (Geoscapheini) or rotted wood (the remainder). They therefore illustrate the range of wing variation possible within an ecologically similar, closely related taxon (Table 2.4). Many species in the subfamily have fully developed tegmina and wings, and are heavy bodied but able flyers (Fig. 2.12A). Male *Panesthia australis*, for example, have been collected at lights in Australia (Roth, 1977; CAN, pers. obs.). Some genera include sexually dimorphic species, with winged males and wingless females (*Miopanesthia*), and a number of species in the genus *Panes-*

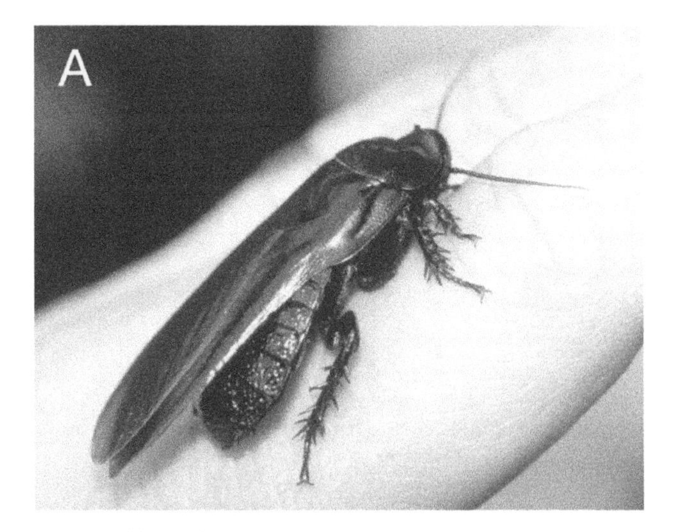

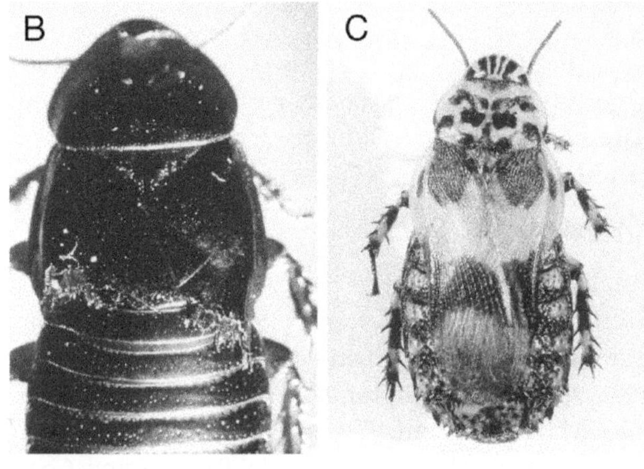

Fig. 2.12 Wing condition in wood-feeding Panesthiinae. (A) Fully winged adult of Australian *Panesthia australis;* photo by C.A. Nalepa; (B) detail of adult Australian *Panesthia cribrata* showing ragged wing bases after dealation; photo courtesy of Douglas Rugg; (C) strikingly patterned winged female of *Caeparia donskoffi* from Vietnam, body length approximately 3.5 cm; photo by L.M. Roth.

Table 2.3. Tryonicinae (Blattidae) illustrate the complete range of wing development, from fully developed wings to completely apterous, with intermediate stages (LMR, pers. obs.).

Wing characters	Genus (no. species)	Country
Fully winged, but wings may not reach the end of the abdomen	*Methana* (10)	Australia
Tegmina reduced, elongated, lateral, completely separated from the mesonotum, reaching a little beyond hind margin of second abdominal tergite, hindwings present, vestigial, lateral, completely covered by the tegmina	*Tryonicus* (3) (female apterous)	Australia
Tegmina small, lateral lobes completely separated from the mesonotum, not reaching the first abdominal tergite, wings absent	*Punctulonicus* (2) *Angustonicus* (2) *Rothisilpha* (2)	New Caledonia
Tegmina lateral, but not completely separated from the mesonotum, wings absent	*Pellucidonicus* (2) *Pallidionicus* (5) *Angustonicus* (1) *Punctulonicus* (1) *Rothisilpha* (1)	New Caledonia
Completely apterous	*Lauraesilpha* (4)	New Caledonia

Table 2.4. Extent of development of tegmina and wings in 10 genera of Panesthiinae; after Table 6 in Roth (1982b). The "reduced" wing category includes brachypterous morphs, micropterous morphs, and those with reduced tegmina and absent wings. One genus includes polymorphic species (*Panesthia*). Sexual dimorphism is found only in the genus *Miopanesthia*.

	Number of species + subspecies with tegmina and wings				
Genus	Fully developed (macropterous)[1]	Fully developed + reduced-wing morphs	Reduced	Absent	Total
Panesthia[2]	23 + 1	5 + 1	15 + 2	11 + 1	54 + 9
Miopanesthia[2]					
Male	6	0	0	2	8
Female	1[3]	0	0	7	8
Ancaudellia[2]	15 + 1	0	3 + 3	0	18 + 4
Salganea[2]	26 + 3	0	12 + 1	4	42 + 4
Caeparia[2]	4	0	0	0	4
Microdina	0	0	1	0	1
Parapanesthia[4]	0	0	0	1	1
Neogeoscapheus[4]	0	0	0	2	2
Geoscapheus[4]	0	0	0	2 + 2	2 + 2
Macropanesthia[4]	0	0	0	4	4

[1]A number of these eventually shed their wings.
[2]Wood-feeding cockroaches; information on the diet of *Miopanesthia, Caeparia,* and *Ancaudellia* from a pers. comm. from K. Maekawa to CAN.
[3]The original description of *M. sinica* Bey-Bienko did not indicate the wing condition of the female; the implication is that they have tegmina and wings (Roth, 1979c).
[4]Soil-burrowing cockroaches (Geoscapheini).

thia are intraspecifically variable. Of these, both males and females may have either well-developed or variably reduced wings. In some species (e.g., *Pane. australis*), the reduced-wing form is uncommon (Roth, 1977).

Uniquely among cockroaches, some macropterous members of this subfamily shed their wings. In some species of *Panesthia, Salganea,* and *Ancaudellia* only the basal region of the tegmina and wings remains intact. The wings are not cleanly snapped at a basal suture, as in termites, but have a raggedy, irregular border (Fig. 2.12B) (Roth, 1979c; Maekawa et al., 1999b). Some early observers thought that dealation resulted from the chewing action of conspecifics (Caudell, 1906), that they "solicit the assistance of their comrades to gnaw them off close to the base." Others, however, suggested that the wings were broken off against the sides of their wood galleries, because dealation occurs even in isolated individuals and because the proposed gnawing action was never observed (McKeown, 1945; Redheuil, 1973). The wings are most likely lost by a combination of both behaviors. In laboratory studies of *Panesthia cribrata,* Rugg (1987) saw adults moving rapidly backward, rubbing the wings against the sides of the cage, and also observed a male chewing the wing of a female, then dragging off a tattered portion and eating it. Rugg illustrates obviously chewed wings, with distinct semicircular portions removed. Individuals are unable to chew their own wings (D. Rugg, pers. comm. to CAN). Like termites and some other insects, Panesthiinae with deciduous wings restrict flight activity to the pre-reproductive stage of their adult life. It would therefore be of interest to determine if flight muscle histolysis accompanies wing loss, and if so, how it relates to fecundity. In crickets, dealation induces histolysis of the wing muscles and a correlated rapid production of eggs (Tanaka, 1994).

A well-corroborated estimate of relationships among 20 species of Panesthiinae inferred from a combined analysis of 12S, COII, and 18S is illustrated in Fig. 2.13 (Maekawa et al., 2003). We mapped four wing-related character states onto the depicted tree: wing morphology (macropterous, reduced wings, or apterous), and in macropterous species, whether the wings are permanent or deciduous. The apterous condition appears to have evolved three times, in *Miopanesthia deplanata, Panesthia heurni,* and the Geoscapheini. Deciduous wings arose twice, in *Salganea* and in the lineage that includes *Panesthia* and *Ancaudellia.* Within *Salganea,* reduced wings seem to be derived from the macropterous, deciduous state. Maekawa et al.'s (2003) phylogeny is not fully resolved and shows the genus *Panesthia* as poly- or paraphyletic. It is nonetheless obvious that the morphological wing condition and the behaviors associated with removing deciduous wings are evolutionarily labile in these cockroaches. Wings are generally dull and uniformly colored in the Panesthiinae that eventually shed them. Un-

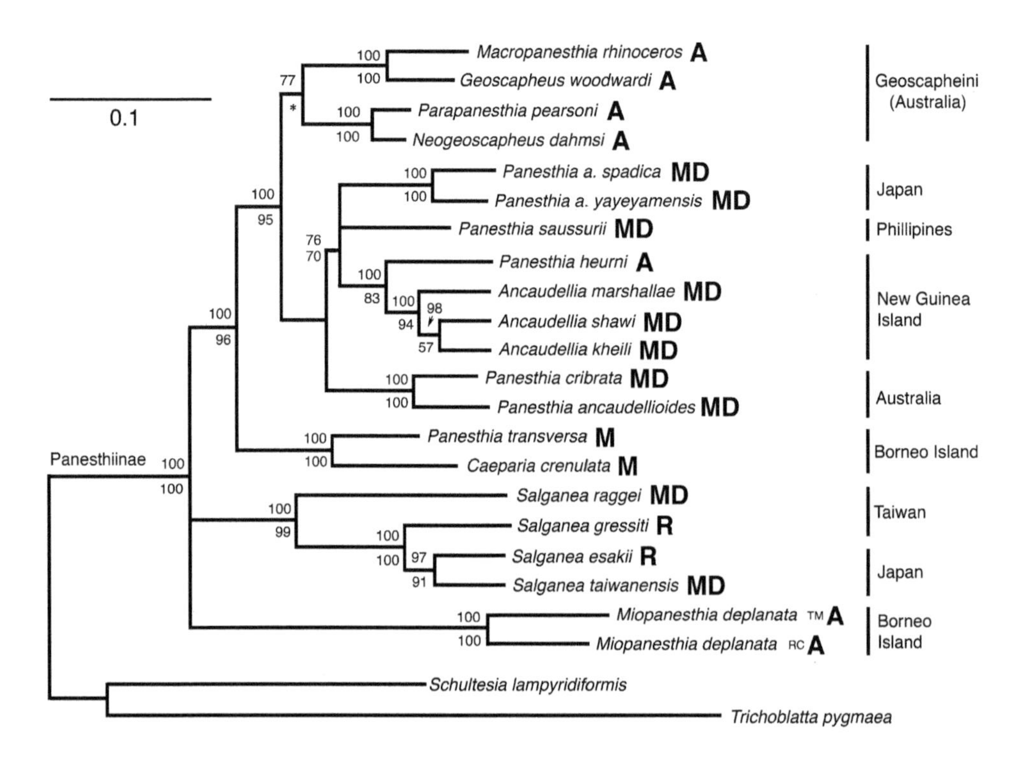

Wing Condition: **M** = macropterous **MD** = macropterous, deciduous
R = reduced **A** = apterous

Fig. 2.13 Phylogenetic distribution of wing condition in the Panesthiinae. The phylogenetic tree is inferred from a combined analysis of 12S, COII, and 18S, obtained using Bayesian inference of phylogeny with the GTR + I + G model of substitution. Posterior probabilities (PP), expressed as percentages, are shown above branches to indicate the level of support for each node. Branches with less than 50% PP were collapsed to form polytomies. Bootstrap values (expressed as percentages) from an MP analysis are shown below the nodes. The asterisk indicates a node that was not supported in more than 50% of bootstrap replicates; however, an analysis in which COII third codon transitions were downweighted by a factor of 4 resulted in 70% support. The scale bar indicates the number of inferred substitutions per site. From Fig. 3 (p. 1305) in Maekawa et al. (2003), courtesy of K. Maekawa and with permission of the Royal Society of London. Wing conditions based on Roth (1979b, 1979c) and the observations of K. Maekawa (pers. comm. to CAN).

like the other macropterous species, *Panesthia transversa* and *Caeparia crenulata* (as well as other species of *Caeparia*) have strongly colored and patterned wings and retain them throughout their adult life (Fig. 2.12C). This reinforces the idea that cockroach wings have functional significance in contexts other than flight; in this case it is likely that retained wings have signal value to predators, conspecifics, or both. A comparison of the population genetics of apterous or brachypterous wood-feeding species to those that have remained flight capable might yield data relevant to dispersal distances.

Intraspecific Wing Variation

A similar reduction in tegmina and wings often occurs in both sexes of a species. Sexual dimorphism is common, however, and it is most often the female that exhibits the greater degree of wing reduction. At one extreme are species with fully winged males and apterous females. Examples include the African genus *Cyrtotria* (= *Agis*) (Rehn, 1932a), *Trichoblatta sericea,* living on and under the bark of *Acacia* trees in India (Reuben, 1988), and many desert Polyphagidae. In *A. investigata,* for example, females are wingless, but at night fully winged males emerge from the sand and fly (Edney et al., 1974). Females of *Escala circumducta* have "almost discarded their organs of flight" and live their entire lives beneath the bark of trees. The fully winged males associate with the females only during a brief pairing season (Shaw, 1918). In cockroaches with extreme wing dimorphism females are often burrowers or crevice fauna, but the habitats of males are unknown, because they have been collected only at lights. Some cases of sexual dimorphism are so extreme that they are problematic to taxonomists trying to associate the two sexes (Roth, 1992). Females of *Laxta* (= *Onisco-*

soma) granicollis are flattened and wingless, resembling "an enormous wood louse," while males are winged and "of more graceful shape" (Swarbeck, 1946). Similarly, males of several species of *Perisphaeria* and *Pseudoglomeris* are slender, winged insects, while the females are apterous and broader (Hanitsch, 1933). More moderate cases of wing dimorphism include species where both sexes have reduced wings but the female more so, and those species discussed above, where both sexes are fully winged, but the female is nonetheless flightless. We are not aware of cases of macropterous females and apterous males, but when wing reduction occurs in both sexes, sometimes the wings of the male are shorter (e.g., *Para. stonei*—Roth, 1990b).

Wing development within a species is not always a fixed character. In some cockroaches, only one sex exhibits variation, for example, *Neotemnopteryx fulva* males are macropterous, but the females may be macropterous or brachypterous (Roth, 1990b). Likewise, *E. africanus* males are macropterous, but female wing reduction varies with altitude (Rehn, 1932b). In other cockroaches, the reduction of tegmina and wings is variable in both sexes. These include at least five species of *Panesthia* (Roth, 1982b), *H. concinna* in the Galapagos (Roth, 1990a), and the Australian *Para. couloniana* (Roth, 1990b). The latter generally has brachypterous tegmina and micropterous wings, but the degree of reduction varies, and there are males whose flight organs are fully developed. This species lives in litter and under bark, but there are also records of it infesting houses (Roach and Rentz, 1998).

Migration

Intraspecific variation in the wing form of insects is usually associated with migratory flight, that is, dispersal or migration from the habitat, as opposed to trivial flight, activity associated with routine behavior such as feeding, mate finding, or escaping from enemies. As such, the environmental cues known to influence wing form are those that signal seasonal habitat deterioration (photoperiod, temperature) or less predictable, density-dependent habitat changes (poor nutrition, stress, crowding) (Travis, 1994; Masaki and Shimizu, 1995). High population density is known to induce a number of morphological and physiological changes in studied cockroach species, for example, *Blab. craniifer* (Goudey-Perriere et al., 1992) and *Eublaberus distanti* (Rivault, 1983), but to date, wing form has not been one of them.

Mass migrations and dispersals have been recorded in cockroaches, though not in wing-polymorphic species. Surface activity in *C. punctulatus* occurs following rainfall, during daylight hours in spring (Nalepa, 2005). Soil-burrowing Australian Geoscapheini undertake spectacular pedestrian migrations after rains—sometimes seen by motorists crossing roads every few yards for 32 km at a stretch (Monteith, pers. com. to LMR). There are two intriguing reports of possible long-distance movement by flight. On a sunny morning in Venezuela at an elevation of 1100 m, Beebe (1951) observed a "flurry" of at least 30 *Blaberus giganteus* fluttering slowly up a gorge used as a flyway for migrating insects. Under the hot sun in an Arizona desert, Wheeler (1911) watched two separate swarms of male *Homoeogamia subdiaphana* alternately flying and quickly running over the sand in a southwesterly direction; he likened their quick movements to those of tiger beetles (Cicindelidae). Overpopulated buildings or sewers have been known to spawn natural migrations in several species of urban pests (Roth and Willis, 1957). It is unusual that many of these movements occur during daylight hours in otherwise nocturnal insects. Stein and Haschemi (1991) report that German cockroaches emigrating from a garbage dump used solar cues for orientation. Most walked directly toward the sun, with their bearing shifting from east to west over the course of the day.

Evolution of Flightlessness

Macropterism is clearly the primitive condition in cockroaches (Rehn, 1932b). Because no fossil cockroaches are known with abbreviated organs of flight (R.J. Tillyard, in Shaw, 1918), it is assumed that Paleozoic cockroaches were swift-flying and diurnal (Brodsky, 1994). Flight may have been advantageous in Carboniferous swamps, as it would allow movement between patches of habitat surrounded by water. On the other hand, the possession of wings does not assure the ability to fly, and apterous and brachypterous cockroaches are less likely to leave fossil evidence than their more volant relatives. There are indications of wing sexual dimorphism in the fossil record. Schneider (1977, 1978) concluded that the wings of Carboniferous females were broader than those of males, and Laurentiaux (1963) demonstrated that there were intersexual differences in both the length and the shape of wings.

It is possible to induce alary reduction experimentally in a normally winged species (e.g., *Blab. craniifer*), but attempts to produce fully developed wings in an apterous cockroach have been unsuccessful; Lefeuvre (1971) therefore concluded that the evolutionary loss of wings is irreversible. On the other hand, Masaki and Shimizu (1995) suggested that wing reduction is possible without elimination of the genetic background for macropterous development, and potential evolutionary reversal of wing loss has been demonstrated in the Hemiptera-Heteroptera (Anderson, 1997) and in the Phasmatodea (Whiting et al., 2003). As robust phylogenetic trees become available for varying cockroach taxa, the possibility of the re-evolution of wings in the Blattaria can be put to the test.

Habitat Factors Associated with Wing Loss

Flight loss in insects is most often associated with environmental stability (Southwood, 1962; Harrison, 1980; Roff, 1990; Denno et al., 1991, 2001b; Wagner and Liebherr, 1992; Zera and Denno, 1997, among others). The logic is that flightless morphotypes are inclined to persist in spatially homogeneous, temporally stable habitats where food, shelter, and mates are continuously accessible to pedestrians. Conversely, flight is retained in insects living in temporary habitats, so that fluctuating levels of resource quality and abundance may be tracked. Although a number of studies support this hypothesis (e.g., Roff, 1990; Denno et al., 1991), the association of cockroaches with their habitat is not as clear as it is in insects such as stenophagous herbivores on annual plants, or waterstriders that live in temporary versus permanent ponds. Few cockroaches are exclusively associated with ephemeral or periodically disturbed habitats, although they may utilize them if available. Some species exhibit seasonal habitat shifts, but there are no known cockroaches with seasonal variation in wing morphology.

Several hurdles to understanding the role of habitat in structuring cockroach wing morphology must be added to those noted earlier. First, there can be a great deal of intraspecific variation in habitat choice. A good example is *Chorisoneura carpenteri* from the Galapagos, a species with both brachypterous and macropterous forms. The fully winged morphs have been collected at elevations of 30–1000 m in agricultural areas, arid zones, pampa, humid forest, and *Scalesia* forest; the brachypterous form has been collected at 120–700 m in all of the listed habitats but one—the agricultural zone (Peck and Roth, 1992). Second, many cockroaches defy being described by just one aspect of their habitat, and it is difficult to tease apart the relative importance of a hierarchy of overlapping ecological levels. Is a canopy cockroach more likely to be wingless if the forest is on a mountain? Is it valid to compare a list of wingless cockroaches found in caves to a list of wingless cockroaches found in Texas (Roff, 1990, p. 395)? Finally, the fact that so many cockroaches in different habitats utilize the same microhabitats confounds analysis. Whether they are found in a desert, grassland, forest, or elsewhere, many cockroaches are associated with a continuum of dark, humid, enclosed spaces that they find or make.

The strength of the association of a given cockroach species with these subterranean and other spaces appears influential in wing development. Cockroaches that live their entire lives in burrows, galleries, or crevices, except for a brief dispersal period at the subadult or young adult stage or when the habitat becomes unsuitable, seem most prone to winglessness. It is apparent from an examination of the Panesthiinae (Fig. 2.13) that the habit of burrowing in wood or soil may be connected to the prevalence of reduced, absent, or deciduous wings in this subfamily. Cockroach species that spend their lives in the loose spaces beneath bark also fall into this category. Shaw (1918) noted that flightless cockroaches are generally cryptic in their habits, and that there was a "definite correlation" between a flattened morphology and the absence of wings. In deserts, cockroach microhabitats include the base of grass tufts and the spaces beneath debris and boulders. The majority of desert cockroaches, however, live a partially or entirely subterranean existence. Half of the 28 desert cockroaches listed by Roth and Willis (1960) live in the burrows of small vertebrates, and additional species burrow into loose sand. It should be noted that obligate cavernicoles are an extreme case of this same continuum. The ecological influences that promote wing loss in all these cockroaches, then, may differ more in degree than in type.

Several characteristics of crevices and burrows may influence wing loss in the cockroaches that permanently or periodically inhabit them. First, these are temporally stable habitats. Logs, leaf litter, and other rotting vegetable matter are continuously or periodically replenished from source plants, and migration to fresh resources, if required, is often a local trip. Second, these are homogeneous microhabitats, in that they are interchangeable dark, moist, protected quarters. If leaf litter on the forest floor loses moisture during the tropical dry season, for example, cockroaches normally found in ground-level litter are known to move into moist, arboreal accumulations of leaves (Young, 1983). Third, these are chiefly two-dimensional microhabitats, particularly for cockroach species that either rarely venture from shelters or have a modest ambit around them. Schal and Bell (1986) found that many of the flightless cockroach species in Costa Rican rainforest ground litter did not move very far in vertical space during their active period. Recent evidence suggests that it is the interaction of habitat dimensionality and habitat persistence that may have the most significant effect on insect wing morphology (Waloff, 1983; Denno et al., 2001a, 2001b). Finally, these cockroaches are able to feed within their shelter (in logs, under bark, in leaf litter, in vertebrate burrows, in social insect nests, in caves), or the shelters are situated in the immediate vicinity of potential food (soil burrowers, under rocks, under logs). The proximity of widespread, persistent, often abundant but low-quality food has two potential implications for the evolution of cockroach wing morphology. First, the insects are less tied to the seasonality of their food source. Flightlessness in insects tends to be positively correlated with their ability to remain throughout the year in their developmental habitat (Anderson, 1997; Denno et al., 2001a). Second, wing reduction and loss is often associated with nutrient limitation (Jarvinen and Vepsalainen, 1976; Kaitala and Hulden, 1990), and cockroaches that rely on rotting vegetable matter as a primary food source may be living close to their nutritional threshold. In caves, wing loss and associated morphological changes occur more frequently in organisms that rely on plant debris than those that rely on bat or bird guano (Culver et al., 1995).

Wing Loss and Life History Trade-offs

Food abundance and quality cannot be divorced from wing morphology because it is costly to produce and maintain the wings and their muscular and cuticular support (Roff and Fairbairn, 1991); insect flight muscle is one of the most metabolically active tissues known (e.g., Weis-Fogh, 1967). Flight behavior is also energetically demanding, and can alter the composition of hemolyph for up to 24 hr afterward in *P. americana* (King et al., 1986). These metabolic expenses place a significant demand on an insect's overall energy budget, and compete with other physiologically demanding life history processes. The best documented of these is egg production. Any easing of the selective pressure to maintain wings allows a female to divert more resources to egg production, increasing her fitness more than if she remained volant ("flight-oogenesis syndrome") (Roff, 1986, 1990; Roff and Fairbairn, 1991). Flight capability can diminish rapidly under the right conditions (Denno et al., 1991; Marooka and Tojo, 1992), and may account for the lack of functional flight muscle in laboratory-reared females of *Periplaneta* (Table 2.1). The flight-oogenesis syndrome also may account for the prevalence of flightless females, rather than males, in cockroach species exhibiting sexual dimorphism in flight ability. The relationship between wing morphology and fecundity has been demonstrated in a number of insect species, including orthopteroids (e.g., Cisper et al., 2000), but is as yet unstudied in cockroaches. The fact that there are numerous cockroach species with males possessing reduced or absent wings suggests that there is a cost to the retention of wings even in males. In some insects, short-winged males have a mating advantage over macropterous males, or a gain in testes and body size (Dingle, 1996; Langellotto et al., 2000). Macroptery in males is most often related to the distribution of females in the habitat, and whether they are accessible to males on foot (Roff, 1990; Denno et al., 2001a). This is likely the case in cockroaches, because in many species females produce volatile sex pheromones; males use these chemical cues to actively seek mating partners (Gemeno and Schal, 2004). The degree of wing development may affect longevity in both sexes (Kaitala and Hulden, 1990; Roff and Fairbairn, 1991). It may be relevant, then, that among the longest-lived of the known cockroaches are apterous species that burrow in wood or soil (Chapter 3).

Wing Loss, Paedomorphosis, and Population Structure

A lack of functional wings is at the heart of two obstacles to understanding the evolutionary biology of some earth-bound cockroaches. First, aptery and brachyptery are associated with a developmental syndrome that reduces morphological complexity, making it difficult to distinguish among closely related taxonomic groups. Second, the loss of mobility associated with aptery can result in complex geographic substructuring of these morphologically ambiguous groups.

Wing reduction or loss is the best indicator of paedomorphosis, defined as the retention of juvenile characters of ancestral forms in the adults of their descendents (Matsuda, 1987; Reilly, 1994). Not all short-winged insects retain juvenile characters, but in other cases, it is clear that many so-called adult characters are absent in short-winged or apterous morphs (Harrison, 1980). The diminishment or loss of structures such as ocelli, compound eyes, antennal and cercal segments, and some integumental structures such as sensilla often accompanies aptery and brachyptery (Matsuda, 1987). These reductions are common in cockroaches (Nalepa and Bandi, 2000), and like other animals (Howarth, 1983; Juberthie, 2000b; Langecker, 2000) occur most often in species that inhabit relatively safe, stable environments, such as caves, burrows, logs, social insect nests, leaf litter, and other cryptic environments. Lefeuvre (1971) found that some cockroach species with reduced wings have fewer developmental stages than macropterous relatives, and that juvenile features can be retained in the tracheal system, peripheral nervous system, and integument. Warnecke and Hintze-Podufal (1990) concluded that the reduced wings of female *Blaptica dubia* are the result of larval characters that persist into maturity, rather than the growth inhibition of adult wings. Other examples include the retention of styles in wingless adult females of *Noc. termitophila* (female cockroaches normally lose their styles prior to the adult stage) (Matsuda, 1979), and the reduced sensory and glandular systems of the myrmecophile *Att. fungicola* (Brossut, 1976). *Cryptocercus* has reduced eyes and cercal segmentation, and exhibits marked paedomorphic traits in its genital morphology (Walker, 1919; Crampton, 1932; Klass, 1995). Females of the desert cockroach *A. investigata* are "generally nymphlike," lack the wings and ocelli seen in the male, and have shorter antennae and cerci (Friauf and Edney, 1969). Because wing loss in cockroaches is female biased, it is most often females that exhibit correlated paedomorphic characters.

The systematics of paedomorphic organisms can be frustrating. Because many structures never develop or develop variably within a group, they cannot be used to delimit taxa, or to infer phylogenetic relationships. Independent losses of ancestral postmetamorphic features is an important source of homoplasy and can confound cladistic analysis (Wake, 1991; Brooks, 1996; Hufford, 1996). The morphological homogeneity of the Polyphagidae has caused quite a few problems with attribution, not only to species but also to genera (Failla and Messina, 1987). Members of the genus *Laxta* "vary so much in color and size and have genitalia so similar as to make distinguishing taxa difficult" (Roth, 1992). Paedomorphic characters and mosaic evolution in the wood-feeding cockroach *Cryptocercus* strongly contribute to problems in determining the phylogenetic relationships of this genus at all taxonomic levels (Klass, 1995, 1998a; Nalepa and Bandi, 1999, 2000; Nalepa et al., 2002). Cave cockroaches, like other cave dwellers (Howarth, 1983; Juberthie, 2000a; Langecker, 2000), are prone to taxonomic problems associated with paedomorphosis. Roth (1990b) noted that *Para. stonei* from different caves all had reduced hindwings but varied in body size, in the development of pulvilli, and in length of tegmina. The genitalia were so similar, however, that he assigned them to different races within the species. A morphometric study by Slaney and Weinstein (1997b) subsequently supported Roth's conclusions.

Molecular and chemical tools are increasingly required to provide characters to distinguish among these morphologically ambiguous cockroach taxa. Humphrey et al. (1998), for example, used protein electrophoresis to propose that morphologically similar populations of *M. rhinoceros* are comprised of three genetic species. Slaney and Blair (2000) used the ITS2 gene region of nuclear ribosomal DNA in the *Para. stonei* group, and their results supported conclusions based on morphology. Molecular phylogenetic relationships, however, are not always completely congruent with relationships based on morphological characters. Basal relationships among species of the wood-feeding blaberid *Salganea* are poorly resolved by molecular analysis, probably because of rapid and potentially simultaneous radiation of the group (Maekawa et al., 1999a, 2001).

In flightless animals the pool of potential mating partners is limited to those that can be found within walking distance, resulting in restricted levels of gene flow. Populations may become subdivided and isolated to varying degrees, resulting in complex genetic substructuring and the formation of local species, subspecies, and races. This is common in caves, where subterranean spaces can be isolated or locally connected via mesocaverous spaces (Barr and Holsinger, 1985). It is also common on mountains, where endemic races and subspecies may be wholly restricted to single peaks (Mani, 1968). *Cryptocerus primarius,* for example, is found in an area of China with a dissected topography characterized by high mountain ridges sandwiched between deep river gorges, forming various partitioned habitats (Nalepa et al., 2001b). This genus of montane cockroaches is also dependent on rotting logs, which ties their distribution to that of mature forests. Any event that has an impact on the distribution of forests, including glaciation (Nalepa, 2001; Nalepa et al., 2002) and deforestation (Nalepa et al., 2001b) will affect the population structure of the cockroach. Consequently the geographic distribution of genetic populations and species groups in both Northeast Asia (Park et al., 2004; Lo et al., 2000b) and the eastern United States (Nalepa et al., 2002) can be unexpected. *Cryptocercus* found in southern Korea, for example, are more closely related to populations in Northeast China than they are to all other Korean members of the genus.

THREE Habitats

Of no other type of insect can it be said that it occurs at every horizon where insects have been found in any numbers.

—S.H. Scudder, "The Cockroach of the Past"

Cockroaches are found in nearly all habitats: tropical and temperate forests, grasslands, heath, steppe, salt marshes, coastal communities, and deserts. They are active in the entire vertical dimension of the terrestrial environment, from the upper forest canopy to deep in the soil, and inhabit caves, mines, hollow trees, burrows, and sub-bark spaces. They are also found in dead leaves, rotting logs, streams and stream edges, epiphytes, arboreal water pools, the nests of social insects, rodents, reptiles, and birds, and human-made structures such as dwellings, ships, and aircraft (Roth and Willis, 1960). Cockroaches occur between latitudes 60°N and 50°S, but most are found between 30°N and 30°S in the warm, humid regions of the Old World (Africa) and tropical America (Guthrie and Tindall, 1968); they are less diverse in the temperate regions. Wolda et al. (1983) cites the number of species captured at various latitudes in Central and North America: 64 in Panama, 31 in Texas, 14 in Illinois, 9 in Michigan, 5 in Minnesota, and 2 in North Dakota. In the high arctic, pest cockroaches readily invade heated structures (Beebe, 1953; Danks, 1981), but several species are physiologically capable of dealing with extremely cold weather in their natural environment (e.g., *Celatoblatta quinquemaculata*—Worland et al., 2004). The general tendency is to live near sea level, where temperatures are higher (Boyer and Rivault, 2003). In his collections on Mt. Kinabalu in Borneo, Hanitsch (1933) found 19 cockroach species up to an altitude of 2135 m, but only three species above it. Light trap catches in Panama also indicate higher diversity in lowland than in mountain sites (Wolda et al., 1983). In Hawaii, *Allacta similis* was found no higher than 1600 m along an altitudinal transect and was thought to be excluded from higher altitudes by the cooler, wetter, montane environment (Gagné, 1979). Nonetheless, the relationship of cockroaches with altitude can be complex. On Volcán Barva in Costa Rica, no cockroaches were found at the lowest elevation sampled (100 m), but they were present at all other elevations (Atkin and Proctor, 1988). There are also montane specialists, such as *Eupolyphaga everestiana* on Mount Everest at 5640 m (Chopard, 1929).

	Dry, ruderal grassland	Scrub	Xeric hammock	Sandhills	Longleaf pine flatwoods	Shrubby longleaf pine flatwoods	Black pine flatwoods	Marginal thickets	Slash pine flatwoods	Pond margins	Moist ruderal grassland	Mesic hammock	Low hammock	Alluvial hammock	Bayhead	Spartina marsh	Sawgrass marsh	Structures	Woodpiles
Cariblatta lutea	X	X	X	X	X	X	X	X	X	X	X	X	X	X	X				
Cariblatta minima	X	X		X	X				X	X	X	X	X			X	X		
Supella longipalpa																		X	
Blattella germanica																		X	
Ischnoptera deropeltiformis	X	X	X	X	X	X			X	X	X	X	X	X	X				
Parcoblatta virginica		X																	
Parcoblatta fulvescens		X	X	X	X							X	X	X					X
Parcoblatta lata													X						
Periplaneta americana																		X	
Periplaneta brunnea																		X	
Periplaneta australasiae																		X	
Eurycotis floridana			X	X								X	X				X		X
Pycnoscelus surinamensis			X									X							
Arenivaga floridensis					X														

Fig. 3.1 Occupation of different habitats by cockroaches in a reserve near the town of Welaka in northeastern Florida. Of the habitats examined, only four contained no cockroaches: ponds, lawns, and dry and moist sparsely vegetated sand. Based on information in Friauf (1953).

HABITAT SPECIFICITY

Sorting out habitat specificity in a secretive taxon like cockroaches is a daunting task. Although some species are known to be habitat specific and have associated morphological, physiological, behavioral, and life history modifications, many are much more flexible in their living conditions. Of 19 examined habitats that contained cockroaches in a reserve in northeastern Florida, *Parcoblatta virginica, Parc. lata,* and *Arenivaga floridensis* were each found in just one habitat, and five cockroach species were found only in structures (Fig. 3.1) (Friauf, 1953). *Cariblatta lutea,* on the other hand, was found in 15 of the habitats, and nymphs of this species have also been recorded from the burrows of small vertebrates (Hubbell and Goff, 1939). In Jamaica *Car. lutea* is found in leaf litter, under debris of every kind, in dead agaves, and in bromeliads (Hebard, 1916a). Even closely related cockroaches may vary widely in habitat choice (Table 3.1), making the detection of phylogenetic trends problematic.

ONTOGENY OF HABITAT USE

Although nymphs generally live in the same habitats as adults (Mackerras, 1970), there are several cockroach species that exhibit ontogenetic niche shifts. The most common pattern is that of females, female-nymph combinations, and groups of young nymphs reported from burrows, shelters, and other protected sites, often in or near a food source. These sheltered sites serve as nurseries, with the habitat of youngest nymphs determined by the partition[3] behavior of the mother; subsequently, nymphs may or may not disperse from their natal area. In all species of *Gyna,* for example, adults are found primarily in the canopy, while nymphs are found at ground level, often burrowing in the dust of treeholes, abandoned insect nests, and caves (Corbet, 1961; Grandcolas, 1997a). Juveniles of *Capucina patula* are restricted to the habitat beneath loose bark of live or fallen trees; adults are occasionally seen on nearby foliage (WJB, pers. obs.). Nymphs of *Car. lutea,* and females and nymphs of *Parcoblatta fulvescens* have been recorded from the burrows of pocket gophers (*Geomys* sp.) (Hubbell and Goff, 1939). Adults of both these species are found in a variety of above-ground habitats. Adults of *Parcoblatta bolliana* are found in grass-

3. Partition is defined as the expulsion by the female of the reproductive product, whether it is an egg or a neonate (Blackburn, 1999).

Table 3.1. New World distribution and microhabitats of *Latiblattella* (Blattellidae). From Willis (1969).

Species	Habitat	Country
Latiblattella inornata	Decaying leaf mold and litter under palms	Canal Zone
Lat. chichimeca	In bromeliads	Mexico
Lat. zapoteca	Under stones at the edge of rivers	Costa Rica
Lat. rehni	In Spanish moss (*Tillandsia usueoides*), under bark of dead pines	Florida
Lat. lucifrons	On *Yucca elata*	Arizona
Lat. angustifrons	On *Inga* spp. trees	Costa Rica
Lat. azteca	On grapefruit trees	Mexico
Lat. vitrea	In dry, curled leaves of corn plants (*Zea zea*)	Mexico, Costa Rica, Honduras

lands, shrub communities, and woods, where they are associated with leaf litter and loose bark. Early instars, however, are consistently found living in nests of *Crematogaster lineolata,* an ant that inhabits the soil beneath large rocks (Lawson, 1967). Females, nymphs, and oothecae of *Escala insignis* have been collected from ant colonies in Australia, but males live in leaf litter (Roth, 1991b; Roach and Rentz, 1998). In Florida, densities of *Blattella asahinai* nymphs and females bearing oothecae are highest in leaf litter of wooded areas; all other adults are more diffusely distributed (Brenner et al., 1988).

SPATIAL DISTRIBUTION

Many factors influence the spatial distribution of a species, and it is difficult to determine whether the arrangement of individuals in a habitat is determined by one, a few, or the combined action of all of them. Individuals may move in response to temporal changes (daily rhythms, weather, season), or to fulfill varying needs (dispersal, mate finding, etc.) (Basset et al., 2003b). The distribution of cockroach individuals is often correlated with the proximity of appropriate food sources. In sparsely vegetated sites, for example, cockroaches are frequently associated with whatever plants (and therefore their litter) are present. This includes deserts (Edney et al., 1978), alpine zones (Sinclair et al., 2001), and other arid or Mediterranean-type habitats such as southwestern Australia, where the number and diversity of ground-dwelling cockroaches depends on the type, percent cover, and depth of the litter present (Abenserg-Traun et al., 1996a). In wood-feeding cockroaches, juvenile food and habitat is set when the parent chooses a log to colonize. The horizontal distribution of cockroaches in caves is often related to the resting positions of bats, which determine the placement of guano and other organic matter. Gautier (1974a, 1974b) calculated the spatial distribution of burrowing *Blaberus* nymphs in caves by counting the number of individuals in 50 cm^2 samples to a depth of 15 cm. He found that nymphs were concentrated in zones where bat guano, fruit, and twigs dropped by the bats accumulated, and were absent from zones of dry soil, stones, or pebbles. In many cave cockroaches, females descend from their normal perches on the cave walls to oviposit or give birth on the cave floor in or near guano (e.g., *Blaberus, Eublaberus, Periplaneta*—Crawford and Cloudsley-Thompson, 1971; Gautier, 1974b; Deleporte, 1976), where the nymphs remain until they are at least half grown. They then climb onto the cave walls, where they complete their development.

CIRCADIAN ACTIVITY

Many species exhibit daily and seasonal movements in response to their dietary, reproductive, and microenvironmental needs; these vary with the individual, sex, developmental stage, species, day, season, and habitat. Activity patterns are expected to differ, for instance, between those cockroaches that forage, find mates, reproduce, and take refuge all in the same habitat (in logs, under bark, in leaf litter) and those that move daily between their harborage and the habitats in which they conduct most other life activities. The most common circadian activity pattern among the latter is for nymphs and adults to rest in harborages during the day, then become active as the sun sets. At dusk, adults climb or fly to above-ground perching sites (Schal and Bell, 1986), while nymphs confine their activities to the leaf litter. Some species are evidently active for short periods just after sunset, whereas others may be observed throughout the night. Within 60 min after sunset, adult males and small nymphs of *Periplaneta fuliginosa* emerge from their harborage, followed by medium and large nymphs and adult females. After feeding, males climb vertical surfaces, while nymphs and most females return to shelter (Appel and Rust, 1986). Males also become active earlier than females in *Ectobius lapponicus.* They begin moving in the late afternoon, while females and nymphs wait until after sunset (Dreisig, 1971). In *Nesomylacris* sp., most females do not become active until just before dawn, while males are active throughout the night. Females of *Epilampra involucris* are active at both dusk and dawn (Fig. 3.2). With few exceptions, temporal overlap among nocturnally active species is large.

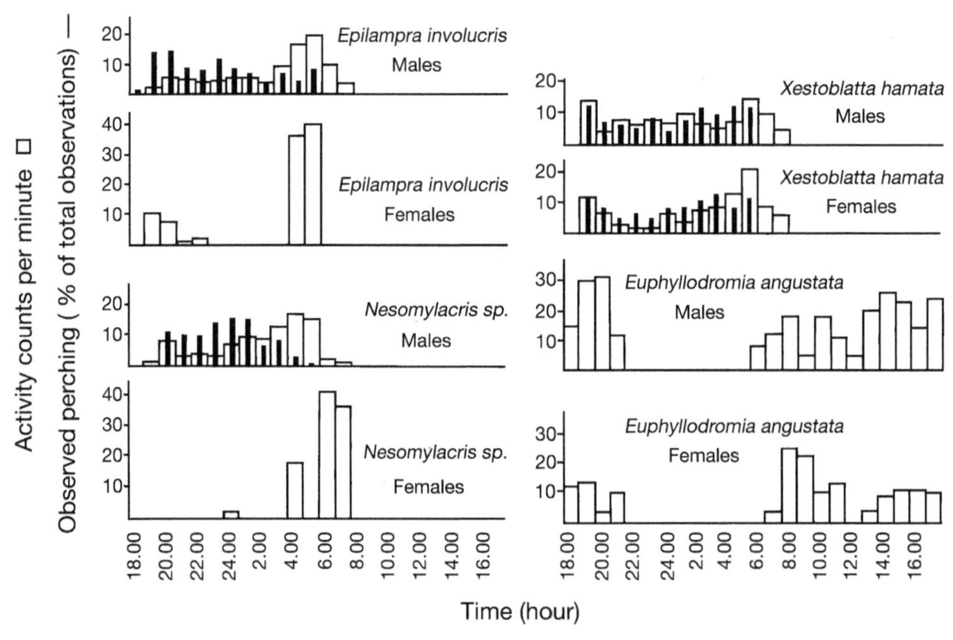

Fig. 3.2 Circadian activity of three nocturnal and one diurnal cockroach species in Costa Rican rainforest. Solid bars are a measure of conspicuousness in the field; open bars indicate locomotor activity in an outdoor insectary. Modified from Schal and Bell (1986).

Not all cockroach individuals are mobile on a nightly basis. Kaplin (1996) found that 40% of individuals of the desert cockroach *Anisogamia tamerlana* are active in a single summer night. In females, locomotor patterns are often associated with the reproductive cycle. In *Blattella germanica,* activity increases when females are sexually receptive and peaks during ovarian development. Locomotion decreases when she is forming or carrying an ootheca (Lee and Wu, 1994; Tsai and Lee, 2000). *Nauphoeta cinerea* females likewise stop locomotor activity shortly after mating; activity rhythms begin again after partition (Meller and Greven, 1996b). In *Rhyparobia maderae* daily activity gradually decreases in parallel with the progressive development of eggs until the level characteristic of pregnancy is reached (Engelmann and Rau, 1965; Leuthold, 1966). This inactivity is correlated with a decreased requirement for locating food and mates; females rarely forage during gestation. An increase in movement prior to partition is associated with locating a suitable nursery for forthcoming neonates. In juvenile cockroaches activity is correlated with the developmental cycle. *Blattella germanica* nymphs are active during the first half of a nymphal stadium. During the last third of the stadium, they remain in the harborage and move very little (Demark and Bennett, 1994). Cockroaches may also "stay home" during adverse weather. The activity of *E. lapponicus* is inhibited by wind (Dreisig, 1971), and *Lamproblatta albipalpus* individuals return to harborage when disturbed by heavy rain (Gautier and Deleporte, 1986).

The distance traveled between shelter and sites of foraging and other activity varies from 28 m in field populations of *Periplaneta americana* (Seelinger, 1984) to no more than a meter or two in female *Macropanesthia rhinoceros* (D. Rugg, pers. comm. to CAN) and *Lam. albipalpus* (Gautier and Deleporte, 1986).

There are a number of day-active cockroach species, but little is known of their biology. Some, such as *Euphyllodromia angustata* (Fig. 3.3), live in tropical rainforest. Others inhabit more arid landscapes; these include

Fig. 3.3 The diurnal species *Euphyllodromia angustata* perching on a leaf, Costa Rica. Note the dead edges of leaf holes and the presence of epiphylls on the leaf surface, both of which are included in the diet of many tropical cockroaches. Photo courtesy of Piotr Naskrecki.

brightly colored Australian species in the blattellid genus *Ellipsidion,* and members of the blattid subfamily Polyzosteriinae (Tepper, 1893; Mackerras, 1965a; Rentz, 1996). In *Platyzosteria alternans,* nymphs are diurnal while adults are nocturnal (Roach and Rentz, 1998).

Activity rhythms in cockroaches are controlled by a circadian master clock in a region of the brain anatomically and functionally connected to the optic system. Light entrains the rhythm and allows for synchronization with environmental light-dark cycles (Foerster, 2004). An absence of cockroach activity rhythms has been observed in deep tropical caves, for example, *Eublaberus posticus* in Trinidad (Darlington, 1970), *Gyna maculipennis* (probably *Apotrogia* n. sp.) in Gabon (Gautier, 1980), but no study has demonstrated free-running activity. *Blaberus colloseus, Blab. atropos,* and *P. americana* positioned close to cave entrances become active when the light intensity falls below 0.7 Lux (Gautier, 1974a; Deleporte, 1976). Adult and older nymphs emerge from their shelters, and younger nymphs crawl onto the surface of the cave floor at nightfall. An intensity change of 1 Lux influences activity rhythms of *Blaberus craniifer* in the laboratory (Wobus, 1966). Observations of cave-dwelling cockroaches in Trinidad suggest that activity rhythms also may be cued by micrometerological events like wind disturbances or an increase in temperature at the beginning of bat activity. Darlington (1968) recorded a 2.5°C increase in temperature in the evening when bats become active in the deep part of Tamana Cave. In the laboratory, Roberts (1960) found that a thermoperiod with variations of 5°C was sufficient to set the rhythm of *R. maderae* in continuous darkness.

VERTICAL STRATIFICATION

In lowland Costa Rican rainforest individuals space themselves in the vertical dimension during their active period (Schal and Bell, 1986). There is intersexual and ontogenetic variation in the behavior, with males tending to perch higher in the vegetation than females (Fig. 3.4). This is not simply a function of perch availability, since many potential perch sites remain unoccupied. Perch height was generally associated with flight ability. Adult females of *E. involucris, Nesomylacris* sp., and *Hyporichnoda reflexa* are either wingless or have very short wings, and they perch close to the ground. *Epilampra unistilata, Xestoblatta hamata,* and *X. cantralli* comprise an intermediate group; all are good fliers and after spending the day in ground litter, fly to higher perches. The arboreal pseudophyllodromiine species (*Imblattella* n. sp. G, and *Cariblatta imitans*) are excellent fliers and perch higher than the intermediate group at night. Except for *Imblattella* spp., early instars are located in ground litter where partition occurs; as nymphs develop they gradually perch higher in the foliage (Schal and Bell, 1986).

Vertical stratification during the active period has been observed in subtropical and temperate cockroaches as well. In the forests and grasslands of eastern Kansas, six species (*Parcoblatta* spp., *Ischnoptera* spp.) are distributed vertically at night among grasses, shrubs, and trees (Gor-

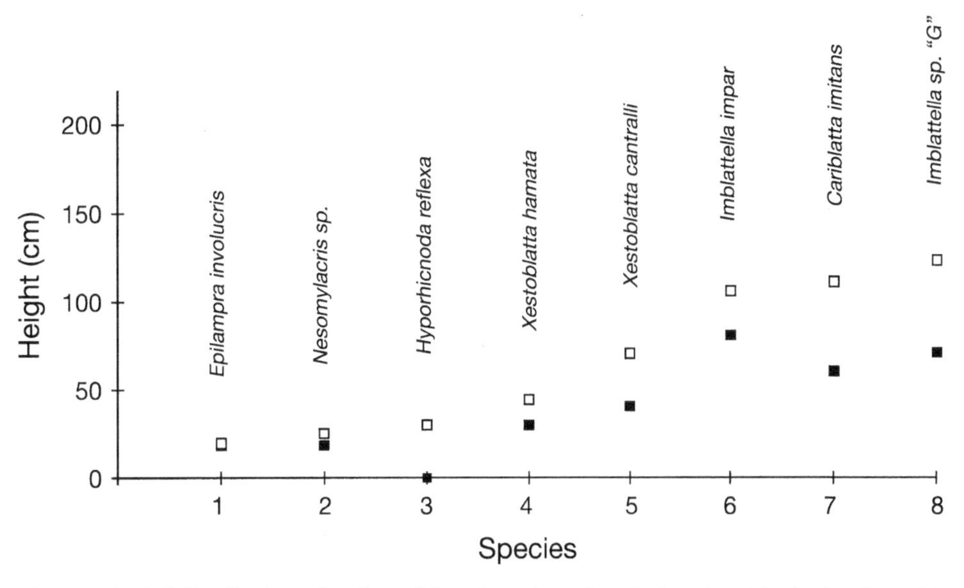

Fig. 3.4 Vertical distribution of male and female cockroaches during the night in the Costa Rican rainforest. Males, open box; females, black box. From Schal and Bell (1986).

ton, 1980). Males are good fliers and are generally located higher than females, most of which remain on or near the ground. Females seldom fly and, except for *Parc. pennsylvanica,* all have reduced wings. The inability to fly, however, is not always correlated with low perch height. Both nymphs and brachypterous females of *Ectobius sylvestris* walk on trunks into tree canopies (Vidlička, 1993).

Cockroaches appear to sort themselves in the vertical dimension via their differential sensitivity to zones of temperature, humidity, and wind currents (Edney et al., 1978; Appel et al., 1983). Schal (1982) found significant differences in these variables up to a height of 2 m in the tropical forest subcanopy. In one experiment, individually marked *E. involucris* were blinded, then placed at heights where they usually do not occur; all individuals migrated back to their typical perch zone. This stratification along micrometeorological gradients relates to the ascent of warm air and pheromone dispersion at night. Females emit sex pheromones while perching, and temperature inversions carry the pheromones aloft. Males perching higher than females would be able to detect rising pheromones and locate receptive females. Perching behavior in adults, then, is primarily a mate-finding strategy (Schal, 1982; Schal and Bell, 1986), a conclusion supported by the observations of Gorton (1980). Among the temperate species he studied in Kansas, males were generally found high, females low, and copulating pairs in between. Vertical stratification may also be related to communication between males and females in desert cockroaches (Hawke and Farley, 1973), but data are lacking to support this idea or to exclude other explanations.

SEASONAL ACTIVITY

Although many cockroach species live in relatively stable environments like tropical caves and lowland rainforests, others contend with the annual rhythmicity of seasonal climates. These include the warm-cold cycles of temperate zones and high mountains, and the alteration of wet and dry seasons in various tropical habitats. Cockroaches cope with environmental extremes and fluctuating availability of food in these environs by using varying combinations of movement, habitat choice, physiological mechanisms, and lifecycle strategies. Cockroaches may track food sources, such as those species that move into the canopy or beneath particular trees coincident with new leaf production or the appearance of spent flowers or rotten fruit. In Puerto Rico, for example, branch bagging indicated that cockroaches were more abundant on *Manilkara* spp. during the wet season, but on *Sloanea berteriana* during the dry season (Schowalter and Ganio, 2003). Cockroaches in seasonal environments may move into more benign microhabitats during harsh climatic conditions, like burrowing into deeper soil horizons or litter piles. In summer when their open woodlands habitat is excessively dry, *Ischnoptera deropeltiformis* can be found clustered in the damp area beneath recumbent portions of sedge-like grass clumps in creek beds (Lawson, 1967). Logs lying on the soil surface also serve as refugia for forest-dwelling cockroaches during dry periods (Lloyd, 1963; Horn and Hanula, 2002). Because of surface contact with the soil and the concomitant higher level of fungal invasion, recumbent logs maintain a higher moisture content than standing wood or the top layers of the forest floor (van Lear, 1996). Log refugia may be particularly important in deciduous forests, where 50–70% of incident radiation penetrates to the forest floor when trees are in their leafless state, as compared to less than 10% when leaves are fully expanded (Archibold, 1995). Likewise, the spaces beneath stones and logs as well as similarly buffered microhabitats may be seasonally occupied. In the high alpine zone of New Zealand, individuals of *Cel. quinquemaculata* burrow deep among buried rock fragments in winter, but in summer are found under surface rocks (Sinclair et al., 2001). In the United Kingdom and most of Western Europe, *Blatta orientalis* can survive normal winters outdoors provided it can avoid short-term extremes of temperature by choosing suitable harborage such as sewers, culverts, and loose soil (le Patourel, 1993). Roth (1995b) noted that cavernicolous *Nocticola brooksi* leave the more open caves of western Australia as these lose moisture during the dry season.

Using light trap collections in Panama, Wolda and Fisk (1981) demonstrated that cockroaches may show cyclic activity even in habitats lacking obvious climatic cycles. In both a seasonal and an aseasonal site, adults were most common between April and July, corresponding to the rainy season in the seasonal site. In follow-up experiments, Wolda and Wright (1992) regularly watered two plots throughout the dry season on Barro Colorado Island in Panama for 3 yr, with two unwatered plots as controls. Windowpane traps were used to monitor cockroaches and other insects. Forty-six cockroach species were captured, with tremendous variation in numbers between years. Seasonal variation was also common but could not be attributed to the experimental treatment. The author concluded that rainfall was not the proximate cause of cockroach seasonal activity. Staggered seasonal peaks suggested strong interactions among some congeneric species (Fig. 3.5) (Wolda and Fisk, 1981).

Withstanding Cold

Cockroaches, like other invertebrates, have a diversity of responses to cold temperatures (Block, 1991). Each strategy entails energetic costs, with many interacting factors,

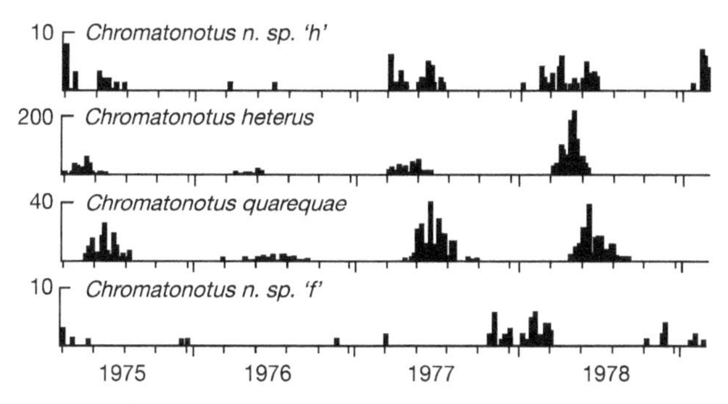

Fig 3.5 The number of individuals of four species of *Chromatonotus* collected per week in a light trap run for 4 yr on Barro Colorado Island, Panama. Modified from Wolda and Fisk (1981).

including the minimum temperature to which they are exposed, the variation in winter temperature, lifecycle stage, body size, habitat, availability of harborage, diet, snow cover, and particularly, water requirements and management (e.g., Sinclair, 2000). Several temperate cockroaches are active throughout winter, including the New Zealand species *Parellipsidion pachycercum, Celatoblatta vulgaris, Cel. peninsularis,* and *Cel. quinquemaculata.* The latter is a tiny (adult weight 0.1 g), brachypterous cockroach inhabiting alpine communities at altitudes greater than 1300 m asl, and is active even when the temperature of its microhabitat is below freezing (Zervos, 1987; Sinclair, 1997). Several North American species of *Parcoblatta* are similarly lively in winter (Horn and Hanula, 2002). Blatchley (1920) wrote of *Parc. pennsylvanica:* "Cold has seemingly but little effect upon them, as they scramble away almost as hurriedly when their protective shelter of bark is removed on a day in mid-January with the mercury at zero, as they do in June when it registers 100 degrees in the shade." Tanaka (2002) demonstrated that in *Periplaneta japonica* the ability to move at low temperature is acquired seasonally. During winter, last instar nymphs recover from being buried in ice in < 100 sec, with some of them moving immediately; in summer, movement was delayed by > 600 sec.

As in other insects, two main physiological responses contribute to winter hardiness in cockroaches: freeze tolerance and the prevention of intracellular ice formation by supercooling. Regardless of the season, *Cel. quinquemaculata* is freeze tolerant, with a lower lethal temperature in winter. Supercooling points fluctuate throughout the year, but the insect uses potent ice nucleators to avoid extensive supercooling. Its level of protection is just adequate for the New Zealand mountains in which it lives, where the climate is unpredictable and temperatures as low as -4°C have been recorded in summer. This cockroach may undergo up to 23 freeze-thaw cycles during the coldest months and remain frozen for up to 21 hr. The added protection of buffered microhabitats is necessary for survival in some winters (Sinclair, 1997, 2001; Worland et al., 1997). The North American montane species *Cryptocercus punctulatus* lives in a more predictable seasonal climate, with the added climatic buffer of a rotting log habitat. It is freeze tolerant only in winter; it uses the sugar alcohol ribitol as an antifreeze in transitional weather, and as part of a quick-freeze system initiated by ice-nucleating proteins when the temperature drops (Hamilton et al., 1985). There was a 76% survival rate among individuals held up to 205 days at -10°C, and winter-conditioned cockroaches that are frozen become active as soon as they are warmed to room temperature. Cold hardiness has also been studied in *P. japonica* (Tanaka and Tanaka, 1997), *Parc. pennsylvanica* (Duman, 1979), *Perisphaeria* spp., and *Derocalymma* spp. (Sinclair and Chown, 2005).

Seasonality and Life Histories

In trapping studies of cockroaches it is usually unknown if the failure to collect a particular species is due to the absence of the taxon in the habitat, the absence of the targeted life stage, or the current inactivity of the targeted life stage. Light traps or windowpane traps, for example, will collect only adult stages of volant cockroaches during the active part of their diurnal and seasonal cycle; taxa absent from these traps may be plentiful as oothecae and juveniles in the leaf litter. It is therefore important to discuss seasonal activity within the framework of a particular taxon's life history strategy (Daan and Tinbergen, 1997). There are complex, multivariate interactions among generation time, the size at maturity, age, lifespan, and growing season length (Fischer and Fiedler, 2002; Clark, 2003). Diapause and quiescence further interact with developmental rates to synchronize lifecycles, determine patterns of voltinism, and regulate seasonal phenology.

In seasonal environments life histories typically balance time constraints, with the synchronization of adult emergence most crucial when nymphal development is extended and adults are relatively short lived (Brown, 1983). Hatching must be timed so that seasonal mortality risks to juveniles are minimized. In *P. japonica*, for example, first-instar nymphs do not recover following tissue freezing, although mid- to large-size nymphs survive (Tanaka and Tanaka, 1997). The most thoroughly studied lifecycles among temperate cockroaches are those of the genus *Ectobius.* All three species in Great Britain spend winter in egg stage diapause, and hatch over a limited period in June after 6–7 mon of dormancy (Fig. 3.6). *Ectobius panzeri* is univoltine, while *E. lapponicus* and *E. pallidus* have semi-voltine lifecycles. Nymphs and eggs of the latter two species diapause in winter in alternate years, but there is complex intrapopulation variability in both species. At the onset of winter the nymphs move to grass tussocks and assume a characteristic posture: the body is flexed ventrally and the legs and antennae are held close to the body. Nymphs may feed during the winter, but no molting occurs from the end of September until the end of April or beginning of May. Adults are short lived; males die shortly after mating in June, but females live until October (Brown, 1973a, 1973b, 1980, 1983). It is also notable that of the three species of *Ectobius* in Great Britain, the smallest species, *E. panzeri* (Brown, 1952), is the only one with a univoltine cycle. *Ectobius duskei,* abundant in the bunch grasses of Asian steppe zones, is also univoltine and endures winters of −30 to −40°C in the egg stage (Bei-Bienko, 1950, 1969). It is thought that short favorable seasons often lead to compressed life histories such as these, characterized by brief developmental times, high growth rates, and smaller adult sizes (Abrams et al., 1996). A radically different life history, however, is exhibited by temperate cockroaches in the genus *Cryptocercus,* and by members of the blaberid subfamily Panesthiinae. Nymphs have extended developmental periods and the full length of the growing season is required to complete a reproductive episode in both *Cryptocercus* and *Panesthia* (Rugg and Rose, 1984b). Female *C. punctulatus* paired with males the previous summer begin exhibiting ovariole and accessory gland activity in April and oviposit in late June and early July. Oothecae hatch in late July and early August, with most neonates reaching the third or fourth instar prior to the onset of winter (Nalepa, 1988a, and pers. obs.). Additional temperate species that have been studied include *An. tamerlana* in the Turkmenistan desert (3-yr lifecycle in males, 4–6 yr in females) (Kaplin, 1995), and *P. japonica,* with a 2-yr lifecycle. The first winter is passed as early instar nymphs, the second one as late-instar nymphs (Shindo and Masaki, 1995).

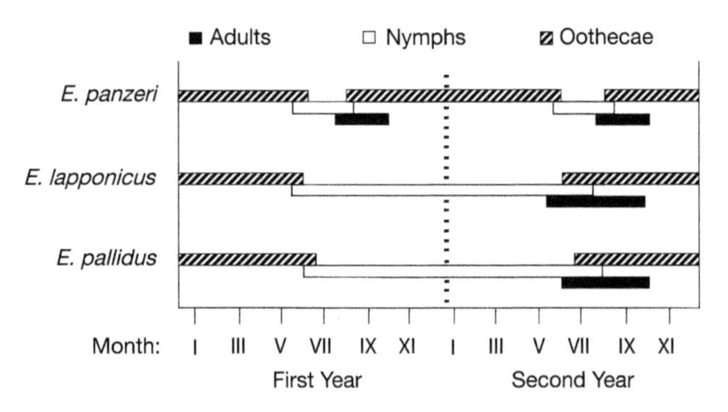

Fig. 3.6 Lifecycle of three species of *Ectobius* in Great Britain. After Brown (1973b), with permission from V.K. Brown.

Recently Tanaka and Zhu have been studying the lifecycles of several species of subtropical cockroaches on Hachijo Island in Japan. *Margattea satsumana* is a univoltine species that overwinters as a non-diapause adult. Nymphs undergo a summer diapause, but develop quickly in autumn under short-day photoperiods. The authors suggest that the summer diapause of nymphs is related to a need for timing reproduction during the following spring (Zhu and Tanaka, 2004b). *Opisthoplatia orientalis* and *Symploce japanica* on this island are both semi-voltine. The latter has a complex 2-yr lifecycle with three kinds of diapause (Tanaka and Zhu, 2003): a winter diapause in mid-size nymphs, a summer diapause in late-stage nymphs, and a winter diapause in adults. *Opisthoplatia orientalis* is a large (25–40 mm) brachypterous species capable of overwintering successfully in any stage without diapause. The ovoviviparous females spend the winter with several different stages of oocytes and embryos held internally, but the growth of these is suppressed. Most of the eggs and embryos do not survive to partition. As a result female ovarian development is reset in spring; there is a synchronized deposition of nymphs in summer, most of which reach the fifth instar prior to winter (Zhu and Tanaka, 2004a). This somewhat odd strategy may be related to the fact that these cockroaches are at the northern limit of their distribution on Hachijo Island, where they are not endemic.

RANGE OF HABITATS

Cockroaches are found in a continuum of dark, humid, poorly ventilated, and often cramped spaces either continuously or when sheltering during their non-active period. Although certain species may be associated with a particular crevice type like the voids beneath rocks or the space beneath loose bark, others are commonly found in more than one of these habitat subdivisions. Many species exploit the interconnectivity of dark, enclosed spaces wherever there is suitable food and moisture, and a distinctive classification of cockroaches as either obligate or facultative inhabitants of caves, litter, or soils is not always a natural one (Peck, 1990). The cave and the forest floor differ far more from the open-air habitat than they do from each other (Darlington, 1970). In closely grown tropical and subtropical forest almost all atmospheric movements are inhibited, surface evaporation of the leaves maintains a high humidity, and the canopy shields the forest floor from the direct rays of the sun. Cockroaches that live in the maze of hiding places that exist in suspended soils or on the forest floor live in a doubly blanketed environment, as moist plant litter further dampens the small fluctuations of light, temperature, and humidity that prevail throughout the forest (Lawrence, 1953). Caves, on the other hand, encompass a continuum of various sized dark, humid voids. To an arthropod, these could range from a few millimeters in size to the largest caverns, and may occur in soil layers, fractured rock, lava tubes, and talus slopes (Howarth, 1983). All of these spaces, whether created by the insect or naturally occurring in soil, leaf litter, guano, debris, rotten wood, or rock are similar in that they are dark, often humid, and buffered from temperature fluctuations.

It is obvious that a crevice-seeking/burrowing lifestyle is suited to a wide range of habitats, as long as dark, humid spaces are present or the substrate allows for their creation. Burrowing, the act of manufacturing or enlarging a space for shelter, is common among Blattaria, but there is a fine line of distinction between a cockroach forcing itself into an existing void, such as one under loose bark, and actually tunneling into the soft, rotted wood beneath. Both photonegativity and positive thigmotaxis predispose cockroaches to burrowing behavior. Beebe (1925, p. 147) offers a vivid definition of positive thigmotaxis: "having the irresistible desire to touch or be touched by something, above, below, and—a thigmotac's greatest joy—on all sides at once" (Fig. 3.7). Additional traits that favor successful colonization of dark, dank habitats include the use of non-visual cues in detecting food, mates, and predators, a lack of highly specialized feeding habits, and physiological adaptations to food scarcity (Darlington, 1970; Culver, 1982; Langecker, 2000).

A subterranean niche offers a relatively simple habitat, with climatic stability and a degree of protection from predators. These benefits are countered by physical and physiological challenges that must be met for successful occupancy. Costs may be incurred in obtaining or constructing burrows and shelters. The insect must cope with an environment that is aphotic, low in production, and high in humidity, endo- and ectoparasites, and pathogens (Nevo, 1999). Suboptimum O_2 and toxic CO_2 levels are also common in burrows, in caves, in wet, decaying logs, at high altitudes, and when insects are encased in snow and ice (Mani, 1968; Cohen and Cohen, 1981; Hoback and Stanley, 2001).

For our discussion of cockroach habitats, we recognize five broad subdivisions: (1) cockroaches that shelter in loose substrates (plant litter, guano, uncompacted soil, dust); (2) crevice fauna (under logs, bark, stones, and clumps of earth, in rolled leaves, leaf bases, bark crevices, scree); (3) those that excavate burrows in a solid substrate (wood, soil); (4) those that make use of existing nests or burrows (active or abandoned nests of social insects and small vertebrates); and (5) those in large burrows: caves and cave-like habitats like sewers and mines. We then address cockroaches found in three rather specialized habitats: deserts, aquatic environments, and the forest canopy. We are aware that there are difficulties in adhering to these distinctions, as the subdivisions grade into each other and species often span categories. Many cockroaches that do not routinely inhabit a burrow, for example, may construct underground chambers for rearing the young, for hibernation, for aestivation, or for molting. Many species travel between shelter and sites of feeding and reproductive activity; others (especially those in categories 3 and 4) live their entire life in shelter, except for brief dispersal periods. Some cockroaches never leave sheltered spaces (some cases of category 5). Those in category 3 actively create their living space, while those in the other four categories generally choose advantageous locations among existing alternatives. In each category, variation exists that is rooted in resource quality, quantity, and location.

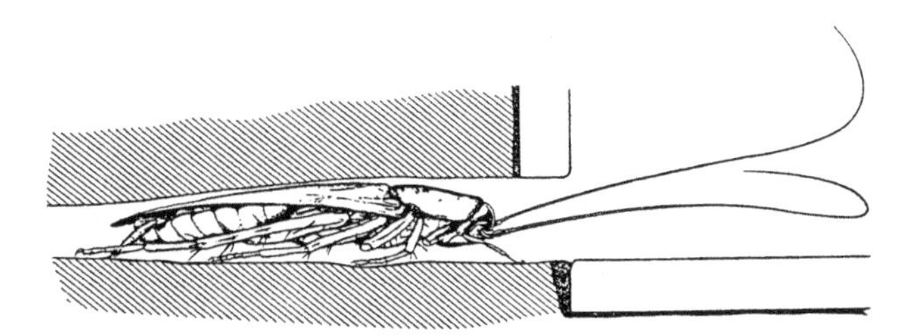

Fig. 3.7 Section through a crevice showing the characteristic rest position of a cockroach. From Cornwell (1968), with permission of Rentokil Initial plc.

In Loose Substrate

Cockroaches in this category either tunnel in uncompacted substrate (loose soil, dust, sand, guano), which may collapse around them as they travel through it, or they utilize small, preexisting spaces (dirt clods, leaf litter, and other plant debris), which their activities may enlarge. Many remain beneath the surface only during inactive periods, although those in guano and leaf litter, particularly juveniles, may conduct all activities there. Certainly the largest class in this category are cockroaches that tunnel in plant litter found on forest floors, in the suspended soils of the canopy (e.g., in epiphytes, treeholes, tree forks), and in piles concentrated by the actions of wind, water, or humans. Some species tunnel only as a defense from predators, or in response to local or seasonal conditions. Substrate categories are often fluid. Those that burrow in guano may also burrow in dirt, and those that tunnel in leaf litter may continue into the superficial layers of soil. Adults of *Therea petiveriana* in the dry, scrub jungles of India burrow in soil, leaf litter, and debris (including garbage dumps) during their non-active period (Livingstone and Ramani, 1978). The nymphs are subterranean and prefer the zone between the litter and the underlying humus, but may descend 30 cm during dry periods (Bhoopathy, 1997). Other versatile burrowers include *Blaberus* spp., which readily bury themselves in dirt or loose guano (Blatchley, 1920; Crawford and Cloudsley-Thompson, 1971), and *Pycnoscelus* spp., found in a wide variety of habitats as long as they can locate appropriate substrate for burrowing (Roth, 1998b; Boyer and Rivault, 2003). All stages of *Pyc. surinamensis* tunnel in loose soil, and are also reported from rodent burrows (Atkinson et al., 1991). The sand-swimming desert cockroaches fall into this category, as well as species such as *Ergaula capensis,* where females and nymphs burrow into well-rotted coconut stumps (Princis and Kevan, 1955), as well as the dry dust at the bottom of tree cavities (Grandcolas, 1997b). *Blattella asahinai* is known to burrow into leaf litter and loose soil; they are sometimes pulled up along with turnips in home gardens (Koehler and Patterson, 1987). Individuals of *Heterogamodes* sp. are known to bury themselves in sand or earth (Kevan, 1962). Several Australian species (*Calolampra* spp., *Molytria vegranda*) seem to spend the daylight hours underground, emerging to feed after dark (Rentz, 1996; D. Rentz, in Roth, 1999b). When collected during their active period or in light traps they usually sport sand grains on their bodies. In caves, *Eu. posticus* nymphs burrow in the surface of loose guano. They may be completely concealed, or may rest with their heads on the surface with their antennae extended up into the air. If the guano is compacted, the cockroaches remain on its surface and are attracted to irregularities such as the edge of a wall, a rock, or a footprint (Darlington, 1970). The recently described species *Simandoa conserfariam* congregates in groups of 20 to 50 individuals of all ages deep within the guano of fruit bats; none have been observed on the surface (Roth and Naskrecki, 2003).

Crevice Fauna

The cockroaches considered crevice fauna are those that insert themselves into preexisting small voids in generally unyielding substrates. These include species found under bark, in bark fissures, in the bases of palm fronds and grass tussocks, in hanging dead leaves, empty cocoons, and hollow twigs, under logs and rocks, in piles of stones, rock crevices, and the excavated galleries of other insects. An example of the latter is the Malaysian cockroach *Margattea kovaci,* which lives in bamboo internodes accessed via holes excavated by boring Coleoptera and Lepidoptera (D. Kovach, pers. comm. to LMR). Burrowing and crevice-dwelling cockroaches can be categorically difficult to separate, particularly species that shelter under rotting logs, in rolled leaves, or in the litter wedged into the base of bunch grasses, spinifex, or the leaf axils of many plants. The spaces under rocks and stones are a particularly important microhabitat for cockroaches in unforested areas. Species of the genera *Deropeltis* and *Pseudoderopeltis,* for example, are abundant under the boulders "bestrewing the Masai steppe country" (Shelford, 1910b). Rock-soil interfaces may also act as corridors between habitats, serving as oases for cockroaches moving between caves, or between patches of forest (Lawrence, 1953). Some cockroach species are morphologically specialized to inhabit the wafer-thin crevices under bark or rocks (Fig. 1.10). The incredibly flattened bodies of tropical Australian *Mediastinia* spp. allow them to slip into the unfolding leaves of gingers, lilies, and similar plants during the day. At night they move to new quarters as the leaves of their previous shelters unfold (D. Rentz, pers. comm. to CAN).

In Solid Substrate

Cockroaches that excavate permanent burrows in solid materials such as wood or compacted soil are more specialized than those that use loose substrate or crevices. They typically exhibit a suite of ecological and behavioral features associated with their fossorial existence, and external morphology tends to converge. There are two major groups that fall into this category, the Cryptocercidae and the Panesthiinae, the latter of which includes the soil-burrowing cockroaches. There are other species whose morphology suggests they are strong burrowers, but little has been published on their field biology. The hissing cockroaches, including *Gromphadorhina portentosa,* have the general demeanor of burrowers. In a recently published book on the natural history of Madagascar, however, the only mention of these cockroaches is as prey for some vertebrates and as hosts for mites (Goodman and Benstead, 2003).

Burrows in solid substrates offer mechanical protection, as well as shelter from some classes of parasites and predators. The fact that dispersal in both the Cryptocercidae and Geoscapheini occurs following rainfall when excavation is likely to be more efficient (Rugg and Rose, 1991; Nalepa, 2005) suggests that burrow creation is energetically costly. Pathogens may accumulate in tunnels, and occupants may not be able to escape if a predator enters the excavated space. It is unknown if burrowing cockroaches have strategies for dealing with flooded burrows, or with the often peculiar O_2 to CO_2 ratios that may occur.

In Wood

Dead wood is a tremendously diverse resource that varies with plant taxon, size (branch to bole), location (forest floor to suspended in canopy), degree and type of rot, orientation (standing versus prone), presence of other invertebrates, and other factors. Cockroach species from

Table 3.2. Examples of cockroaches other than Cryptocercidae and Panesthiinae that have been collected from rotted wood.

Cockroach species	Habitat	Reference
Anamesia douglasi	Under bark, in rotting wood, in fallen timber	Roach and Rentz (1998)
Austropolyphaga queenslandicus	Colonies in preformed chambers in dead logs and stumps	Roach and Rentz (1998)
Lauraesilpha mearetoi	In soft wood of small, dead branches	Grandcolas (1997c)
Lamproblatta albipalpus	Rotten logs and banana trucks, leaf litter	Hebard (1920a) Gautier and Deleporte (1986)
Laxta granicollis *Lax. tillyardi*	Under bark, in rotting wood	Roach and Rentz (1998)
Litopeltis bispinosa	Rotting banana and coconut palms	Roth and Willis (1960)
Methana parva	Under bark, in rotting wood	Roach and Rentz (1998)
Panchlora nivea	Rotting banana and coconut palms, rotten wood	Roth and Willis (1960) Séin (1923)
Panchlora spp.	Rotting logs, stumps, woody vegetation	Wolcott (1950) Fisk (1983)
Paramuzoa alsopi	Juveniles in dead wood	Grandcolas (1993b)
Parasphaeria boleiriana	In soft, rotten wood	Pellens et al. (2002)
Polyphagoides cantrelli	In rotting wood	Roach and Rentz (1998)
Robshelfordia hartmani	In rotting wood, females also collected in caves	Roach and Rentz (1998)
Sundablatta pulcherrima[1]	Abundant in decayed wood	Shelford (1906c)
Ylangella truncata	Adults under bark; juveniles deep in rotten tree trunks	C. Rivault (pers. comm. to CAN)

[1]Described as *Pseudophyllodromia pulcherrima* by Shelford (1906c); LMR's notes on the Shelford manuscript indicate it is in the genus *Sundablatta*.

most families have been collected from rotting logs (Table 3.2), but in the majority of cases it is unknown whether these feed on wood and associated microbes, if they depart to forage elsewhere, or both. This category is more fluid than generally recognized, and divisions in the dietary continuum of rotted leaf litter, soft rotted wood, and wood-feeding are not always easy to make. This is particularly true of the many cockroaches that bore into the well-rotted trunks and stalks of coconut and banana palms, which have been described as "gigantic vegetables with a stalk only a little tougher than celery" (Perry, 1986). Some cockroaches (e.g., *Blaberus*) are found in rotting logs as well as a variety of other habitats, others are not recorded anywhere else. *Tryonicus monteithi, Try. mackerrasae,* and *Try. parvus* are found in rotting wood and under stones and pieces of wood in Australian rainforest, but never under bark or above ground (Roach and Rentz, 1998). *Anamesia douglasi* is found under bark and in rotting wood, but has also been observed on sand ridges (Roach and Rentz, 1998), perhaps sunning themselves like some other Polyzosteriinae. Groups of similar-sized juveniles of *Ylangella truncata,* probably hatched from a single ootheca, live in galleries deep in the interior of large rotting tree trunks. Adults are excellent fliers and are found most often just under the bark of these logs. Attempts to rear nymphs in the laboratory on pieces of rotted wood and a variety of other foodstuffs, however, were not successful (C. Rivault, pers. comm. to CAN). A species of large, reddish, heavy-bodied hissing cockroach has been observed in groups of 40 or 50 inside of rotten stumps and logs in riverine areas of southeastern Madagascar. Groups included both adults and nymphs (G. Alpert, pers. comm. to LMR). The least known cockroaches in this category are those with the elongated, cylindrical body form of many boring beetles. These include *Compsagis lesnei* (Chopard, 1952), found inside of tree branches (Fig. 1.14), and several species of *Colapteroblatta* (= *Poroblatta*) (Roth, 1998a), which Gurney (1937) described as boring into stumps and logs in a manner similar to *Cryptocercus*. There are probably many more wood-boring cockroaches yet to be discovered, particularly in the substantial amount of dead and dying wood suspended in tropical canopies.

Both sexes of all species in the monogeneric family Cryptocercidae are wingless and spend their lives in decaying wood on the floor of montane forests in the Palearctic and Nearctic (Nalepa and Bandi, 1999). As might be expected for insects feeding on dead wood, their distribution and abundance varies in relation to patterns of tree mortality if other habitat requirements are met (Nalepa et al., 2002). Presently *C. punctulatus* in eastern North America is numerous at high elevations in logs of Fraser fir (*Abies fraseri*) killed by balsam wooly adelgid (*Adelges piceae*). Formerly they were easily found in chestnut logs (*Castanea dentata*) abundant on forest floors because of chestnut blight (Hebard, 1945). Occasionally, all families in a log are of the same developmental stage, suggesting that a particular log became suitable for colonizing at a particular point in time. A log may harbor only male-female pairs, for example, or only families with second-year nymphs (CAN, pers. obs.). Both Palearctic and Nearctic species of *Cryptocercus* occur in a wide variety of angiosperms and conifers, with the log host range determined by the plant composition of the inhabited forest. Well-rotted logs as well as those that are relatively sound serve as hosts (Cleveland et al., 1934; Nalepa and Bandi, 1999; Nalepa, 2003). The cockroaches are only rarely collected from wood undergoing the white rot type of decay (Mamaev, 1973; Nalepa, 2003); the conditions associated with white rot generally do not favor many groups of animals (Wallwork, 1976). Inhabited logs can be quite variable in size. Logs harboring *C. primarius* ranged from 10 cm to more than 1 m in diameter (Nalepa et al., 2001b). *Cryptocercus clevelandi* is most often collected in logs of Douglas fir, the large size of which buffers the insects from the warm, dry summers characteristic of southwest Oregon (Nalepa et al., 1997). Large logs provide insulation from winter cold, but *C. punctulatus* is also physiologically equipped to withstand freezing weather (Hamilton et al., 1985).

Wood-feeding cockroaches in the blaberid subfamily Panesthiinae are distributed principally in the Indo-Malayan and Australian regions, with a few species extending into the Palearctic. Six genera live in and feed on rotting wood, and exhibit little variation in morphology and habits. Body size, however, can be quite variable; *Panesthia* spp. range from 15 to more than 50 mm in length (Roth, 1977, 1979b, 1979c, 1982b). The best studied is *Panesthia cribrata* in Australia, found inside of decaying logs but also under sound logs, where they feed on the wood surface in contact with the ground. They are sometimes found in the bases of dead standing trees (Rugg and Rose, 1984a; Rugg, 1987). Host choice in these blaberids is similar to that of *Cryptocercus*. *Panesthia cribrata* in Australia (Rugg, 1987), as well as species of *Panesthia* and *Salganea* in Japan (K. Maekawa, pers. comm. to CAN) utilize softwood as well as hardwood logs. They generally use what is available, and when populations are high, they are found in a greater variety of log types (D. Rugg, pers. comm. to CAN).

All Cryptocercidae and wood-feeding Panesthiinae studied to date are slow-growing, long-lived cockroaches. Development takes about 4 yr in *Cryptocercus kyebangensis* (Park et al., 2002), *C. clevelandi* takes 5–7 yr, and *C. punctulatus* requires 4–5 yr. In the latter two species, adults pair up during the year they mature, but do not reproduce until the following summer. Thus the time from hatch to hatch in *C. clevelandi* is 6–8 yr, and in *C. punctulatus* 5–6 yr. Post reproduction, adults of these two species live for 3 or so yr in the field, females longer than males (Nalepa et al., 1997). Rugg and Rose (1990) calculated that the nymphal period of *Pane. cribrata* was at least 4–6 yr, and that the field longevity of adults exceeds 4 yr. *Panesthia cribrata*, as well as *Pane. australis*, *Pane. matthewsi*, *Pane. sloanei*, and *Pane. angustipennis spadica* live in aggregations, most often comprised of a number of adult females, an adult male, and nymphs of various sizes. Nymphs are also commonly found in groups without adults (Rugg and Rose, 1984a). *Panesthia cribrata* reproduces once per year, but probably gives birth each year (Rugg and Rose, 1989). All species of *Cryptocercus* studied to date live in monogamous family groups, and produce just one set of offspring, with an extensive period of parental care following (Seelinger and Seelinger, 1983; Nalepa, 1984; Nalepa et al., 2001b; Park et al., 2002). The panesthiine genus *Salganea* is also subsocial (Matsumoto, 1987; Maekawa et al., 1999b), but at least one species (*Sal. matsumotoi*) is iteroparous (Maekawa et al., 2005).

In Soil

Those cockroaches known to tunnel in uncompacted media such as leaf litter or loose soil occasionally make forays into more solid substrates. *Periplaneta americana* nymphs and adults have been observed digging resting sites in the clay wall of a terrarium (Deleporte, 1985), and *Pyc. surinamensis* can excavate tunnels that extend up to 13 cm beneath the soil surface. These tubes may end in a small chamber where juveniles molt and females bear young (Roesner, 1940). At least two unstudied blaberids in the subfamily Perisphaeriinae appear to live in permanent soil burrows. Female *Cyrtotria* (= *Stenopilema*) are found in a burrows surrounded by juveniles (Shelford, 1912b). Similarly, a female *Pilema thoracica* accompanied by several nymphs was taken from the bottom of a neat round hole about 15 cm in depth; there were about a dozen such holes in half an acre and all contained families of this species (Shelford, 1908). Cockroaches of a *Gromphadorhina* sp. have been observed in a ground burrow in grassland of the Isalo National Park in Madagascar. The heads and antennae of both adults and nymphs were projecting from the entrance, which was about 5 cm in diameter (G. Alpert, pers. comm. to LMR).

All other cockroaches that form permanent burrows in compacted soil belong to four Australian genera of the subfamily Panesthiinae: *Macropanesthia, Geoscapheus, Neogeoscapheus,* and *Parapanesthia* (Roth, 1991a). They are distributed mainly east of the Great Dividing Range with a concentration in southeast Queensland (Roach and Rentz, 1998). The giant burrowing cockroach *M. rhinoceros* is the best studied (Rugg and Rose, 1991; Matsumoto, 1992), but the biology of the other species is similar (D. Rugg, pers. comm. to CAN). All feed on dry plant litter that they drag down into their burrows. Burrow entrances have the characteristic shape of a flattened semicircle, but may be slightly collapsed or covered by debris during the dry season. Tunnels initially snake along just beneath the soil, then spiral as they descend and widen out; they tend to get narrow again at the bottom. Litter provisions are typically stored in the wider part, and the cockroaches retreat to the narrow blind terminus when alarmed. They are not known to clean galleries; consequently, debris and excrement accumulate (Rugg and Rose, 1991; D. Rugg, pers. comm. to CAN). Species distribution is better correlated with soil type than with vegetation type. Burrows of *M. rhinoceros* may be found in *Eucalyptus* woodland, rainforest, or dry *Acacia* scrub, as long as the soil is sandy. Other species are associated with gray sandy loams, red loam, or hard red soil (Roach and Rentz, 1998). The depth of *Macropanesthia saxicola* burrows is limited by the hard heavy loam of their habitat, and those of *M. mackerrasae* tend to be shallow and non-spiraling because they run up against large slabs of rock. The deepest burrows are those of females with nymphs, the shallowest are those of single nymphs (Rugg and Rose, 1991; Roach and Rentz, 1998). Female *M. rhinoceros* reproduce once per year, and nymphs remain in the tunnel with females for 5 or 6 mon before they disperse, initiate their own burrows, and begin foraging. These mid-size nymphs then enlarge their burrows until adulthood. Development requires a minimum of 2 or 3 yr in the field, but growth rates are highly variable. Adults live an additional 6+ yr (Rugg and Rose, 1991; Matsumoto, 1992). Males are occasionally found in the family during early stages of the nesting cycle. Both sexes emerge from burrows after a rainfall, with females foraging and males looking for females. Surface activity in *M. rhinoceros* occurs from just before midnight to a couple of hours after sunrise; peak of activity is 2 or 3 hr before sunrise. Small nymphs are never observed above ground (Rugg and Rose, 1991).

Recent evidence indicates that among the Panesthiinae, the ecological and evolutionary boundaries between the soil-burrowing–litter-feeding habit, and one of living in and feeding on wood, are more fluid than expected. In 1984, Rugg and Rose (1984c) proposed that the soil-burrowing cockroaches be elevated to the rank of subfamily (Geoscapheinae) on the basis of their unique reproductive biology. Recently, however, a molecular analysis of three genes from representatives of nine of the 10 Panesthiinae and Geoscapheini genera by Maekawa et al. (2003) indicates that these taxa form a well-supported monophyletic group, with the former paraphyletic with respect to the latter (Fig. 2.13). These authors propose that the ancestors of soil-burrowing cockroaches were wood feeders driven underground during the Miocene and Pliocene, when dry surface conditions forced them to seek humid environments and alternative sources of food. This suggestion is eminently reasonable, as there are isolated cases of otherwise wood-feeding cockroach taxa collected from soil burrows or observed feeding on leaf litter. *Ancaudellia rennellensis* in the Solomon Islands lives in underground burrows (Roth, 1982b), even though the remaining species in the genus are wood feeders. There is also a record of a male, a female, and 19 nymphs of *Panesthia missimensis* in Papua New Guinea collected 0.75 m deep in clay, although others in the species were collected in rotten logs (Roth, 1982b). Although the preferred habitat of the endangered *Panesthia lata* is decaying logs, Harley Rose (University of Sydney) has also found them under rocks, sustaining themselves on *Poa* grass and *Cyperus* leaves (Adams, 2004). Even individuals or small groups of *C. punctulatus* are sometimes found in a small pocket of soil under a log, directly beneath a gallery opening (Nalepa, 2005), particularly when logs become dry. These examples are evidence that the morphological adaptations for burrowing in wood also allow for tunneling in soil, and that the digestive physiology of wood-feeding Panesthiinae may be flexible enough to allow them to expand their dietary repertoire to other forms of plant litter when required.

In Existing Burrows and Nests

Some cockroaches specialize in using the niche construction, food stores, and debris of other species. Whether these cockroaches elude their hosts or are tolerated by them is unknown. Of particular interest are the cockroaches that live with insectivorous vertebrates such as rodents and some birds. How do the cockroaches avoid becoming prey?

Insect Nests

A number of cockroaches live in the nests of social insects, although these relationships are rather obscure. Some cockroach species collected in ant and termite colonies have been taken only in this habitat (Roth and Willis, 1960), and are presumably dependent on their hosts. In others, the relationship is more casual, with the cockroaches opportunistically capitalizing on the equable nest climate and kitchen middens of their benefactors. Several species of the genus *Alloblatta,* for example, scavenge the refuse piles of ants (Grandcolas, 1995b). Similar garbage-picking associations are found in *Pyc. surinamensis* with the ant *Campanotus brutus* (Deleporte et al., 2002), and in nymphs of *Gyna* with *Dorylus* driver ants (Grandcolas, 1997a). Occasional collections from insect nests include the Australian polyphagid *Tivia australica,* recorded from both litter and ant nests, and the blattellid *Paratemnopteryx australis,* collected from under bark, in litter, and from termite (*Nasutitermes triodiae*) nests (Roach and Rentz, 1998). In the United States, *Arenivaga bolliana* and *A. tonkawa* have been taken from both nests of *Atta texana* and burrows of small vertebrates (Roth and Willis, 1960; Waller and Moser, 1990). In Africa, *Er. capensis* has been collected in open bush, in human habitations, and in termite mounds, and is just one of several taxa, including *Periplaneta,* that exploit both human and insect societies (Roth and Willis, 1960).

The records we have of more integrated myrmecophiles include the New World genera *Myrmecoblatta* and *Attaphila.* The polyphagid *Myrmecoblatta wheeleri* is associated with nests of *Solenopsis geminata* in Guatemala (Hebard, 1917), and with the carpenter ants *Camponotus abdominalis* in Costa Rica and *C. abdominalis floridanus* in Florida. Deyrup and Fisk (1984) observed at least 20 *Myr. wheeleri* of all sizes when a dead slash pine log was turned over in scrubby flatwoods habitat in Florida. All *Attaphila* spp. (Blattellidae) are associated with leaf-cutting ants in the genera *Atta* and *Acromyrmex* (Kistner, 1982). The best known is *Attaphila fungicola* (Fig. 1.16B), a species that lives in cavities and tunnels within the fungus gardens of *Atta texana.* Both male and female cockroaches have been collected from *A. texana* nests in Texas (Wheeler, 1900), but only females have been collected in Louisiana (Moser, 1964). Within the nest, *Att. fungicola* ride on the backs or the enormous heads of soldiers, which "do not appear to be the least annoyed" (Wheeler, 1900). The cockroach mounts a passing host by grabbing the venter or gaster, then climbing onto the mesonotum; they ride facing perpendicular to the long axis of the ant's body. The weight of the cockroach may cause the ant to topple over (J.A. Danoff-Burg, pers. comm. to WJB). Perhaps for this reason, *Attaphila* chooses for steeds the soldiers, the largest ants in the colony. The cockroaches run along with ants as well as riding on them, and can detect and orient to ant trail pheromone (Moser, 1964), presumably via a unique structure on the maxillary palps (Brossut, 1976). Wheeler (1900) originally thought that the cockroaches fed on the ant-cultivated fungus within the nest, but later (1910) decided that they obtain nourishment by mounting and licking the backs of soldiers. It is, of course, possible that they do both.

Recently, another myrmecophile has been described from jungle canopy in Malaysia, leading us to believe that there are many more such associations to be discovered in tropical forests. The ovoviviparous blattellid *Pseudoanaplectinia yumotoi* was found with *Crematogaster deformis* in epiphytes (*Platycerium coronarium*) exposed to full sunlight 53 m above the ground. The leaves of these stag's horn ferns form a bowl that encloses the rhizome, roots, and layers of old leaves within which the ants and cockroaches live. More than 2800 *Ps. yumotoi* were collected from one nest of about 13,000 ants. The ants protect the cockroaches from the attacks of other ant species. Living cockroaches are not attacked by their hosts, but ants do eat the dead ones (Roth, 1995c; T. Yumoto, pers. comm. to LMR). At least two cockroach species exploit the mutualism between ants and acacias. *Blattella lobiventris* has been found in swollen acacia thorns together with *Crematogaster mimosae* (Hocking, 1970). Female *Nyctibora acaciana* glue their oothecae near *Pseudomyrmex* ant nests on acacias, apparently for the protection provided by the ants against parasitic wasps (Deans and Roth, 2003).

Several species of cockroaches in the genus *Nocticola* have been found within the nests of termites but nothing is known about their biology or their relationship with their hosts (Roth and Willis, 1960; Roth, 2003b). The majority of these are associated with fungus-growing termites (*Macrotermes* and *Odontotermes*), which in the Old World are the ecological equivalents of *Atta.* This strengthens the suggestion that fungus cultivated by social insects may be an important dietary component of cockroach inquilines. Many cockroach species can be found in deserted termite mounds (Roth and Willis, 1960).

Few cockroaches have been found in nests of Hymenoptera other than ants. The minute (3 mm) species *Sphecophila polybiarum* inhabits the nests of the vespid wasp *Polybia pygmaea* in French Guiana (Shelford, 1906b). Apparently the cockroaches feed on small fragments of prey that drop to the bottom of the nest when wasps feed larvae. *Parcoblatta* sp. (probably *Parc. virginica*) are commonly found (68% of nests) scavenging bits of dropped prey and other colony debris in subterranean yellowjacket (*Vespula squamosa*) nests at the end of the colony cycle (MacDonald and Matthews, 1983). Similarly, *Oulopteryx meliponarum* presumably ingest excreta and other debris scattered by the small stingless bee *Melipona*. Additional associations are discussed in Roth and Willis (1960).

Cockroaches living in the nests of social insects profit from protective services, a favorable microclimate, and a stable food supply in the form of host-stored reserves and waste material. The only benefit to the hosts suggested in the literature is the opportunity to scavenge the corpses of their guests. Ants generally ignore live *Attaphila* in the nest (Wheeler, 1900), but the mechanism by which the cockroaches are integrated into colony life has not been studied. Like other inquilines, however, the cuticular hydrocarbons of these cockroaches may mimic those of their hosts. Gas chromatography indicates that the surface wax of *Ps. yumotoi* is similar to that of their ant hosts (T. Yumoto, pers. comm. to LMR), but it is yet to be determined whether these are acquired from the ants by contact or ingestion, or if they are synthesized de novo. Cuticular hydrocarbons are easily transferred by contact between two different species of cockroaches. After 14 days in the same container *N. cinerea* and *R. maderae* merge into one heterospecific group with cuticular profiles that show characteristics of both species (Everaerts et al., 1997). Ants can acquire the hydrocarbons of a non-myrmecophile cockroach (*Supella longipalpa*) via physical contact; these ants are subsequently recognized as foreign by their nestmates and attacked (Liang et al., 2001). Individuals of *Attaphila fungicola* spend so much time licking soldiers (Wheeler, 1910) that these myrmecophiles may be internally acquiring and then reusing epicuticular components of their host.

Vertebrate Burrows

Most records of Blattaria in vertebrate burrows come from deserts (discussed below), as the high moisture content of these habitats is advantageous in arid environments. Cockroach food sources in these subterranean spaces include organic debris, and the feces, cached food, and dead bodies of inhabitants (Hubbell and Goff, 1939). Roth and Willis (1960) indicate that cockroach species found in animal burrows are usually different than those that inhabit caves. Richards (1971), however, suggests that burrows may be important as intermediate stops when cockroaches move between caves, and gives as example the often cavernicolous species *Paratemnopteryx rufa* found in wombat burrows.

Bird Nests

Cockroaches are only rarely associated with the shallow cup-type nest typical of many birds. The one exception known to us is *Euthlastoblatta facies*, which lives in large numbers among twigs in the nests of the gray kingbird in Puerto Rico (Wolcott, 1950). Most records are from the nests of birds that breed gregariously and construct pendulous, teardrop-shaped nests up to 1 m long (Icteridae) or large, hanging apartment houses of dry grass (Ploceinae). Roth (1973a) collected about 10 species of cockroaches in the pendulous nests of an icterid (probably the oriole, *Cassicus persicus*) in Brazil. *Schultesia lampyridiformis* was found in 2 of 7 nests of *Cassicus* about 18 m above ground in the Amazon. Van Baaren et al. (2002) found 5 species in icterid bird nests in French Guiana: *Schultesia nitor, Phoetalia pallida, Pelmatosilpha guianae, Chorisoneura* sp., and *Epilampra grisea*. Immature cockroaches were common in the nests of Ploceinae in Madagascar and the Ivory Coast; all nests of *Foundia* spp. examined in Madagascar harbored cockroaches restricted to this habitat (Paulian, 1948). *Griffiniella heterogamia* lives in nests of a social weaver bird in southwest Africa (Rehn, 1965). Most icterid nests inhabited by the cockroaches were abandoned, and a few carried the remains of dead young birds. The cockroaches are probably scavengers and may also occupy the nests while birds are present (Roth, 1973a).

In Caves and Cave-Like Habitats

Cockroaches are well represented in caves throughout the tropics and subtropics, from 30°N to 40°S of the equator; they are uncommon in temperate caves (Izquierdo and Oromi, 1992; Holsinger, 2000). Except for rare collections of *Arenivaga grata* and *Parcoblatta* sp., no cave cockroaches occur in the continental United States (Roth and Willis, 1960; Peck, 1998). The biology of cave-dwelling cockroaches has been studied most extensively in Trinidad and Australia. In Guanapo Cave in Trinidad, *Eublaberus distanti* is dominant, with *Blab. colloseus* and *Xestoblatta immaculata* also found (Darlington, 1995–1996). These three species, as well as *Eub. posticus*, are also found in the Tamana Caves (Darlington, 1995a). Six cockroach species are reported from caves of the Nullarbor Plain of southern Australia: *Polyzosteria mitchelli,*

Polyz. pubescens, Zonioploca medilinea (Blattidae), *Neotemnopteryx fulva, Trogloblattella nullarborensis,* and *Para. rufa* (Blattellidae). Three are considered accidentals, two are facultative, and one is an obligate cavernicole (Richards, 1971). Cockroaches in the family Nocticolidae are consistent inhabitants of caves throughout the Old World tropics (Stone, 1988; Deharveng and Bedos, 2000). Of the approximately 20 species in the widely distributed genus *Nocticola,* most are cavericolous, a few are epigean or termitophilous, and a few can be found both inside and outside of caves (e.g., *Alluaudellina himalayensis*) (Roth, 1988; Roth and McGavin, 1994). Juberthie (2000a) estimated that worldwide, 31 cockroaches species are known to be obligate cavernicoles, but additional species continue to be described (e.g., Vidlička et al., 2003). Table 3.3 gives examples of cave cockroaches; others are discussed in Asahina (1974), Izquierdo et al. (1990), Martin and Oromi (1987), Martin and Izquierdo (1987), Roth and Willis (1960), Roth (1980, 1988), Roth and McGavin (1994), and Roth and Naskrecki (2003).

It is often difficult to label a given species as a cave cockroach for two reasons. First, many of the described species are based on few collection records. Second, the term *cave* usually refers to an underground space large enough to accommodate a human, but grand expanses such as these are just a small part of the subterranean environment (Ruzicka, 1999). The limits of the hypogean realm are hard to define because cave habitats grade into those of the edaphic environment via smaller-scale subterranean spaces such as animal burrows, tree holes, hollow logs, the area under rocks, and other such dark, humid, organic living spaces. Cockroaches found in many of these non-cave habitats occasionally or consistently exploit caves. Those that are considered "accidentals" are only rarely collected in caves. *Polyz. mitchelli,* for example, is a large ground-dwelling epigean Australian species that has also been taken in caves (Roach and Rentz, 1998). On the other hand, those species that typically inhabit cave entrances may venture outside the cave if the humidity is high enough (e.g., *Para. rufa*—Richards, 1971). Among the cockroaches taken in a range of subterranean-type habitats is the Asian species *Polyphaga aegyptiaca,* found in bat caves, under decaying leaves, and in cliffs along ravines (Roth and Willis, 1960), and *X. immaculata, Eub. distanti, Blaberus giganteus, Blab. atropos,* and *Blab. craniifer.* The latter are all considered cave cockroaches, but are also collected from under decaying litter, in epiphytes, inside rotting logs, and in the rot holes and hollows of trees, particularly those that house bats (Darlington, 1970; Fisk, 1977). Perry (1986) described dozens of adult *Blab. giganteus* in a tree hollow "all sitting, as sea gulls on a beach, evenly spaced and facing upward." *Blatta orientalis, Blattella germanica,* and *P. americana* have all been found in caves, as well as in buildings, wells, sewers, steam tunnels, and mines 660 m below the surface (Roth and Willis, 1960; Roth, 1985) (Fig. 3.8). In one sense, however, these human-made, non-cave habitats may be considered vertebrate burrows. Cockroaches exhibiting morphological correlates of cave adaptation such as elongated appendages and the loss of pigment, eyes, and wings are generally restricted to cave habitats, but even these can be found elsewhere. A species of Australian *Nocticola* with reduced eyes and tegmina and no wings lives beneath rotting logs (Stone, 1988). The troglomorphic *Symploce micropthalmus* lives in the mesocavernous shallow stratum of the Canary Islands, but is also found under stones in humid areas (Izquierdo and Medina, 1992).

Individual caves are commonly divided into zones,

Table 3.3. Examples of cave-dwelling cockroaches.

1. Occur in caves sporadically, and sometimes become established there; show no morphological characters specifically associated with cave dwelling.

Examples: Blattidae: *Periplaneta americana, Polyzosteria mitchelli;* Blaberidae: *Pycnoscelus indicus, Pyc. surinamensis, Blaberus colosseus*

2. Habitually found in caves, but are able to live in or outside of caves; they show no characters adaptive for cave dwelling.

Examples: Blattidae: *Eumethana cavernicola;* Blattellidae: *Blattella cavernicola;* Blaberidae: *Blaberus craniifer, Eublaberus posticus, Aspiduchus cavernicola*

3. Cannot live outside of caves and show marked morphological specializations for the cave habitat (obligate cavernicoles or troglobites).

Examples: Blattidae: *Neostylopyga jambusanensis;* Blattellidae: *Neotrogloblattella chapmani, Loboptera anagae, L. troglobia, Paratemnopteryx howarthi, Para. stonei, Trogloblattella chapmani;* Nocticolidae: *Alluaudellina cavernicola, Typhloblatta caeca, Nocticola simoni, Noc. australiensis, Noc. bolivari, Noc. flabella, Spelaeoblatta thamfaranga*

Fig. 3.8 *Periplaneta* sp. in a sewer manhole in Houma, Louisiana. From Gary (1950).

with each supporting a different community (Juberthie, 2000b). The twilight zone near the entrance is closest to epigean conditions and has the largest and most diverse fauna. Next is a zone of complete darkness with variable temperature, and finally in the deep interior a zone of complete darkness, stable temperature, and stagnant air, where the obligate, troglomorphic fauna appear (Poulson and White, 1969). The degree of fidelity to a zone varies. While the Australian *Para. rufa* is found only from the entrance to 0.4 km into a cave, *Trog. nullarborensis* is found from the entrance to 4.8 km deep; it roams throughout the cave system and is one of the few troglomorphs recorded from the twilight zone (Richards, 1971). *Eublaberus posticus* and *Eub. distanti* may segregate in caves according to their particular moisture requirements. The former prefers the moist inner sections of caves, while the latter is more common in drier guano (Darlington, 1970). The habitable areas of caves, and consequently, populations of cave organisms, are dynamic—they move, expand, and contract, depending on climate and on pulses of organic matter (Humphreys, 1993). After an exceptionally cool night in Nasty Cave in Australia, for example, a common *Nocticola* cockroach could not be found and was thought to have retreated into cracks during the unfavorable conditions (Howarth, 1988). Initially a small species in the subfamily Anaplectinae was sporadically seen in a Trinidadian cave, subsequently formed a thriving colony, then was wiped out when the cave flooded. It did not reappear (Darlington, 1970).

Caves with a source of vertebrate guano support very different cockroach communities than caves that lack such input. Guano caves typically contain very large numbers of few cockroach species able to maintain dense populations and exploit the abundant, rich, but rather monotonous food bonanza (Darlington, 1970). Examples include a population of more than 80,000 *Gyna* sp. in a South African cave (Braack, 1989), more than 43,000 *Eub. distanti* in just one chamber of a cave in Trinidad (Darlington, 1970) (Fig. 3.9), and *Pycnoscelus striatus* found at approximately 2000–3000/m^2 in the Batu Caves of Malaysia (McClure, 1965). A similar scenario is that of approximately 3000 *P. americana* /m^2 in a sewer system more than 27 m beneath the University of Minnesota campus (Roth and Willis, 1957). In guano caves, the distribution of cockroaches usually coincides with that of bats and their excrement (Braack, 1989). Some species are consistently associated with bat guano, wherever it is found. One South African *Gyna* sp. was present in all bat-inhabited caves and cave-like habitats, including the roof of a post office (Braack, 1989).

Highly troglomorphic cockroach species generally support themselves on less rich, less abundant food sources. *Trogloblattella chapmani* is typically found remote from guano beds in passages floored by damp sticky clay or silt (Roth, 1980). *Metanocticola christmasensis* is associated with the often luxuriant tree root systems that penetrate caves (Roth, 1999b), but their diet is unknown (Roth, 1999b). Troglomorphic cockroaches tend to move

Fig 3.9 Habitat stratification in *Eublaberus distanti* in Guanapo Cave, Trinidad. (A) Adults on walls of cave; (B) nymphs on surface of fruit bat guano. Photos courtesy of J.P.E.C. Darlington.

very slowly (e.g., *Nocticola* spp.—Stone, 1988; *Loboptera troglobia*—Izquierdo et al., 1990), and produce few eggs. The oothecae of *Alluaudellina cavernicola* contain only four or five eggs (Chopard, 1919) and those of *Nocticola* (= *Paraloboptera*) *rohini* from Sri Lanka contain just four (Fernando, 1957). Among the seven species of *Loboptera* studied by Izquierdo et al. (1990) in the Canary Islands, reductions in ovariole number paralleled the degree of morphological adaptation to the underground environment. The least modified species had 16–18 ovarioles, while the most troglomorphic had six ovarioles. It is unknown whether troglomorphic cockroaches exhibit the increased developmental time and lifespan, decrease in respiratory metabolism, and loss of water regulatory processes found in many other cave-adapted animals (Gilbert and Deharveng, 2002).

Deserts

While cockroaches are generally associated with humid habitats, there are a number of species that have settled deserts, scrub, grassland, and other arid environments. These habitats vary in temperature, from hot subtropical deserts to colder deserts found at high latitudes or high elevations. In each, however, low precipitation plays a major role in controlling biological productivity. Many polyphagids, some blattellids, and a few blattids inhabit these xeric landscapes. Polyphagidae are most diverse in the deserts of North Africa and South Central Asia (Bei-Bienko, 1950), and best studied in Egypt (Ghabbour et al., 1977; Ghabbour and Mikhaïl, 1978; Ghabbour and Shakir, 1980) and Saudi Arabia (Bei-Bienko, 1950; Grandcolas, 1995a). The cockroaches can be very abundant, comprising nearly a third of the mesofaunal biomass collected in surveys of soil arthropods in the desert of northern Egypt (Ayyad and Ghabbour, 1977). In North America, polyphagid cockroaches occur in the southwestern United States, with one species (*Arenivaga floridensis*) found in Florida.

Desert-dwelling cockroaches exhibit morphological, behavioral, and physiological adaptations for maintaining water balance, avoiding or tolerating extreme temperatures, and finding food in habitats with sparse primary productivity. Behavioral tactics for coping with these extreme conditions include diurnal and seasonal shifts in spatial location and prudent choice of microhabitat. Like many desert arthropods, the sand-swimming Polyphagidae take advantage of the more salubrious conditions beneath the surface of desert soil. *Arenivaga investigata* migrates vertically in loose sand on a diel basis. In spring and summer, activity near the surface commences 2 hr after darkness and continues for most of the night (Edney et al., 1974). In winter, activity corresponds to peaks in nighttime surface temperature (Hawke and Farley, 1973). The insects move about just beneath the sand (Fig. 2.6), making them less susceptible to predators (e.g., scorpions) as they forage for dead leaves, roots, and other food. Throughout the year *A. investigata* can find a relative humidity of about 82% by descending 45 cm in the sand, and can avoid temperatures above 40°C by moving no lower than 15 cm (Edney et al., 1974). The cockroaches descend deeper in the sand in summer than in winter (Edney et al., 1974) (Fig. 3.10). In July, all developmental stages except adult males range 2.5–30 cm below the surface, with a mode at 12.5 cm. In November the insects are found no deeper than 15 cm, with most occurring at 5 cm or less. It is possible that the maximum depth to which these cockroaches burrow may be limited by hypoxia (Cohen and Cohen, 1981).

Although deserts can be very hot, very dry, and sometimes very cold, they have numerous microhabitats where the climate is much less extreme. In addition to the depths of loose sand, these include the burrows of small vertebrates, under boulders, in caves, and amid decaying organic material in dry stream beds, at the base of tussocks, in rock crevices, and under shrubs or trees (Roth and Willis, 1960). Some cockroach species are consistently associated with one of these microhabitats, and others move freely between them. *Arenivaga grata* is found under stones and rocks in scrub oak, oak-pine, and oak manzanita forests in Texas (Tinkham, 1948), but has been reported from bat guano in a cave in Arizona (Ball et al., 1942). Sand-swimming and Australian burrowing cockroaches are frequently found in the root zones of plants. *Arenivaga investigata* is most commonly associated with the shrubs *Larrea tridentata, Atriplex canescens,* and *Croton californicus* (Edney et al., 1978). The burrows of desert vertebrates utilized by some cockroach species are also typically found near desert plants. In the desert, vegetation is a source of shade and food, and subterranean root systems concentrate available moisture (Wallwork, 1976).

About half the desert cockroaches for which we have any information live in the burrows of vertebrates (Roth and Willis, 1960). Various species of *Arenivaga* and *Polyphaga* live in the excavations of desert turtles, prairie dogs, ground squirrels, wood rats, gerbils, and white-footed mice (Roth and Willis, 1960; Krivokhatskii, 1985). In some species, burrows are just one of several utilized microhabitats. The blattellid *Euthlastoblatta abortiva* can be found in both wood rat nests and leaves and dry litter on the ground along the Rio Grande River in Texas (Helfer, 1953). *Arenivaga floridensis* has been observed in the burrows of mice, burrowing freely in loose sand, and amid vegetation in sandhill and scrub communities (Atkinson et al., 1991). Occasionally only females (e.g., *Arenivaga erratica*—Vorhies and Taylor, 1922) or

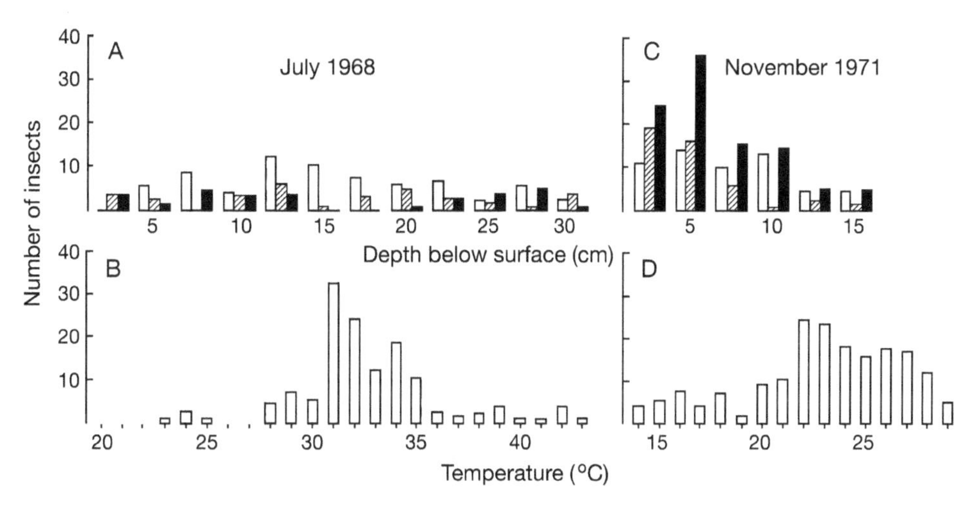

Fig. 3.10 Distribution of *Arenivaga* sp. in relation to depth below the surface (A,C) and temperature (B,D). In (A) and (C) the insects are scored according to size: open columns = 1st–3rd instar; striped columns = 4th–6th instars; solid columns = 7th–9th instars and adults. Adult males were rarely found below the surface and are not included in the data set. After Edney et al. (1974). Reprinted by permission of the Ecological Society of America.

nymphs (e.g., *Car. lutea*—Hubbell and Goff, 1939) are collected from burrows.

Animal burrows generally offer a more favorable microclimate than surface habitats. A higher humidity is maintained by the respiration of the vertebrate occupant (Tracy and Walsberg, 2002), and because of enhanced air circulation in burrows, cockroaches that utilize them avoid the hypoxic conditions that may be encountered by sand-swimming species (Cohen and Cohen, 1981). Richards (1971) indicates that animal burrows have a microclimate that is intermediate between that of caves and that of surface habitats. Recent studies, however, suggest that animal burrows are not always cool and humid refugia from surface conditions. For more than 100 days of the year soil temperatures rose to over 30°C at depths of 2 m in burrows of *Dipodomys* in the Sonoran desert (Tracy and Walsberg, 2002).

In a remarkable case of niche construction, at least one cockroach species mitigates conditions within vertebrate burrows by building a home within a home. In southeastern Arizona *Arenivaga apacha* is a permanent inhabitant of mounds of the banner-tailed kangaroo rat (*Dipodomys spectabilis*) and builds a microenvironment of small burrows ("shelves") within the main burrow of the rat (Cohen and Cohen, 1976). The mini-burrows are tightly packed with the grasses that were dragged into the main burrow by the rat for use as nesting material. Although the rodent burrows extend much deeper, most of the cockroaches were found 30–45 cm below the sand surface. Surface temperatures reached as high as 60°C, burrow temperatures reached 48°C , but the temperature of the grass-lined cockroach shelves averaged 16.5°C. Humidity of the burrows was as low as 20%, but the shelves remained nearly saturated at all times; 91% was the lowest reading. Conditions within the vertebrate burrow were nearly as harsh as the open desert and were made tolerable only by the alterations in the microenvironment made by the cockroaches; the insects died in 3–5 min if subjected to temperatures above 40°C. These cockroaches feed on the stored seeds of their host. "With this stored food available throughout the year and the very stable environmental conditions, the cockroaches have an ideal kind of oasis in the midst of a harsh desert environment" (Cohen and Cohen, 1976).

While *A. apacha* exhibits striking behavioral strategies for living in the harsh desert environment, its closely related congener, the Colorado Desert sand swimming *A. investigata,* relies heavily on well-developed physiological mechanisms. *Arenivaga investigata* has a higher temperature tolerance and lower rates of water loss and oxygen consumption than *A. apacha* (Cohen and Cohen, 1981). This is due in large part to the predominance of long chain wax esters in the cuticle that are effective in waterproofing the insect (Jackson, 1983). *Arenivaga investigata* is also able to tolerate a water loss of 25–30% without lethal effects (Edney, 1967) and is able to absorb water vapor from the surrounding air at ≥ 82% relative humidity (RH) (Edney, 1966). This level of RH is available at 45 cm below the ground surface (Edney et al., 1974). Thus, descending to that level assures the cockroach a predictable source of water. Water vapor is absorbed by means of a unique system of specialized structures on the head and mouthparts (O'Donnell, 1977a, 1977b). A thin layer of hygroscopic fluid is spread on the surface of two eversible

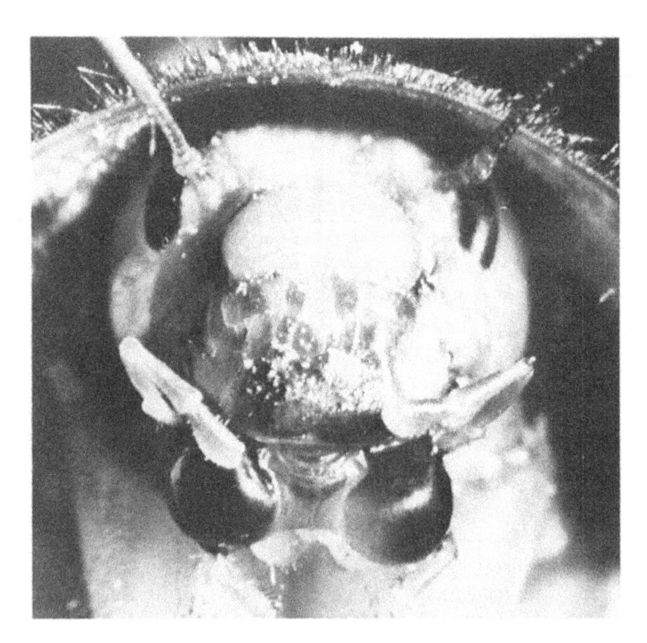

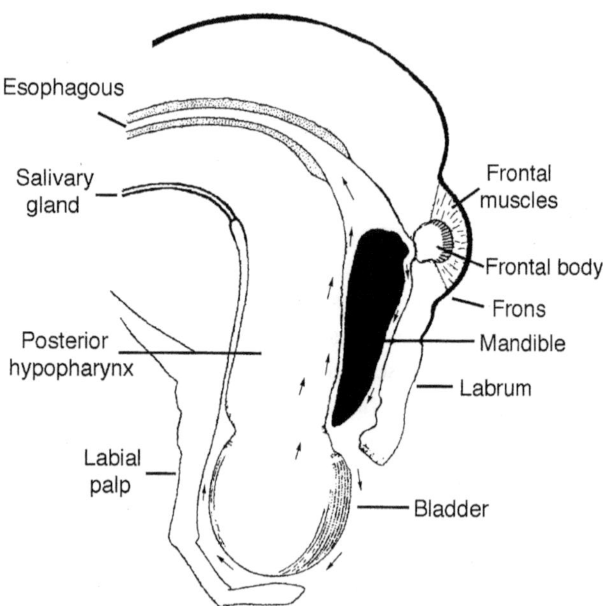

Fig. 3.11 Morphological structures associated with capturing atmospheric water in *Arenivaga investigata. Top,* photograph of head showing the two dark, spherical bladders protruding from the mouth. Note hairs around edge of pronotum. From O'Donnell (1977b), courtesy of M.J. O'Donnell. *Bottom,* sagittal view of the head with portions removed to show details of structures; redrawn from O'Donnell (1981), with permission of M.J. O'Donnell. The frontal body secretes a fluid that spreads over everted hypopharyngeal bladders. Atmospheric water condenses in the fluid and both liquids then flow toward the esophagus and are swallowed. Arrows indicate route of fluid movement from site of production in the frontal bodies to the esophagus.

Table 3.4. Water balance in *Arenivaga.* Data are in mg/100 mg/day at 25°C for a 320 mg nymph. From Edney (1966).

	Dry air	88% RH
Water loss		
Feces	0.19	0.19
Cuticular and spiracular	5.43	0.65
Total	5.62	0.84
Water gain		
Food	0.22	0.44
Metabolism	0.87	0.87
Vapor absorption	0	2.14
Total	1.09	3.45

bladders, one on each side of the mouth (Fig. 3.11). These are coated with a thick layer of cuticular hairs that hold and distribute the fluid via capillary action. The fluid is supplied to the bladders by two glands located on the inside of the labrum and embedded in a massive muscular complex that can be seen oscillating when the glands are secreting fluid. Atmospheric water condenses on the bladders and is then transferred to the digestive system, where it is absorbed. The capture of atmospheric moisture is a solute-independent system, based on the hydrophilic properties of the cuticular hairs on the bladders (O'Donnell, 1981, 1982). As a result of this water uptake system, *A. investigata* can maintain water balance even if no free water is available and food contains only 20% water, provided that air at 82% RH or above is available (Table 3.4). Females and nymphs are capable of absorbing water vapor, but males are not (Edney, 1967). Females are apterous, but males are winged and may be capable of seeking out free water and higher humidity surface habitats.

The Egyptian species *Heterogamisca syriaca* is similarly adapted to desert life. A lipid layer effective up to 56°C protects against evaporation, and the cockroach can extract water vapor from unsaturated air between 20 and 40°C and RH $\geq$ 75% (Vannier and Ghabbour, 1983). Humid air is available at a depth of 50 cm and at the surface during the night. Water absorption presumably occurs via hypopharyngeal bladders, as these have been observed in *H. chopardi* (Grandcolas, 1994a). Under the harshest conditions of water stress, *H. syriaca* may fast to generate metabolic water from fat reserves, which are abundant during the summer months (references in Vannier and Ghabbour, 1983).

Cockroaches that live in arid zones are rich in potential for research into behavioral ecology and physiology. *Thorax porcellana* living in suspended litter in dry forests of India, for example, do not actively seek or drink water when maintained in laboratory culture (Reuben, 1988), and nothing is known about the many diurnal Australian species that enjoy sunbasking. Perhaps as in some birds (Dean and Williams, 1999) the added heat helps speed digestion of a cellulose-based diet. Juvenile *Phyllodromica maculata* live on the dry, grassy hillsides of Bavaria, prefer low humidity, and do not aggregate (Gaim and Seelinger, 1984). Studies of laboratory-bred cockroaches indicate a variety of methods for dealing with heat and water stress. *Periplaneta americana, B. germanica,* and *Blatta orientalis* can withstand a body weight loss of 30% and still recover successfully when given an opportunity to drink water (Gunn, 1935). *Periplaneta fuliginosa* and *R. maderae* nymphs use the salivary glands as water storage organs (Laird et al., 1972; Appel and Smith, 2002). *Gromphadorhina brauneri* and *P. americana* maintain body temperatures below that of surrounding air by evaporative cooling (Janiszewski and Wysocki, 1986), and there is some evidence that *P. americana* can close dermal gland openings to conserve water (Machin et al., 1994). The physiology of water regulation in cockroaches is addressed in detail by Edney (1977), Mullins (1982), and Hadley (1994).

Aquatic Habitats

Most amphibious and quasi-aquatic cockroaches fall into two basic groups: those that live in phytotelmata (small pools of water within or upon plants) and those associated with rivers, streams, and ponds. In both cases, the insects live at the surface of the water or on solid substrate in its immediate vicinity, but submerge to hunt for food or to escape predators. About 62 species (25 genera) of cockroaches have been collected from the leaf bases of bromeliads (Roth and Willis, 1960; Rocha e Silva Albuquerque and Lopes, 1976), but it is unknown how many of these are restricted to this habitat. One example is *Dryadoblatta scotti,* a large, handsome, Trinidadian cockroach found in considerable numbers in epiphytic bromeliads; they rest just above the surface of the water or are partly immersed in it (Princis and Kevan, 1955). Nymphs of *Litopeltis* sp. are encountered during the day at all times of the year in the erect bracts of *Heliconia,* which collect and hold water even during the dry season of Costa Rica. The cockroaches forage at night on the outer and inner surfaces of the bracts, feeding on mold and decayed areas (Seifert and Seifert, 1976).

Numerous species in at least six genera of Epilamprinae live near streams or pools, usually in association with rotting vegetation amid rocks along the edge of the water. *Poeciloderrhis cribrosa verticalis* in Rio de Janeiro (Rocha e Silva Albuquerque et al., 1976) and *Rhabdoblatta annandalei* in Thailand (LMR, pers. obs.) occur near swift-moving streams, and *Rhabdoblatta stipata* in Liberia occurs on logs or mats floating directly in the current (Weidner, 1969). The cockroaches submerge in response to disturbance or when a shadow passes overhead, and swim rapidly below the surface for a minute or two. They then cling to submerged vegetation for up to 15 min before climbing to the surface (e.g., *Epilampra maya* [reported as *Ep. abdomennigrum*] in Panama—Crowell, 1946).

It has been debated as to whether aquatic cockroaches have morphological adaptations that enable underwater respiration. In most species observed to date, it appears that the insects use the abdominal tip as a snorkel, use a bubble of air as an accessory gill, or both. Weidner (1969) writes that individuals of *Rha. stipata* inspire via spiracles located on conical projections adjacent to the cerci, and die in 6–12 hr if the abdominal tip is held under water. *Opisthoplatia maculata* also has spiracular openings at the tip of abdominal projections, and these are protected by long hairs on the ventral surface of the cerci (Takahashi, 1926). Annandale (1906) suggested that the position of these posterior spiracles is an adaptation to an aquatic lifestyle; however, Shelford (1907) and Chopard (1938) point out that this character is present in many terrestrial cockroach species. Scanning electron micrographs of *Ep. abdomennigrum* reveal no unique adaptations of the terminal spiracles; they appear to be identical to those elsewhere on the body (WJB, unpubl. obs.). There are distinct patches of hairs on the ventral side of the cerci in older nymphs that that are absent in other *Epilampra* species examined; however, these hairs are quite distant from the terminal spiracles. The tracheal systems of aquatic and terrestrial cockroaches are morphologically distinct. The tracheae of the latter are thread-like, silvery in appearance, and dilated to their maximum with air. The tracheae of amphibious cockroaches are strap-like, not silvery, and contain just a few scattered air bubbles. Shelford (1916) suggested that the differences are rooted in the need for the amphibious species to be "sinkable," which would be prevented by internal accumulated air.

A large bubble is apparent beneath the pronotal shield of several aquatic species when they are submerged. The air is trapped by easily wetted, long hairs on the underside of the thorax (Takahashi, 1926; Crowell, 1946); these hairs also occur on terrestrial species. Some observers suggest that the bubble is formed by air taken in through the terminal abdominal spiracles, which then issues from the prothoracic spiracles in *Ep. maya* and *O. orientalis* (Shelford, 1907; Takahashi, 1926). Although this may explain the formation of the thoracic air bubble, air usually moves posteriorly through the tracheal system of blaberids (Miller, 1981), and recent observations suggest a different source of the bubble. WJB (unpubl. obs.) observed 48 dives of *Ep. abdomennigrum* nymphs in an aquarium in Costa Rica. When a nymph swimming on the surface is disturbed, it flips 180 degrees, with the venter of the body briefly facing upward. While supine the cockroach envelops an air bubble with its antennae and front legs, and holds the bubble beneath the thorax; the antennae remain extended posteriorly between the legs. As the cockroach dives below the surface, it turns again, righting itself, with the bubble held ventrally. Once underwater, it either grasps vegetation to remain submerged, or floats slowly to the surface. The median time totally submerged was 80 sec (range 20–1507 sec). While floating to the surface, the abdomen is extended upward, lifting the terminal spiracles out of the water. The insect remains motionless while floating on the air bubble for up to 30 min as the abdomen pulses slowly, at 1 or 2 pulses/10 sec.

Arboreal and Canopy Habitats

Rainforest canopies are structurally complex habitats with many niches favorable for maintaining cockroach populations: living and dead leaves, branches, bark crevices, sub-bark spaces, vines, epiphytes, suspended soils, hollow branches, vine-tree interfaces, treeholes, and bird and insect nests, among others. Canopies also contain an exceptionally rich array of organic resources (Novotny et al., 2003) known to be incorporated into cockroach diets. These include nonvascular plants, sap, bird excrement, plant litter, leaves, flowers, and fruit. In most studies of canopy invertebrates cockroaches are a consistent but minor component of the fauna. At times they are relegated to the "other" category (e.g., Nadkarni and Longino, 1990) because of the low number collected. Species-level identification is rarely attempted. In a recent eye-opening review of canopy arthropods worldwide, however, Basset (2001) concluded that while cockroaches represented only 5.3% of the individuals collected, they *dominated* in the amount of invertebrate biomass present. Blattaria represented 24.3% of the biomass, with Hymenoptera (primarily ants) coming in second at 19.8%, and Coleoptera ranking third at 18.8%. The revelation that nearly a quarter of the arthropod biomass in tree canopies may consist of cockroaches is particularly significant because the most commonly used canopy techniques almost certainly under-sample Blattaria. These are fogging, light traps, suspended soil cores, beating foliage, bromeliad bagging, and branch bagging (Table 3.5). Fogging is most effective on insects out in the open and is typically conducted early in the morning when the air is still. At that time, however, nocturnal and crepuscular cockroach species have likely entered harborage for the day. While the insecticide fog might kill them, they may not drop from their shelters. The same is true for cockroaches that live in tree hollows, epiphytes, insect nests, and other enclosed canopy habitats. Light traps, on the other hand, capture only volant cockroaches (Basset et al., 2003b) like *Gyna gloriosa*, taken at a height of 37 m in Uganda (Corbet, 1961). Branch bagging under-represents highly mobile taxa, and must be well timed. More cockroaches were collected at night than during the day using this method (Schowalter and Ganio, 2003), possibly because cockroaches perching on leaves during their active period were included in the night samples. A combination of the above methods may give a clearer picture of cockroach diversity and abundance in the canopy, with the additional use of baited traps and hand collecting from vines, suspended dead wood, treeholes, and other cryptic habitats (Basset et al., 1997). There is evidence that canopy cockroaches are a taxonomically rich group. In a fogging experiment in Borneo cockroaches were about 2% of the catch, but 40 presumed species were represented (Stork, 1991). A difficulty in documenting cockroach diversity, however, is that it is rarely possible to identify cockroach juveniles, and these can make up the bulk of Blattaria collected; 90% of the cockroaches collected by Fisk (1983) in Central American canopies were nymphs. In Venezuela, Paoletti et. al (1991) categorized cockroaches collected in their study as "microinvertebrates" because all were less than 3 mm in size. It is unclear, however, if these were small species or immatures.

Despite the high amounts of precipitation in rainforests, the canopy is a comparatively harsh environment characterized by high mid-day temperatures and low relative humidities, wind turbulence, and intense solar radiation (Parker, 1995; Rundel and Gibson, 1996). Cockroach canopy specialists no doubt have physiological and behavioral mechanisms that allow them to function in these conditions, but we currently have little information on their biology. These taxa are distinct from species commonly collected near the forest floor by light traps and other means (Fisk, 1983), and have been characterized as "smaller, aerial varieties endowed with unexpected beauty" (Perry, 1986). Conspicuously colored beetle mimics like *Paratropes bilunata* live in the canopy; this species imitates both the appearance and behavior of a lycid beetle (Perry, 1986). Fisk (1983) considered the following blattellid genera as canopy indicators in Panama and Costa Rica: *Imblattella, Nahublattella, Chorisoneura, Riatia,* and *Macrophyllodromia.* In Costa Rican lowland rainforest, Schal and Bell (1986) collected *Car. imitans* and two species of *Imblattella* from attached, folded, dead leaves in successional stands, and noted *Nyctibora noctivaga* and *Megaloblatta blaberoides* on trees in mature forest.

Most studies of canopy invertebrates have been con-

Table 3.5. Studies in which cockroaches were collected during canopy sampling.

Method	Location	Habitat	Reference
Beating foliage	Gabon	Lowland rainforest	Basset et al. (2003a)
Branch bagging	Puerto Rico, Panama	Evergreen wet forest	Schowalter and Ganio (2003)
Bromeliad bagging	Venezuela	Cloud forest	Paoletti et al. (1991)
Bromeliad bagging	Mexico	Low, inundated forest, semi-evergreen forest	Dejean and Olmstead (1997)
Fogging	Sabah	Lowland rainforest	Floren and Linsenmair (1997)
Fogging	Australia	Rainforest	Kitching et al. (1997)
Fogging	Japan	Mixed pine stand	Watanabe (1983)
Fogging	Brunei	Lowland rainforest	Stork (1991)
Fogging	Thailand	Dry evergreen forest	Watanabe and Ruaysoongnern (1989)
Fogging	Hawaii	Varied; altitudinal transect	Gagné (1979)
Fogging	Costa Rica, Panama	Lowland forest	Fisk (1983)
Light traps	Sarawak	Lowland mixed dipterocarp forest	Itioka et al. (2003)
Suspended soil cores	Gabon	Lowland forest	Winchester and Behan-Pelletier (2003)

ducted in the tropics. The canopies of temperate forests have proportionately fewer niches available because of the lower occurrence of lianas and epiphytes (Basset et al., 2003b; Novotny et al., 2003). In Japan, no cockroaches were listed in the results of a fogging study on a cypress plantation (Hijii, 1983) but they were recovered from a mixed pine stand (Watanabe, 1983). *Miriamrothschildia* (= *Onychostylus*) *pallidiolus* is an arboreal cockroach in Japan, the Ryuku islands, and Taiwan. The nymphs are very flat and semitransparent, and are found on live or dead tree leaves (Asahina, 1965). In the United States (South Carolina) *Parcoblatta* sp. were present in dead limbs and in and on the outer bark of longleaf pines sampled in winter. All trees had cockroaches on the upper bole, with a mean biomass of 36.2 mg/m^{-2}. Cockroaches were present but variable on other parts of the tree (Hooper, 1996). Additional Blattaria that forage and shelter on live and dead tree boles at various heights include *Aglaopteryx gemma* (Horn and Hanula, 2002) and several species of *Platyzosteria* on tea tree (*Leptospermum*) in Australia (Rentz, 1996).

A number of species that shelter on or near the forest floor spend their active period on trunks or low branches (Schal and Bell, 1986). However, Basset et al. (2003a) reported no difference in the number of cockroaches collected between day and night beat samples in lowland tropical rainforest in Gabon. Seasonal movement into the canopy may occur, coincident with rainfall and its effects on tree phenology. In Central America, Fisk (1983) collected 16 arboreal cockroach species (n = 220) during the dry season, but 24 species (n = 986) during the wet season. Maximum cockroach numbers coincided with peak new leaf production of the early wet season. In a light trapping study in Sarawak, Itioka et al. (2003) monitored cockroach abundance in relation to flowering periods in the canopy. Blattaria were most numerous during the post-flowering stage, and lowest during the non-flowering stage (Fig. 3.12). This seasonal abundance was attributed to the increased amount of humus in the canopy during the post-flowering period, derived from spent flowers, fruits, and seeds. Barrios (2003) found that the number of cockroaches collected by beat sampling comparable leaf areas in Panama was higher in mature trees (n = 237) than in saplings (n = 60). Long-term fluctuations were evident in a study by Schowalter and Ganio (2003). Canopy cockroaches were more abundant in drought years, and least abundant during post-hurricane years in Puerto Rico and Panama.

There are numerous humid microhabitats in treetops, where cockroaches not specifically adapted to the arid conditions of the canopy thrive. Among these are habitats that are little or nonexistent in the understory, such as bird nests and the spaces in and around complex vegetation such as epiphytes, intertwining vines, lianas, tendrils,

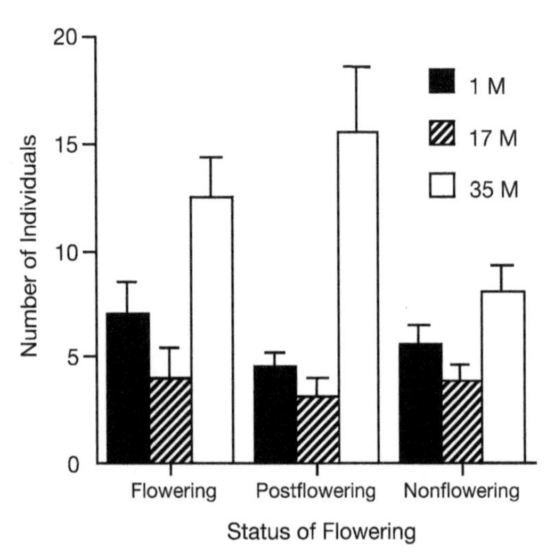

Fig. 3.12 Average monthly numbers of cockroaches in light traps at 1, 17, and 35 m in height during three trapping periods; flowering status of the trees varied during these periods. The study was conducted in tropical lowland dipterocarp forest in Sarawak, Malaysia. After Itioka et al. (2003), with permission of T. Itioka.

and adventitious roots. These provide sheltered resting places and a substantial amount and variety of food, particularly in the form of suspended soils. Fisk (1983) found a general albeit inconsistent correlation between number of cockroaches collected during fogging and the number of lianas per tree. Floren and Linsenmair (1997) fogged trees from which all lianas and epiphytes were removed in Sabah, and found that cockroaches did not exceed 1% of the insects collected, on average. The substantial pool of suspended soil that accumulates in the various nooks and crannies of the canopy may be particularly important in understanding the vertical stratification of cockroach faunas (Young, 1983), yet it is commonly neglected in tropical canopy research (Winchester and Behan-Pelletier, 2003). Suspended soil has a high organic content derived from leaf, fruit and flower litter, epiphyte tissues, decomposing bark, and the feces, food, and faunal remains of canopy-dwelling animals. It also contains a mineral component derived from fine particles carried on wind, rain, and fog (Winchester and Behan-Pelletier, 2003). This above-ground humus in rainforest is often thicker than the rapidly decomposing layer on the ground, and cockroaches that utilize the plant litter on the forest floor may also do so in the litter of the canopy. Leaf litter in plastic cups suspended in the lower branches of cacao trees in Costa Rica attracted cockroaches. Most abundant were species of *Latiblattella* and *Eurycotis;* the latter was also found in ground litter (Young, 1983). Studies of arthropods to date, however, generally indicate that the soil/litter fauna on the forest floor is in large measure distinct from that of the forest above (Basset et al., 2003b). One example among cockroaches is *Tho. porcellana,* which lives in aerial litter caught by the interlaced horizontal branches of plants in scrub jungle in India. The entire lifecycle of this cockroach is confined to suspended soil; they have no direct contact with the substratum (Bhoopathy, 1997). Winchester and Behan-Pelletier (2003) found that unidentified cockroaches collected from suspended soil cores from the crown of an *Ongokea gore* tree in Gabon were stratified; they were more abundant at 42 m than at 32 m above the ground.

Canopy litter is often considered ephemeral, as it can be removed by disturbances such as wind, rain, and arboreal animals (Coxson and Nadkarni, 1995). That is not true of the suspended soil trapped in some of the container epiphytes, such as the bird's nest *Asplenium* ferns and species of *Platycerium* with basal, clasping structures. In both, the litter mass acts as a sponge to retain water and nutrients (Rundel and Gibson, 1996). In the Neotropics epiphytes and hemiepiphytes may comprise greater than 60% of all individual plants, individual trees may support several hundred bromeliads, and a single bromeliad can contain more than 100 gm of soil (Gentry and Dodson, 1987; Paoletti et al., 1991). This is a substantial resource pool for cockroaches that feed on the accumulated debris and microorganisms contained within. Dejean and Olmsted (1997) found cockroaches in 67–88% of collected bromeliads (*Aechmea bracteata*) examined on the Yucatan peninsula of Mexico. Rocha e Silva Albuquerque et al. (1976) identified more than 30 cockroach species in bromeliads and list additional ones from the literature.

FOUR

Diets and Foraging

Timid roach, why be so shy?
We are brothers, thou and I.
In the midnight, like yourself,
I explore the pantry shelf!

—C. Morley, "Nursery Rhymes for the Tender-Hearted"

Cockroaches are typically described as omnivores, scavengers, or "classic generalists" (Dow, 1986), insects that feed on most anything they encounter. Indeed, the success of pest cockroaches in human habitations may be based largely on their ability to feed on soap, glue, wire insulation, and other materials that they certainly did not encounter during their evolution and do not encounter while living in more natural habitats. Our knowledge of cockroach diets stems largely from studies of these domestic pests, and it is assumed that their dietary habits are the norm (Bell, 1990). Some non-pest species (e.g., certain cave cockroaches) do appear omnivorous, but the term is not an adequate descriptor for the majority of Blattaria. Outside the man-made environment, the cockroach diet typically contains more refractory material than is generally appreciated (Mullins and Cochran, 1987). They can be selective eaters, and in some cases, specialized. There are several reasons for this rather biased image of cockroach diets. Some species will eat almost anything in urban or laboratory settings, but are highly selective in the wild. Few feeding observations or gut analyses from cockroaches in natural habitats have been conducted; in existing studies the picture is far from complete. We may have an indication of the menu at a particular point in time; however, we do not know if the food item in question is a small or large component of the diet. Further, the menu may vary with availability of certain foods, and with the age, sex, and reproductive or developmental status of the consumer.

FORAGING BEHAVIOR

With some exceptions, three feeding syndromes characterize the cockroaches that can be observed from ground level in tropical rainforest. First, nymphs of most species become active at nightfall, and begin to forage in the leaf litter on the forest floor. They can be seen skeletonizing wet, dead leaves, leaving harder veins and similar tissue. Leaf chips or dead leaf mush dominate the gut contents, but nematodes, fungi, insect larvae, and oligochaetes are also found. This feeding strategy was confirmed in the laboratory, where cockroach nymphs were observed ingesting the entire "sandwich": the leaf and everything on it (WJB, pers. obs.). Second, adults emerge from tree holes, leaf litter, and other harborages, and begin a vertical migration up into the canopy; the heights reached are species specific and probably relate to nutritional preferences (Schal and Bell, 1986). When the adults have reached the "correct" height, they move onto leaves and begin feeding on materials that have fallen or grow on the leaves. Third, a subset of species, mostly blattellids, shelter in curled dead leaves at a height of 1.5 to 2 m; palm fronds are commonly chosen as harborage. At night the cockroaches flit about leaves in the canopy, scraping algae and other microvegetation from the phylloplane. These species do not feed at a preferred height. Other foraging strategies include feeding on bark and epiphylls of rotting logs (*Capucina*) and feeding in rotting wood (nymphs of *Megaloblatta*). Some species have never been observed feeding, such as the green cockroach *Panchlora nivea*, but their guts contain a sweet-smelling substance that may be nectar from the upper canopy (WJB, pers. obs.)

Locating Food

Individually marked cockroaches in the rainforest generally home in on food via exploration and olfactory cues, sometimes arriving at fruit falls from quite long distances (Schal and Bell, 1986). Once near the food item, the cockroach's antennae and palps are used to inspect the resource; the information gathered is then used as basis to decide whether ingestion should proceed (WJB, unpubl. data). In domestic species (*Blattella germanica*), food closest to the harborage is exploited first (Rivault and Cloarec, 1991); this is probably also the case for cockroaches in natural habitats.

Individuals of *Diploptera punctata* in Hawaii are attracted to moist, dead leaves (WJB and L.R. Kipp, unpubl. obs.). Experiments were conducted on a large (2 m tall) croton bush in the late afternoon, during the inactive period of the cockroach. The insects previously had been seen foraging in the bush at 9:00 the same morning. Dead, wet leaves were placed on a branch about 1.2 m from the ground, and within 5 min individuals appeared near the bait leaves, apparently lured from their harborages at the base of the plant by the leaf odor. When "activated" by the odor they scurried about, waving their antennae. When a branch route took them near, but not to the dead leaf, they would get to the end of the branch, antennate rapidly, then turn and run down the branch to seek another route. Sometimes an individual made several attempts, over various routes, before locating the wet leaf. They were never observed flying to the bait. In Hawaii, *D. punctata* foraged from early evening (6:00 p.m.) to midmorning (10:00 a.m.), with two peaks in activity at 8:00 a.m. and 10:00 p.m. Nonetheless, the cockroaches could be activated to return to the above-ground portions of the plant at any time by hanging new decaying leaves within the canopy. Members of this population survived and reproduced for 6 mon in WJB's laboratory in Kansas on a diet consisting solely of dead oak and hackberry leaves.

Relocating Food

Urban cockroaches (*B. germanica*) search individually and independently for food. Items are not transported back to shelter, but eaten where they are found (Durier and Rivault, 2001a). In at least two cockroach groups the place where food is acquired differs from where it is utilized. Obtaining food and using it are thus separated in time and space, and the obtainer and the user are not necessarily the same individual (Zunino, 1991). Both groups that employ this "grocery store" strategy live in excavated underground chambers. The Australian soil-burrowing cockroaches forage during the night and the early morning hours of the wet season. After a rain and above a certain threshold temperature, they emerge, transport a quantity of dead leaves down into the burrow, and then do not emerge again until the next rain. Females grasp a food item in their mandibles and drag it backward down into the burrow. If they are approached when they are on the surface they will drop whatever they are carrying and "get a fair scuffle up" running back to their burrow (D. Rugg, pers. comm. to CAN). Gathered leaves are eaten by both the forager and any young offspring in the nest. Nymphs begin provisioning their own burrow when they are about half-grown. The food cache accumulated during the rains must sustain burrow inhabitants throughout the dry season (Rugg and Rose, 1991, pers. comm. to CAN). Other cockroaches known to transport and store food live in the tunnels of small vertebrates. *Arenivaga apacha* in the burrows of kangaroo rats in Arizona can be found nesting amid *Yucca, Ephedra, Atriplex,* and grass seeds that they have filched from the supply gathered and stored by the host rodent. "Our suspicion that the cockroaches gather and hoard provisions was confirmed when we saw the cockroaches carry dried dog food and sesame seeds that were sprinkled over the top of the aquaria soil into small caches underground" (Cohen and Cohen, 1976).

There are records of other cockroach species transporting food, but in these cases it occurs only in competitive situations. Rivault and Cloarec (1990) discovered that *B. germanica* began to "steal" food items from a dish as the items became small enough to carry and as food became scarce. Adults and larger nymphs stole more food than younger nymphs, and more stealing occurred when two or more individuals were present at a food source than when a lone individual was feeding. Similarly, when LMR fed crowded laboratory cultures with rice, he observed young nymphs position individual pieces of it between their front legs and mouthparts and run off on their hind legs (identity of species is lost to memory). Annandale (1910) documented *Periplaneta americana* using the mandibles to seize, hold, and transport termite alates in Calcutta.

Competition at food sources can trigger intraspecific aggression in *B. germanica.* The insects vary their tactics with age, and tailor them to the developmental stage of the opponent. Most agonistic interactions are between individuals of the same developmental stage. Young nymphs are primarily biters, but begin kicking more often as they develop; a good boot becomes more effective with the increased body weight characteristic of older stages (Rivault and Cloarec, 1992c). Young nymphs are generally tolerated by older stages and often reach food by crawling beneath larger conspecifics (Rivault and Cloarec, 1992a, 1992b). The relative amount of food required by large and small nymphs lowers the cost of benevolence for older insects.

Food relocation and aggression are both proximate mechanisms for obtaining and securing food from competitors. In burrow dwellers, relocation also allows them to feed at leisure in a location relatively safe from predators. Resource competition also may influence life history strategies, resulting in the distribution of competitors within a guild either in time (Fig. 3.5) or in space.

Learning

In many species, the location of the night harborage is spatially separated from other resources such as food and water. In the laboratory and in urban settings, individuals of *B. germanica* learn the position of their shelter and of stable food sources in relation to visual landmarks; however, olfactory information, which provides more reliable information about the presence of food, can override the visual cues. The insects learned to associate a certain type of food with a specific site, and were "disturbed" (exhibited complex paths) when the association between food type and food position was modified (Durier and Rivault, 2001b). Young nymphs of this species tend to explore smaller areas, cover shorter distances, and remain longer at depleted food sources than older cockroaches, eventually learning that "there is no point in waiting near a depleted patch, as it will not be renewed immediately" (Cloarec and Rivault, 1991). *Periplaneta americana* is differentially attracted to various dietary nutrients, and learned to associate certain odors with a proteinaceous food source, particularly when they were protein deprived. No such association between odor and carbohydrate could be established (Gadd and Raubenheimer, 2000). Watanabe et al. (2003) demonstrated that *P. americana* can be classically conditioned to form olfactory memories. The species also begins including novel foods in its diet after nutrient imbalances (Geissler and Rollo, 1987). It is probable that similar associations occur in nature; cockroach species known to have a wide dietary repertoire may both acquire knowledge of food-associated odors and benefit from past experience.

FEEDING VARIATION AND FOOD MIXING

Urban pest cockroaches (*Supella longipalpa*), like many omnivores (Singer and Bernays, 2003), balance their diet by selecting among available foods rather than by trying to obtain all nutrients from one food type (Cohen et al., 1987). *Periplaneta fuliginosa* is described as a "cafeteria-style eater" that will sample several types of food before concentrating on one (Appel and Smith, 2002). Other species known to have a varied diet in natural habitats, like *Parcoblatta* (Table 4.1), may do the same thing. Laboratory studies indicate that cockroaches are capable of selecting their diet relative to nutrient demand at every point in the lifecycle. Within a species, foraging behavior and dietary preferences vary with sex and ontogeny, and undergo dramatic changes correlated with reproductive and developmental cycles. In the field, it is possible that these predilections are also influenced by the seasonal availability of specific foods. Just after a local mast fruiting, for example, their diet may be higher in sugars and yeasts, and lower in fiber. When fruit is not available or their needs change, they may rely on less nutritious, higher-fiber foods such as litter or bark, or seek items that provide specific nutrients.

Age

As in most young animals (Scriber and Slansky, 1981; White, 1985) the dietary requirements of young cockroach nymphs differ from those of older nymphs and adults. Cochran elegantly demonstrated this in his studies of *Parcoblatta* spp., cockroaches that void urates to the exterior in discrete pellets if dietary nitrogen levels exceed a certain "break even" point with respect to nitrogen demands. In nymphs less than 1 mon old, a diet of 4% nitrogen results in only minimal urate excretion. On the same diet, nymphs 1–2 mon old void urates at a rate of 8–13% of excreta by weight and large nymphs reach an equilibrium at less than 1.5% nitrogen in the diet (Cochran, 1979a; Cochran and Mullins, 1982). In numerous species, this high requirement for nitrogen is reflected in the behavior of neonates, whose first meals are largely derived from animal or microbial sources. In many species the first meal consists of the embryonic membranes and the oothecal case. The female parent may provide bodily secretions originating from glands in or on the body, or from either end of the digestive system (Chapter 8). The few studies of coprophagy to date indicate that this behavior is most prevalent in early instars, suggesting that microbial protein is a crucial dietary component (Chapter 5). The need for animal or microbial protein may help explain why it is difficult to rear many cockroaches in the laboratory. While adults may thrive, "nymphs are more difficult to rear, starving to death in the midst of a variety of food stuffs" (Mackerras, 1970).

As they develop, juveniles may adopt the same diet as adults (e.g., wood, guano in caves) or feed on different materials, such as the rainforest species in which nymphs feed on litter but adults have a more varied menu. Studies in laboratory and urban settings indicate ontogenetic changes in foraging behavior, as well as variation in feeding behavior and food choice within a stadium. Immediately after hatch nymphs of *B. germanica* are able to find food and return to shelter, but they improve their foraging performance as they age (Cloarec and Rivault, 1991). *Periplaneta americana* nymphs take large meals during the first three days post-molt, then feed very little until the next (Richter and Barwolf, 1994). Juveniles of *Su. longipalpa* change their dietary preferences within a stadium. Protein consumption remains relatively low and constant, whereas carbohydrate consumption is highest during the first week, then declines gradually until the end of each instar (Cohen et al., 1987) (Fig. 4.1A). When given a wide range of protein:carbohydrate choices, *Rhyparobia maderae* nymphs consistently selected a ratio of approximately 25:75, suggesting that they have the ability to balance their diet (Cohen, 2001). Subadults of *B. germanica* are impressively capable of compensating dietary imbalances by choosing foods that redress deficiencies (Raubenheimer and Jones, 2006).

Sexual Differences

Current evidence suggests that male foraging behavior and food choice differs from that of females; generally, male cockroaches feed less and on fewer food types. In the Costa Rican rainforest male cockroaches always have less food in their guts than do females after the usual nightly foraging period (WJB, unpubl. data). This is particularly true for seven species of blattellids, in which 50–100% of males had empty guts. In more than 30 male *Latiblattella* sp. examined, none had any food in the gut. In contrast, males of four species of blaberids often had medium to full guts, although females had still fuller guts. This difference may be due to the active mate searching required of blattellid males as compared to blaberids. Male cockroaches tend to have narrower diets than females (Table 4.2), which may relate to the nutrients required for oogenesis. A similar pattern was obvious in *D. punctata* in Hawaii; 44% of females had guts filled to capacity, whereas male guts were never full. Nymphal guts were variable (19% full, 81% not full). It appeared that first instars had not fed at all, suggesting that they were relying on fat body reserves developed *in utero* while being fed by their viviparous mother. Older nymphs had fed to repletion. In all stages, the gut content was homogeneous material resembling dead leaf mush (WJB, unpubl. data). The amount of food consumed by male *P. americana* varies greatly on a daily basis, with the insects fasting on approximately one-third of days (Rollo, 1984b). Male German cockroaches did not exhibit cyclical feeding patterns, but the degree of sexual activity appears influential.

Table 4.1. Diet of four species of *Parcoblatta*, based on 45 nocturnal observations of feeding adults (Gorton, 1980). Note that two species were not observed ingesting animal food sources.

	Parc. pennsylvanica	*Parc. uhleriana*	*Parc. lata*	*Parc. virginica*
Mushrooms	+	+		+
Cambium			+	
Flower petals			+	
Moss		+		
Sap	+		+	
Cercropid spittle	+			
Live insect	+			
Bird feces		+		
Mammalian feces	+			
Mammalian cartilage		+		

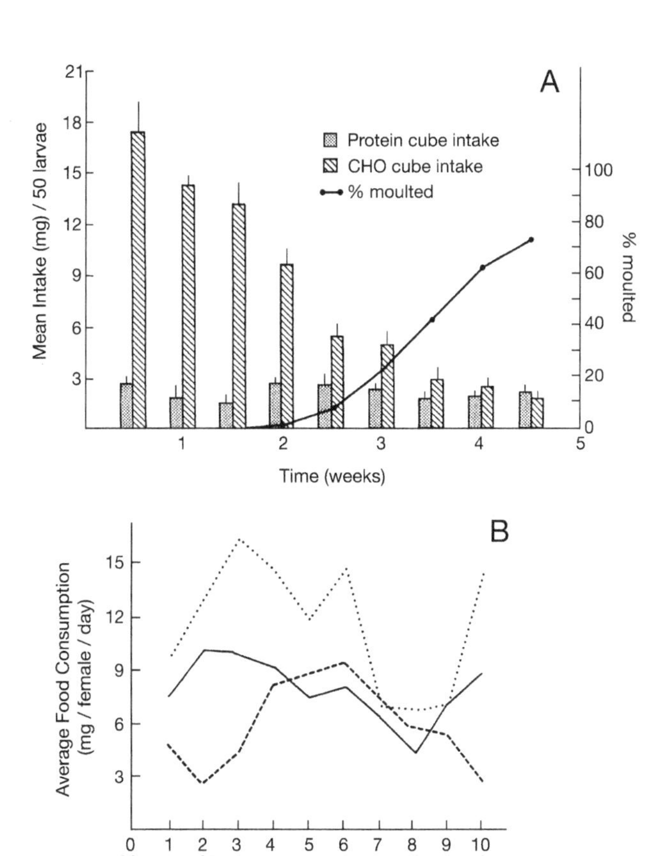

Fig. 4.1 Dietary self-selection in cockroaches. (A) Mean intake of protein and carbohydrate (CHO) cubes and cumulative percent molting in *Supella longipalpa* first instars over the course of the stadium. From Cohen et al. (1987), courtesy of Randy W. Cohen. (B) Food consumption by adult female *Parcoblatta fulvescens* over the course of the reproductive cycle when given a dietary choice. Dashed line, 5% protein-cellulose diet; dotted line, 5% protein-dextrose diet; solid line, 42% protein diet. EC, egg case formation; ECD, egg case deposition. From Lembke and Cochran (1990), courtesy of Donald G. Cochran. Both graphs reprinted with permission of Elsevier Press.

The food intake of *B. germanica* males mated twice per week was greater than that of males allowed to mate only once (Hamilton and Schal, 1988).

In many oviparous females, food intake and meal type is correlated with the ovarian cycle. Food intake falls to a low level a few days prior to ovulation and remains low until the ootheca is deposited in *P. americana* (Bell, 1969), *Parcoblatta fulvescens, Parc. pennsylvanica* (Cochran, 1986b), *Su. longipalpa* (Hamilton et al., 1990), and *B. germanica* (Cochran, 1983b; Cloarec and Rivault, 1991; Lee and Wu, 1994). Water intake is also cyclical (Fig. 4.2) (Cochran, 1983b, 1986b). In the ovoviviparous *R. maderae,* food intake declines at the time of ovulation and remains at a relatively low level until partition; neural input from mechanoreceptors in the wall of the brood sac directly inhibits feeding (Engelmann and Rau, 1965). In pregnant females of *D. punctata,* gut fullness varies relative to embryo length, with a trend toward full guts when embryos are small (2–5 mm) and empty guts when embryos are large (6–8 mm) (WJB, unpubl. data).

Females in at least two blattellid species select among various food types according to their vitellogenic requirements. In choice experiments with *Xestoblatta hamata,* Schal (1983) found that high-nitrogen foods were consumed mainly on nights 3 and 4 of the ovarian cycle. Females of *Parc. fulvescens* given one high-protein and two low-protein diets fed so that they remained in nitrogen balance; relative proportions of the different nutrients varied over the reproductive cycle (Fig. 4.1B). Females with access to only high-protein diets excreted urates, an indication that ingested protein levels exceeded their needs. Ovarian cycles of the self-selecting individuals were similar in length to those of the females fed a high-protein diet (Cochran, 1986b; Lembke and Cochran, 1990).

STARVATION

Willis and Lewis (1957) determined the mean survival times of 11 species of cockroaches deprived of food, water, or both (Table 4.3). When deprived of food and water, the insects can live from 5 days (male *Blattella vaga*) to 42 days (female *P. americana*). When given dry food

Table 4.2. Gut contents of cockroaches collected between 20:00 and 4:00 at La Selva Research Station, Costa Rica, between January and May 1992 (WJB and J. Aracena, unpub. data).

Cockroach species	n	Material in foregut
Blaberidae		
Capucina rufa		
Male	5	Epiphylls
Female	2	Epiphylls, bark scraps
Nymph	6	Epiphylls, bark scraps
Epilampra rothi		
Male	64	Dead leaf chips
Female	20	Algae, green plant, dead leaf, trichomes
Nymph	80	Dead leaf chips, insect parts
Blattellidae		
Xestoblatta hamata		
Male	16	Dead leaf, bird dung
Female	11	*Inga* bark chips, algae, dead leaf chips, fruit, leaf debris
Nymph	25	Finely ground dead leaf, insect parts
Cariblatta imitans		
Male	16	Algae
Female	10	Algae
Nymph	4	Algae

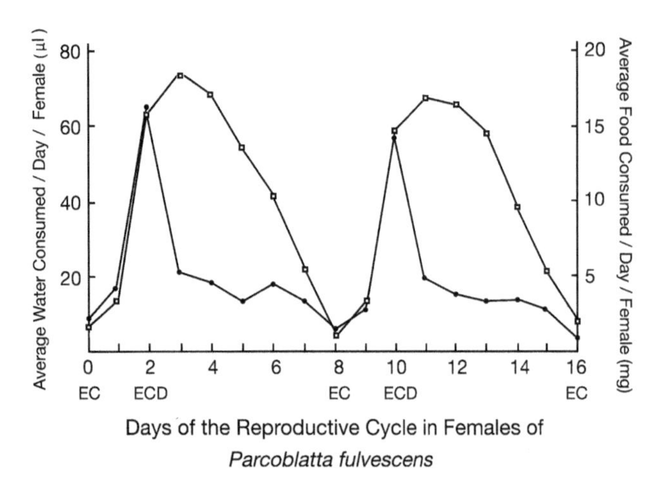

Fig. 4.2 Feeding and drinking cycles in relation to the reproductive cycle of the wood cockroach *Parcoblatta fulvescens.* Filled circles, water consumption; open squares, food consumption; EC, egg case formation; ECD, egg case deposition. From Cochran (1986b), courtesy of Donald G. Cochran, with permission from Elsevier Press.

but no water, they lived for about the same period of time as those deprived of both. If they are provided with water, most lived longer. Some species can live for 2 to 3 mon on water alone, and others significantly longer. Virgin females of *Eublaberus posticus* live an average of 360 days on water alone, whereas starved but mated females can live an average of 8 mon and are even able to produce 1 or 2 litters, yielding about 26 young. One female mated at emergence was starved for 252 days, during which time she produced 2 litters totaling 50 nymphs. She was then given food on day 252 (and thereafter), mated again 4 days later, and lived an additional 525 days, producing 5 more oothecae from which 24, 18, 5, 1, and 0 nymphs hatched. Although this female had been starved for the first 8 mon of adult life, after food was made available she managed to give birth to a total of 98 offspring, which is about normal for this species (Roth, 1968c).

There is a significant difference in starvation resistance between males and females in cockroach species exhibiting sexual dimorphism in body size. In Table 4.3, males and females are of similar size only in *Neostylopyga rhombifolia, Eurycotis floridana,* and *Nauphoeta cinerea;* in these cases, survival of males and females is similar. In the remaining species males are significantly smaller than females and are more vulnerable to starvation. A larger body size is correlated with bigger fat bodies and their accumulations of carbohydrates, lipids, and uric acid; these reserves can be rapidly mobilized on demand (Mullins and Cochran, 1975b; Downer, 1982). The nutrients and water housed in developing oocytes are additional resources available to starving females. The strategy for a food-deprived female of *P. americana* seems to involve resorption of yolk-filled eggs, storage of their yolk proteins, and then rapid incorporation of protein into eggs when feeding re-ensues (Roth and Stay, 1962b; Bell, 1971; Bell and Bohm, 1975).

A variety of digestive attributes help cockroaches buffer food shortages. The large crop allows an individual to consume a substantial quantity of food at one time. This bolus then acts as a reservoir during periods of fasting. When fully distended with food, the crop is a pear-shaped organ about 1.5 cm in length and 0.5 cm at its widest part (in *Periplaneta australasiae*). It extends back to the fourth or fifth abdominal segment, crowding the other organs and distending the intersegmental membranes. A meal may be retained in the crop for several days (Abbott, 1926; Cornwell, 1968). Solid food is also retained in the hindgut of starving *P. americana* for as long as 100 hr, although the normal transit time is about 20 hr (Bignell, 1981); this delay likely allows microbial biota to more thoroughly degrade some of the substrates present, particularly fiber. The functional significance of intestinal symbionts increases in times of food deficiency and helps to maintain a broad nutritional versatility (Zurek, 1997). A starving cockroach is thus indebted to its microbial partners on two counts: first, for eking out all possible nutrients in the hindgut, and second, for mobilizing uric acid stored in the fat body (Chapter 5). When food is again made available, starved *P. americana* binge. After starving for 13 days the amount of food consumed rose to five times the normal level, then leveled off after approximately 20 days. Greater consumption was accomplished by larger and longer meals, not by increasing the number of foraging trips (Rollo, 1984a).

PLANT-BASED FOOD

There is little evidence that any cockroach species is able to subsist solely on the mature green leaves of vascular plants. There are reports of occasional herbivory, such as that of Crowell (1946), who noted that the small, round leaves of the aquatic plant *Jussiaca* are included in the diet of *Epilampra maya.* Often, cockroaches that appear to be feeding on green leaves are actually eating either a small, dead portion at the leaf edge or around a hole, or other material on the leaf (WJB, unpubl. obs.). To test the extent to which tropical cockroaches include fresh vegetation in their diets, WJB set up a series of two-choice tests in laboratory cages at La Selva Biological Station in Costa Rica. Ten species of cockroaches were tested: *Capucina* sp., *Cariblatta imitans, Epilampra involucris, Ep. rothi, Ep. unistilata, Latiblattella* sp., *Imblattella impar, Nahublattella* sp., *Nesomylacris* sp., and *X. hamata.* The insects were offered a choice of green leaves versus dead leaves of the same plant species; only leaves eaten readily by local Or-

Table 4.3. Longevity of cockroaches on starvation diets. Tests were performed at 36–40% relative humidity, except for tests with *R. maderae*, which were run at 70%. Note that controls (+ food, + water) are not adult lifespans; controls were terminated when all the experimental insects of the species died. Modified from Willis and Lewis (1957).

Species	Sex	Mean length of survival (days)			
		+ food + water	+ food − water	− food + water	− food − water
Blattidae					
Blatta orientalis	Female	64	16.8	32.1	14.2
	Male	40	11.5	20.0	11.9
Neostylopyga rhombifolia	Female	108	25.4	26.7	22.1
	Male	128	24.6	29.3	21.9
Periplaneta americana	Female	190	40.1	89.6	41.7
	Male	97	27.3	43.7	28.1
Eurycotis floridana	Female	86	26.6	43.0	26.7
	Male	70	21.8	29.7	21.1
Blattellidae					
Blattella germanica	Female	85	11.9	41.9	12.8
	Male	54	8.8	9.6	8.2
Blattella vaga	Female	95	7.9	32.4	8.5
	Male	69	5.4	16.8	4.8
Supella longipalpa	Female	80	12.8	14.3	14.5
	Male	74	11.5	10.1	9.0
Blaberidae					
Diploptera punctata	Female	102	18.7	42.9	18.7
	Male	119	14.5	28.9	15.8
Rhyparobia maderae	Female	181	160.0	54.3	51.3
	Male	150	84.0	56.0	35.1
Nauphoeta cinerea	Female	98	24.3	61.1	27.0
	Male	94	22.8	46.1	27.3
Pycnoscelus surinamensis	Female	139	18.8	73.2	24.3
	Male	74	9.9	39.8	10.6

thoptera were used. The feeding behavior of the cockroaches was observed throughout the night, and their guts dissected the next day. Without exception, no cockroach ate fresh vegetation. Individuals that nibbled the greenery appeared repelled and on occasion could be observed jumping away from the leaf. When offered a choice of paper versus green leaves, the cockroaches ate the paper. When only green leaf was offered, they refused to feed.

Nonetheless, there are numerous records of cockroaches as plant pests (Roth and Willis, 1960). In 1789, Captain William Bligh had to wash down his ships with boiling water so that cockroaches would not destroy the breadfruit trees he was transporting from Tahiti to the West Indies (Roth, 1979a). One of the more frequently reported plant pests is *Pycnoscelus surinamensis,* which destroyed the roots of 300,000 tobacco plants in Sumatra. In greenhouses, it is known to girdle rose bushes, eat the bark and stems of poinsettias, and damage orchids, cucumbers, and lilies. It was responsible for the destruction of 30,000–35,000 rose plants in one Philadelphia greenhouse, and regularly hollows the hearts of palms and ferns in the southern United States (Roth, 1979a). Apparently, it managed to sneak into Biosphere 2 and took a strong liking to every kind of living plant. Tomatoes, sweet potato leaves, flowers and fruit of squash plants, rice seedlings, ripe papayas and figs, and green sorghum seeds were each included on the bill of fare (Alling et al., 1993). While the culprit cockroach was never identified, both *Pyc. surinamensis* and *P. australasiae* were found in the beehives brought in to pollinate crops (Susan C. Jones, pers. comm. to CAN).

The most commonly reported type of plant damage by cockroaches is to seedlings, new leaves, and growing root and shoot tips. These are likely preferred because their actively growing tissues have physically tender, thin-walled cells, lower levels of secondary compounds, and higher levels of nitrogen than mature leaves (Chown and Nicolson, 2004). Examples include *P. americana* destroying 30% of the freshly planted seeds of the quinine-producing plant *Cinchona pubescens* in Puerto Rico (Roth, 1979a), and *Shelfordina* (= *Imblattella*) *orchidae* damaging developing roots and shoots of orchids in Australian greenhouses (Rentz, 1987). *Calolampra elegans* and *Cal. solida* (Blaberidae) are pests requiring control measures in a variety of Australian crops, including sunflower, soybean, sorghum, cotton, navy beans, wheat, and maize. The cockroaches live in litter and the upper layers of soil, and emerge at night to chew the stems of seedlings at or near ground level (Robertson and Simpson, 1989; Murray and Wicks, 1990; Roach and Rentz, 1998). Cockroach herbivory in tropical forests is probably more common than generally realized; damage to newly flushed leaves in the canopy of Puerto Rican rainforest has been correlated with the abundance of cockroaches (Dial and Roughgarden, 1995).

Overt herbivores are not limited to feeding on green leaves of vascular plants; the category includes organisms that feed on other plant parts as well (Hunt, 2003). Many cockroach species, then, are at least partly herbivorous, because they include pollen, nectar, sap, gum, roots, bark, twigs, flowers, and fruit in their diet. Among those known to feed on pollen are *Sh. orchidae* (Lepschi, 1989), *Paratropes bilunata* (Perry, 1978), *Latiblattella lucifrons* (Helfer, 1953), and *Ellipsidion* sp. (Rentz, 1996). *Balta bicolor* is commonly found on the leaves and spent flower heads of *Gahnia* sp. in eucalypt woodlands (Rentz, 1996) and both males and females are attracted to pollen placed on a tree branch (Fig. 4.3). In a survey of insects captured by the pitcher plant *Sarracenia flava* in North Carolina, CAN (unpubl. data) collected males of four species of *Parcoblatta* (*Parc. fulvescens, Parc. uhleriana, Parc. virginica,* and *Parc. lata*), and both sexes of *Cariblatta lutea.* Since all these are winged as adults, while females of the *Parcoblatta* species are brachypterous, the cockroaches may be seeking nectar as an easily harvested source of energy to fuel flight. This suggestion is strengthened by the observation that volant *Blattella asahinai* adults, but not nymphs, feed on aphid honeydew (Brenner et al., 1988). *Trichoblatta sericea* in India feeds on the gum exuded from the bark of *Acacia* trees, and less commonly on gum from other trees (*Azadirachta, Moringa, Enterolobium*) (Reuben, 1988). Since individuals lived twice as long and had four times the reproductive output when fed a diet of powdered gum arabic when compared to a diet of biscuit crumbs or wheat flour, gum may be providing essential nutrients. The digestive physiology of this species would be of interest, as most gums are carbohydrate polymers that require microbial degradation if they are to be assimilated (Adrian, 1976). A number of cockroaches are noted as feeding "on flowers" (e.g., *Opisthoplatia orientalis*—Zhu and Tanaka, 2004a; *Ectobius pallidus*—Payne, 1973), but it is unclear as to whether the individuals were actually feeding on flower petals, or standing on the flower ingesting pollen or nectar. *Arenivaga apacha* (Cohen and Cohen, 1976) and possibly other cockroaches that dwell in vertebrate burrows feed on the stored seeds of their host, while sand-swimming species of *Arenivaga* include the roots of desert shrubs in their diet (Hawke and Farley, 1973). Many species feed on ripe fruit, an energy-rich, seasonally available food source. *Diplotera punctata,* for example, feeds on mangoes, papayas, and oranges, as well as on the outer covering of *Acacia* pods (Bridwell and Swezey, 1915) and the bark of *Cypress,* Japanese cedar, citrus, and *Prosopis* spp. (Roth, 1979a).

Leaf Foraging

In tropical rainforests leaf surfaces are "night habitat" for many crepuscular and nocturnal cockroaches. It is the only time and place that the majority of cockroaches that live in rainforests of Queensland, Australia (D. Rentz, pers. comm. to CAN), and Costa Rica (WJB, pers. obs.) can be seen. The insects emerge from harborage on the forest floor, move up the plants, then out onto foliage, or they move onto leaves from the innumerable hiding places in the different strata of the forest canopy. Adhesive footpads (arolia and euplantae) help the cockroaches negotiate sleek planes of vegetation, but it is only young leaves that commonly have smooth, simple surfaces. As leaves age they become elaborate, textured habitats rich in potential food sources (Walter and O'Dowd, 1995) (Fig. 4.4). In general, leaves provide two menu categories for cockroaches (WJB, unpubl. obs.). First, leaves act as *serv-*

Fig. 4.3 *Balta bicolor* feeding on pollen applied to a branch; male (*left*), female (*right*). Photo courtesy of David Rentz.

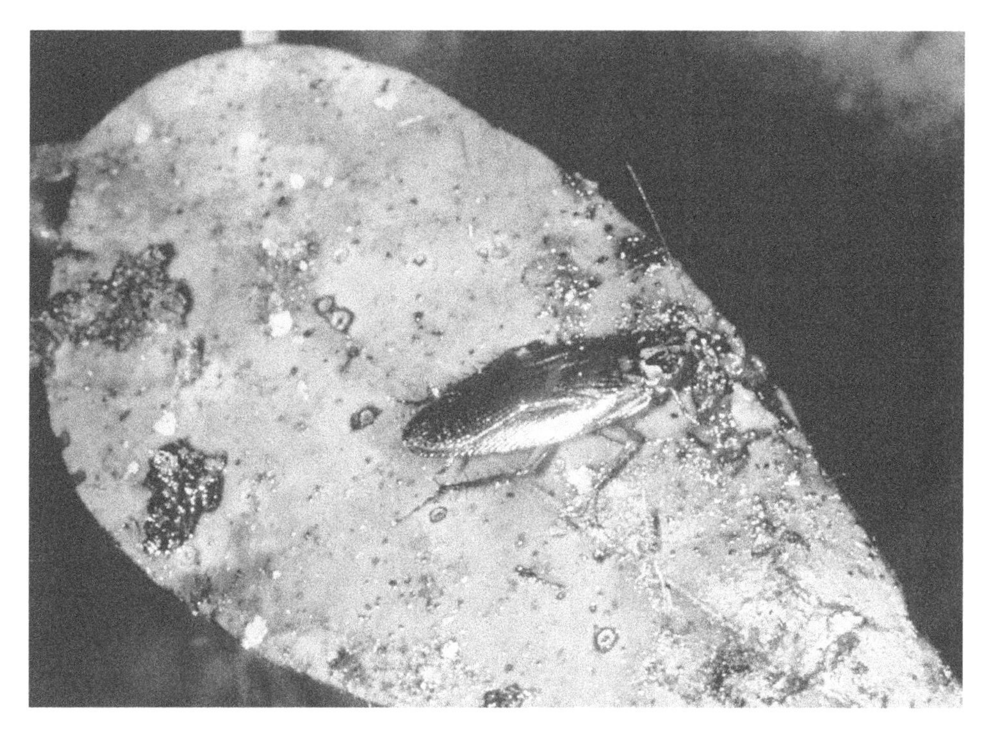

Fig. 4.4 *Beybienkoa* sp., night foraging on leaf surface material, Kuranda, Queensland. Photo courtesy of David Rentz.

ing trays for the intercepted rain of particulate organic matter that falls perpetually or seasonally from higher levels of the forest. This includes bird and other vertebrate feces, pollen, spores, leaves, twigs, petioles, sloughed tree bark, flower parts, and pieces of ripe fruit originating from the plant and from sloppy vertebrate eaters. Also offered on these leaf trays are dead leaf material around herbivore feeding damage, and the excreta, honeydew, silk webbing, eggshells, exuvia, and corpses of other arthropods. Live mites, aphids, and other small vulnerable arthropods on leaves are potential prey items. The second menu category on leaves in tropical forests is the salad course: leaves are *gardens* that support a wide range of nonvascular plants (epiphylls) and microbes. These include lichens, bryophytes, algae, liverworts, mosses, fungi, and bacteria.

Cockroaches in Costa Rican rainforest have been observed feeding on the majority of items listed above (WJB and J. Aracena, unpubl. obs.). Dissections of the cockroaches and inspection of their gut contents, however, indicate that ingestion of the different food types can be rather specific. Those cockroaches for which fairly large sample sizes are available are listed in Table 4.2. *Capucina rufa* and *Cap. patula* forage on dead logs, feeding on epiphylls, fungi, and bark scraps. *Epilampra involucris* females perch near the ground, where they feed on ground litter and the materials that fall onto it. Males of this species, which perch on leaves at heights of up to 50 cm, eat algae, bryophytes, lichens, pollen, spores, fruit, and flakes of shed bark. A subset of small, mobile species fly about in the canopy and scrape epiphylls from leaf surfaces at night. *Imblattella* and *Cariblatta* feed primarily on leaf trichomes, blue-green algae, liverworts, and spores. Only algae were found in the guts of male, female and juvenile *Car. imitans.* Trichomes, which normally interfere with foraging by small herbivores and carnivores (Price, 2002), are ingested by several cockroach species (WJB, unpubl. obs.). The many tropical cockroaches that fulfill their nutritional requirements by feeding on the broad variety of materials offered on leaf laminae may, like ants (Davidson et al., 2003), be categorized as leaf foragers. Those that specialize on the epiphylls and other plant products (trichomes, pollen, honeydew) found in this habitat may be described as cryptic herbivores (Hunt, 2003).

Detritus

Many cockroaches feed on detritus (Roth and Willis, 1960; Mullins and Cochran, 1987), a broad term applied to nonliving matter that originates from a living organism (Polis, 1991). A unique feature of detritivores is that there is no co-evolutionary relationship between the consumer and the ingested substrate. This is in stark contrast with the relationship between herbivores and higher plants, and in predator-prey systems. A consequence of this lack of co-evolutionary interaction is that detritivores are less specialized than predators and herbivores, and they defy classification into straightforward food chains (Anderson, 1983; Price, 2002; Scheu and Setälä, 2002). The food of detritivores is nutritionally very different from feeding on living plants or animals because it has been colonized and altered by microbes. Litter is a "resource unit" comprised of recently living material, degraded litter, dissolved organic matter, complex consortia of fungi, bacteria, nematodes, and protozoa, and the metabolic products of these (Nalepa et al., 2001a; Scheu and Setälä, 2002). The notion that detritivores may ingest a large amount of living microbial material, and may develop co-evolutionary relationships with these organisms, is not typically considered (Chapter 5).

Dead plant material in varying states of decay is known to be the primary food source for cockroach taxa in a variety of habitats. This is particularly true for species living at or near ground level in tropical forests, which have an unlimited supply of decaying litter within easy reach. Plant detritus is constantly accumulating on the forest floor, either seasonally or constantly. In the rainforest canopy, detritivores have access to suspended litter and the dead material that typically edges herbivore damage on live leaves (Fig. 3.3). Many cockroaches feed on leaf litter (Table 4.4), which in general is of higher resource quality and decomposes more quickly than twigs and other woody materials (Anderson and Swift, 1983); however, decayed wood may serve as a food source more commonly than is generally appreciated (Table 3.2). In rainforests, practically all wood is rotten to some extent, and the division between decayed wood, rotted plant litter, and soil organic matter is difficult to assess (Collins, 1989). Many cockroach detritivores live within their food source—"a situation reminiscent of paradise" (Scheu and Setälä, 2002).

Physically tough substrates like leaf litter and wood are macerated by a combination of mandibular action and

Table 4.4. Examples of cockroaches subsisting largely on leaf litter.

Habitat	Cockroach taxon	Reference
Rainforest	*Epilampra irmleri*	Irmler and Furch (1979)
	6 species (Malaysia)	Saito (1976)
	20 species of nymphs (Costa Rica)	WJB (pers. obs.)
Dry forest, scrub	Geoscapheini	Rugg and Rose (1991)
	Thorax porcellana	Reuben (1988)
Desert	*Arenivaga investigata*	Hawke and Farley (1973) Edney et al. (1974)
	Heterogamisca chopardi	Grandcolas (1995a)
Aquatic	*Litopeltis sp.*	Seifert and Seifert (1976)
	Poeciloderrhis cribrosa verticalis	Rocha e Silva Albuquerque et al. (1976)
	Opisthoplatia maculata	Takahashi (1926)

Fig. 4.5 Proventriculus of *Blattella germanica*, transverse section. From Deleporte et al. (1988), courtesy of Daniel Lebrun. Scale bar = 100 μm. When the "teeth" are closed the inward pointed denticles almost occlude the lumen. Hairs on the pulvilli may help filter the coarse food from the fine (Cornwell, 1968).

passage through the proventriculus, a strongly muscled and often toothed armature that lies just behind the crop (Fig. 4.5). It might be expected that the morphology of this organ is functionally related to diet, but that does not appear to be the case. The various folds, denticles, and pulvilli on the structure are, in fact, useful characters in phylogenetic studies of cockroaches (McKittrick, 1964; Klass, 1998b). The proventriculus of the wood-feeding taxa *Cryptocercus* (Cryptocercidae) and *Panesthia* (Blaberidae), for example, are completely different; that of *Cryptocercus* resembles that of some termites, and *Panesthia* has the flaccid, wide proventriculus of a blaberid. *Macropanesthia rhinoceros,* which feeds on dead, dry leaves, lacks a proventriculus (Day, 1950). This species, as well as *Geoscapheus dilatatus, Panesthia cribrata,* and *Cal. elegans* are known to ingest sand, probably to aid in the mechanical fragmentation of their food (Zhang et al., 1993; Harley Rose, pers. comm. to CAN).

ANIMAL-BASED FOOD

Like a large number of herbivores and detritivores (e.g., Hoffman and Payne, 1969), many cockroaches incorporate animal tissue into their diet when the opportunity arises. *Parcoblatta uhleriana* has been observed feeding on mammalian cartilage (Gorton, 1980), but most records of cockroaches feeding on living and dead vertebrates come from species that dwell in caves (discussed below) and from pest cockroaches. The latter can eat a great deal of flesh, particularly of human corpses. They also nibble on the calluses, wounds, fingernails and toenails, eyelashes, eyebrows, earwax, dandruff, eye crust, and the nasal mucus of sleeping individuals, particularly children. At times they "bite savagely," leaving permanent scars (Roth and Willis, 1957; Denic et al., 1997). Most reports are from ships, nursing homes, unhygienic urban settings, and primitive tropical living quarters. See Roth and Willis (1957) for a full roster of these horror stories.

Many cockroaches are equipped for predation: they are agile, are aggressive in other contexts, have powerful mandibles, and possess spined forelegs to help secure prey. The recorded victims of cockroaches include ants, parasitic wasps, *Polistes* larvae, centipedes, dermestids, aphids, leafhoppers, mites, and insect eggs (Roth and Willis, 1960). Both *B. vaga* and *B. asahinai* eat aphids and are considered generalist predators (Flock, 1941; Persad and Hoy, 2004). *Periplaneta americana* has been observed both catching and eating blowflies in a laboratory setting (Cooke, 1968), and pursuing and capturing termite dealates in and around dwellings. They pounced on termites from a distance of 5 cm, and followed them into crevices in the floor (Annandale, 1910; Bowden and Phipps, 1967). Cockroaches that feed on guano, leaf litter, or epiphylls also ingest the invertebrate microfauna that inhabit their primary food source (WJB, pers. obs). Dead invertebrates are scavenged by *Blattella karnyi* (Roth and Willis, 1954b), *Parcoblatta pennsylvanica* (Blatchley, 1920), and *P. fuliginosa* (Appel and Smith, 2002), among others. "The insect collector will often find that cockroaches, particularly in the tropics, will play sad havoc with his dead specimens" (Froggatt, 1906).

There are a few instances of cockroaches harvesting the secretions and exudates of heterospecific insects. Several are known to feed on honeydew (e.g., *Eurycotis* spp. sipping it from fulgorids—Naskrecki, 2005). *Parcoblatta pennsylvanica* has been observed feeding on cercopid spittle (Gorton, 1980). Recently two species of Costa Rican *Macrophyllodromia* were observed grazing the white, waxy secretion on the tegmina of at least two species of Fulgoridae (Fig. 4.6) (Roth and Naskrecki, 2001).

Conspecifics as Food Sources

The remaining cases of animal-based food pertain to fellow cockroaches. This fits the profile of other detritivores, as intraguild predation and cannibalism are widespread within decomposer food webs (Scheu and Setälä, 2002). There are a few cases of cockroaches preying on other cockroach species, like *N. cinerea* killing and eating *D. punctata* (Roth, 2003a). A more significant source of animal tissue, however, originates from same-species interactions (Nalepa, 1994). Most records of cockroach cannibalism come from domestic pests in lab culture (e.g., *Periplaneta* spp.—Pope, 1953; Roth, 1981a; *B. germanica*—Gordon, 1959), and it is the vulnerable that are most often taken as prey. Hatchlings, freshly molted nymphs, and the weak or wounded are the most frequent victims. It is usually the abdomen that is eaten first, to take advantage of the uric acid pool stored in the fat body (Cochran, 1985). Adult cockroaches in culture (Abbott, 1926) and in caves (Darlington, 1970) often have their wings extensively nibbled (although this may also be the result of aggressive interactions). The most ubiquitous ecological factor favoring cannibalism is the quality and quantity of available food, which depends to varying degrees upon population density (Elgar and Crespi, 1992).

Egg eating is a form of cannibalism, although in some cases the ingested eggs may be unfertilized or unviable (Joyner and Gould, 1986). In cockroaches, oothecae may be partially or entirely eaten prior to hatch (Roth and Willis, 1954b; Nalepa, 1988a), and oothecae carried by fe-

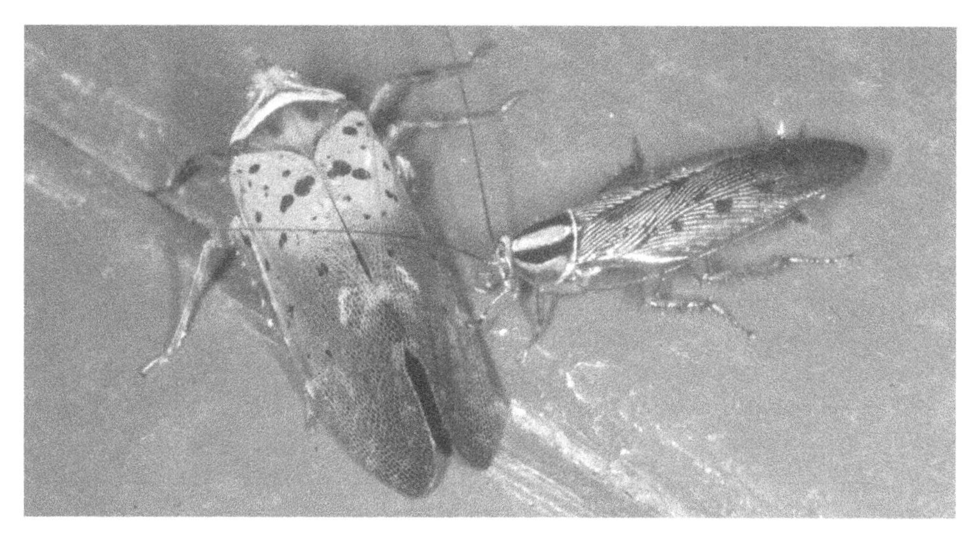

Fig. 4.6 The Costa Rican cockroach *Macrophyllodromia maximiliani* palpating the elytron of the fulgorid *Copidocephala guttata*. From Roth and Naskrecki (2001), courtesy of Piotr Naskrecki, with permission from the Journal of Orthoptera Research.

Table 4.5. Organic composition of exuvia from adult ecdysis and oothecae from several cockroach species, as determined by ^{13}C-NMR analyses. Reprinted from Kramer et al., "Analysis of cockroach oothecae and exuvia by solid state ^{13}C-NMR spectroscopy," *Insect Biochemistry* 21 (1991): pp. 149–56; copyright (1991), with permission from Elsevier.

	Relative amount (%) in/on exuvia			
Species	Protein	Chitin	Diphenol	Lipid
Periplaneta americana	49	38	11	2
Blattella germanica	59	30	9	2
Gromphadorhina portentosa	53	38	8	1
Blaberus craniifer	52	42	5	1
Rhyparobia maderae	61	35	4	1
	Relative amount (%) in/on post-hatch oothecae			
Species	Protein	Oxalate	Diphenol	Lipid
Periplaneta americana	87	8	4	1
Periplaneta fuliginosa	86	7	6	1
Blatta orientalis	88	7	4	1
Blattella germanica	95	<1	3	1

males are not immune to biting and cannibalism by conspecifics (Roth and Willis, 1954b; Willis et al., 1958). After hatch, neonates of ovoviviparous cockroaches eat the embryonic membranes and the oothecal case (Nutting, 1953b; Willis et al., 1958); the sturdier oothecal cases of oviparous species are probably eaten by older nymphs or adults. After hatch in *Cryptocercus,* for example, oothecal cases are occasionally found still embedded in wood, but chewed flush with the surface of the gallery; hatching oothecae isolated from adults always remain intact (Nalepa and Mullins, 1992). It is estimated that females of *Cryptocercus* may be able to recover up to 59% of the nitrogen invested into a clutch of eggs by consuming the oothecal cases after hatch, but it is unknown how much of this nitrogen is assimilated (Nalepa and Mullins, 1992). Cannibalism may be part of an evolved life history strategy in young families of *Cryptocercus* (Nalepa and Bell, 1997; Chapter 8).

Cast skins are a prized food source and are eaten quickly by the newly molted nymph or by nearby individuals. In *P. americana* the cast skin is usually consumed within an hour after molt (Gould and Deay, 1938), and the older the nymph, the more quickly the skin is eaten (Nigam, 1932). Nymphs of *B. germanica* are known to force newly emerged individuals away from their cast skins and "commence to eat the latter with great gusto" (Ross, 1929). A nymph of *E. posticus* usually eats its exuvium immediately after molt, before the new cuticle has hardened. Nearby cockroaches also eat fresh exuvia, and occasionally the molting cockroach as well (Darlington, 1970). Competition to feed on exuvia has been observed in both *Macropanesthia* (M. Slaytor, pers. comm. to CAN) and *Cryptocercus* (CAN, unpubl. obs.). In the latter, "snatch and run" bouts can occur where an exuvium changes ownership a half dozen times or more before it is completely consumed. The competition is understandable in that a cast skin is a considerable investment on the part of a growing nymph; exuvia from young instars of *E. posticus,* for example, comprise nearly 16% of their dry weight (Darlington, 1970). The cuticle is made up of chains of a polysaccharide, chitin, embedded in a protein matrix. Protein and chitin are 17% and 7% nitrogen by mass, respectively (Chown and Nicolson, 2004), and together these may account for 95% or more of the organic materials in an exuvium or oothecal case (Table 4.5).

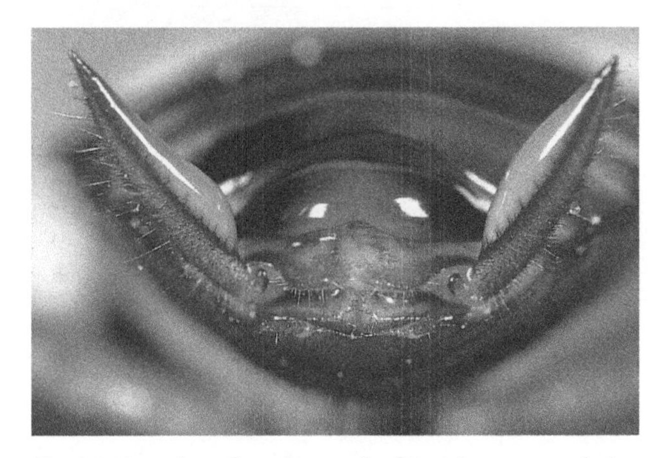

Fig. 4.7 Rear view of a male nymph of *Periplaneta australasiae,* showing the proteinaceous secretion that accumulates on the cerci and terminal abdominal tergites. Photo courtesy of Thomas Eisner.

Cockroaches apparently have the enzymes required to break down the chitin polysaccharide chain; endogenous chitinase is distributed throughout the gut of *P. americana* (Waterhouse and McKellar, 1961). Exuvium consumption appears directly related to nitrogen budget in *P. americana;* the behavior occurs more commonly in females, in insects reared on a low-protein diet, and in those deprived of their fat body endosymbionts (Mira, 2000).

In addition to the direct consumption of bodies, body parts, and reproductive products, cockroaches feed on materials exuded from the body of conspecifics in several contexts (Table 4.6). A form of nuptial feeding occurs in most cockroach species whose mating behaviors have been studied. Tergal glands are common in mature male cockroaches (Chapter 6). The secretions they produce attract the female during courtship, and as she climbs onto the male's back to feed on them she is properly positioned for genital contact (Roth, 1969; Brossut and Roth, 1977). Tergal secretions are general phagostimulants, and gravid, unreceptive females as well as males and nymphs feed on the gland of a courting male (Roth and Willis, 1952a; LMR, unpubl. obs.; Nojima et al., 1999b). In at least two blattellid species, *B. germanica* and *X. hamata,* males use the secretion of the uricose (accessory) gland as a nuptial gift (Mullins and Keil, 1980; Schal and Bell, 1982). During auto- and allogrooming cockroaches may ingest cuticular waxes, as well as anything else on the body surface; they spend a significant amount of time grooming antennae, legs, feet, and wings (Bell, 1990). Females and nymphs of both sexes in a variety of oviparous species produce a grayish viscous secretion on the cerci and terminal abdominal segments (Fig. 4.7). The material reappears 5–10 min after molt or the removal of the secretion. During autogrooming of the glandular area, the upper layer of the secretion is removed by the hind tibia; the leg is then cleaned by drawing it through the mouthparts (Naylor, 1964). The material is primarily (90%) proteinaceous and may serve as supplemental food (Roth and Stahl, 1956). Nymphs have been observed ingesting it from each other (D. Abed and R. Brossut, pers. comm. to CAN). Newly molted cockroaches eat their exuvium together with the glandular material accumulated on it (Roth and Stahl, 1956). The secretion also serves in defense, by mechanically impairing small predatory arthropods (Roth and Alsop, 1978; Ichinosé and Zennyoji, 1980). Allogrooming has been observed in *Pane. cribrata* (Rugg, 1987) and *Cryptocercus punctulatus* (Seelinger and Seelinger, 1983), neither of which produce this type of exudate. Neonates in at least six cockroach subfamilies feed on body fluids or glandular secretions of the mother (Chapter 8). These originate from a variety of locations on the adult body and have been analyzed only in the viviparous *Diploptera punctata* (Chapter 7).

Table 4.6. Conspecifics as food sources (modified from Nalepa, 1994).

Feeding behavior	Selected references
Cannibalism/necrophagy	Gordon (1959), Roth (1981a)
Oophagy (oothecae/ oothecal cases)	Nutting (1953b), Roth and Willis (1954b), Willis et al. (1958), Nalepa (1988a)
Consumption of exuvia	Roth and Willis (1954b), Willis et al. (1958)
Male-female transfer	
Tergal glands	Nojima et al. (1999b), Kugimiya et al. (2003)
Accessory glands	Mullins and Keil (1980), Schal and Bell (1982)
Cuticular secretions (from grooming and cercal exudates)	Roth and Stahl (1956), Seelinger and Seelinger (1983)
Parental feeding	Stay and Coop (1973), Roth (1981b), Perry and Nalepa (2003)
Coprophagy	Cruden and Markovetz (1984), Lembke and Cochran (1990)

CAVES

Caves are almost entirely heterotrophic; they depend on the transfer of energy and nutrients from the surface environment. Food is brought in with plant roots, water (i.e., organic material brought in with percolating rainwater, flooding, streams), and animals, particularly those that feed in the outside environment but return to the cave for shelter during their inactive period (Howarth, 1983; Gnaspini and Trajano, 2000; Hüppop, 2000). Although caves are generally considered food deficient, there is tremendous variation among and within caves. Food scarcity may be considered general, periodic (variation in time), or patchy (variation in space) (Hüppop, 2000). The best examples of the latter are guano beds that can be several meters deep and support tremendous populations of invertebrates. These islands of life, however, "are surrounded by desert, as most of the underground space is severely oligotrophic and sparsely populated" (Gilbert and Deharveng, 2002).

Guano

Vertebrate excrement is by far the most important nutritional base for cave Blattaria; cockroaches that feed on guano are apparently found on all main continents (Gnaspini and Trajano, 2000). If the vertebrates use the same roosting areas year round, then guano deposition is predictable in space as well as time and can support very large, persistent groups of cockroaches (guanobies). This occurs primarily in the tropics, because there food is available for bats throughout the year (Poulson and Lavoie, 2000). Cave cockroaches feed on the droppings of birds and of frugivorous, insectivorous, and haematophagous bats, but not carnivorous bats (Table 13.1 in Gnaspini and Trajano, 2000). The abundance and quality of guano varies not only in relation to the diet of a vertebrate guano source, but also seasonally, depending on roosting sites and the availability of food items (Darlington, 1995a). Communities that develop on guano can be very distinct. In one Australian cave, guano may be inhabited by mites, pseudoscorpions, beetles, and maggots, while in a nearby cave the guano is dominated by cockroaches (*Paratemnopteryx* sp.) and isopods (Howarth, 1988). *Eublaberus distanti* living in Tamana Cave, Trinidad, wait nightly buried under the surface of guano, with their antennae extended above the surface. When the insectivorous bat *Natalus tumidirostris* begins to return from foraging at about 3:00 a.m., the cockroaches emerge to feed on the fresh droppings raining from above. The frugivorous bat *Phyllostomus hastatus hastatus* is found in the same cave, and though *Eub. distanti* may burrow through their droppings, the cockroaches do not feed on them (Hill, 1981). None of the six cockroach species found in the caves of the Nullarbor Plain in south Australia are associated with bat guano, but *Paratemnopteryx rufa* and *Trogloblattella nullarborensis* utilize bird droppings (Richards, 1971).

Most cockroaches that live on the surface of guano appear highly polyphagous (Richards, 1971) and will take advantage of any animal or vegetable matter present in the habitat. Indeed, species able to benefit from all types of food present in caves have more aptitude for colonizing the subterranean environment (Vandel, 1965). The gut contents of *Eub. posticus* are indistinguishable from guano, but Darlington (1970, 1995a) considers both *Eub. distanti* and *Eub. posticus* primarily as scavengers on the guano surface. These cockroaches are not indiscriminant feeders, however, as they will pick out the energy-rich parts of food presented to them (Darlington, 1970). The cave floor in Guanapo is covered with bat droppings, dead bats, live and dead invertebrates, as well as fruit pulp, seeds, nuts, and other vegetable fragments defecated by the bats (Darlington, 1995–1996). In cave passages remote from guano beds the choices are much more restricted. Leaves, twigs, and soil that wash or fall into caves generally form the food base for troglobites (Poulson and White, 1969). There also may be occasional bonanzas of small mammals that blunder into caves but cannot survive there (Krajick, 2001). The ability of many cockroaches to endure long intervals without food, particularly if water is available (Table 4.3), may allow for exploitation of the deep cave environment. This starvation resistance is based at least in part on the capacity to binge at a single meal when food is available, together with the bacteroid-assisted ability to mete out stored reserves from the fat body when times are lean.

Plant Food in Caves

Cavernicolous cockroaches that depend on plant litter transported by water (Roth and McGavin, 1994; Weinstein, 1994) are attracted to traps baited with wet leaves (Slaney and Weinstein, 1996). While sinking streams may be continual, low-level sources of flotsam, seasonal flood debris supplies the bulk of the plant litter in most tropical caves (Howarth, 1983; Gnaspini and Trajano, 2000). In Australia, some caves may receive an influx of water and associated organic matter only once every 5 yr (Humphreys, 1993). Seeds defecated by frugivorous bats and the seeds of palm and other plants regurgitated by oilbirds commonly sprout in guano beds (Darlington, 1995b). The "forests of etiolated seedlings" (Poulson and Lavoie, 2000) that emerge may serve as food to cave cockroaches, but this is unconfirmed. *Periplaneta, Blaberus,* and other genera that feed on the guano of frugivorous bats also take advantage of fruit pieces dropped onto the floor (e.g., Gautier, 1974a). Fruit bats in Trinidad bring the fruit back to the caves, eat part of it, and then drop the remainder (Brossut, 1983, p. 150).

Live/Dead Vertebrates as Food in Caves

Those cockroaches that live in bat guano opportunistically feed on live, dead, and decomposing bats. Juveniles in maternity roosts that lose their grip and fall to the cave floor are particularly vulnerable (Darlington, 1970). *Blaberus* sp. have been observed rending the flesh of a freshly fallen bat, starting with the eyes and lips (D.W., 1984). Among the species recorded as feeding on dead bats are *Blattella cavernicola* (Roth, 1985), *Gyna caffrorum, Gyna* sp., *Hebardina* spp., *Symploce incuriosa* (Braack, 1989), and *Pycnoscelus indicus* (Roth, 1980). Cockroaches that live in the guano of oilbirds are treated to fallen eggs and occasional bird corpses (Darlington, 1995b). LMR once placed a dead mouse into a large culture of *Blaberus dytiscoides* and it was skeletonized overnight; he suggested to his museum colleagues that the cockroaches might be used to clean vertebrate skeletons.

Live/Dead Invertebrates as Food in Caves

Many cave cockroaches scavenge dead and injured invertebrates including conspecifics, and several have been reported to take live victims. Both *B. cavernicola* (Roth, 1985) and *Pyc. indicus* (Roth, 1980) prey on the larvae of tinead moths; *Pyc. indicus* also appears to be the main predator of a hairy earwig (*Arixenia esau*) found on the guano heap. Crop contents of both *Trog. nullarborensis* and *Para. rufa* consisted of numerous small chitinous particles and setae. In *Trog. nullarborensis* it was possible to identify small dipterous wing fragments and lepidopterous scales (Richards, 1971).

Geophagy in Caves

True troglobites are rarely associated with guano but little information is available regarding their food sources. At least two cockroach species appear geophagous. Roth (1988) found clay in the guts of five nymphs of *Nocticola australiensis,* and suggested that *Neotrogloblattella chapmani* subsists on the same diet (Roth, 1980). The latter is confined to remote passages away from guano beds. Clays and silts in caves contain organic material, protists, nematodes, and numerous bacteria that can serve as food for cavernicoles. Chemoautotrophic bacteria may be particularly important in that they are able to synthesize vitamins (Vandel, 1965). Cave clay is a source of nutrition in a number of cave animals, including amphipods, beetles, and salamanders (Barr, 1968). One species of *Onychiurus* (Collembola) survived over 2 yr on cave clay alone (Christiansen, 1970).

Microbivory in Caves

As with detritivores in the epigean environment, the primary food of cave cockroaches may be the decay organisms, rather than the organic matter itself (Darlington, 1970). This may be particularly true for cockroaches that spend their juvenile period or their entire lives buried in guano. In *Sim. conserfarium,* for example, groups of all ages are found at a depth of 5–30 cm in the guano of fruit bats in West African caves (Roth and Naskrecki, 2003). What better microbial incubator than a pile of feces, leaf litter, or organic soil in a dark, humid environment in the tropics? In addition to ingesting microbial cytoplasm and small microbivores together with various decomposing substrates, it is possible that some cave cockroaches directly graze thick beds of bacteria and fungi that live off the very rocks. These include stalactite-like drips of massed bacteria, and thick slimes on walls (Krajick, 2001). In Tamana cave, fungi dominate the guano of insectivorous bats. The low pH combined with bacteriocides produced by the fungi is responsible for the low number and diversity of bacteria. The pH of frugivorous bat guano, on the other hand, favors bacterial growth, which supports a dense population of nematodes (Hill, 1981). Recent surveys using molecular techniques indicate that even oligotrophic caves support a rich bacterial community able to subsist on trace organics or the fixation of atmospheric gases (Barton et al., 2004).

FIVE Microbes: The Unseen Influence

on the
back of a cockroach
no larger than
myself millions of
influenza germs may lodge i
have a sense of responsibility
to the public and i
have been lying for two weeks
in a barrel of moth
balls in a drug store
without food or water

—archy, "quarantined"

Why are cockroaches almost universally loathed? One of the primary reasons is because of the habitats they frequent in the human environment. Cockroaches are associated with sewers, cesspools, latrines, septic tanks, garbage cans, chicken houses, animal cages, and anywhere else there are biological waste products. Their attraction to human and animal feces, rotting food, secretions from corpses, sputum, pus, and the like gives them a well earned "disgust factor" among the general public (Roth and Willis, 1957). Why, however, are they are attracted to environments reviled by most other animals? It is it is obvious to us that the common denominator in all these moist, organic habitats is the staggeringly dominant presence of bacteria, protozoa, amoebae, fungi, and other microbial material. While these consortia are rarely if ever discussed as food for macroarthropods (e.g., Coll and Guershon 2002), in the case of cockroaches, that may be a glaring oversight. The main source of nourishment for cockroaches in mines and sewers, for example, is human feces (see Roth and Willis, 1957, plate 4), which can be 80% bacterial, by fresh weight (Draser and Barrow, 1985). *Blattella germanica* has been observed feeding on mouth secretions of corpses riddled with lung disease; these secretions contained infectious bacteria in almost pure culture (Roth and Willis, 1957). Granted, the above cases refer to cockroaches associated with the man-made environment, while the main focus of this book is on the 99%+ species that live in the wild. We contend, however, that microbes are an essential influence in the nutrition, ecology, and evolution of all cockroaches; indeed, it can be difficult to determine the organismal boundaries between them. Here we address microbes as gut and fat body mutualists, as part of the external rumen, the food value of microbes, various mechanisms by which cockroaches may ingest them, and some non-nutritional microbial influences. Finally, we discuss some strategies used by cockroaches to evade and manage disease in their microbe-saturated habitats.

MICROBES IN AND ON FOODSTUFFS

Because of the intimate association of microbial consortia and the substrate they are decomposing, both are ingested by detritivores. It is the microbial material, rather than the substrate that may serve as the primary source of nutrients (Berrie, 1975; Plante et al., 1990; Anduaga and Halffter, 1993; Gray and Boucot, 1993; Scheu and Setälä, 2002). Scanning electron micrographs show that millipedes, for example, strip bacteria from the surface of ingested leaf litter (Bignell, 1989), and similar to cockroaches, they can be found feeding on corpses in advanced stages of decay (Hoffman and Payne, 1969). Most foods known to be included in the diet of cockroaches in natural habitats are profusely covered with microbes. Bacteria and fungi are present on leaves before they are abscised, and their numbers increase rapidly as soon as the litter has been wetted on the ground (Archibold, 1995). The floor of a tropical rainforest is saturated with microbial decomposers, and as decay is successional, different species of microbe are associated with different parts of the process. A square meter of a tropical forest floor may contain leaves from 50 or more plant species, and each leaf type may have a different microflora and microfauna. Microbial populations may also vary with season, with climate, with soil, and with the structure of the forest; there is no simple way to recognize all of the variables (Stout, 1974). Dead logs, treeholes, bird and rodent nests, bat caves, and other such cockroach habitats are also microbial incubators. Bacteria are ubiquitous, but flagellates, small amoebae, and ciliates are also important agents of decomposition, and are associated with every stage of plant growth and decline, from the phylloplane to rhizosphere (Stout, 1974). Fermenting fruits and plant exudates (e.g., oozing sap) support the growth of yeasts, which are exploited as a source of nutrients in many insect species (Kukor and Martin, 1986). Cockroaches in culture favor overripe fruit, with the rotted part of the fruit eaten first, and fruit fragments intercepted by leaves in tropical forests are far from fresh. *Blattella vaga* has been observed in large numbers around decaying dates on the ground (Roth, 1985). Vertebrate feces are obviously rich sources of microbial biomass, particularly in bat caves, and, as discussed in Chapter 4, some cave cockroaches apparently assimilate bacteria from ingested soil.

THE ROLE OF MICROBES IN DIGESTION

The success of cockroaches within their nutritional environment results in large part from their relationship with microorganisms (Mullins and Cochran, 1987) at three levels: the microbes that comprise the gut fauna, the microbes found on ingested foodstuffs and fecal pellets, and the intracellular bacteria in the fat body.

Hindgut Microbes

The guts of all cockroach species examined house a diverse anaerobic microbiota, with ciliates, amoebae, flagellates, and a heterogeneous prokaryotic assemblage, including spirochetes (Kidder, 1937; Steinhaus, 1946; Guthrie and Tindall, 1968; Bracke et al., 1979; Bignell, 1981; Cruden and Markovetz, 1984; Sanchez et al., 1994; Zurek and Keddie, 1996; Lilburn et al., 2001). Methanogenic bacteria, a good indicator of microbial fermentative activity (Cazemier et al., 1997b), are found both free in the gut lumen and in symbiotic association with ciliates and mastigotes in most cockroach species tested (Bracke et al., 1979; Gijzen and Barugahare, 1992; Hackstein and Strumm, 1994). *Nyctotherus* (Fig. 5.1) can host more than 4000 methanogens per cell (Hackstein and Strumm, 1994), and hundreds to thousands of the ciliate can be found in full-grown cockroaches (van Hoek et al., 1998). Microbes are densely packed within the gut, but in a predictable spatial arrangement; food is processed sequentially by specific microbial groups as it makes its way through the digestive system. Volatile fatty acids (VFAs) are present in the hindgut, further suggesting the degradation of cellulose and other plant polysaccharides (Bracke and Markovetz, 1980). The hindgut wall of cockroaches is permeable to organic acids (Bignell, 1980; Bracke and Markovetz, 1980; Maddrell and Gardiner, 1980), indicating that the host may directly benefit from the products of microbial fermentation. Long cuticular spines and extensive infolding of the hindgut wall increase surface area and provide points of attachment for the microbes (Bignell, 1980; Cruden and Markovetz, 1987; Cazemier et al., 1997a). Finally, redox potentials indicate conditions are more reducing than in other insect species, with the exception of termites (Bignell, 1984). These features of cockroach digestive physiology support the notion that plant structural polymers play a significant role in the nutritional ecology of Blattaria; however, we currently lack enough information to appreciate fully the subtleties of the interactions in the hindgut. It is known to be a fairly open system, with a core group of mutualists, together with a "floating" pool of microbes recruited from those entering with food material (Bignell, 1977b, pers. comm. to CAN). Populations of the microbial community shift dynamically in relation to the food choices of the host. Whatever rotting substrate is ingested, a suite of microbes responds and proliferates (Gijzen et al., 1991, 1994; Kane and Breznak, 1991; Zurek and Keddie, 1998; Feinberg et al., 1999).

Cellulases are distributed throughout the cockroach

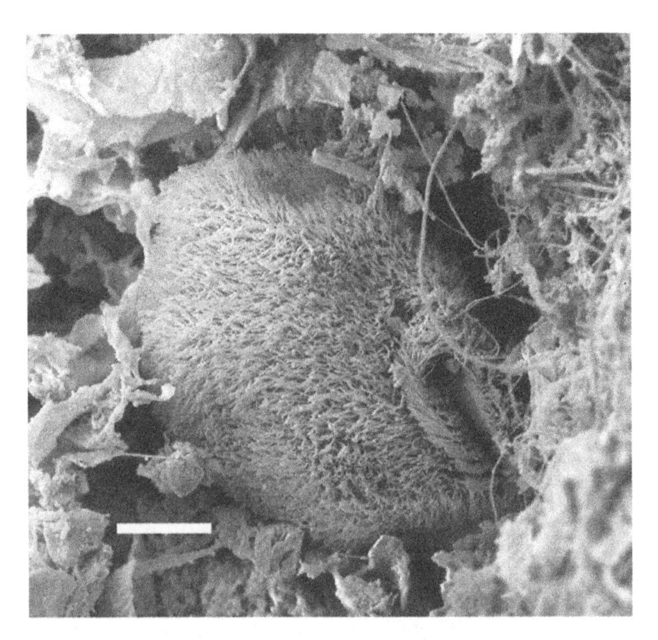

Fig 5.1 Scanning electron micrograph of the ciliate *Nyctotherus ovalis* from the hindgut of *Periplaneta americana.* Scale bar = 20 μm. From van Hoek et al. (1998); photo courtesy of J. Hackstein, with permission of the journal Molecular Biology and Evolution.

digestive system, and these enzymes are both endogenous and microbial in origin (Wharton and Wharton, 1965; Wharton et al., 1965; Bignell, 1977a; Cruden and Markovetz, 1979; Gijzen et al., 1994; Scrivener and Slaytor, 1994b). The nature of the contribution of cellulose to cockroach nutritional ecology, however, has been difficult to determine; in most cases no obvious nutritional benefit can be detected (Bignell, 1976, 1978), even in some wood-feeding cockroaches. Zhang et al. (1993), for example, found that *Geoscapheus dilatatus,* which feeds on dead, dry leaves, was able to utilize cellulose and hemicellulose more efficiently than the wood-feeding species *Panesthia cribrata.* The latter was surprisingly inefficient in extracting both cellulose (15%) and hemicellulose (3%) from its diet. In omnivorous domestic species, cellulose digestion may be a backup strategy, to be used when other available foods are inadequate (Jones and Raubenheimer, 2001). This is supported by evidence that solids are retained longer in the gut of starving *Periplaneta americana* (Bignell, 1981), allowing more time for processing the less digestible components. Retention time in animals with hindgut fermentation is directly related to digestive assimilation and efficiency (Dow, 1986; van Soest, 1994). The fact that so many cockroaches feed on cellulose-based substrates in the field but there is so little evidence for it playing a significant metabolic role suggests another possible function: the breakdown of cellulose may primarily provide energy for bacterial metabolism (Slaytor, 1992, 2000). Fibrous materials, then, may be ingested because they serve as fuel for microbial growth on the ingested substrate, on feces, and in the gut, and it is the microbes and their products that are of primary nutritive importance to the cockroach (Nalepa et al., 2001a).

Ontogeny of Microbial Dependence

Although it is often tacitly assumed that hosts derive net advantage from their mutualists throughout their lifecycle, in a number of associations it is only at key stages in the host lifecycle that exploitation of symbionts is important (Smith, 1992; Bronstein, 1994). Regardless of the exact nature of the benefits, young cockroaches depend more than older stages on gut microbiota. If the hindgut anaerobic community is eliminated, adequately fed adults are not affected. The overall growth of juvenile hosts, however, is impeded, and results in extended developmental periods. The weight of antibiotic-treated *P. americana* differed by 33% from controls at 60 days of age. Defaunation also lowered methane production and VFA concentrations within the hindgut, and the gut itself became atrophied (Bracke et al., 1978; Cruden and Markovetz, 1987; Gijzen and Barugahare, 1992; Zurek and Keddie, 1996).

The nutritional requisites of young cockroaches also differ from those of adults (*P. americana*), and are reflected in the activities of hindgut anaerobic bacteria, including methanogens (Kane and Breznak, 1991; Gijzen and Barugahare, 1992; Zurek and Keddie, 1996). Juvenile *P. americana* produce significantly more methane than adults, particularly when on high-fiber diets (Kane and Breznak, 1991), and demonstrable differences occur in the proportions of VFAs in the guts of adults versus juvenile stages (*Blaberus discoidalis*) fed on the same dog food diet (McFarlane and Alli, 1985).

Coprophagy

Although coprophagy simply means feeding on fecal material, it is an extremely complex, multifactorial behavior (Ullrich et al., 1992; Nalepa et al., 2001a). Fecal ingestion can be subdivided into several broadly overlapping categories, depending on the identity of the depositor, the nature of the fecal material, the developmental stage of the coprophage, and the degree to which feces are a mainstay of the diet. Many cockroaches feed on the feces of vertebrates, such as *Periplaneta* spp. in sewers or caves, desert cockroaches attracted to bovine and equine dung (Schoenly, 1983), and a variety of species attracted to bird droppings (Fig. 5.2). Here we highlight the feces of invertebrate detritivores (including conspecifics) as a source of cockroach food, and divide the behavior into three, not mutually exclusive categories.

Fig. 5.2 Unidentified nymph feeding on bird excrement, Ecuador. Photo courtesy of Edward S. Ross.

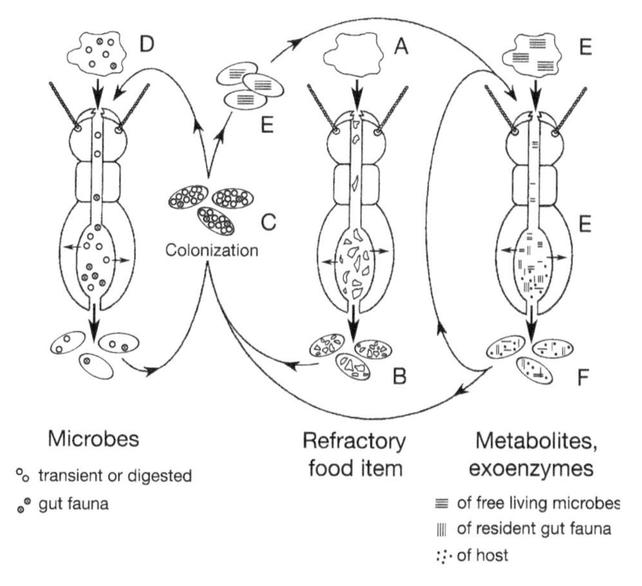

Fig. 5.3 Detritivore-microbial interactions during coprophagy. When a cockroach feeds on a refractory food item (A), any starches, sugars, lipids present are digested, and endogenous cellulases permit at least some structural polysaccharides to be degraded as well. Much of the masticated litter, however, may be excreted relatively unchanged (B), and serve as substrate for microbial growth (C). Ingested microbes, whether from the substrate (D) or from the fecal pellets of conspecifics (C), may be digested, passed in the feces, or selectively retained as mutualists. Microbes on the food item, on the feces, and in the hindgut are sources of metabolites and exoenzymes of possible benefit to the insect (E). Metabolites of the insect and of the gut fauna excreted with the feces (F) may be used by microbes colonizing the pellets or reingested by the host during coprophagy. Various authors shift the balance among these components, depending on the arthropod, its diet, its environment, and its age. From Nalepa et al. (2001a), with the permission of Birkhäuser Verlag.

Coprophagy as a Source of Microbial Protein and Metabolites

As food, the feces of detritivores are not fundamentally different from rotting organic matter; the feces of many differ very little from the parent plant tissue (Webb, 1976; Stevenson and Dindal, 1987; Labandeira et al., 1997). The differences that do occur, however, are important ones: feces are higher in pH, have a greater capacity to retain moisture, have increased surface to volume ratios, and generally occur in a form more suitable for microbial growth (McBrayer, 1973). Fecal pellets are colonized by a succession of microbes immediately after gut transit, with microflora increasing up to 100-fold (Lodha, 1974; Anderson and Bignell, 1980; Bignell, 1989). Fragmentation of litter is particularly important for bacterial growth, for unlike fungi, whose hyphae can penetrate tissues, bacterial growth is largely confined to surfaces (Dix and Webster, 1995; Reddy, 1995). The process is similar to gardeners creating a compost pile: microbially mediated decomposition occurs best when plant litter is moist and routinely turned. Coprophagy exploits the microbial consortia concentrated on these recycled cellulose-based foodstuffs (Fig. 5.3); the microorganisms serve not only as a source of nutrients and gut mutualists, but they also "predigest" recalcitrant substrates. Microbial dominance is so pronounced that fecal pellets may be considered living organisms. They consist largely of living cells, they consume and release nutrients and organic matter, and they serve as food for animals higher on the food chain (Johannes and Satomi, 1966).

Coprophagy as a Mechanism for Passing Hindgut Mutualists

All developmental stages feed on feces, but coprophagy is most prevalent in the early instars of gregarious domestic cockroaches (*B. germanica, P. americana, P. fuliginosa*) (Shimamura et al., 1994; Wang et al., 1995; Kopanic et al., 2001). Feces contain protozoan cysts, bacterial cells, and spores, and are the primary source of inoculative microbes (Hoyte, 1961a; Cruden and Markovetz, 1984). Very young cockroaches, with a hindgut volume of 1 μl, already show significant bacterial activity (Cazemier et al., 1997a). Repeated ingestion of feces is no doubt required, however, because a successional colonization of the various gut niches by microbes is the norm (Savage, 1977). Obligate anaerobes have to be preceded by facultative anaerobes, and a complex bacterial community has to precede protozoan populations (Atlas and Bartha, 1998). Because cockroach aggregations are generally species specific, horizontal transmission of microbial mutualists from contemporary conspecifics may be considered typical. Mixed-species aggregations are occasionally reported (Roth and Willis, 1960). Neonates, then, may also have sporadic access to interspecific fecal material. Analysis of rDNA repeats from the cockroach hindgut ciliate *Nyctotherus* indicates that there is a significant phylogenetic component to the distribution of the ciliates among hosts, but transpecific shifts do occur (van Hoek et al., 1998). The longevity of cysts and spores in fecal pellets would contribute to transmission across species; cysts of *Nyctotherus* are estimated to survive 20 weeks under favorable conditions (Hoyte, 1961b).

We have little information on transmission of gut mutualists in non-gregarious species. In subsocial species of cockroaches or those with a short period of female brooding, transmission is probably vertical, via filial coprophagy (Nalepa et al., 2001a). In *Cryptocercus* spp. intergenerational transfer occurs via proctodeal trophallaxis (Seelinger and Seelinger, 1983; Nalepa, 1984), the direct transfer of hindgut fluids from the rectal pouch of a donor to the mouth of a receiver (McMahan, 1969). We do not know the mechanism of microbial transmission in oviparous species that abandon the egg case. Perhaps the female defecates in the vicinity of the ootheca, or the eggs are preferentially deposited near conspecific feces. Alternatively, neonates may acquire their gut biota directly from ingested detritus. Metabolically complementary consortia of microbes are always present on ingested organic material, because the microorganisms are themselves using it as a food source (Costerton, 1992; Shapiro, 1997). The mode of transmission of gut microbes in cockroaches is related to the degree of host-microbe interdependence and to host social behaviors; these three comprise a co-varying character suite (Troyer, 1984; Ewald, 1987; Nalepa, 1991; Chapter 9).

Coprophagy as a Mechanism for Passing Cockroach-Derived Substances

A coprophage has access to the metabolites, soluble nutrients, exoenzymes, and waste products of microbes both proliferating on feces and housed in the host digestive system, but also to products that originate from the insect host itself. The excretion of urate-containing fecal pellets by some blattellids can be a mode of intraspecific nitrogen transfer (Cochran, 1986b; Lembke and Cochran, 1990), discussed below. There are behaviorally distinct defecation behaviors in *P. americana* associated with physically different feces, and certain types of feces are eaten by early instars more frequently than others. Young nymphs were the only developmental stage observed feeding on the more liquid feces smeared on the substrate (Deleporte, 1988). Adult *Cryptocercus punctulatus* occasionally produce a fecal pellet that provokes a feeding frenzy in their offspring, while other pellets are nibbled or ignored (Fig. 5.4) (Nalepa, 1994). This behavior was also noted in *C. kyebangensis* as "clumping behavior" (Park et al., 2002). The basis of the appeal of these pellets is unknown.

Fig. 5.4 First instars of *Cryptocercus punctulatus* massed on and competing for a fecal pellet recently excreted by the adult female. Only certain pellets induce this behavior. Photo by C.A. Nalepa.

MICROBES AS DIRECT FOOD SOURCES

It is extremely difficult to characterize the degree to which microbes are used as food. Ingested microbes may be digested, take up temporary residence, or pass through; many live as commensals and symbionts. Studies of cockroaches as disease vectors indicate that some bacteria fed to cockroaches are passed with feces, while others could not be recovered even if billions were repeatedly ingested (Roth and Willis, 1957). A mushroom certainly qualifies as food, but so does any microbe that dies within the digestive system, releasing its nutrients to be assimilated by the cockroach host, other microbes resident in the gut, or a coprophage feeding on a subsequent fecal pellet. We do not know the degree to which cockroaches feeding on dead plant material handle the substrate/microbe package in bulk (the gourmand strategy) versus pick through the detrital community, ingesting only the relatively rich microbial biomass (the gourmet strategy). If the latter, they are not detritivores, because they feed primarily on living matter and on material of high food value (Plante et al., 1990). The gourmet strategy may be common among the youngest cockroach nymphs in tropical rainforests. Many of them never leave the leaf litter (WJB, pers. obs.), and small browsers can be highly selective

(Sibley, 1981). Even if a cockroach is a gourmand, however, it may only digest and assimilate the microbial biomass, and pass the substrate in feces relatively unchanged, "like feeding on peanut butter spread on an indigestible biscuit" (Cummins, 1974).

Regardless of the strategy, it is generally agreed that for most detritivores microorganisms are the major, if not sole source of proteinaceous food, and are assimilated with high efficiency, 90% or more in the case of bacteria (White, 1985, 1993; Bignell, 1989; Plante et al., 1990). On a dry weight basis, fungi are 2–8% nitrogen, yeasts are 7.5–8.5%, and bacteria are 11.5–12.5% (Table 5.1). These levels are comparable to arthropod tissue and may exceed cockroach tissue (about 9.5% in *C. punctulatus* adults) (Nalepa and Mullins, 1992). In addition to being rich sources of nitrogen, microbes contain high levels of macronutrients such as lipids and carbohydrates, and critical micronutrients, such as unsaturated fatty acids, sterols, and vitamins (Martin and Kukor, 1984). Even if the ingested biomass is small, the nutrient value may be highly significant (Seastedt, 1984; Ullrich et al., 1992). Irmler and Furch (1979), for example, pointed out that a litter-feeding cockroach in Amazonia would need to consume impossible amounts (30–40 times its energy requirement) of litter to satisfy its phosphorus requirement; it is known, however, that microbial tissue is a rich source of this element (Swift et al., 1979).

The External Rumen

The importance of microbial tissue to an arthropod may reside as much in its metabolic characteristics while on recalcitrant substrates as in its nutrient content once ingested. The bacteria and fungi responsible for decay predigest plant litter in a phenomenon known as the "external rumen." The microbes remove or detoxify unpalatable chemicals (e.g., tannins, phenols, terpenes), release carbon sources for assimilation, and physically soften the substrate. These changes improve the palatability of plant litter and increase both its water-holding capacity and its nutritional value (Wallwork, 1976; Eaton and Hale, 1993; Scrivener and Slaytor, 1994a; Dix and Webster, 1995). As a result, decay organisms can guide food choice in cockroaches. Both *Cryptocercus* and Panesthiinae are collected from a wide variety of host log taxa, as long as the logs are permeated with brown rot fungi (Mamaev, 1973; Nalepa, 2003). It is the physical softening of wood that was suggested as the primary fungal-associated benefit for *Pane. cribrata* by Scrivener and Slaytor (1994a). Ingested fungal enzymes did not contribute to cellulose digestion, and fungal-produced sugars were not a significant source of carbohydrate. Microbial softening of plant litter may be particularly important for juveniles (Nalepa, 1994). Physically hard food is known to affect cockroach development (Cooper and Schal, 1992) and young cockroach nymphs preferentially feed on the softer parts of decaying leaves on the forest floor (WJB, pers. obs.)

Table 5.1. Nitrogen levels of various natural materials exploited as food by invertebrates. Compiled by Martin and Kukor (1984).

Material	Nitrogen content (% dry weight)
Bacteria	11.5–12.5
Algae	7.5–10
Yeast	7.5–8.5
Arthropod tissue	6.2–14.0
Filamentous fungi	2.0–8.0
Pollen	2.0–7.0
Seeds	1.0–7.0
Cambium	0.9–5.0
Live foliage	0.7–5.0
Leaf litter	0.5–2.5
Soil	0.1–1.1
Wood	0.03–0.2
Phloem sap	0.004–0.6
Xylem sap	0.0002–0.1

Microbes on the Body

Omnivores and detritivores contact microbes at much higher rates than do herbivores or carnivores (Draser and Barrow, 1985). In cockroaches, a high frequency of encounter is obvious from the habitats they frequent and from the abundant literature on their role as vectors. A large number and variety of bacteria, parasites, and fungi are carried passively on the cuticle of pest cockroaches (Roth and Willis, 1957; Fotedar et al., 1991; Rivault et al., 1993). Despite being nonfastidious feeders with regard to bacteria, however, cockroaches are scrupulous in keeping their external surfaces clean (Fig. 5.5). More than 50% of their time may be spent grooming (Bell, 1990) and in many species the legs are morphologically modified with comb-like tubercles, spines, or hairs to aid the process (Mackerras, 1967b; Arnold, 1974). Mackerras (1965a) described the concentration of hairs on the ventral surfaces of the fore and hind tibiae of *Polyzosteria* spp. as "long handled clothes brushes" used to sweep both dorsal and ventral surfaces of the abdomen. The final stage of the grooming process is to bring the leg forward to be cleansed by the mouthparts (Fig. 1.18). It seems reasonable to assume that microbes and other particulate matter concentrated on the legs during grooming activities are ingested at this point and may be used as food. This suggestion is strengthened by studies of the wood-feeding cockroach *Cryptocercus*. An average of 234 microbial colony-forming units/cm^2 cuticle have been detected on *C. punctulatus* (Rosengaus et al., 2003), and the insects are known to allogroom, using their mouthparts to directly graze the cuticular surface of conspecifics. Young nymphs spend 8% of their time in mutual grooming (Fig. 5.5B) and 15–20% of their time grooming adults. Grooming decreases with increasing age, and allogrooming was never observed in adults (Seelinger and Seelinger, 1983). Grooming has a number of important functions, and high levels of autogrooming may be related primarily to the prevention of cuticular pathogenesis in their microbe-saturated habitats. Digestion of some of the gleaned bacteria may be an auxiliary benefit, particularly if resident gut bacteria play a role in neutralizing ingested pathogens. Intense allogrooming in developmental stages with high nutrient requirements is suggestive that there may be a nutritional reward for the groomer, in the form of microbes, cuticular waxes, or other secretions. Starvation is known to increase grooming interactions in termites (Dhanarajan, 1978), and the observation that young *Cryptocercus* nymphs spend up to a fifth of their time grooming the heavily sclerotized adults, presumably the most pathogen-resistant stage, further supports this hypothesis. However, young nymphs also may be acquiring antimicrobials or other non-nutritive beneficial substances from adults during grooming, and keeping nest mates free of infection is in the best interest of the groomer as well as the groomee. Radiotracer studies are necessary to confirm the assimilation of ingested microbes.

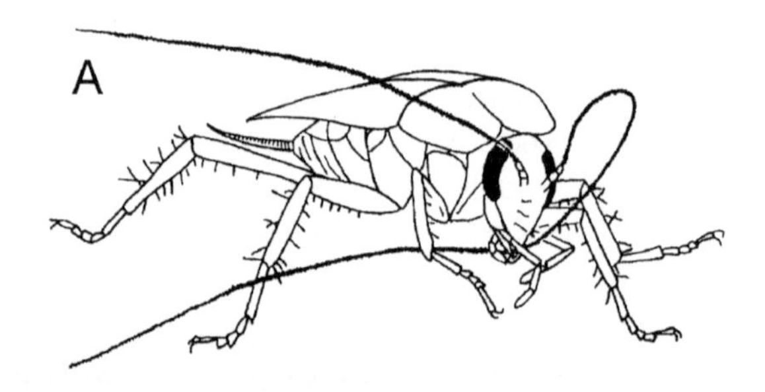

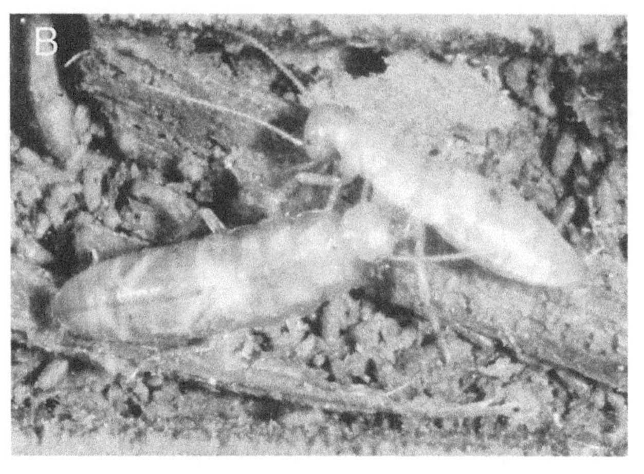

Fig 5.5 Grooming behavior. (A) *Periplaneta americana* passing an antenna through its mouth during autogrooming. Modified from Jander (1966), courtesy of Ursula Jander. (B) Fourth-instar *Cryptocercus punctulatus* allogrooming a sibling. Photo by C.A. Nalepa.

Flagellates as Food

Trophic stages of protozoans are vulnerable when they are passed from adult to offspring during proctodeal trophallaxis in the wood-feeding cockroach *Cryptocercus*. Some flagellate species are extremely large—*Barbulanympha* may be up to 340 μ long (Cleveland et al., 1934), and first instars of *Cryptocercus* are unusually small (Nalepa, 1996). Consequently, large flagellates may not be able to pass through the proventriculus of early instars without being destroyed; the phenomenon has been reported in termites. Remnants of the flagellate *Joenia* were observed in the gizzards of all young *Kalotermes* examined by Grassé and Noirot (1945). It may take several molting cycles before the gizzard of the young host is of a diameter to allow passage of the largest flagellates. Typically, the large protozoans are the last ones established in *Cryptocercus;* they are not habitually found in the hindgut until the third instar (Nalepa, 1990). Until then, the numerous flagellates passed from adult to offspring in the proctodeal fluids are a high-quality, proteinaceous food (Grassé, 1952) available at low metabolic cost to the consumer (Swift et al., 1979). The normal death of protozoans within the gut may also contribute to microbial protein in the hindgut fluids. Cleveland (1925) indicated that "countless millions of them must die daily" in a single host.

Fungi as Food

Many animals feed on fungal tissue by selectively grazing on fruiting bodies and mycelia. Others consume small quantities of fungal tissue along with larger amounts of the substrate on which the fungus is growing (Kukor and Martin, 1986). Cockroaches as a group span both categories, using fungi as food either incidentally or specifically.

Among the more selective feeders are species like *Parcoblatta,* which include mushrooms in their diet (Table 4.1), and *Lamproblatta albipalpus,* observed grazing on mycelia covering the surface of rotten wood and dead leaves (Gautier and Deleporte, 1986). The live and dead plant roots used as food by the desert cockroach *Arenivaga investigata* are sheathed in mycorrhizae, and numerous fungal hyphae can be found in the crop (Hawke and Farley, 1973). *Shelfordina orchidae* eats pollen, fungal hyphae, and plant tissue (Lepschi, 1989), and gut content analyses have clearly established that many species in tropical rainforest consume fungal hyphae and spores (WJB, unpubl. obs.). Australian *Ellipsidion* spp. are often associated with sooty mold, although it is not known if they eat it (Rentz, 1996). No known cockroach specializes on fungi, although species that live in the nests of fungus-growing ants and termites may be candidates.

All types of decaying plant tissues, whether foliage, wood, roots, seeds, or fruits, are thoroughly permeated by filamentous fungi (Kukor and Martin, 1986). The fungal contribution to the nutrient budget of cockroaches, however, is unknown. Chitin is the major cell wall component of most fungi and constitutes an average of 10% of fungal dry weight (range = 2.6–26.2) (Blumenthal and Roseman, 1957). Although chitinases are apparently rare in the digestive processes of most detritus-feeding insects (Martin and Kukor, 1984), it is distributed throughout the digestive tract of *P. americana.* The enzyme is related to cannibalism and the consumption of exuvia (Waterhouse and McKellar, 1961), but may also play a role in breaking down fungal polysaccharides.

BACTEROIDS

Bacteroids are symbiotic gram-negative bacteria of the genus *Blattabacterium* living in the fat body of all cockroaches and of the termite *Mastotermes darwiniensis.* The endosymbionts reside in specialized cells, called mycetocytes or bacteriocytes, with each symbiont individually enclosed in a cytoplasmic vacuole (Fig. 5.6A,C). They are transmitted between generations vertically, via transovarial transmission, a complex, co-evolved, and highly coordinated process (Sacchi et al., 1988; Wren et al., 1989; Lambiase et al., 1997; Sacchi et al., 2000). DNA sequence analyses indicate that the phyletic relationships of the bacteroids closely mirror those of their hosts, with nearly equivalent phylogenies of host and symbiont (Bandi et al., 1994, 1995; Lo et al., 2003a) (Fig. 5.7). Bacteroids synthesize vitamins, amino acids, and proteins (Richards and Brooks, 1958; Garthe and Elliot, 1971) but the symbiotic relationship appears grounded on their ability to recycle nitrogenous waste products and return usable molecules to the host (Cochran and Mullins, 1982; Cochran, 1985; Mullins and Cochran, 1987). The establishment of the urate-bacteroid system in the cockroach-termite lineage occurred at least 140 mya (Lo et al., 2003a), and was an elaborate, multi-step process. It involved the regulation or elimination of urate excretion, the intracellular integration of the bacteroids, the evolution of urate and mycetocyte cells in the fat body, and the coordination of the intricate interplay between host and symbionts during transovarial transmission (Cochran, 1985).

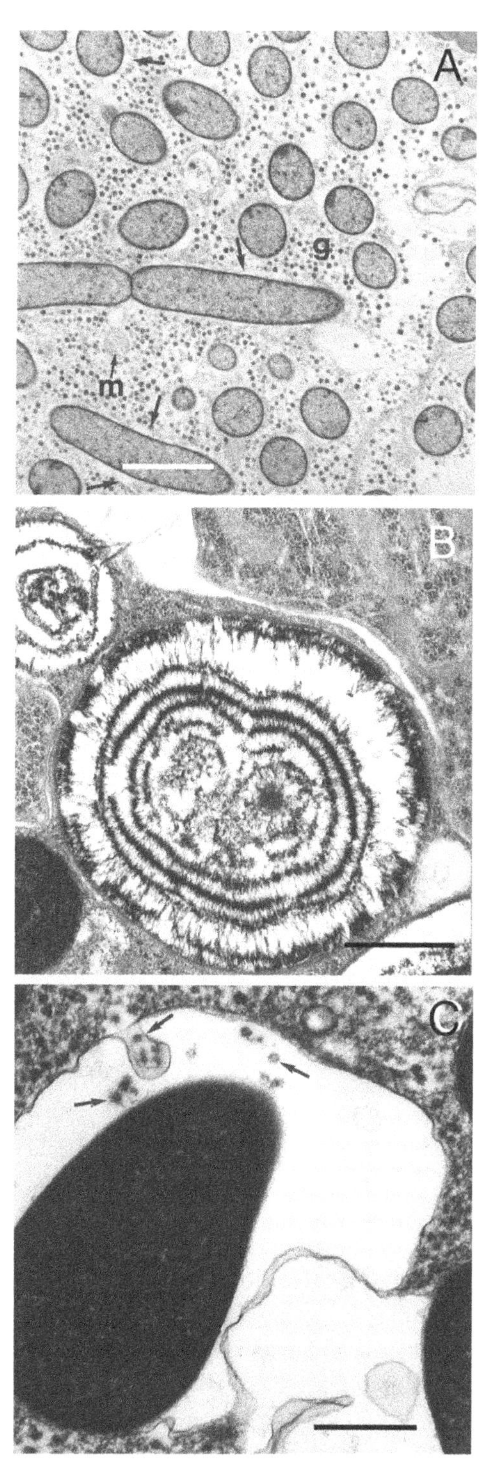

Fig. 5.6 Transmission electron micrographs of the fat body of *Cryptocercus punctulatus.* (A) Bacteriocyte with cytoplasm filled by symbiotic bacteria (g = glycogen granules; m = mitochondria; arrows = vacuolar membrane). Scale bar = 2.2 μm. (B) Urocyte of *C. punctulatus.* Note the crystalloid subunit arranged concentrically around dark cores of urate structural units. Scale bar = 0.8 μm. (C) Detail of a bacteriocyte showing glycogen particles (arrows) both enclosed in a vacuolar vesicle and within the vacuolar space surrounding the bacteroid, suggesting exchange of material between host cell cytoplasm and the endosymbiont. Scale bar = 0.5 μm. From Sacchi et al. (1998a); photos courtesy of Luciano Sacchi.

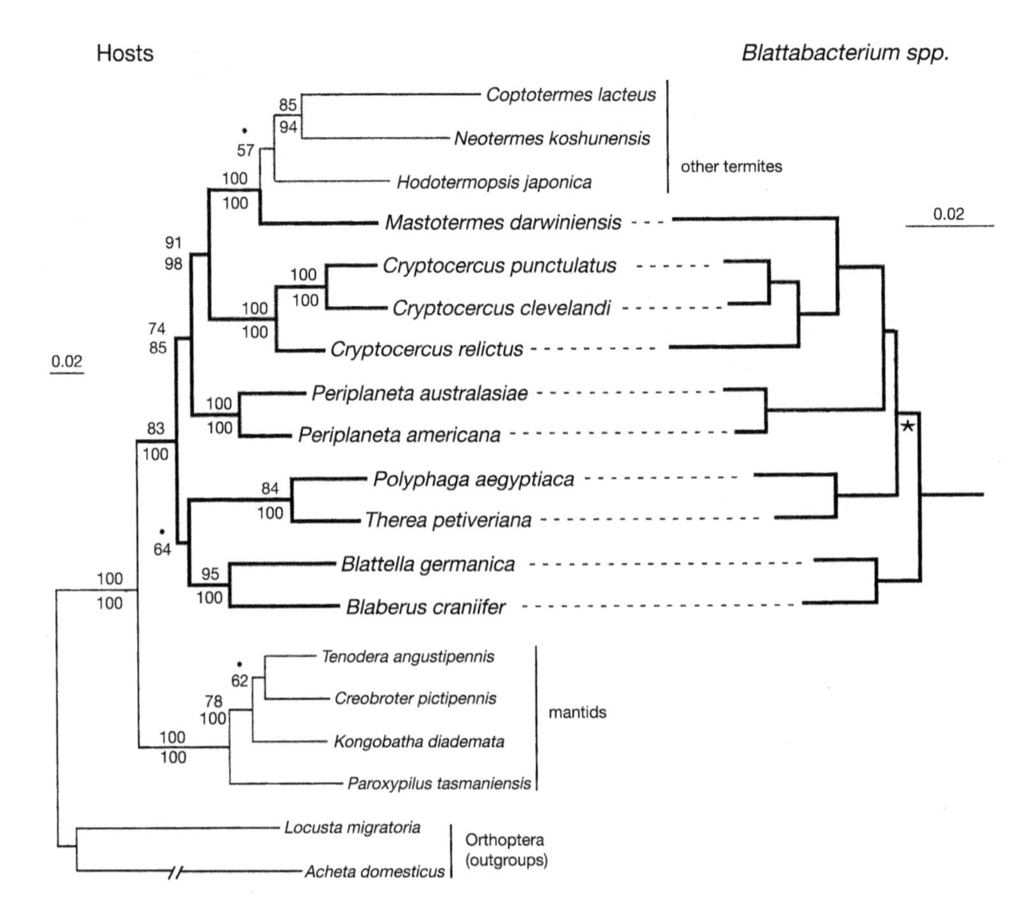

Fig. 5.7 Phylogeny of dictyopteran species and a comparison with the phylogeny of endosymbiotic *Blattabacterium* spp. The host phylogeny was based on a combined analysis of 18S rDNA and mitochondrial COII, 12S rDNA, and 16S rDNA sequences. Tree length: 2901, consistency index: 0.55. Bold lines indicate those dictyopteran taxa that harbor *Blattabacterium* spp., and that were examined in host endosymbiont congruence tests. The asterisk indicates the only node in the topology that was in disagreement with that based on host phylogeny. From Lo et al. (2003a), reprinted with permission from Nathan Lo and the journal Molecular Biology and Evolution.

Urate Management

Nitrogen excretion in cockroaches is a complex phenomenon that differs from the expected terrestrial insect pattern of producing and voiding uric acid. Several different patterns are apparent. The majority of species studied (thus far > 80) do not void uric acid to the exterior even though they may produce it in abundance (Cochran, 1985). When cockroaches are placed on a diet high in nitrogen, urates accumulate in their fat body (Mullins and Cochran, 1975a); they are typically deposited in concentric rings around a central matrix in storage cells (urocytes) adjacent to bacteriocytes (Cochran, 1985) (Fig. 5.6B). When the diet is deficient in nitrogen or individual nitrogen requirements increase, bacteroids mobilize the urate stores for reuse by the host, and the fat body deposits become depleted. Uric acid storage thus varies directly with the level of dietary nitrogen and is not excreted under any conditions. Even when fed extremely high levels of dietary nitrogen, American and German cockroaches continue to produce and store uric acid in the fat body and other tissues, ultimately leading to their death (Haydak, 1953; Mullins and Cochran, 1975a). At least three other patterns of urate excretion are found in the family Blattellidae. In the Pseudophyllodromiinae, the genera *Euphyllodromia, Nahublattella, Imblattella,* and probably *Riatia* sparingly void urate-containing pellets, with urates constituting 0.5–3.0% of total excreta by weight (Cochran, 1981). Feeding experiments showed that high-nitrogen diets did not change urate output in *Nahublattella nahua,* but did increase it in *N. fraterna* in a dose-dependent manner. In both cases diets high in nitrogen content led to high mortality. The genus *Ischnoptera* (Blattellinae) excretes a small amount of urates (2% by weight) mixed with fecal material; this pattern is similar to that of other generalized orthopteroid insects, except for the very small amount of urates voided (Cochran and Mullins, 1982; Cochran, 1985).

The most sophisticated pattern of nitrogen excretion occurs in at least nine species in the Blattellinae (*Parcoblatta, Symploce, Paratemnopteryx*), which void discrete, formed pellets high in urate content. These pellets are distinct from fecal waste (Fig. 5.8), suggesting that the packaging does not occur by chance. The cockroaches store urates internally as well (Cochran, 1979a). The level of dietary nitrogen in relation to metabolic demand for nitrogen is the controlling factor in whether uric acid is voided (Cochran, 1981; Cochran and Mullins, 1982; Lembke and Cochran, 1990). This is nicely illustrated in Fig. 5.9, which shows urate pellet excretion in female *Parcoblatta fulvescens* on different diets over the course of a reproductive cycle. Excreted urate pellets serve as a type of external nitrogen storage system, which may be accessed either by the excretor or by other members of the social group in these gregarious species. Reproducing females have been observed consuming the urate pellets, and they do so primarily when they are on a low-nitrogen, high-carbohydrate diet. A female carrying an egg case was even observed eating one, although they do not normally feed at this time. This system allows the cockroaches to deal very efficiently with foods that vary widely in nitrogen content. High nitrogen levels? The cockroaches store urates up to a certain level, and beyond that they excrete it in the form of pellets. Nitrogen limited? They mobilize and use their urate fat body reserves. Nitrogen depleted? They scavenge for high-nitrogen foods, including bird droppings and the urate pellets of conspecifics. Nitrogen unavailable? They slow or stop reproduction or development until it can be found (Cochran, 1986b; Lembke and Cochran, 1990).

Implications of the Bacteroid-Urate System

The bacteroid-assisted ability of cockroaches to store, mobilize, and in some cases, transfer urates uniquely allows them to utilize nitrogen that is typically lost via excretion in the vast majority of insects (Cochran, 1985). These symbionts thus have a great deal of power in structuring the nutritional ecology and life history strategies of their hosts. Bacteroids damp out natural fluctuations in food availability, allowing cockroaches a degree of independence from the current food supply. An individual can engorge prodigiously at a single nitrogenous bonanza, like a bird dropping or a dead conspecific, then later, when these materials are required for reproduction, development, or maintenance, slowly mobilize the stored reserves from the fat body like a time-release vitamin. The legendary ability of cockroaches to withstand periods of starvation is at least in part based on this storage-mobilization physiology. The beauty of the system, however, is that stored urates are not only recycled internally by an individual, but, depending on the species, may be transferred to conspecifics, and used as currency in mating and parental investment strategies. Any individual in an ag-

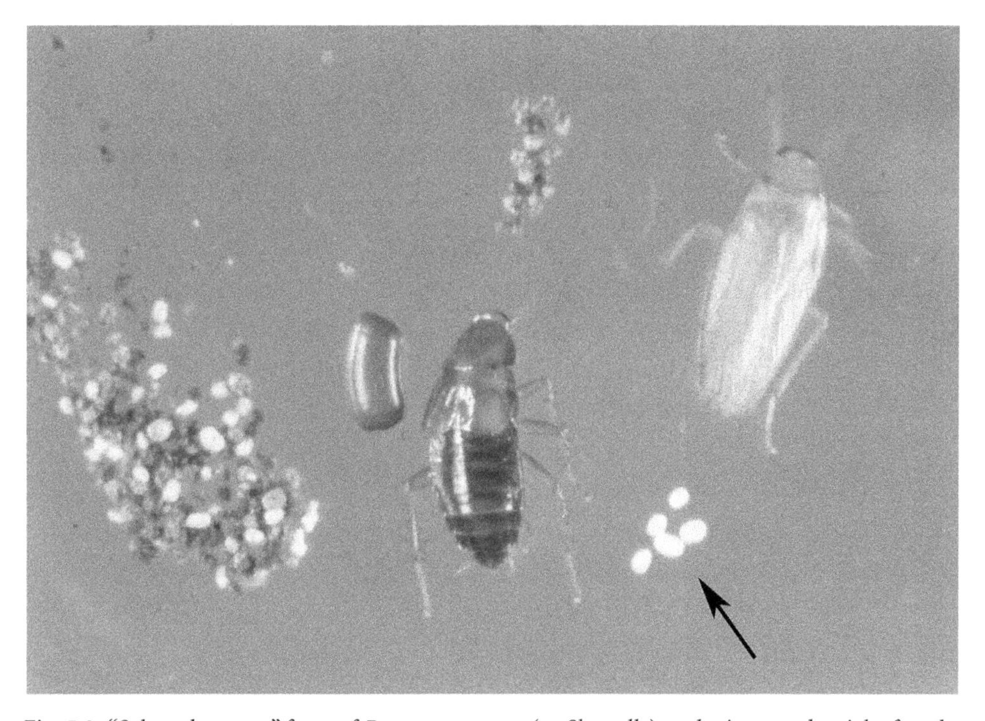

Fig. 5.8 "Salt and pepper" feces of *Paratemnopteryx* (= *Shawella*) *couloniana;* male, *right;* female and ootheca, *left.* The pile of feces to the left of the ootheca shows the variation in color of the pellets. Some of these have been separated into piles of the dark-colored fecal waste pellets (above the female) and the white, urate-filled pellets (arrow). Photo courtesy of Donald G. Cochran.

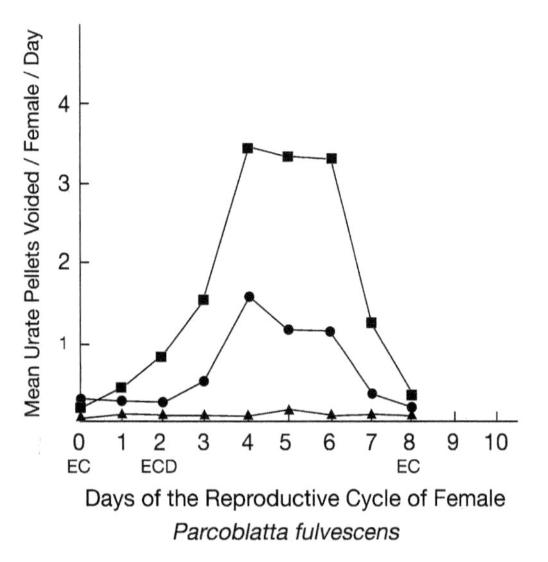

Fig. 5.9 Urate pellet excretion by adult female *Parcoblatta fulvescens* in relation to the reproductive cycle and level of dietary nitrogen. Filled triangles, 4.0% nitrogen diet; filled circles, 5.4% nitrogen diet; filled squares, 6.7% nitrogen diet. EC, egg case formation; ECD, egg case deposition. From Cochran (1986b), courtesy of Donald G. Cochran, with permission from Elsevier Press.

gregation of the cockroaches that excrete urate pellets (like *Parcoblatta*) potentially benefits when just one of them exceeds its nitrogen threshold (Lembke and Cochran, 1990). In cockroach species in which the male transfers urates to the female during or after mating (Mullins and Keil, 1980; Schal and Bell, 1982), it would not be surprising to discover that female mate or sperm choice decisions are based on the size or quality of the nuptial gift (Chapter 6). The diversity of modes of post-ovulation provisioning of offspring observed in cockroaches (brood milk, gut fluids, exudates) is likely to be rooted in the ability of a parent to mobilize and transfer stored reserves of nitrogen (Nalepa and Bell, 1997). Finally, cockroaches are able to use the uric acid scavenged from the feces of birds, reptiles, and non-blattarian insects, adding to the list of advantages of a generalized coprophagous lifestyle (Schal and Bell, 1982).

Bacteroids as Food

There is some evidence that fat body endosymbionts in cockroaches and in the termite *Mastotermes* may be a direct source of nutrients to developing embryos. During embryogenesis a portion of the bacterial population degenerates, with a concomitant increase in glycogen granules in the cytoplasm as the symbionts degrade (Sacchi et al., 1996, 1998b). Bacteroids are also reported to shrivel in size, then disappear when a postembryonic cockroach is starved (Steinhaus, 1946; Walker, 1965).

ADDITIONAL MICROBIAL INFLUENCES

There is a general under-appreciation of the ubiquity of microorganisms and the varied roles they play in the biology and life history of multicellular organisms. Microbes can affect their hosts and associates in unexpected ways, often with profound ecological and evolutionary consequences (McFall-Ngai, 2002; Moran, 2002). If this is true for organisms that are *not* habitually affiliated with rotting organic matter, shouldn't microbial influence be exponentially higher in cockroaches, insects that seek out habitats saturated with these denizens of the unseen world? Our focus so far has been primarily on the role of microbes in the nutritional ecology of cockroaches. The diverse biosynthetic capabilities of microbes, however, allow for wide-ranging influences in cockroach biology.

Microbes may alter or dictate the thermal tolerance of their host. Hamilton et al. (1985) demonstrated that the sugar alcohol ribitol acts as an antifreeze for *C. punctulatus* in transitional weather, and as part of a quick freeze system when temperatures drop. Because microbes produce significantly more five-carbon sugars than animals and because ribitol had not been previously reported in an insect, the authors suggested that microbial symbionts might be responsible for producing the alcohol or its precursors. Cleveland et al. (1934) indicated that the effects of temperature on the cellulolytic gut protozoans of *Cryptocercus* confine these insects to regions free from climatic extremes. These effects differ between the eastern and western North American species. If the insects are held at 20–23°C, the protozoans of *C. clevelandi* die within a month, whereas those of *C. punctulatus* live indefinitely.

Microbial products may act like pheromones. Because cockroach aggregation behavior is in part mediated by fecal attractants in several species, it is possible that gut microbes may be the source of at least some of the components. Such is the case in the aggregation pheromone of locusts (Dillon et al., 2000) and in the chemical cues that mediate nestmate recognition in the termite *Reticulitermes speratus* (Matsuura, 2001).

Microbes may influence somatic development. There is a "constant conversation" between host tissues and their symbiotic bacteria during development, with the immune system of the host acting as a key player (McFall-Ngai, 2002). Aside from their profound effect on cockroach development via various nutritional pathways, bacterial mutualists may directly influence cockroach morphogenesis. It is known that gut bacteria are required for the proper postembryonic development of the gut in *P. americana* (Bracke et al., 1978; Zurek and Keddie, 1996); normal intestinal function may depend on the induction of host genes by the microbes (Gilbert and

Bolker, 2003). The highly complex and tightly coordinated interactions of *Blattabacterium* endosymbionts with their hosts during transovarial transmission and embryogenesis (Sacchi et al., 1988, 1996, 1998b) suggest that these symbionts may influence the earliest stages of cockroach development.

MICROBES AS PATHOGENS

Microbes can be formidable foes. Most animals battle infection throughout their lives, and devote substantial resources to responding defensively to microbial invaders (e.g., Irving et al., 2001). Cockroaches, like other animals that utilize rotting organic matter (Janzen, 1977), must fend off pathogenesis and avoid or detoxify the chemical offenses of microbes. Most Blattaria lead particularly vulnerable lifestyles. They are relatively long-lived insects that favor humid, microbe-saturated environments; many live in close association with conspecifics, particularly during the early, vulnerable part of life. They also have a predilection for feeding on rotting material, conspecifics, feces, and dead bodies. Pathogens and parasites such as protozoa and helminths (e.g., Fig. 5.10) are no doubt a strong and unrelenting selective pressure, but cockroach defensive strategies must be delicately balanced so that their vast array of mutualists are not placed in the line of fire. An example of these conflicting pressures lies in cockroach social behavior. On the one hand, beneficial microbes promote social behavior. Transmission of hindgut microbes requires behavioral adaptations so that each generation acquires microflora from the previous one, and consequently selects for association of neonates with older conspecifics. On the other hand, pathogenic microbes exploit cockroach social behavior, in that their transmission occurs via inter-individual transfer. Oocysts of parasitic *Gregarina*, for example, are transmitted via feces (Lopes and Alves, 2005), and the biological control of urban pest cockroaches with pathogens is predicated largely on their spread via inter-individual contact in aggregations (e.g., Mohan et al. 1999; Kaakeh et al.,1996). Roth and Willis (1957) document inter-individual transfer of a variety of gregarines, coccids, amoebae, and nematodes via cannibalism, coprophagy, or proximity.

Fig. 5.10 Hairworm parasite (*Paleochordodes protus*) of an adult blattellid cockroach (in or near the genus *Supella*) in Dominican amber (15–45 mya). From Poinar (1999); photo courtesy of George Poinar Jr.

Cockroaches have a variety of behavioral and physiological mechanisms for preventing and managing disease. At least two cockroach species recognize foci of potential infection and take behavioral measures to evade them. Healthy nymphs of *B. germanica* are known to avoid dead nymphs infected with the fungus *Metarhizium anisopliae* (Kaakeh et al., 1996). The wood-feeding cockroach *Cryptocercus* sequesters corpses and controls fungal growth in nurseries (Chapter 9). The former behavior may function to shield remaining members of the family from infection. Vigilant hygienic behavior or fungistatic properties of their excreta or secretions may also play a role throughout the gallery system. Fungal overgrowth of tunnels is never observed unless the galleries are abandoned (CAN, pers. obs.).

The glandular system of cockroaches is complex and sophisticated, with seven types of exocrine glands found in the head alone (Brossut, 1973). The mandibular glands of two species (*Blaberus craniifer* and *Eublaberus distanti*) secrete an aggregation pheromone; otherwise the function of cephalic glands is unknown (Brossut, 1970, 1979). The secretion of some of these may have antimicrobial properties, and could be spread over the surface of the body to form an antibiotic "shell" during autogrooming, particularly if the cockroach periodically runs a leg over its head or through its mouthparts during the grooming behavioral sequence. Autogrooming therefore may function not only to remove potential cuticular pathogens physically, but also to disseminate chemicals that curtail their growth or spore germination. Dermal glands are typically spread over the entire abdominal integument of both males and females (200–400/mm^2) (Sreng, 1984), and five types of defensive-type exocrine glands have been described (Roth and Alsop, 1978) (Fig. 5.11). Most of the latter produce chemical defenses effective against an array of vertebrate and invertebrate predators (Fig. 1.11A), but the influence of these chemicals on non-visible organisms is unexplored. They may well function as "immediate effronteries" to predators as well as "long term antagonists" to bacteria and fungi (Roth and Eisner, 1961; Duffy, 1976), and act subtly, by altering growth rates, spore germination, virulence, or chemotaxis (Duffy, 1976). Most cockroach exocrine glands produce multi-component secretions (Roth and Alsop, 1978). The man-

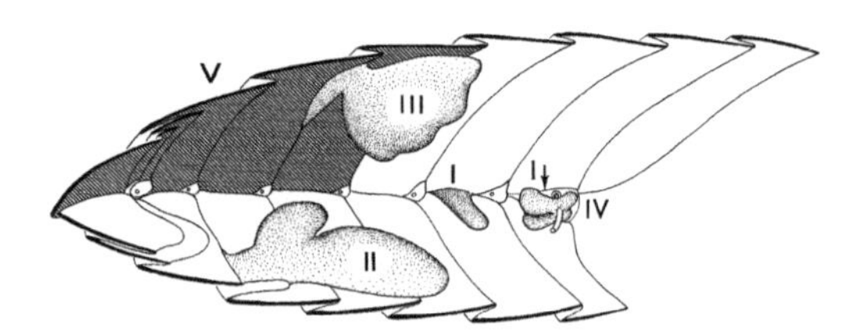

Fig. 5.11 Diagrammatic sagittal section of a cockroach abdomen, showing gland types I–IV and location of the secretory field for gland type V. One of the two type I glands has been omitted and its position indicated by an arrow. Only half of the medially opening Type III gland is shown. From Roth and Alsop (1978), after Alsop (1970), with permission from David W. Alsop.

dibular glands of *Eub. distanti,* for example, is a blend of 14 products (Brossut, 1979). Brossut and Sreng (1985) list 93 chemicals from cockroach glands, some of which are known to be fungistatic in other systems, for example, phenols (Dillon and Charnley, 1986, 1995), naphthol, *p*-cresol, quinones (Brossut, 1983), and hexanoic acid (Rosengaus et al., 2004). Phenols have been identified from both the sternal secretions and the feces of *P. americana,* and neither feces nor the filter paper lining the floor of rearing chambers exhibit significant fungal growth (Takahashi and Kitamura, 1972). Other cockroaches also produce a strong phenolic odor when handled (Roth and Alsop, 1978). It is of interest, then, that phenols in the fecal pellets and gut fluids of locusts originate from gut bacteria, and are *selectively* bacteriocidal (Dillon and Charnley, 1986, 1995). Given the extraordinarily complex nutritional dynamics between cockroaches and microbes in the gut and on feces, these kinds of probiotic interactions are probably mandatory. It is a safe assumption that cockroaches engage in biochemical warfare with microbes, but they have to do so judiciously.

Blattaria have both behavioral and immunological mechanisms for countering pathogens that successfully breach the cuticular or gut barrier. Wounds heal quickly (Bell, 1990), and cockroaches are known to use behavioral fever to support an immune system challenged by disease. When *Gromphadorhina portentosa* was injected with bacteria or bacterial endotoxin and placed in a thermal gradient, the cockroaches preferred temperatures significantly higher than control cockroaches (Bronstein and Conner, 1984). The immune system of cockroaches differs from that of shorter-lived, holometabolous insects, and mimics all characteristics of vertebrate immunity, including both humoral and cell-mediated responses (Duwel-Eby et al., 1991). *Blaberus giganteus* synthesizes novel proteins when challenged with fungi (Bidochka et al., 1997), and when American cockroaches are injected with dead *Pseudomonas aeruginosa,* they respond in two phases. Initially there is a short-term, nonspecific phase, which is superseded by a relatively long-term, specific response (Faulhaber and Karp, 1992). When challenged with *E. coli, P. americana* makes broad-spectrum antibacterial peptides. Activity is highest 72–96 hr after treatment, and newly emerged males respond best (Zhang et al., 1990). Cellular immune responses are mediated by hemocytes, primarily granulocytes and plasmatocytes (Chiang et al., 1988; Han and Gupta, 1988) whose numbers increase in response to invasion and counter it using phagocytosis and encapsulation (Verrett et al., 1987; Kulshrestha and Pathak, 1997).

Sexual contact carries with it the risk of sexually transmitted diseases (e.g., Thrall et al., 1997), but no cockroaches were listed in an extensive literature survey on the topic (Lockhart et al., 1996). *Wolbachia,* a group of cytoplasmically inherited bacteria that are widespread among insects (including termites—Bandi et al., 1997) have not yet been detected in cockroaches, but few species have been studied to date (Werren, 1995; Jeyaprakash and Hoy, 2000). Further surveys of Blattaria may yet detect *Wolbachia,* but because they are transmitted through the cytoplasm of eggs, these rickettsiae may have trouble competing with transovariolly transmitted bacteroids (Nathan Lo, pers. comm. to CAN).

The cost of battling pathogens likely has life history consequences for cockroaches, since it does in many animals that inhabit more salubrious environments (Zuk and Stoehr, 2002). Immune systems can be costly in that they use energy and resources that otherwise may be invested into growth, reproduction, or maintenance, thus making them subject to trade-offs against other fitness components (Moret and Schmidt-Hempel, 2000; Møller et al., 2001; Zuk and Stoehr, 2002). It may be possible, for example, that the prolonged periods of development typical of many cockroaches may be at least partially correlated with an increased investment in immune function. The life of a cockroach has to be a fine-tuned balancing act between exploiting, cultivating, and transmitting microbes, while at the same time suppressing, killing, or avoiding the siege of harmful members of the microbial consortia that surround them. Until recently, these relationships have been difficult to study because the microbes of interest are poorly defined, many have labile or nondescript external morphology, and most cannot be cultured in vitro. The availability of new methodology that allows insight into the origins, nature, and functioning of microbes (Moran, 2002) in, on, and around cockroaches portends a bright future for studies on the subject. Until then, it should be considered that the ability of cockroaches to live in just about any organic environment may have its basis in their successful management of the varied, sophisticated, cooperative, and adversarial relationships with "inconspicuous associates" (Moran, 2002).

SIX

Mating Strategies

The unfortunate couple were embarrassed beyond all mortification, not simply for having been surprised in the act by the minister, but also for their inability to separate, to unclasp, to unlink, to undo all the various latches, clamps and sphincters that linked them together, tail to tail in opposite directions.

— D. Harington, *The Cockroaches of Stay More*

The genitalia of male cockroaches are frequently used as an example of the extreme complexity that may evolve in insect reproductive structures (e.g., Gwynne, 1998). They have been likened to Swiss army knives in that a series of often-hinged hooks, tongs, spikes, and other lethal-looking paraphernalia are sequentially unfolded during copulation. Marvelous though all that hardware may be, it has not yet inspired research on its functional significance. Seventy years ago Snodgrass (1937) stated that "we have no exact information on the interrelated functions of the genital organs" of cockroaches, and the situation has improved only slightly since that time. While there is a vast literature on pheromonal communication, reproductive physiology, male competition, and behavioral aspects of courtship in cockroaches, we know surprisingly little about the "nuts and bolts" of the copulatory performance, and in particular, how the male and female genitalia interact.

Here we briefly describe cockroach mating systems, and the basics of mate finding, courtship, and copulation. We then focus on just a few topics that are, in the main, relevant to the evolution of cockroach genitalia. We make no attempt to be comprehensive. Our emphasis is on male and female morphological structures whose descriptions are often tucked away in the literature on cockroach systematics and are strongly suggestive of sperm competition, cryptic mate choice, and conflicts of reproductive interest. One goal is to shift some limelight to the female cockroach, whose role in mating dynamics is poorly understood yet whose morphology and behavior suggest sophisticated control over copulation, sperm storage, and sperm use.

MATING SYSTEM

In nearly all cockroach species studied, males will mate with multiple females even if the exhaustion of mature sperm and accessory gland secretions preclude the formation of a spermatophore (Roth, 1964b; Wendelken and Barth, 1987); cockroach mating systems are therefore best classified on the basis of female behavior. However, it is difficult to determine how many mating partners a female has in the wild, and, as might be expected for insects that are mostly cryptic and nocturnal, field studies of mating behavior are rare.

One Male, One Copulation

Females of at least two cockroach species are reported to be monandrous in the strictest sense of the word. Once mated, *Neopolyphaga miniscula* (Jayakumar et al., 2002) and *Therea petiveriana* (Livingstone and Ramani, 1978) females remain refractory to subsequent insemination for the rest of their lives; the latter repel suitors by kicking with their hind legs.

One Male, Multiple Copulations

Wood-feeding cockroaches in the genus *Cryptocercus* may be described as socially monogamous; males and females establish long-term pair bonds and live in family groups. Genetic monogamy is yet to be determined, but opportunities for extra-pair copulations are probably few. When paired with a female, males fight to exclude other males from tunnels (Ritter, 1964), and adults of both sexes in families defend against intruders (Seelinger and Seelinger, 1983). In the two copulations observed in *C. punctulatus,* one lasted for 34 min and the other for 42 min (Nalepa, 1988a); sneaky extra-pair copulations therefore seem unlikely. The best opportunity for cheating, if it occurs, would be after adult emergence but prior to establishment of a pair bond. Adult males and adult females each can be found alone in galleries, particularly during spring and early summer field collections (Nalepa, 1984).

Typically, males and females pair up during summer, overwinter together, and produce their sole set of offspring the following summer. Although sperm from a single copulation are presumably sufficient to fertilize these eggs (average of 73), pairs mate repeatedly over the course of their association. There is evidence of sexual activity the year before reproduction, immediately prior to oviposition, during the oviposition period, after the hatch of their oothecae, and 1 yr after the hatch of their single brood (Nalepa, 1988a). Prior to oviposition, repeated copulation may function as paternity assurance or perhaps nutrient transfer, but mating after the eggs are laid is more difficult to explain. Rodríguez-Gironés and Enquist (2001) note that mating frequency is particularly high in species where males associate with females and assist them in parental duties. Superfluous copulations evolve in these pairs because females attempt to sequester male assistance and males are deprived of cues about female fertility. It would be of interest to determine if this pattern of repeated mating behavior occurs in other socially monogamous, wood-feeding cockroaches like *Salganea;* these also live in family groups with long-term parental care (Matsumoto, 1987; Maekawa et al., 2005).

Multiple Males, One Copulation per Reproductive Cycle

In most studied cockroaches female receptivity is cyclic. It declines sharply after copulation and is not restored until after partition. In some species it takes several reproductive cycles before another mating partner is accepted, in others receptivity is restored following each reproductive event. Females, then, may be described as monandrous within each period that they are accepting mates, but polyandrous over the course of their reproductive life. Because they store sperm, it is only during the formation of the first clutch of eggs that their partners are under little threat from sperm competition. The pattern of cyclic receptivity occurs in both oviparous and live-bearing cockroaches. Both *Blattella germanica* (Cochran, 1979b) and *B. asahinai* (Koehler and Patternson, 1987) may copulate repeatedly, although a single mating usually provides sufficient sperm to last for the reproductive life of the female. *Periplaneta americana* females alternate copulation with oothecal production, and may mate as soon as 3–4 hr after depositing an egg case (Gupta, 1947). A pair of *Ellipsidion humerale* (= *affine*) were observed copulating four times within a month, alternating with oothecal production (Pope, 1953). Similarly, blaberid females ordinarily mate just once prior to their first oviposition. After eclosion of the nymphs, they may then enter another cycle of receptivity, mating, oviposition, and egg incubation (Engelmann, 1960; Roth, 1962; Roth and Barth, 1967; Grillou, 1973). Once mated, female *Eublaberus posticus* are fertile for life, and remating does not improve reproductive performance (Roth, 1968c); nonetheless, remating has been observed (Darlington, 1970).

Multiple Males, Multiple Copulations per Reproductive Cycle

Reports of multiple mating by a female within a single reproductive cycle exist, but they are the exception rather than the rule among examined species. In his study of more than 200 female *B. germanica,* Cochran (1979b) recorded just a single instance of a female mating twice prior to her first egg case. In their extensive studies of the same species, Roth and Willis (1952a) noted one pair that copulated twice within a 24-hr period. Hafez and Afifi (1956) report that in *Supella longipalpa* "copulation may occur once or twice a day" but give no further details. On rare occasions, a female of *Diploptera punctata* may be found carrying two spermatophores; however, one of these is always improperly positioned (Graves, 1969). Sperm are likely transferred only from the one correctly aligned with the female's spermathecal openings (discussed below).

MATE FINDING

Most cockroaches that have been studied rely on chemical and tactile cues to find their mates in the dark (Roth and Willis, 1952a). In many cases volatile sex pheromones mediate the initial orientation; these have been demonstrated in 16 cockroach species in three families. The pheromones are most commonly female generated and function at a variety of distances, up to 2 m or more, depending on the species (Gemeno and Schal, 2004). Females in the process of releasing pheromone ("calling") often assume a characteristic posture (Fig. 6.1): they raise the wings (if they have them), lower the abdomen, and open the terminal abdominal segments to expose the genital vestibulum (Hales and Breed, 1983; Gemeno et al., 2003). In some species the initial roles are reversed, with males assuming a characteristic stance while luring females (Roth and Dateo, 1966; Sreng, 1979a). A calling male may maintain the posture for 2 or more hr, with many short interruptions (Sirugue et al., 1992). Based on the limited available data, the general pattern appears to be that in species where the male or both sexes are volant, females release a long-range volatile pheromone. Males release sex pheromones in species where neither sex can fly (Gemeno and Schal, 2004).

Non-chemical Cues

While research has focused primarily on chemical cues (and justly so), mate finding and courtship may be multimodal in a number of species, that is, they integrate chemical, visual, tactile, and acoustic signals. Vision apparently plays little or no significant role in sexual recognition, courtship, or copulation in the species typically studied in laboratory culture (Roth and Willis, 1952a). However, in many cockroaches the males have large, well-developed, pigmented eyes, suggesting the possibility that optical cues may be integrated with pheromonal stimuli during mate seeking and mating behavior. Visual orientation seems particularly likely in Australian Polyzosteriinae and in brightly colored, diurnally active blattellids. The delightful discovery of pronotal headlights on males of *Lucihormetica fenestrata* suggests that even nocturnally active cockroaches may use sight in attracting or courting mates (Zompro and Fritzsche, 1999). This species lives in bromeliads in the Brazilian rainforest and has two elevated, kidney-shaped, strongly luminescent organs on the pronotum (Fig. 6.2). These protuberances are highly porous (probably to allow gas exchange) and absent in nymphs and females. Males of several related species sport similar structures, but because live material had never been examined, their function as lamps was unknown.

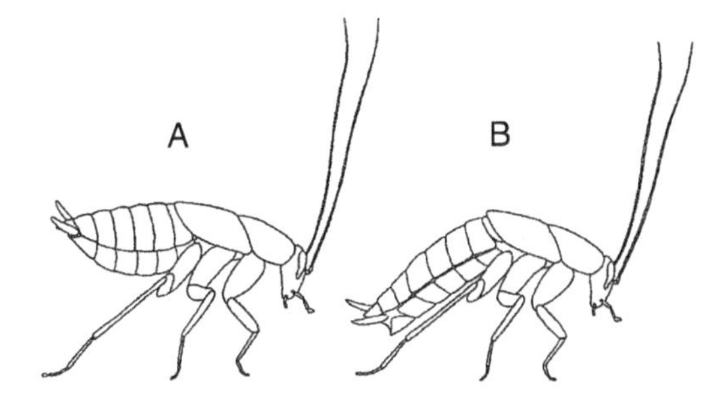

Fig. 6.1 Calling behavior in female *Parcoblatta lata*. Females in the calling posture raise the body up from the substrate and alternate between two positions: (A) upward with longitudinal compression, and (B) downward with longitudinal extension. From Gemeno et al. (2003), courtesy of César Gemeno, with permission of Journal of Chemical Ecology.

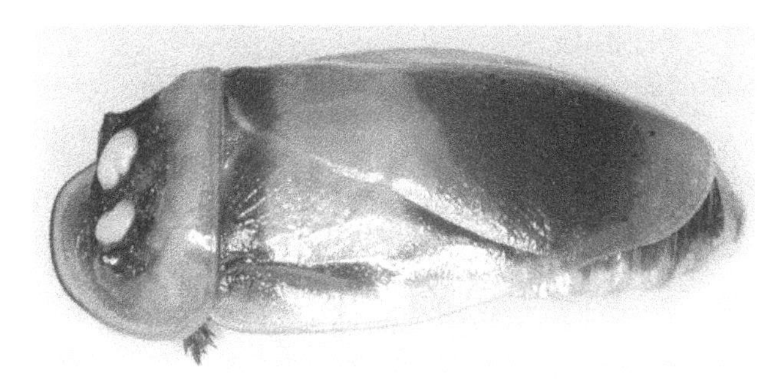

Fig. 6.2 Male *Lucihormetica fenestrata* Zompro & Fritzsche, 1999 (holotype) exhibiting its pronotal "headlights." Copyright O. Zompro, courtesy of O. Zompro.

COURTSHIP AND COPULATION

Once in the vicinity of a potential mate, contact pheromones on the surface of the female and short-range volatiles produced by the male facilitate sexual and species recognition and coordinate courtship. Recently the topic was comprehensively reviewed by Gemeno and Schal (2004). Developments in the field worth noting include the finding that short-range and contact pheromones not only mediate mate choice and serve as behavioral releasers during courtship, but may regulate physiological processes as well. The phenomenon is best studied in *Nauphoeta cinerea*, where male pheromones may influence female longevity, the number and sex ratio of offspring, and their rate of development in the brood sac (Moore et al., 2001, 2002, 2003).

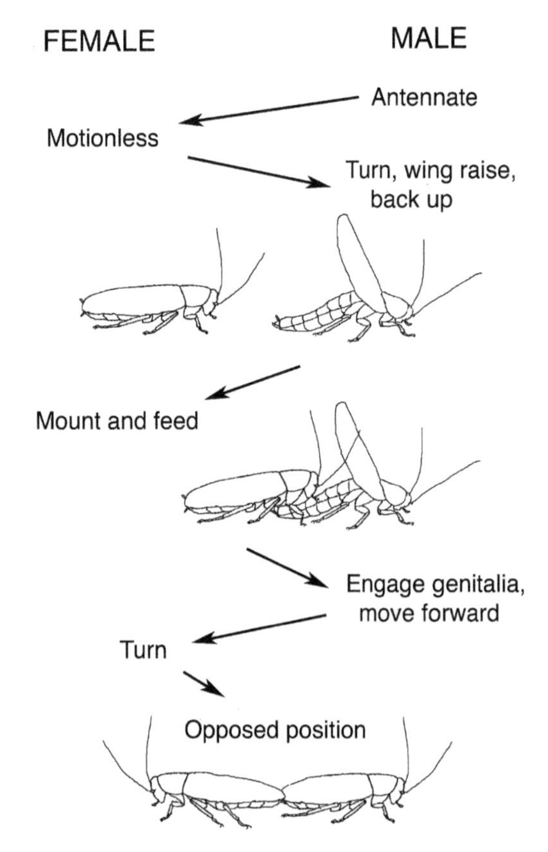

Fig. 6.3 "Basics" of type I courtship and copulation in cockroaches, after initial orientation to a potential mate.

With few exceptions, pre-copulatory behavior is remarkably uniform among cockroaches (Roth and Willis, 1954b; Roth and Dateo, 1966; Roth and Barth, 1967; Roth, 1969; Simon and Barth, 1977a). Antennal contact with the female usually instigates a male tergal display (Fig. 6.3); he turns away from her and presents the dorsal surface of his abdomen. The female responds by climbing onto his back and "licks" it, with the palps and mouthparts closely applied and working vigorously. The "female above" position lasts but a few seconds before the male backs up and extends a genitalic hook that engages a small sclerite in front of her ovipositor. Once securely connected, he moves forward, triggering the female to rotate 180 degrees off his back. The male abdomen untwists and recovers its normal dorsoventral relationship almost immediately. The pair remains in the opposed position until copulation is terminated.

Although the final position assumed by cockroaches in copula is invariably end to end, there are two additional behavioral sequences that may precede it. Both are characterized by the lack of a wing-raising display and female feeding behavior.

Type II mating behavior is characterized by the male riding the female, and is known in *Pycnoscelus indicus* and *Jagrehnia madecassa.* After the male contacts the female he crawls directly onto her back. He twists the tip of his abdomen down and under that of the female, engages her genitalia, then dismounts and assumes the opposed position (Roth and Willis, 1958b; Roth, 1970a; Sreng, 1993). In type III pre-copulatory behavior, neither sex mounts the other. After contact is made between the sexes, the male typically positions himself behind the female with his head facing in the opposite direction, then moves backward until genitalic contact is established. Cockroaches that fall into this category include *Gromphadorhina portentosa* (Barth, 1968c), *Panchlora nivea* (Roth and Willis, 1958b), *Pan. irrorata* (Willis, 1966), *The. petiveriana* (Livingstone and Ramani, 1978), *Panesthia australis* (Roth, 1979c), and the giant burrowing cockroach *Macropanesthia rhinoceros.* Mating in the latter has been described as being "like two Fiats backing into each other" (D. Rugg, pers. comm. to CAN) (Fig. 6.4). In *Epilampra involucris,* the male arches his abdomen down and then up in a sweeping motion until he contacts the female's genitalia (Fisk and Schal, 1981). In *Panesthia cribrata,* the two sexes start out side by side. The female raises the tip of her abdomen and the male bends toward the female until the tips of their abdomens are in close proximity. The male then turns 180 degrees to make genital contact (Rugg, 1987). It is of interest that type III pre-copulatory behavior occurs in the Polyphagidae (*Therea*), and in four different subfamilies of Blaberidae. A common thread is that most of these cockroaches are strong burrowers, suggesting that the behavior may be an adaptation to some aspect of their enclosed lifestyle. It is also notable that termites initiate copulation by backing into each other (Nutting, 1969).

Acoustic Cues

In some cockroach species mating behavior is highly stereotyped, with an internally programmed, unidirec-

Fig. 6.4 Copulating pair of *Macropanesthia rhinoceros,* a species with type III mating behavior. Photo courtesy of Harley Rose.

tional sequence of acts (Bell et al., 1978); in others, male-female interaction is more flexible (Fraser and Nelson, 1984). Variations that do occur often take the form of behaviors that produce airborne or substrate-borne vibrations, particularly when males are courting reluctant females (Fig. 6.5). These signals typically occur after antennal contact but prior to full tergal display, and include rocking, shaking, waggling, trembling, vibrating, pushing, bumping, wing pumping, wing fluttering, "pivot-trembling," anterior-posterior jerking, hissing, whistling, tapping, and stridulation. Although Barth (1968b) suggested that vibrating and wing fluttering during courtship produce air currents that serve to disseminate pheromone, very little is known regarding the role of these behaviors in influencing female receptivity. Hissing during courtship is best known in *G. portentosa* (Fraser and Nelson, 1984), but occurs in other species as well. Males of Australian burrowing cockroaches pulse the abdomen during courtship, and the behavior is accompanied by an audible hiss in the larger species (D. Rugg, pers. comm. to CAN). *Elliptorhina chopardi* males produce broad-band, amplitude-modulated hisses like *G. portentosa,* but also complex, bird-like whistles; dual harmonic series warble independently from the left and right fourth spiracle (Fraser and Nelson, 1982; Sueur and Aubin, 2006). The common name of *Rhyparobia maderae* is the "knocker" cockroach, because of the male habit of tapping the substrate with his thorax in the presence of potential mates (Fig. 6.5B). Highly developed stridulating organs are found on the pronotum and tegmina of some Blaberidae (Oxyhaloinae and Panchlorinae) (Roth and Hartman, 1967; Roth, 1968c). Males of *Nauphoeta cinerea* use the structures to produce characteristic phrases consisting of complex pulse trains and chirps if a female is unresponsive to his overtures (Hartman and Roth, 1967a, 1967b). There is currently no evidence, however, that the male's distinctive song (Fig. 6.5D) influences her response. Sounds produced by *N. cinerea* during courtship can be recorded from the substrate on which they are standing as well as by holding a microphone at close range (Roth and Hartman, 1967). Given the evidence that cockroaches can be sensitive to vibration as well as airborne sound (Shaw, 1994a), substrate-borne courtship signals may be more common than is currently appreciated. This is especially relevant for tropical cockroaches that perch at various levels in the canopy during their active period. Bell (1990) noted that cockroaches on leaves can detect the vibrations of approaching predators. These cockroach species also have potential for communicating with each other via leaf tremulation. The cockroach "ear" is the subgenual organ on the metathoracic legs, a fan-shaped structure lying inside and attached to the walls of the tibiae. The subgenual organ of *P. americana* is one of the most sensitive known insect vibration detectors (Autrum and Schneider, 1948; Howse, 1964).

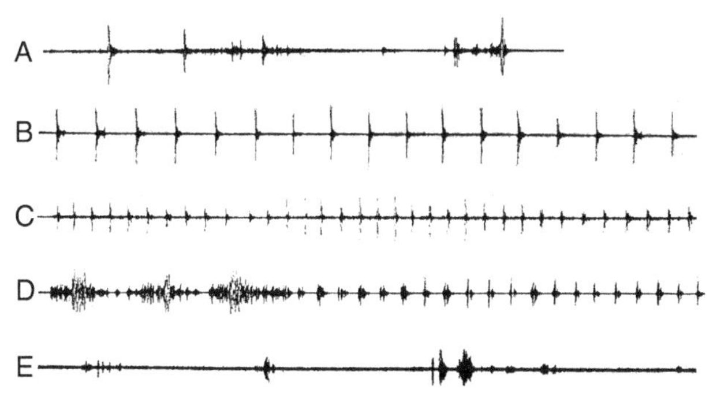

Fig. 6.5 Oscilloscope records of sounds in cockroaches. (A) Arrhythmic rustling sound made by a courting male *Eublaberus posticus;* (B) sound produced by a male *Rhyparobia maderae* tapping upon the substrate, which in this case, was a female on which the male was standing; (C) courting sounds produced by a male *Diploptera punctata* by striking the wings against the abdomen; (D) phrase produced by stridulation during courtship in male *Nauphoeta cinerea;* compare to (E) disturbance sound made by male *N. cinerea.* After Roth and Hartman (1967); see original work for reference signals and sound levels.

Length of Copulation

The length of copulation is variable in cockroaches, both within and between species. In successful matings, the male and female commonly remain in the linear position for 50–90 min, but length can vary with male age, the time since his last mating, and his social status. The shortest recorded copulations are in the well-studied *N. cinerea.* A male's first copulation is his shortest, ranging from 9.5 (Moore and Breed, 1986) to 17 (Roth, 1964b) min. Dominant males of this species copulate significantly longer than do their subordinates (Moore and Breed, 1986; Moore, 1990). If males 14–15 days old are consecutively mated to a series of females, they remain in copula 22 min during the first mating, 100 in the second, and 141 in the third (Roth, 1964b). The most extended matings reported from natural settings are those of *Xestoblatta hamata,* where copulation in the rainforest may last for up to 5 hr (Schal and Bell, 1982), and *Polyzosteria limbata,* where copulation occurs in daylight and pairs sometimes remain linked for over 24 hr (Mackerras, 1965a).

Spermatophores

In all cockroach species the male transfers sperm to the female via a spermatophore; it begins forming in the male as soon as the mating pair is securely connected (Khalifa, 1950; van Wyk, 1952; Roth, 2003a). When it is complete, the spermatophore in *Blattella* descends the ejaculatory duct and is pressed by the male's endophallus against the female genital sclerites (Khalifa, 1950). In *Periplaneta* the spermatophore is not discharged until at least an hour from the beginning of copulation (Gupta, 1947). In *N. cinerea,* where copulation length is typically short, mating pairs detached after 10–12 min can be separated into three groups. In some, only a copious secretion is present; in others a spermatophore has been transferred but is not secured. A third group has a spermatophore firmly inserted (Roth, 1964b).

Three spermatophore layers can be distinguished in *Blattella:* a clear, transparent section covering the ventral surface, a lamellated portion that forms the dorsal wall, and at its core, suspended in a milky white mass, are two sacs containing the sperm (Khalifa, 1950). *Periplaneta*'s spermatophore has just one sperm sac (Jaiswal and Naidu, 1976). In *Blaberus craniifer* the spermatophore consists of four heterogeneous layers, and is invested with a variety of enzymes including proteases, esterases, lipases, and phosphatases (Perriere and Goudey-Perriere, 1988). Several mechanisms exist for fixing the spermatophore in the female (Graves, 1969): (1) the soft outer layer hardens against the female genital sclerites (Blattinae); (2) a thick, wax-like shell holds it in place (most Blattellidae); (3) a large quantity of glue-like secretion secures it (Blaberinae, one Zetoborinae); (4) a uniquely shaped, elongated spermatophore is enclosed in a large membranous bursa copulatrix in the female (Diplopterinae, Oxyhaloinae, Panchlorinae, Pycnoscelinae, one Zetoborinae).

When transferring the spermatophore, the male orients its tip so that the openings of the sperm sacs are aligned directly with the female spermathecal pores (Khalifa, 1950; Roth and Willis, 1954b; Gupta and Smith, 1969); this is apparently unusual among insects (Gillott, 2003). The sperm do not migrate from the spermatophore until copulation is terminated. When first transferred, the spermatophore of *N. cinerea* contains non-motile, twisted sperm; they became active about 2 hr later. Two to three days after mating only a few sperm remain in the spermatophore but the spermathecae are densely filled with them (Roth, 1964b; Vidlička and Huckova, 1993). If the spermatophore is removed 25 min after the male and female detach in *B. germanica,* "a thin thread of spermatozoa, hair-like in appearance, may extend from the female's spermathecal opening" (Roth and Willis, 1952a). It takes about 5 hr for sperm to migrate into the spermathecae of *D. punctata* (Roth and Stay, 1961). The stimulus for sperm activation may be in male accessory gland secretions transferred along with the sperm (Gillott, 2003), produced by the female in the spermathecae or spermathecal glands (Khalifa, 1950; Roth and Willis, 1954b), or both. Little is known regarding the mechanism by which sperm move from the spermatophore to the spermatheca. Among the nonexclusive hypotheses are the active motility of sperm, migration in chemotactic response to spermathecal or spermathecal gland secretions, contractions of visceral muscles associated with the female genital ducts, and aspiration by pumping movements of the musculature of the spermatheca (Gupta and Smith, 1969). Male accessory gland secretions may play a role in stimulating female muscle contraction (Davey, 1960). The activity and morphology of sperm may change once they reach the spermatheca. In *Periplaneta,* alterations were noted chiefly in the acrosome (Hughes and Davey, 1969).

Sperm Morphology

Cockroaches have extremely thin sperm, with long, actively motile flagellae (Baccetti, 1987). The sperm head and the tail are indistinguishable in some species, such as *B. germanica,* but can be distinct and variable among other examined cockroaches. The sperm head in *Arenivaga boliana,* for example, is helical, and that of *Su. longipalpa* is extremely elongated (Breland et al., 1968). Total sperm length varies considerably, with *B. germanica* and *P. americana* at the extremes of the range in 10 examined cockroach species (Breland et al., 1968). The limited data we have suggest that body size and sperm length may be negatively correlated (Table 6.1), but the relative influences of body size, cryptic choice mechanisms, and sperm competition have not been studied.

Dimorphic sperm have been described in *P. americana* (Richards, 1963). A small proportion are "giants," sperm that have big heads and tails that are similar in length but two or more times the diameter of typical sperm. These chunky little gametes swim at approximately the same speeds as the "normal," more streamlined, sperm, and are thought to be the result of multinucleate, diploid, or

Table 6.1. Sperm length relative to body length in cockroaches. Sperm data from Jamieson (1987) and Vidlička and Huckova (1993).

Species	Approximate[1] body length length (mm)	Sperm length (μ)	Ratio body length:sperm length
Blattella germanica	12.0	450	27:1
Pycnoscelus indicus	~ 21.0[2]	250	84:1
Nauphoeta cinerea	27.0	300	90:1
Periplaneta americana	37.5	85	441:1
Blaberus craniifer	55.0	180	306:1

[1]Body length can range fairly widely within a species, for example, male *B. germanica* ranges from 9.6 to 13.8 mm in length (Roth, 1985).
[2]Body length based on its sibling species, *Pyc. surinamensis.*

higher degrees of heteroploidy. Giants range from 0–30% of the total in testes; smears from either seminal vesicles or spermathecae of females, however, yield a much lower percentage, just 0–2%. Most never leave the male gonads, and it is unknown whether those that do are capable of effecting fertilization. Alternate sperm forms are fairly common among invertebrates, and in some cases are specialized for functions in addition to or instead of fertilization (Eberhard, 1996). These include acting as nuptial gifts, suppressing the female's propensity to remate, and creating a hostile environment for rival sperm (e.g., Buckland-Nicks, 1998). The topic is thoroughly discussed in Swallow and Wilkinson (2002).

Sperm Competition

When the probability of female remating is high, selection should favor adaptations in males that allow them to reduce or avoid competition with the sperm of another male. This can lead to rapid and divergent evolution of traits that function in sperm competition and its avoidance. These traits may be manifest in behavior (e.g., mate guarding), genital morphology (e.g., structures that deliver sperm closer to the spermatheca), and physiology (e.g., chemicals in the ejaculate that enhance the success of sperm). Selection may also act at the level of the sperm itself, in that some may be adapted to outcompete others for access to eggs (Ridley, 1988; Eberhard, 1996; Simmons, 2001).

In studies of sperm competition paternity is typically reported as P_2, the proportion of offspring sired by the last male to mate with a female in controlled double mating studies (Parker, 1970). A P_2 between 0.4 and 0.7 indicates sperm mixing. A P_2 higher than 0.8 suggests that sperm are either lost prior to the second mating, or that second-male sperm precedence or displacement is in operation. Values of < 0.4, where the first male is favored, are rare (Simmons, 2001).

Classical studies of sperm competition have been conducted in two cockroach species: *B. germanica* and *D. punctata.* Cochran (1979b) studied the phenomenon in the German cockroach and used the genetic mutant rose eye to recognize paternity. In the single instance of a female mating twice prior to the first egg case, the second male sired 95% of the eggs. Just over 20% of females remated between egg cases; Gwynne (1984), using Cochran's data, calculated the P_2 of these to be 0.43. Using a slightly different approach with the same data, Simmons (2001, Table 2.1) calculated the P_2 as 0.69 when mutant males were the first to mate and 0.33 when wild-type males were the first to mate. The P_2 calculated using mixed broods only was < 0.37 (Simmons, 2001, Table 2.3). *Blattella* is exceptional, then, in that the general

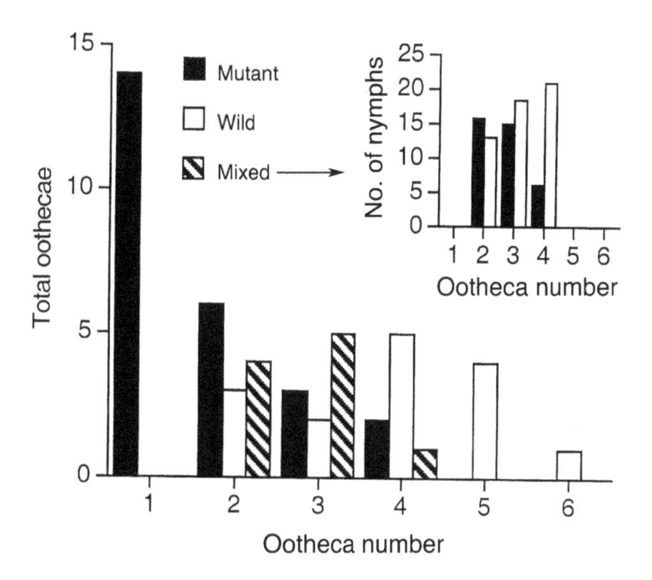

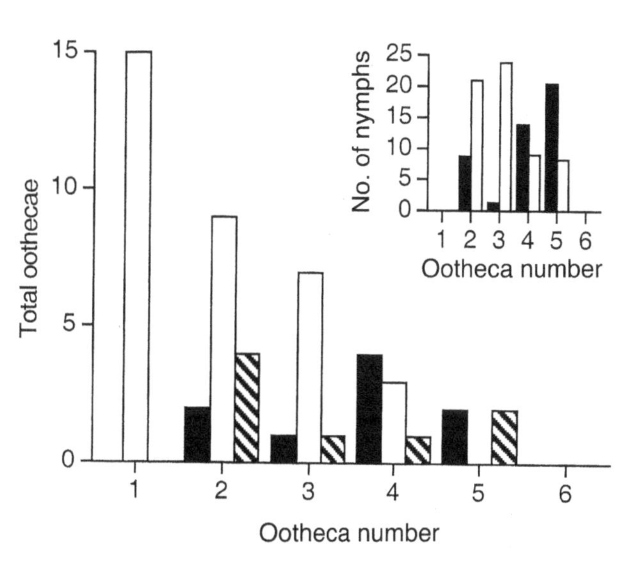

Fig. 6.6 Sperm competition in *Blattella germanica.* Virgin females with the recessive eye color mutation rose eye were initially mated to a mutant male, then to a wild-type male (top graphs), or first to a wild-type male, and subsequently to another mutant (bottom graphs). In each case the female was exposed to the second male only after her first egg case began protruding; progeny of the first egg case were thus sired exclusively by the initial male. Inset graphs detail the paternity of nymphs from oothecae of mixed parentage, that is, those containing eggs fertilized by both males. After data in Cochran (1979b), with permission of D.G. Cochran.

trend is first-male precedence. A focus on average P_2 values can be misleading, however, because variation within a species can be extreme (Lewis and Austad, 1990; Eberhard, 1996). A detailed examination of Cochran's data indicates that in most reproductive episodes, the eggs of some oothecae were exclusively fathered by the first male, some were exclusively fathered by the second male, and some were of mixed parentage (Fig. 6.6). In the waning stages of the female's reproductive life sperm from the second male sired a higher proportion of the offspring, suggesting that remating may occur in response to declining sperm supply (Cochran, 1979b). Maternal influence may account for some variation in paternity. Females have four spermathecae, each with a separate opening, and thus potential for selective sequestration and release of sperm (discussed below). It is noteworthy, based on the P_2 values cited above, that either the sperm of mutant males are somewhat inferior competitors, or that females exhibit some preference for the sperm of wild-type males.

Woodhead (1985) used irradiated males to examine sperm competition in *D. punctata,* a viviparous cockroach that remates only after partition of the first brood. The P_2 averaged 0.67 but was higher when the second male was the normal male (0.89), rather than the irradiated male (0.46). Plots of the position of viable versus sterile eggs in individual oothecae suggested sperm mixing; there was no consistent spatial pattern of egg fertilization by the two sires. The spermatheca in *Diploptera* females is tubular, a shape usually associated with sperm stratification (Walker, 1980).

Variation in Ejaculates

A number of studies indicate that males increase the size of their ejaculate in the presence of rival males (summarized in Wedell et al., 2002). Harris and Moore (2004) tested the idea in *N. cinerea* by exposing adult males during their post-emergence maturation period to the chemical presence of potential competitors (other males) or mates (females); spermatophore size, testes size, and sperm numbers were then determined and compared to isolated male controls. The authors could not demonstrate an influence of male competitors on testes size or sperm number. Spermatophore size increased in the presence of either sex, suggesting the possibility of a group effect on this reproductive character. Males did transfer significantly more sperm during copulation when, after adult emergence, they matured in the presence of females rather than males. One caution in interpreting this study is that the development of the testes and the production of sperm in *Nauphoeta* may be largely complete prior to adult emergence, as it is in *G. portentosa, Byrsotria fumigata* (Lusis et al., 1970), *Blatta orientalis* (Snodgrass, 1937), and *P. americana* (Jaiswal and Naidu, 1972).

Hunter and Birkhead (2002) addressed the relationship between sperm competition and sperm quality by comparing the viability of male gametes in species pairs with contrasting mating systems. They found a higher percentage of dead sperm in *N. cinerea,* which the authors considered monandrous, than in *D. punctata,* which they considered polyandrous. It is unclear, however, as to how much the mating systems in these two species differ. Female *Diploptera* typically mate just after adult emergence, then carry the spermatophore until shortly before the ootheca is formed. They readily remate after partition of the first brood (Stay and Roth, 1958; Woodhead, 1985). Similarly, virgin female *Nauphoeta* are unreceptive after their first copulation; after partition, they may or may not mate again (Roth, 1962). Females of both species, then, may be considered monandrous during their first reproductive period, but polyandrous over the course of their lifetime.

MALE INVESTMENT: TERGAL GLANDS

"Tergal gland" is a generalized term describing a great variety of functionally similar glandular structures that have evolved on the backs of males (Roth, 1969). Male tergal glands occur in almost all cockroach families, but are rare in Polyphagidae and Blaberidae. Within the latter, the glands are restricted to the Epilamprinae and Oxyhaloinae. The most complex and morphologically varied glands occur in male Blattellidae, but at least 73 blattellid genera have species that lack these specializations (Roth, 1969, 1971a; Brossut and Roth, 1977).

Males display their tergal glands to potential mates during the wing-raising (or in wingless species, "back-arching") phase of courtship. The female responds by approaching the male, climbing on his dorsum, and feeding on the gland secretion. The glands thus serve to maneuver the female into the proper pre-copulatory position and arrest her movement so that the male has an opportunity to clasp her genitalia (Roth, 1969; Brossut and Roth, 1977). The extraordinary morphological complexity of the glands in some taxa, however, suggests that they may serve additional roles in courtship and mating.

Morphology and Distribution

When present in the Blattidae, tergal glands almost always occur on the first abdominal tergite. In Blattellidae as many as five segments may be specialized, but most genera in this family have just one tergal gland, usually on segment 1, 2, or 7 (Roth, 1969; Brossut and Roth, 1977). There are many genera where males either have or lack tergal glands. Among species of *Parcoblatta,* for example, males may have glands on the first tergite only, on the first and second tergites, or they may be absent (Hebard, 1917). In Australian *Neotemnopteryx fulva,* the tergal gland on the seventh tergite ranges from a pair of dense tufts to a few, nearly invisible, scattered setae; Roth (1990b) illustrates four variations. Uniquely among cockroaches, the gland of *Metanocticola christmasensis* is on the metanotum (Roth, 1999b). The "best" positions are

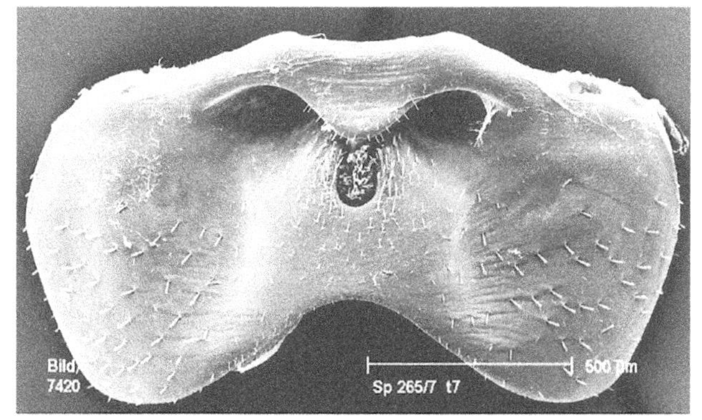

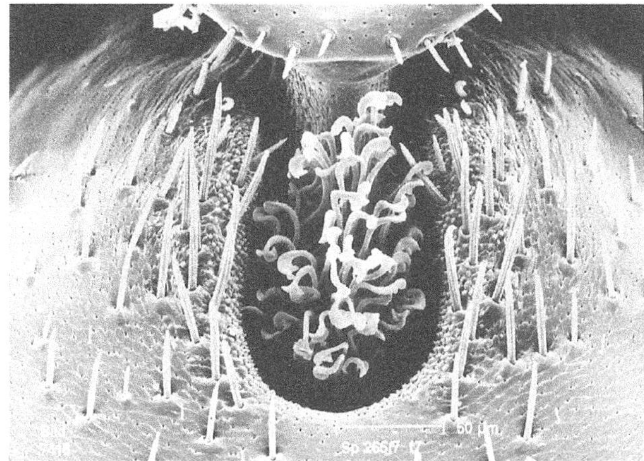

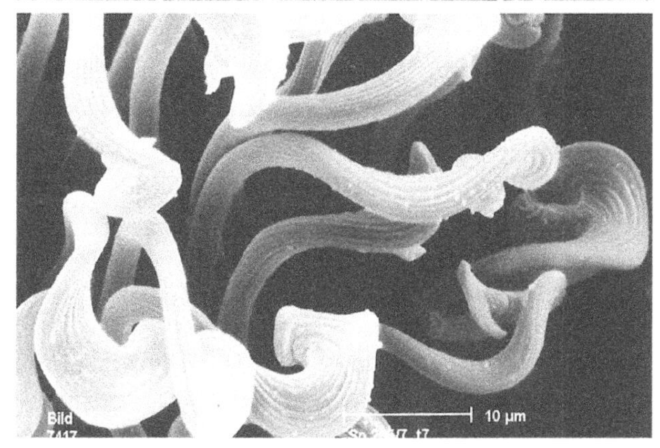

Fig. 6.7 Scanning electron micrographs of the tergal gland of male *Phyllodromica delospuertos* (Blattellidae), in increasing detail. *Top,* tergite 7, *middle,* tergal gland, *bottom,* bristles of the gland. From Bohn (1999), courtesy of Horst Bohn, with permission from the journal Spixiana.

considered to be the more anterior ones, because they draw the female forward, bringing her genitalia into closer alignment with those of the male (Roth, 1969). The Anaplectinae and Cryptocercidae have tergal modifications of unknown functional significance because they occur in unusual locations. In the former the tergal gland is on the supra-anal plate (Roth, 1969). In *C. punctulatus* the gland is located on the anterior part of the eighth tergite, completely concealed beneath the expanded seventh tergite (Farine et al., 1989). Because of its relatively inaccessible position, it is unlikely that it functions to elicit mounting by the female. Nonetheless, females of *C. punctulatus* have been observed straddling the male prior to assuming the opposed position (Nalepa, 1988a).

Because tergal glands are often markedly different among different genera and species, they can be useful characters in cockroach taxonomy (Brossut and Roth, 1977; Bohn, 1993). Morphologically they range from very elaborate cuticular modifications to the complete absence of visible structures. The glands may take the form of shallow or deep pockets containing knobs, hairs, or bristles (Fig. 6.7), fleshy protuberances, cuticular ridges, groups of agglutinated hairs, tufts or concentrations of setae, or just a few setae scattered on the tergal surface. In species with no externally visible specializations, internal cuticular reservoirs nonetheless may be present (Roth, 1969; Brossut and Roth, 1977). Sometimes secretory cells are merely distributed in the epithelium beneath the cuticle, opening to the exterior via individual pores, and the presence of pheromone-producing cells is inferred from female mounting and feeding behavior (e.g., *Blaberus, Archimandrita, Byrsotria*—Roth, 1969; Wendelken and

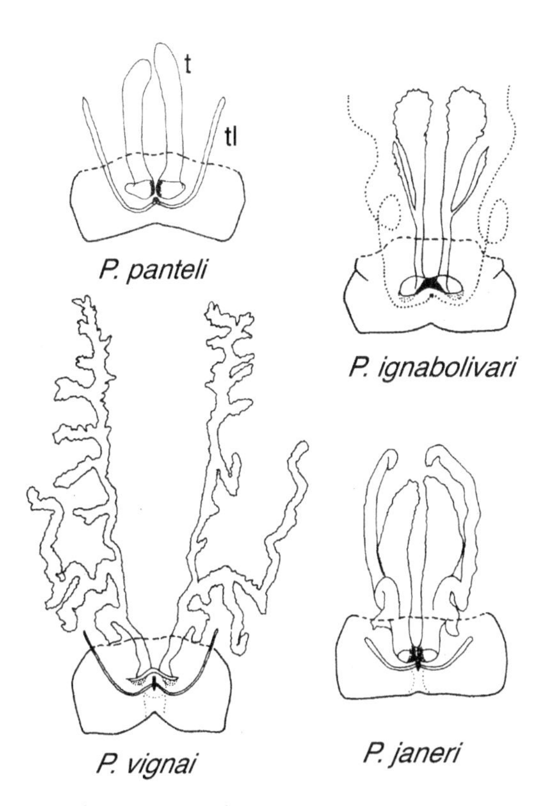

Fig. 6.8 Male tergite 7 of representative species of *Phyllodromica* (Blattellidae: Ectobiinae) showing two sets of tubular pouches underlying the tergal gland. The anterior pair of tubes ("t") are thick and sometimes branched; the posterior pair of tubules ("tl") are very thin and unbranched. The "tl" tubules of *Phy. ignabolivari* were lost during preparation and are indicated by dotted lines. From Bohn (1993), courtesy of Horst Bohn, and with permission from the Journal of Insect Systematics and Evolution (= Entomologica Scandinavica).

Barth, 1985). In some blattellids the internal glandular apparatus is enormous. *Blattella meridionalis* has glands that form elongate sacs extending well into the next abdominal segment (Roth, 1985). In the *panteli* group of *Phyllodromica* the internal reservoirs consist of two pairs of long tubular pouches (Fig. 6.8). The anterior pair is thick, branched in some species, and open to the exterior via an open bowl or pocket. The posterior pair of tubules is very thin and unbranched, with small openings that lie behind the larger openings of the anterior glands (Bohn, 1993).

Functional Significance

External pits, "bowls," or depressions function as reservoirs for the tergal secretion oozing up from underlying glandular cells (Roth, 1969; Brossut and Roth, 1977; Sreng, 1979b). In some instances, drops of liquid can be seen forming at the opening of the gland as the female feeds (e.g., *R. maderae*—Roth and Barth, 1967). The secretion produced by the tergal glands is a mixture of short-range volatile and non-volatile fractions, the latter including protein, lipids, and carbohydrates (Brossut et al., 1975; Korchi et al., 1999). The best-studied, that of *B. germanica,* is a complex synergistic mixture of polysaccharides, 17 amino acids, and lipids, including lecithin and cholesterol. Maltose, known from baiting studies to be a potent phagostimulant for the species, is one of the primary sugars (Kugimiya et al., 2003; Nojima et al., 1999a, 1999b). There is little relationship between response to the secretion and sexual receptivity. Both sexes and all stages are attracted (Nojima et al., 1999b). Because tergal secretions exploit a female's underlying motivation to feed, they can be classified as "sensory traps" (Eberhard, 1996). They mimic stimuli that females have evolved, under natural selection, for use in other contexts.

It is uncertain to what degree tergal secretions provide a nutritional boost to grazing females. The behavior is most often described as "licking" or "palpating," but the action of the female's mandibles and the manner in which she presses her mouthparts against the male's gland indicate that she actually eats the secretion. The male typically lets her feed 3–7 sec before attempting to make genitalic connection (Roth and Willis, 1952a; Barth, 1964; Roth, 1969). Females of *Eurycotis floridana* may graze for nearly a minute, longer than any other studied species (Barth, 1968b). Feeding may also be "quite prolonged" in *Periplaneta* spp., with the female vigorously biting the tergite. The male gland in *Rhyparobia maderae* can be extensively scarred (Simon and Barth, 1977b), attesting to female enthusiasm for the fare. Roth (1967c) suggested that in species with very deep, well-developed tergal glands located near the base of the male's wings, females may feed on tergal secretions during the entire period of copulation, that is, they may not rotate off the male's back into the opposed position. The extent to which tergal glands provide females with a significant source of nourishment is in need of examination, particularly in species with large glandular reservoirs. In many insects with courtship feeding the food gift provides no significant nutritional benefit to the female (Vahed, 1998). The amount of secretion ingested by *B. germanica* does seem negligible. On the other hand a female may feed on the tergal secretion of the male 20 times in a half hour without resultant copulation (Table 6.2), and courtship activities can deplete the gland (Kugimiya et al., 2003).

Blattella germanica is a good example of the concept that in species utilizing sensory traps, males are selected to exaggerate the attractiveness of the signal while minimizing its cost (Christy, 1995). The German cockroach has double pouches on the seventh and eighth tergites, with the ducts of underlying secretory cells leading to the lumen of the pouch (Roth, 1969). During courtship, the female feeds on the secretions in the cavities on the eighth tergite. After 2–5 sec, the male slightly extends his abdomen, causing the female to switch her feeding activities to the gland on the seventh tergite, triggering genitalic extension on the part of the male. The female can contact the tergal secretions with her palps, but the cuticular openings of the glands are too small to permit entry of the mandibles and allow a good bite. She plugs her paraglossae into the cavities and ingests the tiny amount of glandular material that sticks to them. The forced lingering as she repeatedly tries to access the secretions keeps her positioned long enough for a copulatory attempt on the part of the male (Nojima et al., 1999b). The tergal glands in *B. germanica* are akin to cookie jars that allow for the insertion of your fingers but not the entire hand. The design encourages continued female presence, but frugally dispenses what is presumably a costly male investment. Males of other species may take a more direct approach to "encouraging" females to maintain their position. In a number of *Ischnoptera* spp., the tergal gland is flanked by a pair of large, heavily sclerotized claws, each of which has four stout, articulated setae forming the "fingers." When the female is feeding on the tergal gland she must place her head between these claws "and probably applies pressure to the articulated setae" (Roth, 1969, Figs. 47–53; Brossut and Roth, 1977, Figs. 18–19). These structures, however, are quite formidable for simple mechanoreceptors, and may function in restraining the female rather than for just signaling her presence.

Because tergal secretions are sampled by the female prior to accepting a male or his sperm, they may provide a basis for evaluating his genetic quality, physiological condition (Kugimiya et al., 2003), or in some species, his

Table 6.2. Summary of sexual behavior of 10 pairs of *Blattella germanica* observed for 30 min; from Roth and Willis (1952a), LMR's first published study on cockroaches.

Behavior of cockroaches	Pair number									
	1	2	3	4	5	6	7	8	9	10
Number of times male courted female[1]	20	44	4	14	27	17	37	48	33	17
Number of times female fed on tergal gland[2]	10	19	1	0	2	9	10	9	20	3
Time (sec) male in courtship position	679	1385	59	169	576	698	997	1106	916	576
Copulation successful?	+	—	—	—	—	—	—	—	—	—

[1]Courting defined as the male elevating and holding his wings and tegmina at a 45 to 90 degree angle.
[2]This figure also indicates the number of times the male tried to engage the female's genitalia, which almost invariably occurs after she has fed on the tergal gland for several seconds.

ability to provide a hearty postnuptial gift in the form of uric acid (discussed below). Oligosaccharides in the tergal secretion of *B. germanica* do vary individually and daily (unpublished data in Kugimiya et al., 2003). Perhaps repeated tasting by the female (Table 6.2), then, is an evaluation process. Alternatively, females may need to exceed a certain threshold of contact with or ingestion of the tergal secretion before accepting genitalic engagement (Gorton et al., 1983). Finally, she may simply be trying to maximize her nutritional intake. Repeated instances of a female applying her mouthparts to a male tergal gland but leaving without copulation is particularly prevalent in starved females (Roth, 1964b). Nojima et al. (1999a) suggested that tergal secretions may be indirect nutritional investment in progeny, but the nutritional value to the female and her offspring remains to be demonstrated.

MALE INVESTMENT: URIC ACID

Roth and Willis (1952a) were the first to note that in *B. germanica*, a chalk-white secretion composed of uric acid oozes from male uricose glands (utriculi majores) and covers the spermatophore just before copulating pairs separate (Fig. 6.9). Subsequent surveys made evident that uricose glands are unique to a relatively small subset of Blattaria. Within the Blaberoidea, the glands are common in the Pseudophyllodromiinae, less frequent in the Blattellinae, and in the Blaberidae occur only in some Epilamprinae. They are absent in Blattoidea (Roth and Dateo, 1965; Roth, 1967c).

Several hypotheses addressing the functional significance of uric acid expulsion via uricose glands have been offered. Because uric acid is the characteristic end product of nitrogen metabolism in terrestrial insects (Cochran, 1985), initially it was thought that mating served as an accessory means of excretion in these species (Roth and Dateo, 1964). The glands of males denied mating partners become tremendously swollen with uric acid (Roth and Willis, 1952a), like cows that need milking. These excessive accumulations can result in increased male mortality (Haydak, 1953; Roth and Dateo, 1965). Field observations of cockroaches seeking out and ingesting uric acid from bird and reptile droppings (Fig. 5.2), however, weaken the excretion hypothesis. It would also be unusual for males of a species to have a waste elim-

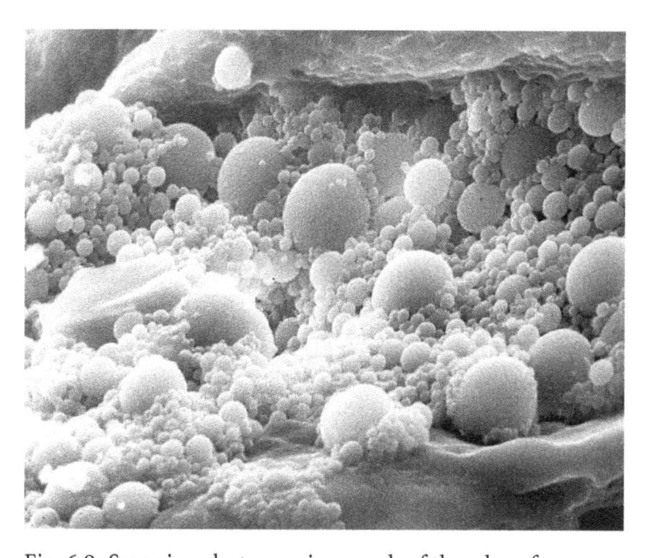

Fig. 6.9 Scanning electron micrograph of the edge of an emptied spermatophore with adhering spherical urate granules of varying diameter (*Blattella germanica*). From Mullins and Keil (1980), courtesy of Donald Mullins, and with copyright permission from the journal Nature (www.nature.com/).

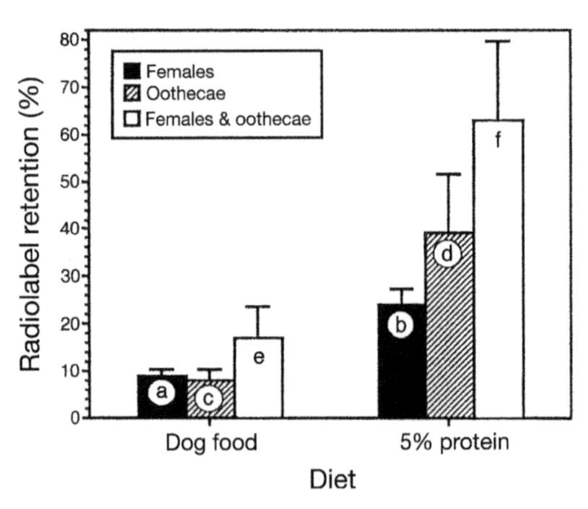

Fig. 6.10 Comparison of total radiolabel content of *Blattella germanica* females and the oothecae they produced while feeding on either a dog food (25% crude protein) or a 5% protein diet; these females were mated to virgin males that had been simultaneously injected with ^{3}H leucine (a representative amino acid) and ^{14}C hypoxanthine (a purine converted to uric acid in vivo). Dog food fed-females and their oothecae contained 17% of the male contributed radiolabel. Those on the low-protein diet contained 63% of the radiolabel made available to them at mating. Values are mean ± SEM. a vs. b, $p = 0.005$; c vs. d, $p = 0.027$; e vs. f, $p = 0.007$ (Student's t-test). From Mullins et al. (1992), courtesy of Donald Mullins and with permission from the Journal of Experimental Biology.

ination system unavailable to females and juveniles. From the female perspective, it was suggested that a spermatophore slathered with an excretory product would be an unattractive meal, and prevent her from consuming it before the sperm moved into storage (Roth, 1967, 1970a). An alternative suggestion was that the uric acid may function as a mating plug that deters additional inseminations (Cornwell, 1968). In species such as *Miriamrothschildia* (= *Onychostylus*) *notulatus*, *Lophoblatta* sp., *Cariblatta minima*, *Amazonina* sp., and *Dendroblatta sobrina*, so much uric acid is applied by males that the female genital segments gape open (Roth, 1967c).

The most strongly supported hypothesis is that the uric acid transferred during mating acts as a nuptial gift. In *B. germanica*, radiolabeled uric acid can be traced from the male to the female, and subsequently to her oocytes; the transfer occurs more readily when the female is maintained on a low-nitrogen diet (Fig. 6.10). The urates are probably ingested by the female, along with the spermatophore, but it is possible that a small fraction may enter via her genital tract (Mullins and Keil, 1980). An analogous urate transfer and incorporation occurs in *X. hamata*. In this case, the female turns, post-copulation, and feeds on a urate-containing slurry produced and offered by the male (Schal and Bell, 1982). After copulation the male raises his wings, telescopes his abdomen, widens the genital chamber, exposes a white urate secretion, and directs it toward the female, who ingests it. Females feed for about 3.5 min. As in *B. germanica*, females on nitrogen-deficient diets transfer to their maturing oocytes more male-derived uric acid than do females on high-protein diets. The magnitude of the gift offered by males of these two species depends on a combination of male age, size, diet, and frequency of mating. The uricose glands of newly emerged male *B. germanica* contain little or no secretion; they become filled in one or two days (Roth and Dateo, 1964). The glands are nearly emptied at each copulation (Roth and Willis, 1952a).

Male to female transfer of uric acid probably occurs in all cockroach species that possess male uricose glands. A recently mated female *Blattella humbertiana* was observed removing excess uric acid with her hind legs, then eating some of the material before it hardened (Graves, pers, comm. to LMR in Roth, 1967c). In three species of *Latiblattella* the male's genitalia and posterior abdominal segments are covered with "chalky white secretion" after mating, and females of *Lat. angustifrons* have been observed applying their mouthparts to it after mating (Willis, 1970).

Paternal Investment or Mating Effort?

A nuptial gift can benefit a male in two ways. The gift can function as paternal investment, where transferred nutrients or defensive compounds increase the number or quality of resultant offspring, or it can function as mating effort, which increases the male's fertilization success with respect to other males that mate with the same female (Eberhard, 1996). The hypotheses are not mutually exclusive, and there is debate on the distinction between them, centering mainly on the degree to which a donating male has genetic representation in the offspring that benefit from the gift. The latter is dependent on female sperm-use patterns, the length of her non-receptive period following mating, and the time delay until the female lays the eggs that profit from the male's nutritional contribution (reviewed by Vahed, 1998).

The incorporation of male-derived urates into oothecae of *B. germanica* suggests paternal investment, supported by three lines of evidence (Mullins et al., 1992). First, urate levels in eggs steadily decrease during development. This strongly suggests that the uric acid is metabolized during embryogenesis (Mullins and Keil, 1980), presumably via bacteroids transmitted transovarially by the female (e.g., Sacchi et al., 1998b, 2000). Second, ^{14}C radioactivity not attributable to ^{14}C urate is present in tissue extracts of oothecae (Mullins and Keil, 1980; Cochran and Mullins, 1982; Mullins et al., 1992). As pointed out by Mullins and Keil (1980), however, the ^{14}C radiolabel reflects pathways involving carbon atoms and not necessarily the path of nitrogen contained in urates. In subsequent work, however, Mullins et al. (1992) demonstrated that, third, ^{15}N from uric acid fed to females did find its way into the nitrogen pool of oothecae, and was incorporated into four different amino acids. The question nonetheless remains as to whether the uric acid derived from a particular male ends up in the offspring that he sires (Vahed, 1998). Female *B. germanica* expel the empty spermatophore with the adhering urates about 24 hr after mating, then consume them between 4 and 18 days later, depending on her nutritional status (Mullins and Keil, 1980). Females typically transfer 90% of the food reserves accumulated during the pre-oviposition period into the next ootheca (Kunkel, 1966). It seems reasonable to assume, then, that the majority of the uric acid transferred during a given copulation is incorporated into the eggs of the next reproductive bout, particularly in unsated females. Young females rarely mate more than once prior to their first ootheca (Cochran, 1979b), so during the first oviposition period a male can be reasonably certain that his nuptial gift will benefit his own offspring. Females may, however, mate between ovipositions. Paternity of subsequent oothecae is variable, but there is a tendency for first-male precedence (Fig. 6.6). The nuptial gifts of male consorts following the first male, then, may benefit some nymphs fathered by other males.

Gwynne (1984) argued that uric acid donation should not be classified as paternal investment, because, as a waste product, uric acid is likely to be low in cost. Vahed (1998) countered that it is likely to be a true parental investment precisely because of the low cost. If a gift is cheap, just a small resultant benefit to offspring will maintain selection for the investment. Neither author appreciated the fact that males deplete their uricose glands with each copulation, and actively forage for uric acid by seeking it out in bird and reptile droppings. The degree to which this foraging activity entails a cost in predation risk and energetic expense is an additional consideration.

Although a demonstration that male-derived nutrients are incorporated into eggs supports the paternal investment hypothesis, it does not necessarily rule out the mating effort hypothesis (Vahed, 1998). Because female cockroaches feed on male-provided urates after spermatophore transfer, the nuptial gift cannot influence overt mate choice. The possibility remains that after copulation, females may bias sperm use based on the size or quality of the urate gift. In many species, females preferentially use the sperm of males that provide the largest nuptial gifts (reviewed by Sakaluk, 2000). With four separate chambers for sperm storage (discussed below), female *B. germanica* certainly have potential for exercising choice. The existence of substantial variation in sperm precedence suggests that she may be doing so.

MALE GENITALIA

The genitalia of most male cockroaches are ornate, strongly asymmetrical, and differ, at times dramatically, among species. Because they are among the primary characters used in cockroach taxonomy, some beautifully detailed drawings are available, but we have little understanding as to the functional significance of most components. The genital sclerites are usually divided into the left, right, and median (also called ventral) phallomeres. These can be relatively simple and widely separated, or form groups of convoluted, well-muscled structures elaborately subdivided into movable rods, hooks, knobs, spines, lobes, brushes, flagellae, and other sclerotized processes (Fig. 6.11).

Several male genital sclerites are associated with the process of intromission and insemination; these include "tools" for holding the female, positioning her, and orienting her genitalia to best achieve spermatophore transfer. In *Blatta orientalis,* for example, all five lobes of the left phallomere, together with the ventral phallomere, serve to stabilize the ovipositor valves of the female, while a sclerite of the right phallomere spreads the valves from the center so that the spermatophore can be inserted (Bao and Robinson, 1990). Nonetheless, phallomeres are nearly absent in some blaberids, suggesting that elaborate hardware is not always a requisite for successful copulation.

Mate-Holding Devices

Some male genital structures function as mate-holding devices, allowing him to stay physically attached to the female during copulation. If the female mounts the male prior to genitalic connection (type I mating behavior), the male has a greatly extensible, sclerotized hook ("titillator"), used to seize and pull down her crescentic sclerite and to maintain his grasp on her when she rotates off his back into the opposed position. After the pair is end to end the male inserts the genital phallomeres. In *B. germanica* a pair of lateral sclerites, the paraprocts, grip the ovipositor valves from each side, and parts of the right phallomere (cleft sclerite) hold the valves from the ventral side (Fig. 6.12). The location of the genital hook in cockroach males varies, and distinguishes the Pseudophyllodromiinae (hook on right—Fig. 6.11A) from the Blattellinae (hook on left—Fig. 6.11C). The hook is always on the right in the Blaberidae (Fig. 6.11D) (Roth, 2003c).

Besides maintaining his grasp during positional changes, there are two basic reasons why a male needs a secure connection to the female during copulation: male competition and female mobility. In several species of

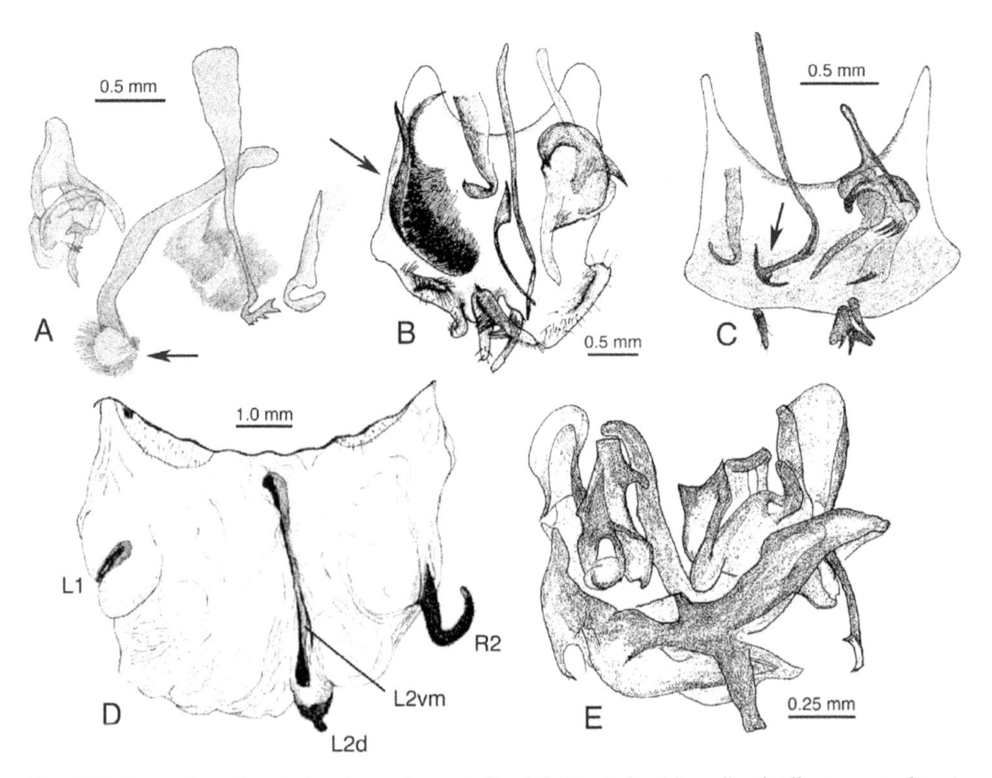

Fig. 6.11 Examples of variation in male genitalia. (A) Genitalia (dorsal) of *Allacta australiensis* (Blattellidae: Pseudophyllodromiinae). Accessory median phallomere is broad, with an apical brush-like modification (arrow). From Roth (1991d). (B) Subgenital plate and genitalia (dorsal) of *Hemithyrsocera nathani* (Blattellidae: Blattellinae). A huge, sclerotized, densely setose brush-like structure is found on the left side (arrow). From Roth (1995a). (C) Subgenital plate and genitalia (dorsal) of *Parasigmoidella atypicalis* (Blattellidae: Blattellinae). Note distally curved median phallomere with a pick-axe-like apex (arrow) and three-fingered "claw" on *right.* From Roth (1999b). (D) Highly reduced phallomeres on the extruded aedeagal membrane of *Panesthia cribrata* (Blaberidae: Panesthiine). From Walker and Rose (1998). Phallomeres are labeled according to McKittrick's (1964) classification. (E) Extraordinarily complex genitalia (dorsal) of *Homopteroidea nigra* (Polyphagidae). From Roth (1995d).

blaberids, rivals disturb or attack courting or mating males. Copulations may be broken off because of interference in *N. cinerea* (Ewing, 1972). In *B. craniifer* males assault copulating pairs by jumping on their backs and attacking their point of juncture. The interference may cause separation of the pair, but only if it occurs during the first few seconds after they assume the opposed position. The copulating male "shows no reluctance in fighting with the intruder," and "the trio may careen about the mating chamber" (Wendelken and Barth, 1985, 1987). A tight grasp of the female is also required because the pair may travel during copulation. Pairs are usually quiescent unless disturbed, in which case they move away. It is invariably the female that is responsible for the locomotion, dragging the passive male along in her wake (Roth and Barth, 1967). She can move with astonishing speed, pulling the "furiously backpedaling" male behind her (Simon and Barth, 1977a). *Blattella germanica* (Roth and Willis, 1952a), *Byr. fumigata* (Barth, 1964), *Ell. humerale* (= *affine*) (Pope, 1953), *Latiblattella* spp. (Willis, 1970), *Parcoblatta fulvescens* (Wendelken and Barth, 1971), and *P. americana* (Simon and Barth, 1977a) are among the species in which this behavior has been reported. It also occurs in *G. portentosa,* even though the male is much heavier than the female (Barth, 1968c).

Intromittent Organs

The need for a secure connection, then, may account for some of the claspers, hooks, and spines in the male's genitalic assemblage but cannot explain the bewildering complexity (Fig. 6.11E) of many components. The similarity of some cockroach structures to those of other, better-studied insects, however, allows us in some cases to make inferences from genitalic design. In particular, brushes and slender, elongate spines, rods, and flagellae, especially those with modified tips, may be sexually selected structures that increase a copulating male's fertilization success. This may be accomplished in one of three basic ways: via the manipulation of rival sperm, by the circumvention of female control of sperm use, or via internal courtship of the female (Eberhard, 1985; Simmons, 2001).

A number of intromittent structures in male cockroaches have been called a penis, pseudopenis, phallus, or pseudophallus. Although these structures may be associated with the ejaculatory duct or have the appearance of organs specialized for penetration, sperm is transferred indirectly in cockroaches, via a spermatophore. Penis-like organs therefore function in some capacity other than to convey sperm directly from the testes of the male to the sperm storage organs of the female. In *P. americana* the pseudopenis, a structure of the left phallomere, is characterized as having a blunt, hammer-like tip and a thin dark ridge along its length (Bodenstein, 1953). According to Gupta (1947) the expanded tip of the pseudopenis enters the female gonopore (entry to the common oviduct) during copulation, and rotates 90 degrees on its own axis. In some Blattellidae (including *Blattella*) a conical membranous lobe between the right and left phallomeres is considered a penis. It is a posterior continuation of the ejaculatory duct and projects into the female genital chamber during copulation. A free spine, or virga, extends through the membranous wall of the penis above the gonopore. Snodgrass (1937) noted that males insert the virga into the female's spermathecal groove during copulation, and suggested that it functioned to guide the sperm of the copulating male to their storage destination. Because sperm remain in the spermatophore until after the pair disengages, however, the functional basis of the virga must be sought elsewhere. In Pseudophyllodromiinae, R3, a sclerite of the right phallomere, has an expanded anterior edge that is elongate, in some genera extraordinarily so. Most often it is curved and flat, but in *Supella* it is

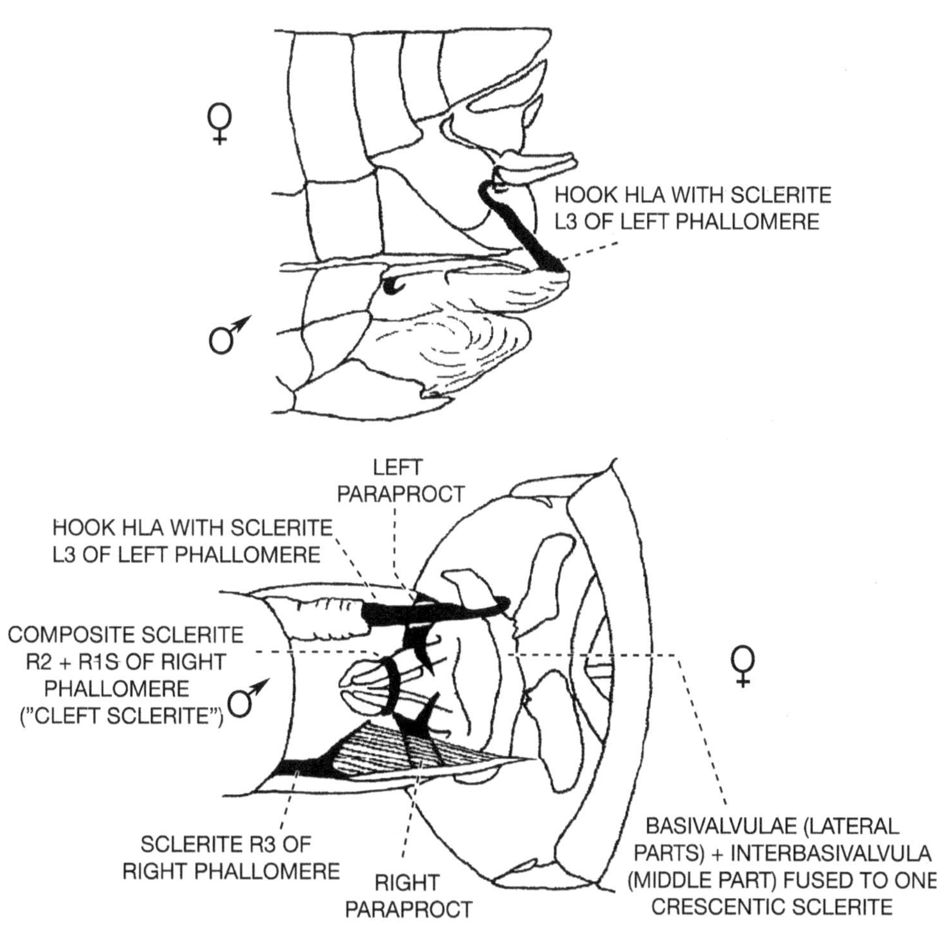

Fig. 6.12 Diagrammatic representation of the external genitalia of *Blattella germanica* during copulation. (A) Side view of the initial position, female superior. The hooked left phallomere is extended and inserted into the genital chamber of the female. (B) The insects in the end-to-end position, ventral view. The paraprocts are holding the ovipositor from each side and the cleft sclerite is holding it from the ventral side. The last sternite in both insects and the endophallus have been removed. After Khalifa (1950), with permission from the Royal Entomological Society. Labels of the various structures courtesy of K.-D. Klass.

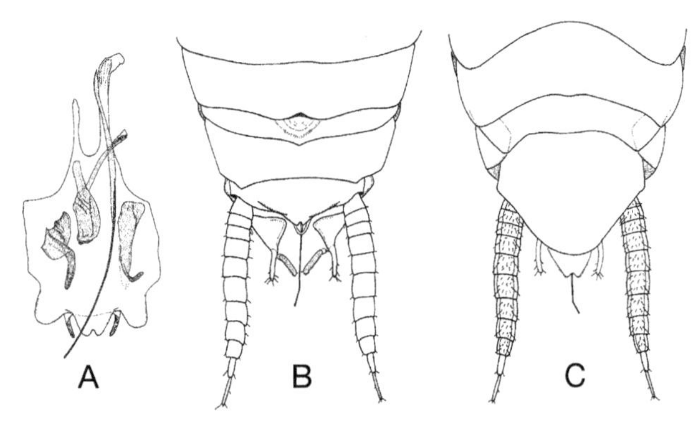

Fig. 6.13 Male *Chorisoserrata jendeki* (Pseudophyllodromiinae) (A) genitalia, (B) dorsal view of abdomen, and (C) ventral view of abdomen, demonstrating the genitalic filament, or whip, that projects from the abdomen. From Vidlička (2002), courtesy of the author and with permission from the journal Entomological Problems.

flat and horseshoe shaped, and in *Lophoblatta* it forms a long whip-like structure (Fig. 113 in McKittrick, 1964). Male *Chorisoserrata jendecki* have a genitalic filament that dangles from the abdomen, like a tail (Fig. 6.13) (Vidlička, 2002). *Nahublattella,* in the same subfamily, has a long whip as part of the left phallomere complex (Klass, 1997). *Loboptera* (Bohn, 1991a), *Neoloboptera,* and *Nondewittea* (Roth, 1989b) (Blattellinae) have elongated filaments associated with the median phallomere complex. In males of the tortoise beetle *Chelymorpha alternans,* whips similar to these are threaded up the female's spermathecal duct during the early stages of copulation, and the length of the whip is related to the probability of fathering offspring (Rodriguez et al., 2004).

Sperm Removal

At the conclusion of a successful copulation in cockroaches the transferred sperm are housed within a spermatophore in the female genital tract. The male is long gone before his gametes move to the female spermathecae, and is likely to have little direct influence on where, how, and if his sperm are stored. If his female consort is not a virgin, however, there is potential for a copulating male to increase his fertilization success by using genital appendages to move or remove the stored sperm of a rival. Male intromittent organs are known to extract stored sperm in one of three basic ways (Eberhard, 1996; Miller, 1990). First, a genital structure may be inserted into or near a spermatheca and the ejaculate issued with enough force to flush out a rival's sperm. This mechanism is unlikely in cockroaches since sperm transfer is indirect, via a spermatophore. Second, male genital appendages may be used to induce the female to discard the sperm of other males. When a female cockroach oviposits, eggs emerging from the oviduct pass over sensory hairs that trigger a contraction in the peripheral muscle layer of the spermathecal bulb and sperm are discharged to fertilize the egg (Roth and Willis, 1954b; Lawson and Thompson, 1970). Copulating males may take advantage of this reflex by using genital armature to tickle the mechanoreceptors, causing the female to expel the sperm of rivals before the male deposits his own. Third, the male may directly remove rival sperm using backward-facing hooks, spines, barbs, or brushes at the tip of elongate appendages (e.g., Yokoi, 1990; Kamimura, 2000). These structures enter the spermatheca, then scrape out, scoop out, or snag and drag the sperm present. This is possible in cockroaches, as in several species genital sclerites have the appearance of organs used for sperm removal or displacement in other insect species; these include brushes (Fig. 6.11A) and hooks (Fig. 6.11C) at the tip of intromittent-type organs.

Copulatory Courtship

If a female cockroach mates with more than one male during her reproductive lifetime, the manner in which she subsequently handles the sperm received from each partner plays a key role in determining the paternity of her offspring. After a copulation is terminated and the male leaves, the fate of his gametes is primarily under female control as they move from the spermatophore to the spermatheca(e), while they are being stored, while traveling from the spermatheca to egg, and at the site of fertilization (Eberhard, 1994, 1996). Female control of sperm use and the resultant potential to bias paternity is called cryptic mate choice, so named because it occurs within the recesses of the female body and is difficult to observe or investigate directly (Thornhill, 1983).

If female post-copulatory sperm-use decisions are cued on particular types of stimuli, it will favor the male to elaborate structures and behaviors that produce those stimuli (Eberhard, 1985, 1994, 1996, 2001). Complex genital sclerites, then, may function to increase a male's fertilization success indirectly, via internal courtship of the female. Internal thrusting is known to have a stimulatory function in copulating animals (Eberhard, 1996), and has been noted in a few cockroach species; however, the behavior also may be associated with the deep insertion of the genitalia, the transfer of the spermatophore, or the direct removal of rival sperm. Males of *B. fumigata* often make rhythmic pumping motions during the first few moments of copulation (Barth, 1964). Likewise, abdominal contractions of male *N. cinerea* occur throughout copulation but are most frequent in the initial stages (Vidlička and Huckova, 1993). Late in copulation the male of *Eub. posticus* "raises up on his forelegs and makes rhythmic pushing movements of his abdomen in a pulsating fashion" (Wendelken and Barth, 1987). *Diploptera punctata* males move their abdomen from side to side just prior to releasing the female (Roth and Stay, 1961). Conversely, females of *Parc. fulvescens* assume an arched posture during copulation, and rhythmical movements were observed for which the female appeared responsible (Wendelken and Barth, 1971). In addition to internally stimulating the female with genital structures, males may sing, tap, rub, hit, kick, wave, lick, wet with secretions, bite, feed, rock, and shake females in attempting to influence cryptic choice decisions (Eberhard, 1996). The production of oral liquid during mating by male *Parc. fulvescens* was listed by Eberhard (1991) as a form of copulatory courtship. A repeating sequence of pronotal butting, abdominal wagging, and circling behavior has been observed in *C. punctulatus* after genital disengagement (Nalepa, 1988a) and has been interpreted by Eberhard (1991) as post-copulatory courtship.

Reduction and Loss of Genitalic Structures

The genital phallomeres of some blaberid cockroaches are lightly sclerotized, considerably reduced, or in some cases, altogether absent. The Panchlorinae are characterized by the absence of a genital hook, and if the remaining two phallomeres are present, they are markedly reduced (Roth, 1971b). Likewise, one or more phallomeres may be reduced or absent in many Panesthiinae (including Geoscapheini) (Fig. 6.11D) (Roth, 1977). *Macropanesthia rhinoceros* and *M. heppleorum* males completely lack a genital hook, and sclerites L1 and L2d are also missing. Some of the Australian soil-burrowing cockroaches exhibit intraspecific variation in the reduction of phallomeres (Walker and Rose, 1998). The occurrence of poorly developed male genitalia in cockroaches corresponds very well with copulatory behavior. A reduced or absent genital hook is strong evidence of type III mating behavior, that is, the male backs into the female to initiate mating (Roth, 1971b, 1977).

Simple genital structure in males is predicted by the cryptic choice hypothesis if females are monandrous, because sexual selection by female choice is possible only if females make genitalic contact with more than one male (Eberhard, 1985, 1996). In monandrous females, the choice of sire is settled prior to copulation, via mechanisms such as premating courtship or male-male contests. The mating strategy in cockroaches with reduced genitalia is not known well enough to determine if that is the case here; however, one male is usually present in social groups of *Panesthia*. *Panesthia cribrata* typically lives in aggregations, often (29%) comprised of a single adult male, a number of adult females, and nymphs of various sizes (Rugg and Rose, 1984a).

Additional correlates of reduced male genitalia in cockroaches also must be considered. Among the Panesthiinae species studied, the absence of an oothecal covering around the eggs is correlated with the absence or reduction of male genital structures (Walker and Rose, 1998). All of the species for which we have information also exhibit a burrowing lifestyle, tunneling in soil, rotted wood, or rotted palms. How all these threads connect (burrowing lifestyle, mating system, copulatory behavior, male genital morphology, and absence of egg case) awaits further study. It is of interest (Chapter 9), however, that termites are monogamous (Nalepa and Jones, 1991) and that isopteran males are largely unencumbered by genitalia (Roonwal, 1970). Termites also live in burrows, mate by backing into each other, and except for *Mastotermes,* have lost the casing around their eggs. Species in the Cryptocercidae, the sister group of termites, live in burrows and are apparently monandrous, but male genitalia are not markedly reduced; they do, however, exhibit a number of paedomorphic characters (Klass, 1997).

THE FEMALE PERSPECTIVE

A variety of female traits can bias paternity, including the premature interruption of copulation and the acceptance or rejection of matings from additional males. Females may also accept a male for copulation but reject him as a father. This is possible because insemination and fertilization are uncoupled in space and time (Eberhard, 1985), and because females have many opportunities to modify the probability that a given copulation will result in egg fertilization. There are at least 20 different mechanisms that can result in cryptic female choice (Eberhard, 1994, 1996), many of which may apply to cockroaches. These include sperm transport to storage sites, sperm nourishment during storage, the ability to discharge or digest stored sperm, and the biased use of stored sperm to effect fertilization, particularly in females with multiple spermathecae. Sperm selection may even occur at the site of fertilization; Eberhard (1996) gives as an example *Periplaneta,* which has up to 100 micropyles for sperm entry at one end of the egg (Davey, 1965). After fertilization ovoviviparous females may abort the egg case. The multiplicity of female mechanisms reduces the likelihood that males will be able to evolve overall control of female reproductive processes, even if males try to prevent further matings via genital plugs, mate guarding, or induced unreceptivity (Eberhard, 1996). While there are no available studies that directly address cryptic choice in female cockroaches, we do have anatomical data from the taxonomic literature from which we can make some inferences. Here we summarize some of the relevant information in the hope that it may serve as a springboard for future investigation.

Female Receptivity

Female cockroaches have strong control of the courtship and mating process; there are several points in the behavioral sequence when she can terminate the transaction. In those cockroach species where females produce volatile pheromones, she may not call; if males produce the pheromones, she may not respond. Females may refuse to mount and feed on the tergites of a displaying male, but if she does, she may not allow genitalic engagement. If she does allow genitalic engagement, she may terminate copulation prematurely. A female's attractiveness to potential mates and her response to sexual overtures from them may or may not be congruent (Brousse-Gaury, 1977). Males of *Su. longipalpa,* for example, begin courting females 8 or 9 days after the female's imaginal molt. Females of this age do not respond to male sexual displays nor do they mate. Female calling and sexual receptivity are initiated 11 to 15 days after adult emergence. A lack of calling behavior in mature females, however, does not necessarily mean that they are unreceptive; 8% will mate if courted (Hales and Breed, 1983).

Response to Courtship

In most species newly emerged females require a period of maturation before they will accept mates. Virgin females of *N. cinerea, R. maderae,* and *Byr. fumigata* become receptive at an average of 4, 9, and 15 days, respectively (Roth and Barth, 1964). *Eublaberus posticus* females mate just after emergence, after their wings have expanded but before the cuticle has hardened (Roth, 1968c). *Jagrehnia madecassa* (Sreng, 1993), *Neostylopyga rhombifolia* (Roth and Willis, 1956), and *D. punctata* (Roth and Willis, 1955a) females are receptive when they are freshly emerged, pale, and teneral. The latter have a narrow window of opportunity for copulation; most that are isolated for several days following emergence do not mate when they are eventually exposed to males (Stay and Roth, 1958). In *N. cinerea,* younger females require longer periods of courtship prior to copulation than do older ones (Moore and Moore, 2001).

Females display their lack of receptivity to courting males in a variety of ways. A *Parc. fulvescens* female uninterested in mating decamps immediately upon contacting the male (Wendelken and Barth, 1971). Unreceptive blaberid females commonly flatten themselves against the substratum with their antennae tucked under their body (e.g., *Byr. fumigata*—Barth, 1964). *Blaberus* females will lower the pronotum or the entire body (Grillou, 1973), tilt the body down on the side facing the male, or kick at courting males (Wendelken and Barth, 1987). Some blattid females can be aggressively unreceptive, and escalate their belligerent behavior when courted by highly motivated males. Occasionally persistence pays off; females sometimes gradually shift to a less aggressive, more receptive pattern of behavior (Simon and Barth, 1977b). Aggression by males directed against unreceptive females is infrequent. *Blaberus giganteus* males occasionally bite an unreceptive female's wings (Wendelken and Barth, 1987), but forced copulation by males cannot occur in species where mating is dependent on female mounting and feeding behavior (Roth and Barth, 1964).

Copulation Refusal

Females often mount and feed on the tergal glands of courting males, but refuse to allow genitalic engagement. The nature of tergal secretions may be at least in part responsible; in the German cockroach the secretions smell like food and thus may lure hungry females regardless of their interest in mating. After mounting and feeding, a cooperative female orients her abdomen and opens her genital atrium to facilitate interaction with male genitalia (Roth and Willis, 1952a). Alignment of the two abdominal tips can require considerable female adjustment, particularly in species where she is larger than the male. *Byrsotria fumigata* females flex the abdominal tip forward ventrally so that genital connection can be made (Barth, 1964) and *Blab. craniifer* females may partially dismount in an attempt to improve the orientation of the genitalia (Wendelken and Barth, 1987). Cooperative females also open wide to allow full genital access. In *Eur. floridana* the gape of a receptive female's genital atrium is so impressive that the male can insert the entire tip of his abdomen (Barth, 1968b). Species in which the sexes back into each other also require female cooperation to copulate successfully. *Panesthia cribrata* females raise the tip of the abdomen and open the posterior plates (O'Neill et al., 1987).

After the genitalia are engaged, there are three major points at which a pair may separate: during turning to the opposed position, a few seconds after turning, and during the first 15 min of copulation. The signal to assume the opposed position comes from the male. He moves slightly forward, and the female responds by rotating off his back. If the female initiates the turning, it invariably results in separation of the pair (Simon and Barth, 1977a). After assuming the opposed position, brief genitalic connections of 4–7 sec are not uncommon in *B. germanica* (Roth and Willis, 1952a). *Eublaberus posticus* females frequently kick at the point of intersexual juncture with their metathoracic legs (Wendelken and Barth,

1987). In 12% of copulations of *D. punctata* observed by Wyttenbach and Eisner (2001), the teneral female pushed at the male with her hind legs until he disengaged; in each case the female subsequently accepted a second male. Females of *N. cinerea* require a longer period of courtship prior to copulation if they can detect the chemical traces of former female consorts on a male (Harris and Moore, 2005)—the cockroach equivalent of lipstick on his collar. After genitalic engagement, they can apparently determine if a male is depleted of sperm or seminal products because of those recent matings. After the first copulation "males are less adept at grasping the female," and pairs often remained joined for only a few seconds or minutes; no spermatophore is transferred. The female pushes the male with her hind legs, forcing him to release her (Roth, 1964b). Further evidence of female control of copulation in *N. cinerea* comes from transection experiments. When female genitalia were denervated males could not grasp the female properly and they stayed connected for only a few seconds (Roth, 1962).

Copulatory Success

Several studies report that male *B. germanica* have an abysmal record of successfully courting and copulating with females provided to them. Curtis et al. (2000) exposed each of 9 virgin males to serial batches of 2–10 virgin females throughout their lifetime (total of 341 females). Only 27 females were successfully inseminated. One-third of the males sired no offspring, and a further third inseminated just a single female. In a study of 55 virgin pairs by Nojima et al. (1999b), 84% of males courted females, 65% of the females responded by tergal feeding, but only 37% made the transition to copulation. Roth and Willis (1952a) did a detailed analysis of courtship and copulation in 10 pairs of German cockroaches (Table 6.2). Males courted rather vigorously in most cases; male 8, for example, courted the female 48 times in 30 min. Four females (pairs 3, 4, 5, 10) were nearly or completely unresponsive to male courtship, and 5 females responded by tergal feeding but refused to mate (pairs 2, 6–9). Just one of the 10 observed pairs successfully copulated. This puzzling lack of copulatory success has been noted in at least 2 other cockroach species. O'Neill et al. (1987) reported that in the majority of observed courtships, females of *Pane. cribrata* (Blaberidae) were not receptive. Males of *P. americana* (Blattidae) are rarely readily acceptable to the female (Gupta, 1947); only one in 20 attempted matings appeared successful in Rau's (1940) study of the species.

Female Loss of Receptivity

Although female sexual receptivity is inhibited as a result of mating in all cockroach species studied (Barth, 1968a), the fine points of its physiological control are far from straightforward. Not only do details of regulation differ among species, but the various components of mating behavior are controlled in distinct ways within a species (Roth and Barth, 1964). "It is essential to be wary of generalization" (Grillou, 1973). Mechanical cues are of primary importance in examined cockroaches, but chemical influences cannot always be ruled out (Engelmann, 1970). Interaction with male genitalia, the presence of the spermatophore in the female genital tract, and sperm or seminal fluid in the spermathecae have all been reported as mechanical cues influential in the initial or sustained loss of receptivity in cockroaches following mating (Roth and Stay, 1961; Roth, 1964b; Stay and Gelperin, 1966; Smith and Schal, 1990; Liang and Schal, 1994). The phenomenon is best studied in three cockroach species, the blattellids *B. germanica* and *Su. longipalpa,* and the blaberid *N. cinerea.* In the blattellids, one aspect of female receptivity, calling, is turned off by two successive mechanical cues provided by males during copulation. First, the insertion of a spermatophore results in the immediate cessation of calling. The behavior can be suppressed in experimental females by a spermatophore in the genital tract, by the insertion of a fake spermatophore, and by copulation with vasectomized males. The spermatophore effect, however, is transient. The presence of sperm or seminal fluids in the spermathecae is the stimulus that maintains the suppression of calling behavior in the first as well as the second ovarian cycles. The ventral nerve cord plays a crucial role in the transmission of the inhibitory signals (Smith and Schal, 1990; Liang and Schal, 1994). Signals transferred via the nerve cord also decrease locomotor activity in females (Lin and Lee, 1998).

The suppression of receptivity in *N. cinerea* following mating requires a single cue: mechanical stimulation caused by the insertion of the spermatophore into the bursa copulatrix (Roth, 1962, 1964b). The insertion of glass beads into the bursa results in the same loss of receptivity, manifested as a lack of a feeding response to male tergal displays. Spermatophore removal experiments indicate that female receptivity is lost immediately after the male reproductive product is firmly inserted into the bursa but prior to the migration of sperm into the spermatheca. Cutting the nerve cord above the last abdominal ganglion in *N. cinerea* renders the female "permanently" receptive. However, it is curious that the ventral nerve cord in most females must remain intact for two days for female receptivity to be inhibited. Vidlička and Huckova's (1993) finding that female *N. cinerea* become unresponsive to male sex pheromone about 2 days after mating is consistent with the results of these transection studies. Roth (1970b) suggests the possibility that mating stimuli are transmitted rapidly to the last abdominal ganglion but require a longer period to reach the brain, or that there is another source of stimulation in the genital region. If firmly inserted spermatophores are removed from mated females, about 15% will mate again (Roth, 1964b). After copulation, females remain unreceptive until after partition, at which time most remate. The absence of sperm in the spermatheca does not influence the return of receptivity after the first oviposition (Roth, 1962, 1964a, 1964b).

Mating Plugs

In cockroaches, the physical presence of a spermatophore in the genital tract of a female may play a dual role in preventing sperm transfer from other males. Besides acting as mechanical triggers in turning off female receptivity, they may also serve as short-term physical barriers to the placement of additional spermatophores. Copulating males typically deposit spermatophores directly over the spermathecal openings. If a female accepts an additional male and a second spermatophore is inserted, it is doubtful that the second male's sperm could access female sperm storage organs. Additional spermatophores are usually improperly positioned (Roth, 1962; Graves, 1969). Spermatophore shape and its mechanism of attachment vary among cockroach taxonomic groups and some types are probably more refractory to dislodgment than others. In some blaberids the spermatophore has a dorsal groove that fits closely against the female genital papilla (Graves, 1969). In blattellids with uricose glands, uric acid deposited on the spermatophore can fill the genital atrium of the female (Roth, 1967c).

The spermatophore is discarded by the female after 20–24 hr in *P. americana* (Jaiswal and Naidu, 1976), after 2–3 days in *Blatta orientalis* (Roth and Willis, 1954b), after 4–9 days in *Eub. posticus* (Roth, 1968c), by the 5th day in *Blab. craniifer* (Nutting, 1953b), by the 6th day in *D. punctata* (Engelmann, 1960), and after 6–13 days in *R. maderae* (Roth, 1964b). Young females of *N. cinerea* extrude the spermatophore after 5 or 6 days, but older females may retain it for over a month (Roth, 1964b). The mechanism by which cockroach females eject the spermatophore is not altogether clear. In *B. germanica*, the spermatophore remains in place about 12 hr and then shrinks; the shriveled remains may adhere to the female for several days (Roth and Willis, 1952a). Jaiswal and Naidu (1976) indicate that shrinkage of the outermost layer also causes spermatophore separation in *P. americana*, but Hughes and Davey (1969) thought that it disintegrated as a result of exposure to spermathecal secretions. Disintegration of the spermatophore is also reported in *Blab. craniifer* (Hohmann et al., 1978). A secretion from the spermathecal glands apparently facilitates spermathecal extrusion in four examined Blaberidae (*D. punctata, R. maderae, N. cinerea, Byr. fumigata*). The secretion is under the control of the corpora allata, and loosens the spermatophore by softening the material covering it (reviewed by Roth, 1970b). Nonetheless, a few experimental females of *R. maderae* were able to extrude their spermatophores despite surgical removal of the spermathecal glands (Engelmann, 1957).

Mechanical Stimulation and "Imposed Monogamy"

Roth's (1964b) demonstration that the suppression of female receptivity results from the physical insertion of the spermatophore into the bursa in *N. cinerea* has been interpreted as evidence that males force monandry on females during their first reproductive cycle. The bursa and the brood sac are in close physical proximity within the female genital tract. This serves as the basis for the argument that males are co-opting the physiological mechanism evolved to suppress female receptivity during pregnancy, and so females are precluded from evolving countermeasures to this manipulation (Harris and Moore, 2004; Montrose et al., 2004). Several points must be carefully considered before accepting this interpretation.

First, while the brood sac is spatially proximate to the genital papilla on which the spermatophore is secured, there is no evidence that the two structures share a mechanism for suppressing female receptivity. The highly distensible brood sac is situated at the anterior end of the vestibulum. It is separated from the genital papilla by the laterosternal shelf (McKittrick, 1965) (Fig. 6.14A). When the female is incubating an ootheca, the genital papilla is forced to stretch as the egg case projects into the vestibulum (Fig. 6.14B). Nonetheless, engaging the mechanoreceptors in the brood sac of a virgin has little to no effect on her receptivity. When glass beads were inserted into the brood sac without applying pressure to the bursa, 72% of virgins subsequently mated. Some physiological change occurs *after ovulation* that makes females responsive to inhibitory stimuli from the stretched brood sac (Roth, 1964b, p. 925). The loss of receptivity after the first copulation of her adult life, and the loss of receptivity in response to an ootheca stretching the brood sac, then, do not have a shared control mechanism.

Second, the imposed monogamy scenario is predicated on the assumption that multiple copulations within the first reproductive cycle confer benefits on female *N. cinerea*. In many insects, females profit from multiple matings because they can increase fitness via increased egg production and fertility (Arnqvist and Nilsson, 2000). A male, on the other hand, benefits by rendering females sexually unreceptive after mating, thus increasing the probability that his sperm will fertilize the majority of the female's eggs (Cordero, 1995; Eberhard, 1996; Gillott,

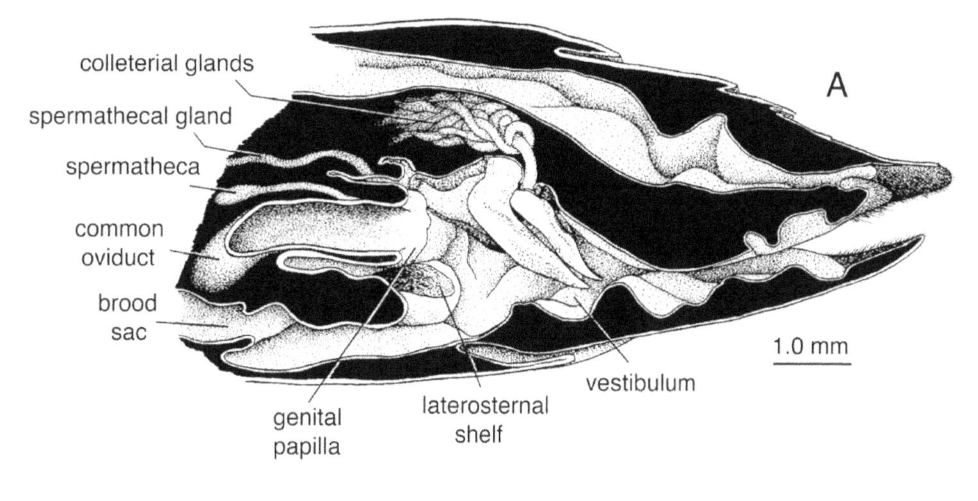

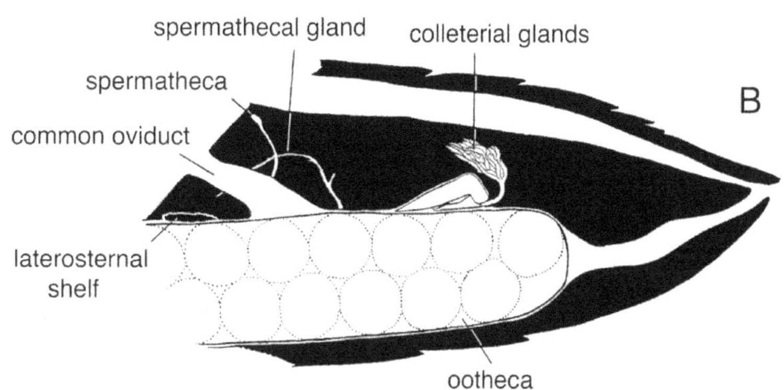

Fig. 6.14 (A) Sagittal section of the female genitalia of *Gromphadorhina portentosa* (Blaberidae). (B) Diagrammatic sagittal section of blaberid female genitalia with ootheca in brood sac. From McKittrick (1964).

2003). If multiple matings do increase female fitness, it follows that the control of female sexual receptivity is a source of conflict between the sexes, and females are expected to evolve resistance to the stimuli males use to induce receptivity loss (Arnqvist and Nilsson, 2000). That does not appear to be the case in *N. cinerea.* Copulation is known to confer numerous fitness benefits on female cockroaches (discussed below), but within the framework of cyclic receptivity typical of *N. cinerea* there is currently no evidence that more than one mate within the first reproductive cycle is advantageous. Moreover, morphological and experimental evidence suggests that spermatophore placement and therefore loss of receptivity in *N. cinerea* is likely under female control, suggesting that there is no conflict of reproductive interest between the sexes on this issue. Not only do females have morphological features specialized for proper spermatophore placement and retention, these features are regulated by her nervous system. Receptivity in *N. cinerea* is suppressed only if the spermatophore is firmly placed and properly positioned (Roth, 1964b). While in some blaberids a large amount of glue-like secretion cements the spermatophore into place, in *Nauphoeta* and several related genera the bursa is largely responsible for spermatophore retention (Graves, 1969). The bursa is deep, is extensively membranous, and almost completely wraps around the correspondingly elongated spermatophore. If the nerve cords are severed prior to mating in female *R. maderae,* another species with a deep, membranous bursa, 70% of males were not able to insert the spermatophore properly. They were placed elsewhere in the genital atrium or dropped by the male without being transferred. In many cases the male had pierced the wall of the brood sac and the spermatophore was in the female's body cavity. "It seems the female takes an active role in the proper positioning of the spermatophore in the bursa copulatrix, and an intact nerve cord is needed for proper muscular movements of the female genitalia" (Roth and Stay, 1962a).

Loss of Receptivity during Gestation

Pregnant blaberid females typically do not respond to courting males. The physical presence of an ootheca in the brood sac inhibits mating behavior, and its removal leads to the return of receptivity (*N. cinerea, Byr. fumigata*) (Roth, 1962, 1964b; Grillou, 1973). The suppression of receptivity appears to be the direct result of sensory stimulation via mechanoreceptors that are abundant within the brood sac (Brousse-Gaury, 1971a, 1971b; Roth, 1973b; Greenberg and Stay, 1974). Internal gestation of eggs, then, leads to potentially large differences between oviparous and ovoviviparous species in the sexual availability of females (Wendelken and Barth, 1987). Live-bearing females are removed from the mating pool for extended periods of time; gestation lasts 35–50 days in *N. cinerea* (Roth, 1964a), 51 days in *R. maderae* (Roth, 1964b), and 55–65 days in *Blab. craniifer* (Grillou, 1973). *Blattella germanica,* a species that externally carries the ootheca for about 21 days before the young hatch (Roth and Stay, 1962c), is intermediate. Oviparous females that drop their oothecae shortly after their formation lack the lengthy gestation periods of ovoviviparous cockroaches (Chapter 7) and so have relatively high rates of "recidivist receptivity" (Wendelken and Barth, 1987). Potentially, then, these females mate more frequently and presumably with a greater number of males.

Secondary Effects of Copulation

The primary role of copulation is egg fertilization, but a variety of secondary effects also occur. In cockroaches these include the suppression of female receptivity, but also diverse processes that facilitate female reproduction, such as the acceleration of oocyte growth, the prevention of oocyte degeneration, an increase in the number of oocytes matured and oviposited, the appropriate construction of the egg case, and, in ovoviviparous species, its proper retraction. The degree to which mating influences these processes as well as the details of their physiological control vary among studied species (Griffiths and Tauber, 1942a; Wharton and Wharton, 1957; Roth and Stay, 1961, 1962a, 1962c; Engelmann, 1970; Roth, 1970b; Adiyodi and Adiyodi, 1974; Hales and Breed, 1983; Goudey-Perriere et al., 1989). These secondary effects clearly promote female reproductive fitness, but are also considered beneficial to the male because they increase the likelihood that his sperm will be used by the female to sire her eggs (reviewed by Cordero, 1995; Gillott, 2003).

Mating has been shown to stimulate oocyte maturation in all cockroach species studied to date (Holbrook et al., 2000b), but the instigating stimuli differ. The physical presence of the spermatophore, stimulation from male genitalia, mechanical pressure from a filled spermatheca, and the chemical presence of the spermatophore all have varying degrees of influence on female reproductive processes. The action of these stimuli also may be moderated, sometimes strongly, by nutritional and social factors. The mechanical stimulation caused by the firm insertion of the spermatophore in *N. cinerea* not only suppresses female receptivity, but is also responsible for stimulating oocyte development and for ensuring the normal formation and retraction of the ootheca during the first reproductive cycle (Roth, 1964b). The physical presence of the spermatophore has been similarly demonstrated to be sufficient stimulus for accelerating oocyte maturation in oviparous *Su. longipalpa;* an artificial spermatophore is a reasonable substitute (Schal et al., 1997). *Diploptera punctata* females are dependent on spermatophore insertion for rapid development of their oocytes. However, the act of mating alone, without passage of a spermatophore, may be sufficient for oocyte maturation in some females. The physical stimulus of the spermatophore together with the action of the male genitalia appear to produce maximum reproductive effects (Roth and Stay, 1961). The acceleration of oocyte growth that occurs after mating in *P. americana* can be prevented by removing the spermatophore prior to the movement of sperm into the spermatheca, or by mating the female to males whose spermatophores are of normal size and shape but lack sperm. Pipa (1985) concluded that the stimulus for oocyte growth in this species originates from the deposition of sperm or other seminal products into the spermatheca. The proper formation and retraction of the ootheca into the brood sac in *N. cinerea* (Roth, 1964b) and *Pyc. indicus* is dependent on the presence of sperm in the spermatheca. After spermatheca removal, severance of spermathecal nerves, or mating with castrated males, females produced abnormal egg cases or scattered the eggs about (Stay and Gelperin, 1966).

Male accessory glands typically contain a variety of bioactive molecules that, when transferred to the female during mating, influence her reproductive processes (Gillott, 2003). The spermatophore of *Blab. craniifer* is richly invested with enzymes whose activities change during the three days subsequent to mating; the longer the spermatophore remains in place (from 0–24 hr), the sooner oviposition occurs. Acetone extracts of the spermatophore topically applied to the female induce the same increases in vitellogenesis as do juvenile hormone mimics. Nonetheless, the physical presence of the spermatophore is also required for the full expression of reproductive benefits, and both mechanoreceptors and chemoreceptors are found in the bursa (Brousse-Gaury and Goudey-Perriere, 1983; Perriere and Goudey-Perriere, 1988; Goudey-Perriere et al., 1989).

In many cockroach species the female either internally digests and incorporates, or removes and ingests the spermatophore sometime after it is transferred to her (Engelmann, 1970). However, there is currently little evidence that spermatophores are of nutritional value, aside from the uric acid that covers them in some species. Mullins et al. (1992) injected ^{3}H leucine into male *B. germanica.* The males transferred it to females during mating, who subsequently incorporated it into their oothecae. The source of the leucine-derived materials is unknown, but the authors suggested that it may have originated from the spermatophore or seminal fluids.

Spermathecae

Our understanding of the functional anatomy of the female cockroach reproductive tract in relation to cryptic mate choice languishes behind that of some other insect groups. The shape, number, elasticity, duct length, coiling pattern, musculature, presence of valves or sphincters, and chemical milieu of spermathecae play a strong role in sperm selection by females (Eberhard, 1996). Multiple sperm storage sites are particularly important in allowing females to cache and use the ejaculates of different males selectively (Ward, 1993; Hellriegel and Ward, 1998). Sperm storage organs in cockroaches have not received much consideration since McKittrick (1964), who demonstrated a great deal of variety in the form, number, and arrangement of spermathecae (Fig. 6.15). In *Cryptocercus* the spermatheca is forked, with the branches terminally expanded; the single spermathecal opening lies in the roof of the genital chamber. The spermatheca of *Lamproblatta* has a wide, sclerotized basal portion and a slender forked distal region. Within the Polyphagidae, *Arenivaga* has a single, unbranched spermatheca, but *Polyphaga* has a small tubular branch coming off about halfway up the main duct. In the Blattellidae the spermathecal opening is shifted to a more anterior position on the roof of the genital chamber, far in advance of the base of the ovipositor. Some species of *Anaplecta* have, in addition, a pair of secondary spermathecae that open separately on the tip of a small membranous bulge, the genital papilla, that lies at the anterior end of the floor of the genital chamber (Fig. 6.15F). The cockroaches of this genus thus have either one or three spermathecae. The Pseudophyllodromiinae, Blattellinae, Ectobiinae, Nyctiborinae, and Blaberidae have secondary spermathecae only. The spermathecal pores in these may be widely spaced (Fig 6.15G—Pseudophyllodromiinae except *Supella*) or more closely situated within a spermathecal groove (Fig 6.15H—*Supella, Pseudomops*), thought by Snodgrass (1937) to function as a sperm conduit. One pair of spermathecae, each with a separate opening, is typically present in Pseudophyllodromiinae, but the Blattellinae may have two (Fig. 6.15I) or more pairs, each with a separate opening. *Xestoblatta festae* averages 10 or 11 spermathecal branches, but these converge into just two exterior openings (Fig. 6.16K). *Nyctibora* sp. (Fig. 6.15J) and *Paratropes mexicana* have three pairs of spermathecae. All Blaberidae have a single pair of spermathecae that open on the genital papilla or directly into the common oviduct; in most species they are accompanied by a conspicuous pair of spermathecal glands (McKittrick, 1964).

Spermathecal Glands

Initially, the energy necessary for sperm maintenance and motility is provided in the semen. The seminal fluid of *P. americana* contains small amounts of protein, substantial glycogen, and some glucose, phospholipid, and other PAS-positive substances (Vijayalekshmi and Adiyodi, 1973). Females are presumably responsible for fueling the long-term metabolic needs of sperm, as well as for creating a favorable environment for extended storage. In *Periplaneta,* for example, a female mated during her first preoviposition period can produce fertile eggs for 346 days subsequent to her first ootheca (Griffiths and Tauber, 1942a). *Parcoblatta fulvescens* females can produce more than 30 oothecae without remating (Cochran, 1986a). It is possible, however, that at times stored sperm are neglected, digested, or destroyed; Breland et al. (1968) noted that the sperm in cockroach spermatheca are sometimes degenerated.

Spermathecal walls are typically glandular, a trait functionally associated with providing for the maintenance requirements of the enclosed sperm. In some species the storage and secretory functions are largely separated via the development of one or more spermathecal glands (Gillott, 1983). Because cockroach spermathecae are also secretory, however, it has been difficult to make a distinction between spermathecae and spermathecal glands without direct observation of the location of stored sperm. An example is *P. americana,* whose spermatheca has two branches, both of which are muscular and secretory. The first spermatheca ("A" of Lawson and Thompson, 1970) is an *S*-shaped capsular branch that terminates in a large swelling lined with a dense and deeply pigmented cuticular intima. It has a thick, underlying muscular layer and a smooth surface facing the lumen. Spermatheca "B" is a long, slender, tightly coiled branch with a thinner lining and strongly rugose inner surface. Secretory cells with collection centers fed by microvilli are far more numerous in the former than in the latter. The two spermathecae join basally to form a common duct. For many years, the slender, coiled branch was thought to be a spermathecal gland, until sperm were found in both branches following copulation (Marks and Lawson, 1962; Lawson and Thompson, 1970). Lawson thought that "B" served as a secondary storage reservoir for sperm. Hughes and Davey (1969) noted that the tubular branch seemed to release sperm more slowly than the capsular branch, or only after the capsular branch had finished discharging them. If so, sperm from the capsular branch may fertilize the majority of the female's eggs, and a multiply mated female may bias paternity via differential sperm storage.

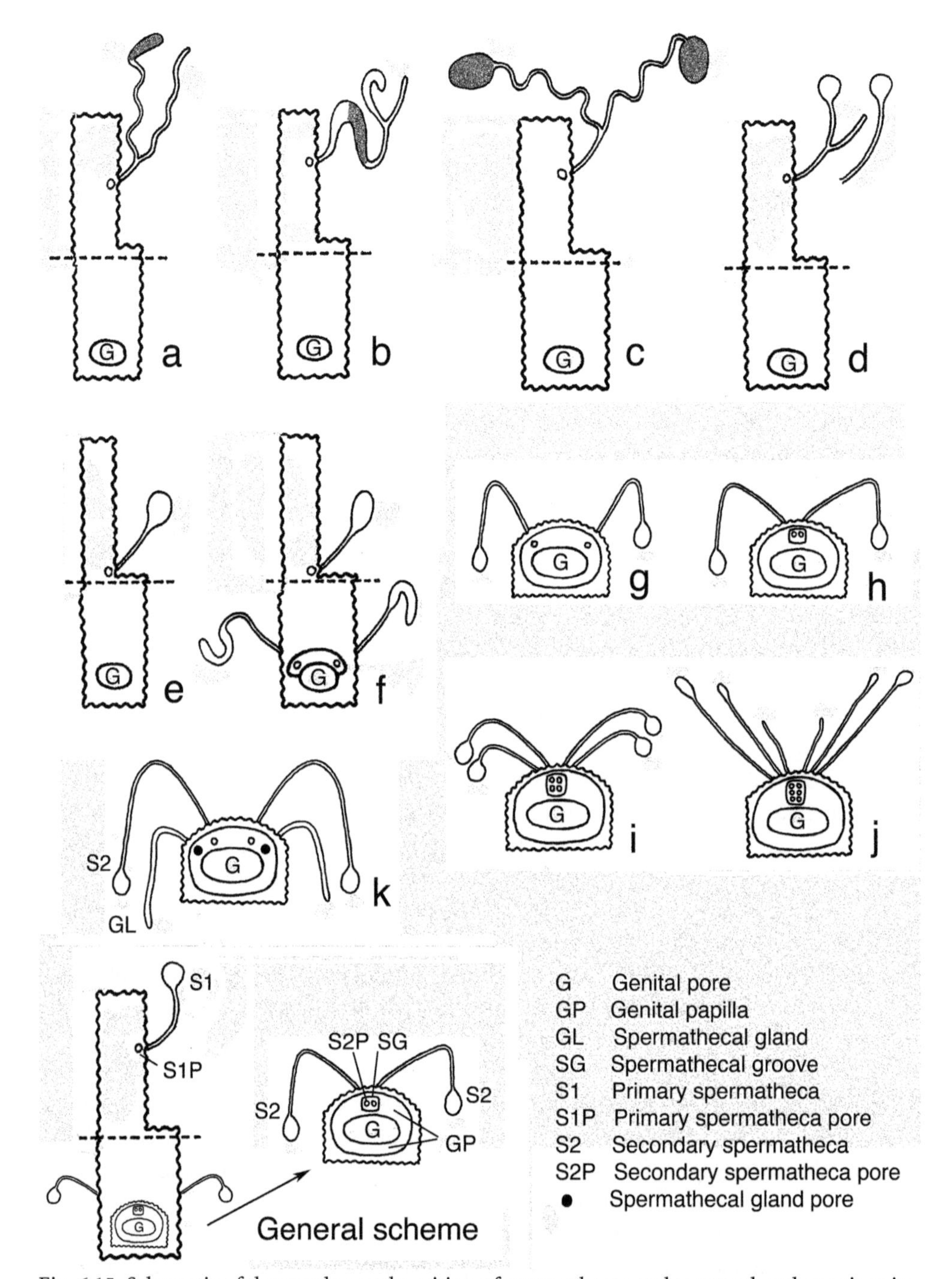

Fig. 6.15 Schematic of the number and position of spermathecae and spermathecal openings in representative cockroaches. (A) Blattinae, Polyzosteriinae; (B) *Lamproblatta;* (C) *Cryptocercus;* (D) *Polyphaga* (*left*), *Arenivaga* (*right*); (E) *Anaplecta* sp. A, B; (F) *Anaplecta* sp. C; (G) Pseudophyllodromiinae (except *Supella*); (H) *Supella, Pseudomops;* (I) Ectobiinae, Blattellinae (except *Pseudomops, Xestoblatta*); (J) *Nyctibora;* (K) Blaberidae. Area above the dashed line represents the dorsal wall of the genital chamber, area below the dashed line represents the ventral wall of the genital chamber. Shaded portions of the spermathecae are sclerotized areas. (A) to (E) have primary spermathecae only; (F) has both primary and secondary spermathecae; (G) to (K) have secondary spermathecae only. After Klass (1995), from data in McKittrick (1964), with permission of K.-D. Klass.

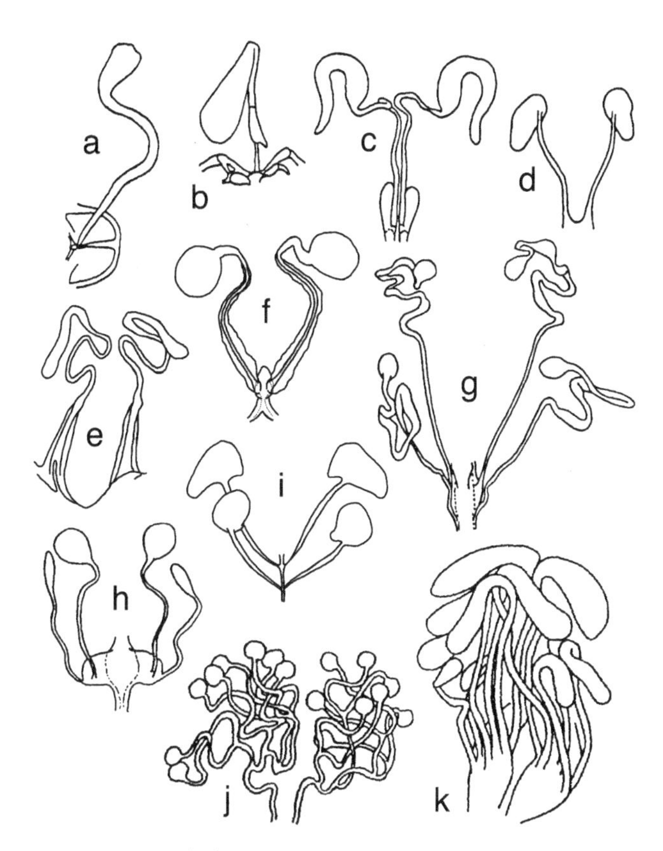

Fig. 6.16 Morphological variation in cockroach spermathecae (A) *Arenivaga bolliana;* (B) *Hypercompsa fieberi;* (C) *Neoblattella* sp.; (D) *Plecoptera* sp.; (E) *Miriamrothschildia notulatus;* (F) *Pseudomops septentrionalis;* (G) *Parcoblatta virginica;* (H) *Blattella germanica;* (I) *Ectobius pallidus;* (J) *Loboptera decipiens;* (K) *Xestoblatta festae.* From McKittrick (1964) and Gurney and Roth (1966).

In those cockroaches that apparently possess both spermathecae and spermathecal glands, ambiguity as to whether all branches function in sperm storage has implications for species in the Blaberidae. Based on morphological observations, most species in this family have been described as having a pair of spermathecae and a pair of spermathecal glands, some of them quite elaborate (McKittrick, 1964). In *R. maderae,* for example (Fig. 6.17), the glands are large, slender, highly branched, and open posterior to the openings of the spermathecae (van Wyk, 1952). Spermathecal glands in *Diploptera* entwine each spermatheca, and are "constantly filled with an intensely basophilic secretion" (Hagan, 1941). Marks and Lawson (1962), however, reported four paired spermathecae in *Blab. craniifer,* with the posterior member of each pair coiled, slender, and unbranched, and the anterior member sparsely branched. A functional analysis of these organs is necessary given their potentially influential role in sperm handling by the female. Spermathecal glands are thought to stimulate spermatozoa to enter the spermathecae (Khalifa, 1950), activate sperm, provide "lubrication" (van Wyk, 1952), and facilitate the extrusion of the spermatophore after mating (Engelmann, 1959, 1960).

Spermathecal Shape

Two "basic" spermathecal shapes are represented in cockroaches: the tubular form, with little difference in width between the duct and the spermatheca proper (= ampulla), and the capitate form, shaped like a lollipop. Shape varies widely across cockroach species and sometimes within a species. In *Agmoblatta thaxteri* each spermatheca has a double terminal bulb, like a figure 8 (Gurney and Roth, 1966). The genus *Tryonicus* can be inter- and intraspecifically polymorphic (Fig. 6.18) (Roth, 1987b); however, some apparent variation in spermathecal shape may be due to the amount of ejaculate stored or to the preservation of specimens at different stages of muscular activity. Both the ampulla and ducts are surrounded by a sheath of profusely innervated striated muscle (Gupta and Smith, 1969). The sheath is best developed at the base, where it consists mainly of circular fibers and functions as a sphincter in opening and closing the entry (van Wyk, 1952).

It has been suggested that spermathecal shape can pre-

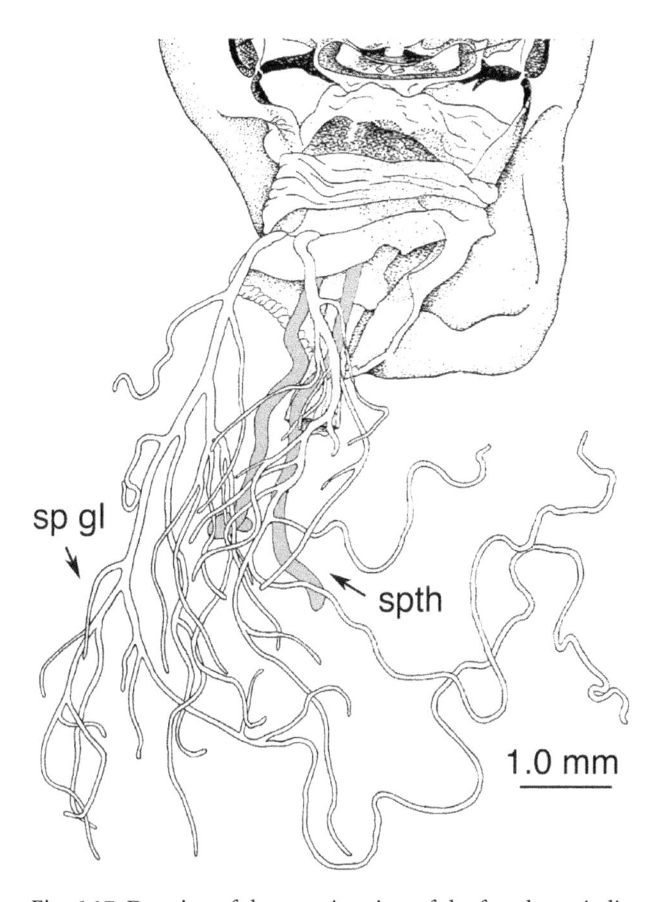

Fig. 6.17 Drawing of the anterior view of the female genitalia of *Rhyparobia maderae,* showing the tubular spermathecae (spth, shaded gray) and extensive, branched spermathecal gland (sp gl). Slightly modified from McKittrick (1964).

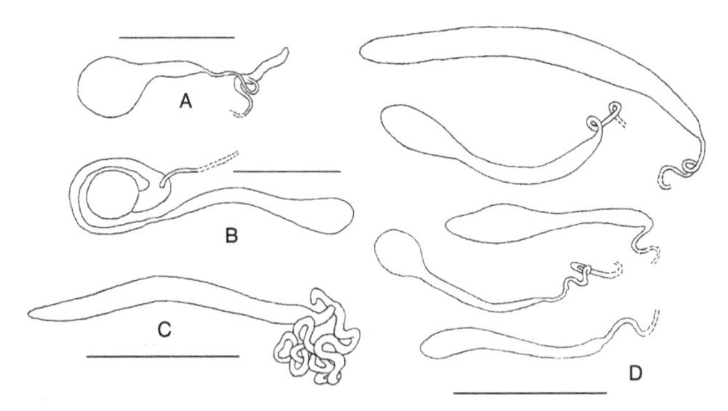

Fig. 6.18 Inter- and intraspecific variation in spermathecae of cockroaches in the genus *Tryonicus* (Blattidae: Tryonicinae). (A) *Tryonicus parvus;* large, bulbous reservoir arising preapically from a convoluted duct. (B) *Tryonicicus angusta;* reservoir spherical, sclerotized at one end and club-shaped on the other. (C) *Tryonicus* sp. 1; large spermathecal duct is same diameter as the spermathecal branch beyond the point of insertion of main reservoir. (D) *Tryonicus monteithi* from five locations in Queensland, Australia. After Roth (1987b). Scale bar is 0.5 mm in all cases.

dict sperm use patterns (Walker, 1980), but the functional significance of spermathecal shape is complex (Otronen, 1997) and not yet clear (Ridley, 1989). Large, globular ampullae may be associated with sperm mixing. Long tubular spermathecae may promote the layering of ejaculates, enhancing the "last in, first out" pattern of sperm precedence, or may serve as "sperm traps" to imprison the sperm of less favored males. Spermatozoa of *R. maderae* are apparently stored chiefly in the distal portion of the female's tubular spermatheca. The proximal portion is filled with a granular secretion, which, according to van Wyk (1952), probably serves as food for the sperm.

Multiple Storage Sites

Most examined cockroaches have just one or two spermathecal lobes. The Blattellidae are extraordinary, however, in that some species have two, others, including *Blattella,* have four, and in some, the spermathecae look like a fistful of balloons (Fig. 6.16J). Each spermatheca may have its own opening (i.e., multiple spermathecae) (*Nyctibora*—Fig. 6.15J), or multiple branches may share a common orifice. In the latter case, the ducts may be arborescent (Fig. 6.16J), or branch from a single point (Fig. 6.16K).

Multiple storage sites offer potential for allowing a female to separate the sperm of different males spatially, giving her greater scope for choosing among potential sires and for postponing mate choice until oviposition. The bias can take the form of differential transport to storage sites, biased sperm survival in different spermathecal lobes, or differential transport from storage to the site of fertilization. Multiple spermathecae may also prevent male genitalic structures from accessing previously stored sperm, and allow specialization for more than one function, such as long- versus short-term storage (Eberhard, 1996; Otronen et al., 1997; Hellriegel and Ward, 1998; Pitnick et al., 1999). It is known, for example, that in the fly *Dryomyza anilis,* sperm movements in and out of individual spermathecae occur independently (Otronen, 1997). Differential sperm storage is also known in the fly *Scatophaga stercoria,* and is mediated by female muscular activity (Hellriegel and Bernasconi, 2000). A detailed examination of the fates of different ejaculates within blattellid cockroaches is clearly indicated. The only relevant information known to us is from *B. germanica.* When the spermatophore is transferred to the female, the two sperm sac openings align directly with two of the spermathecal pores (Khalifa, 1950); nonetheless, sperm can be found in all four spermathecae of mated females (van Wyk, 1952; Marks and Lawson, 1962). Cochran's (1979b) study of sperm precedence in the species suggests that selective use of sperm may be possible in multiply mated females (Fig. 6.6).

SEXUAL CONFLICT OVER SPERM USE

Male and female reproductive interests do not always coincide, and the conflict may be evident in their genital morphology. "Disagreement" over the removal or repositioning of stored sperm can select for male genitalia better designed to penetrate the female's sperm storage organs, as well as female organs that are more resistant to male intrusion (Eberhard, 1985, 1996; Chapman et al., 2003). There is a potential example of such antagonistic co-evolution among cockroaches in the Moroccan and Spanish species of *Loboptera* (Blattellinae) studied by Horst Bohn (1991a, 1991b). As noted above, males have a genital whip as part of the left phallomere complex. Females have spermathecae that are multiply lobed with long, convoluted ducts and as many as 10 branches on each side (*L. glandulifera*). In some species, the length of spermathecal ducts appears correlated with whip length in the male (Fig. 6.19), suggesting that as the female receptacle elongates, so does the adaptive value of a long whip in potential sires (and vice versa). Some males additionally have a sclerite densely covered with bristles, or membranes covered with long, narrow, hair-like scales in the vicinity of the intromittent organ (also occurring in other genera—Fig. 6.11B). In some *Loboptera* species the whip itself is covered in small bristles (*L. delafrontera*) or is densely hairy (*L. juergeni*). Spermathecae appear to have valves, sphincters, or other adaptations that serve to control sperm movement or to interact with male intromittent organs. Ducts can have accordion-like walls (*L. truncata, L. cuneilobata*), or a series of irregular swellings,

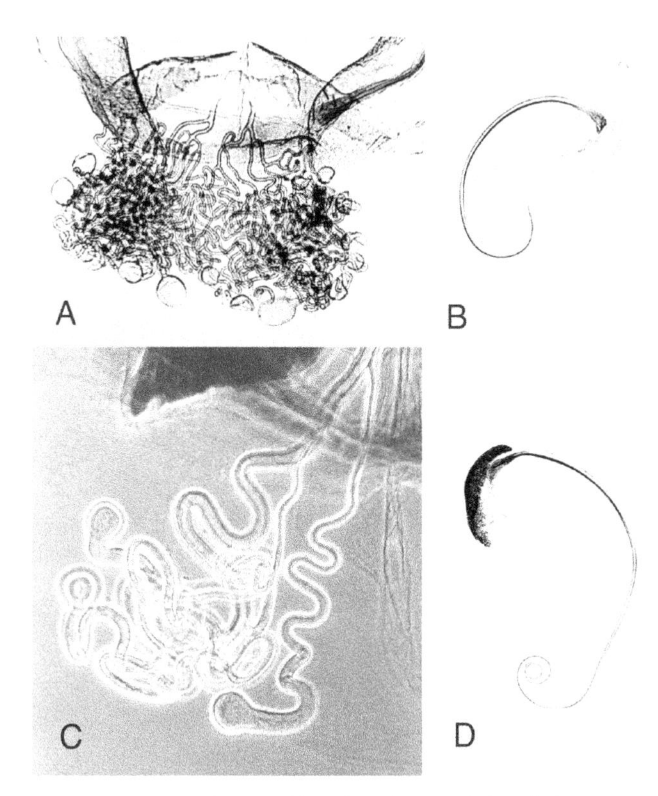

Fig. 6.19 Spermathecae of female *Loboptera* (Blattellidae: Blattellinae) and corresponding genitalic structure in male. (A) Multi-branched spermathecae of *L. decipiens nevadensis;* (B) whip in male of the same species; (C) multi-branched spermathecae of *L. barbarae* (phase contrast); (D) whip in male of the same species. From Bohn (1991b), courtesy of Horst Bohn, and with permission from the Journal of Insect Systematics and Evolution (= Entomologica Scandinavica).

giving them the appearance of a string of pearls (*L. minor minor*). Multiple reversals in the coiling direction of long thin, spermathecal ducts are common in the genus. Terminal ampullae may be globular, club shaped, or the same width as the spermathecal duct; branch points of ducts may be widely separated or originate from a single point. The morphological evidence for co-evolution of genital structures in male and female *Loboptera* is compelling; nonetheless, sexual biology and behavior in the genus are largely unknown.

OPPORTUNITIES

The literature to date suggests the taxa with the most promise for potentially productive studies of sexual selection occur within the Blattellidae, the largest but least known family of cockroaches. Males in this family variably possess diverse complex intromittent genital structures, elaborate tergal glands, uricose glands, and the most variable testes of examined species (Ph.D. thesis by E.R. Quiaoit, cited by Roth, 1970a). Females can have multiple spermathecae; furthermore, their reproduction can be closely tied to food availability, as they invest a high proportion of their bodily reserves into each reproductive event. The existence of these elaborate morphological structures, together with both prenuptial feeding via tergal glands and postnuptial feeding via uricose glands may be red flags signaling that male and female reproductive interests do not coincide. The potential for reproductive conflict is great when males provide nuptial gifts, because females are selected to obtain an optimal supply of nutrients, while males are selected for those traits that assure she uses his sperm (Thornhill and Alcock, 1983). The possession of morphologically complex, multiple spermathecae in females and a variety of intromittent-type structures in males suggest that control of sperm use in some blattellids may be an evolutionary chess game played out inside the female body during and after copulation. Blattellids as well as other cockroach taxa, then, are potentially rich sources of research material for a wide range of studies on insect mating strategies. Can the number of spermathecae or their structure be correlated with the morphology of any of the "blades" on the male's Swiss army knife? Do elaborate spermathecae occur only in species with male uricose glands? Do complex male genital structures influence female sperm use, and if so, how do they do it? Does the quantity or composition of tergal secretion influence female choice? Are complex tergal glands and the possession of uricose glands correlated? Does the amount of uric acid transferred after copulation influence female sperm acceptance and use? It is clear that the scope of research needs to be expanded beyond the domestic pets and pests typically kept in laboratories, with an increased emphasis on bringing field and laboratory work into closer alignment. Even so, the study of sexual selection in cockroaches is in its early stages, despite the opportunities offered by even the most easily obtained and studied species. What is the function of giant sperm in *Periplaneta?* Do female American cockroaches preferentially use sperm from the capsular branch of the spermatheca? Is there differential use of the sperm from the four spermathecal chambers of German cockroaches? If so, is the male virga involved in influencing female sperm choice decisions? A creative scientist capable of overcoming the technical challenges inherent in these kinds of studies could be amply rewarded.

SEVEN

Reproduction

Be fruitful, and multiply, and replenish the Earth.

—Genesis 1:28

Perhaps no aspect of cockroach biology has been studied as extensively as the range of mechanisms by which they replenish the earth. Understandably so, given that their variation in this arena is a rich source of comparative material and that reproduction in many species is amenable to laboratory study. Several reviews of cockroach reproduction are available, including Roth and Willis (1954b, 1958a), Roth (1970a, 1974a), and Bell and Adiyodi (1982b), among others.

In the majority of cockroaches, reproduction is characterized by the formation of an ootheca: eggs are released from the ovaries, move down the oviducts, are oriented into two rows by the ovipositor valves, then surrounded by a protective covering. Three general reproductive categories are recognized, with two of these broken into subcategories (Table 7.1) (Roth, 1989a, 1991a, 2003c; Roth and Willis, 1954b, 1958a). In oviparity type A, females drop the egg case shortly after formation. In oviparity type B, females carry the ootheca externally throughout embryonic gestation, then drop it immediately prior to hatch; eggs also may hatch while the ootheca is attached to the mother. Ovoviviparous females gestate eggs internally, but the embryos rely primarily on yolk nutrients to fuel and support development. In category A ovoviviparous females, the ootheca is first extruded, as in oviparous taxa, but it remains attached and is retracted a short time later into a brood sac. When the nymphs are ready to hatch, the ootheca is fully extruded and the neonates emerge from their embryonic membranes. The eggs are deposited directly from the oviducts into the brood sac in ovoviviparous type B species; there is no oothecal case. In viviparous forms, oviposition is similar to the ovoviviparous type A cockroaches, but the embryos are nourished within the brood sac on a proteinaceous fluid secreted by the mother.

OVIPARITY

Oviparous type A cockroach species characteristically produce an ootheca, a double row of eggs completely enclosed by a protective outer shell (Stay, 1962; Roth, 1968a). A raised

Table 7.1. Modes of reproduction in cockroaches. After Roth (1989a, 2003c).

Characters	Oviparity A	Oviparity B	Ovoviviparity A[1]	Ovoviviparity B[2]	Viviparity[3]
Handling of ootheca	Dropped shortly after formation	Carried externally throughout gestation	After it is formed, retracted into the brood sac	No ootheca; eggs pass directly into brood sac	After it is formed, retracted into the brood sac
Physical properties of egg case	Hard and dark, completely enclosing eggs	Proximal end is permeable	In most, variably reduced and incomplete	—	Incomplete membrane
Water handling	Sufficient water in eggs, or additional water absorbed from substrate	Obtains water from the female during embryogenesis	Obtains water from the female during embryogenesis	Obtains water from the female during embryogenesis	Obtains water from the female during embryogenesis
Pre-partition non-yolk nutrients from mother?	No	Water-soluble material	Probably water-soluble material	Probably water-soluble material	Proteinaceous secretion from walls of brood sac
Taxa	All but Blaberidae and some Blattellidae	A few Blattellidae	A few Blattellidae, most Blaberidae	One tribe of Blaberidae (Geoscapheini)	One known species of Blaberidae
Examples	*Periplaneta, Eurycotis*	*Blattella, Lophoblatta*	*Blaberus, Nauphoeta*	*Macropanesthia, Geoscapheus*	*Diploptera punctata*

[1]"False" ovoviviparity of earlier studies.
[2]"True" ovoviviparity.
[3]"False" viviparity.

crest, the keel, runs along the mid-dorsal line of the egg case, and at hatch, the nymphs swallow air, forcing open this line of weakness (as in the opening of a handbag). The hatchlings generally exit en masse, and the keel snaps shut behind them (Fig. 7.1). If some eggs are lost due to unviability, parasitism, or disease, the entire brood may fail to hatch, because opening the keel typically requires a group effort. The ootheca is structurally sophisticated (Lawson, 1951; D.E. Mullins and J. Mullins, pers. comm. to CAN), and functions in gas exchange, water balance, and mechanical protection.

The oothecae of oviparous type A cockroaches vary in their ability to prevent water loss from the eggs (Roth and Willis, 1955c). In some species the ootheca and eggs at oviposition do not contain sufficient moisture for embryogenesis; in these the ootheca must be deposited in a humid or moist environment where the eggs absorb water (e.g., *Ectobius pallidus, Parcoblatta virginica*). Alternatively, if the ootheca and eggs contain sufficient moisture for the needs of the embryos at the time of oviposition, the ootheca possesses a protective layer that retards water loss (e.g., *Blatta orientalis, Periplaneta americana, Supella longipalpa*). The eggs of *Blatta orientalis* hatch even if oothecae are kept at 0% relative humidity during development. When physically abraded, however, the oothecae lose 60% or more of their water within 10 days, while controls lose only 5% (Roth and Willis, 1955c, 1958a).

Fig. 7.1 Unidentified neonate cockroaches freshly hatched from an ootheca attached to a leaf, Bukit Timah, Malaysia. Note that the keel has snapped shut behind them. Photo courtesy of Edward S. Ross.

Oothecal Deposition and Concealment

The majority of oviparous type A cockroaches select and prepare a site for egg case deposition with some care (Chapter 9; Roth and Willis, 1960; Roth, 1991a), and the stereotyped behavioral sequences involved have been used as taxonomic characters (McKittrick, 1964). *Therea petiveriana* simply deposits oothecae randomly in dry leaves (Ananthasubramanian and Ananthakrishnan, 1959). Other species attach them to the substrate (with saliva or genital secretions), and many find or construct a crevice,

Fig. 7.2 The diurnal Australian cockroach *Polyzosteria mitchelli* digging a hole for hiding her ootheca. It is a beautiful species, with a bronze dorsal surface spotted and barred with orange or yellow, a pale yellow ventral surface, and sky-blue tibiae. The lively colors fade after death. Photo by E. Nielsen, courtesy of David Rentz.

glue the ootheca in a precise position inside it, then conceal it with bits of debris, pieces of the substrate, or excrement (Fig. 7.2). Ootheca concealment is known in blattids (e.g., *Blatta orientalis, Eurycotis floridana, Methana marginalis, Pelmatosilpha purpurascens, Periplaneta americana, P. australasiae, P. brunnea, P. fuliginosa*), blattellids (*Ectobius sylvestris, Parcoblatta pennsylvanica, Supella longipalpa, Loboptera decipiens, Ellipsidion affine, Ell. australe*), and cryptocercids (*Cryptocercus punctulatus*). In the latter, wood and saliva are used to pack oothecae into slits carved in the ceilings of their wood galleries; the keels of the oothecae are left uncovered (Nalepa, 1988a). Concealment behavior may vary among closely related cockroach species. Female *Ectobius pallidus*, for example, carefully bury their oothecae after deposition; *E. lapponicus* and *E. panzeri* seldom do (Brown, 1973a). Intraspecific variation in this behavior may depend to some extent on the substrate on which the insects are found or maintained. *Nyctibora noctivaga* simply drops its ootheca in the laboratory, but in Panama, oothecae were found glued to leaves and in crevices of the piles supporting a house (McKittrick, 1964). Although females whose eggs absorb water from the substrate have to be exceptionally discriminating in where they place oothecae, they do not always make wise choices. In five species of *Parcoblatta*, it is common to find shrunken oothecae, as well as oothecae that have burst and extruded material from the keel (Cochran, 1986a). A great many unhatched and shriveled oothecae of *Parc. pennsylvanica* were found under the bark of pine logs in an early stage successional forest by Strohecker (1937); mortality was attributed to the high temperature of logs exposed to direct sunlight. In species that leave oothecae exposed, the egg case may be cryptically colored. Shelford (1912b) described the ootheca of an unknown species from Ceylon (now Sri Lanka) that was attached to the upper surface of a leaf. It was white, mottled with brown, and looked "singularly like a drop of bird's excrement."

External Egg Retention

In cockroaches displaying oviparity type B, the egg cases are carried externally for the entire period of embryogenesis with the end of the ootheca closely pressed to the vestibular tissues of the female's genital cavity. The proximal end of the egg case is permeable, allowing for transport of water from the female to the developing eggs (Roth and Willis, 1955b, 1955c; Willis et al., 1958). Recently, Mullins et al. (2002) injected radiolabeled water into female *Blattella germanica* carrying egg cases. The water was detected moving from the female to the proximal end of her ootheca, then spreading throughout the egg case following a concentration gradient (Fig. 7.3). A variety of water-soluble materials were also transferred across the female-ootheca divide, including glucose, leucine, glycine, and formate. Preliminary experiments of these authors indicate that the labeled materials also can be detected in nymphs after hatch. Scanning electron microscopy and the use of fluorescent stains pinpointed the structural basis of flow into the ootheca (Fig. 7.4). Small pores completely penetrating the oothecal covering are

Time post injection (h)	Radiolabel content (%) Female	Oothecal quadrants	Oothecal total
0.5	99.59	0.08 / 0.00; 0.32 / 0.00	0.41
2	97.59	0.73 / 0.02; 1.63 / 0.03	2.41
6	95.81	2.05 / 0.27; 1.56 / 0.31	4.19
24	81.93	5.22 / 3.09; 6.54 / 3.21	18.07

Fig. 7.3 Distribution of radiolabel in oothecae attached to *Blattella germanica* females at four time intervals after injection of ^3H_2O into the females. See original paper for sample sizes and variation. After Mullins et al. (2002), with permission from The Journal of Experimental Biology. Image courtesy of Donald and June Mullins.

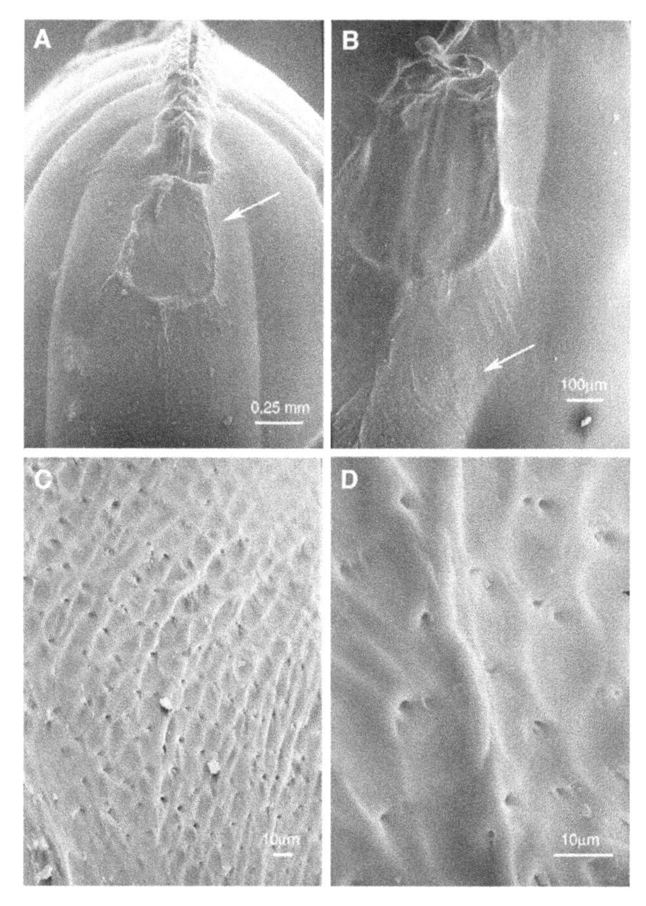

Fig. 7.4 Scanning electron microscopy images of *Blattella germanica* oothecae, demonstrating the morphological basis of their permeability. (A) Proximal end of an ootheca showing the "escutcheon-shaped" vaginal imprint (arrow). (B) Magnification of the ventro-lateral escutcheon region; arrow indicates the pore field area. (C) Magnification of the pore-field area. (D) Pores. From Mullins et al. (2002), with permission from The Journal of Experimental Biology. Images courtesy of Donald and June Mullins.

found in the wrinkled region surrounding the "escutcheon-shaped" vaginal imprint on the proximal end (Mullins et al., 2002).

Because the barrier between mother and developing embryos is permeable, females that externally carry egg cases throughout gestation have the advantage of parceling water and other soluble materials to the embryos on an "as needed" basis. They also have some degree of behavioral control over the embryonic environment. Nymphs of *B. germanica* are known to settle in microhabitats where temperatures are favorable to their development (Ross and Mullins, 1995); it is probable that a female carrying an egg case acts similarly on behalf of her embryos. In most instances, hatch of the egg case is initiated while it is still attached to the mother. The activity level of the female increases significantly prior to hatch, indicating either that she can detect impending hatch, or that her increased activity level initiates it (D. E. Mullins and K. R. Tignor, pers. comm. to CAN).

Oviparity type B occurs in two subfamilies of Blattellidae. In the Blattellinae, at least nine species of *Blattella* and one species of the closely related *Chorisia* exhibit this reproductive mode (Roth, 1985). In the Pseudophyllodromiinae two species of *Lophoblatta* carry their oothecae externally throughout gestation. The first of these was found by LMR in the Amazon basin in 1967; a female *Loph. brevis* carrying an ootheca was collected on a banana plant, and the eggs hatched the following day. A second species with external egg retention, *Loph. arlei,* was taken from a bird nest. All other known *Lophoblatta* deposit their oothecae shortly after they are formed (Roth, 1968b).

OVOVIVIPARITY

Ovoviviparity occurs in all Blaberidae except the viviparous *Diploptera punctata,* and in four genera of Blattellidae: *Sliferia, Pseudobalta* (Pseudophyllodromiinae) (Roth, 1989a, 1996), *Stayella,* and *Pseudoanaplectinia* (Blattellinae) (Roth, 1984, 1995c). As in oviparous cockroaches, type A ovoviviparous species extrude the ootheca as it is being formed. When oviposition is complete, however, the egg case is retracted back into the body and incubated internally in a type of uterus, the brood sac, throughout development. The brood sac is an elaboration of the membrane found below the laterosternal shelf in oviparous cockroaches and is capable of enormous distension during gestation (Fig. 6.14). The eggs have sufficient yolk, but must absorb water from the female to complete development. At hatch, the nymphs are expelled from this maternal brood chamber, and quickly shed their embryonic cuticle. There is some evidence that pressure exerted by the female on the ootheca during extrusion supplies the hatching stimulus (Nutting, 1953a).

Ovoviviparous females are thought to provide only water and protection to embryos during gestation, with the yolk serving as the main source of energy and nutrients. This is supported by data indicating that in ovoviviparous *Rhyparobia maderae* and *Nauphoeta cinerea,* water content increases and dry weight decreases during embryogenesis, just as it does in oviparous *P. americana* (Roth and Willis, 1955c; Roth, 1970a). Even if it is not reflected as weight gain, however, ovoviviparous cockroaches may be supplying more than water to their retained embryos. This is suggested by the physiological intimacy of the embryonic and maternal tissues, and the evidence that maternal transfer of materials occurs in oviparous *B. germanica.* Based on morphological evidence, Snart et al. (1984a, 1984b) suggested that *Byrsotria fumigata* and *Gromphadorhina portentosa,* two Blaberidae commonly considered ovoviviparous, should in fact be classified as viviparous. The surface of the brood sac in these two cockroaches is covered with numerous, closely packed papillae. Pores in the apical region of each papilla exude material thought to result from secretory activity of the brood sac, and the brood sac wall has ultrastructural features characteristic of insect integumentary glands. These authors suggest that the brood sac in these two ovoviviparous cockroaches is sufficiently similar to that of the viviparous *D. punctata* to make it likely that the brood sacs of all three function in the same manner. Depriving female *Byr. fumigata* and *G. portentosa* of food and water resulted in smaller nymphs, but the relative effects of food and water deprivation are unknown. Recent behavioral observations of *G. portentosa* indicate that the brood sac indeed may be producing secretions that serve as nutrition to young cockroaches; however, the material is expelled and ingested by neonates immediately after hatch instead of while they are embryos developing inside their mother (Chapter 8). Until demonstrated otherwise, then, *G. portentosa* should be considered ovoviviparous, with post-hatch parental feeding.

Four genera of Blaberidae, *Macropanesthia, Geoscapheus, Neogeoscapheus,* and *Parapanesthia* (Rugg and Rose, 1984b, 1984c), are classified as ovoviviparous type B and deposit their eggs directly into the brood sac, where they form a jumbled mass (Fig. 7.5B) rather than the two rows

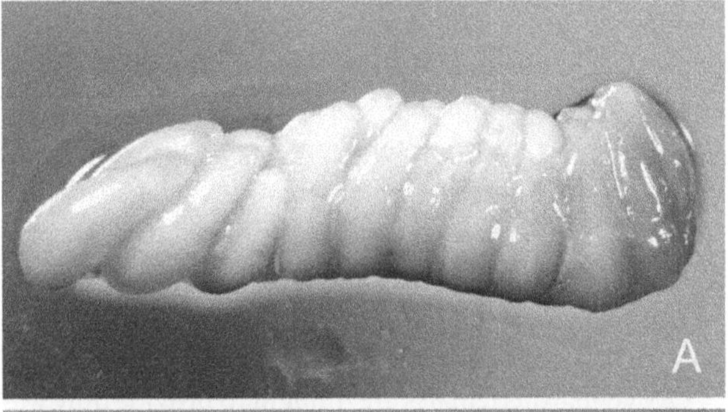

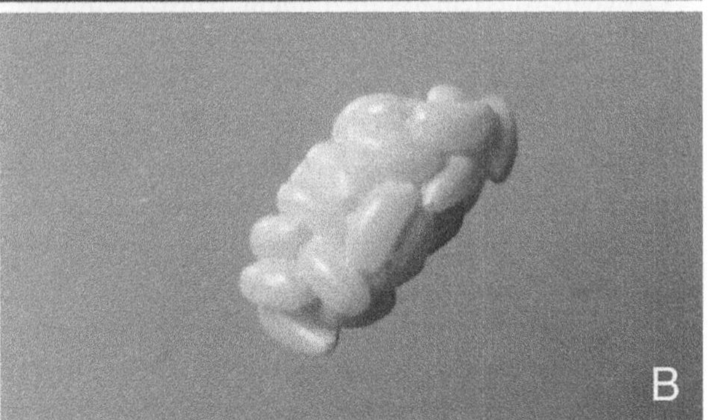

Fig. 7.5 Oothecae of two Panesthiinae. (A) Thin, membranous, incomplete oothecal case of *Panesthia cribrata* (ovoviviparity A). (B) Massed eggs of *Geoscapheus dilatatus,* a species that lacks an oothecal case (ovoviviparity B). Photos courtesy of Harley Rose.

typical of other cockroaches (Fig. 7.5A). These are the only cockroach taxa known to deposit eggs without forming an ootheca. Some species in the same subfamily (*Panesthia australis, Pane. cribrata*) exhibit an apparent intermediate stage, where some eggs occur in parallel rows within an incomplete oothecal membrane, while others are applied haphazardly to its outer surface as the ootheca is retracted. In *Pane. australis,* 90% of examined oothecae had eggs externally attached to the egg case (Rugg and Rose, 1984b, 1984c; D. Rugg, pers. comm. to CAN).

VIVIPARITY

Diploptera punctata is the only known viviparous species of cockroach. Its ootheca contains about a dozen small eggs and has an incomplete oothecal membrane (Roth and Hahn, 1964). Initially the eggs lack sufficient yolk and water to complete development (Roth and Willis, 1955a), but embryos ingest water and nutritive material synthesized and transported by the walls of the brood sac at a rate paralleling embryonic growth (Stay and Coop, 1973, 1974; Ingram et al., 1977). The brood sac "milk" is composed of about 45% protein, 5% free amino acids, 25% carbohydrates, and 16–22% lipids. The milk proteins are encoded by a multigene family that arose via the modification of genes preexisting in ovoviviparous species (Williford et al., 2004). Embryos begin oral intake of the milk just after closure of their dorsal body wall and continue until shortly before partition. The ultimate source of nutrition for the embryos is the food intake of the mother; females normally double their body weight during gestation, and the embryos of starved females die.

Diploptera nymphs are large and well developed when they emerge, requiring fewer molts to adulthood than any studied cockroach. Egg fresh weight increases more than 73 times during gestation (Table 7.2) (Roth and Willis, 1955a), while the fresh weight of the ovoviviparous species *N. cinerea* doubles. In the latter, the weight increase is correlated solely with the absorption of water; solids are slowly lost until partition (Roth and Willis, 1955c). Neonates of *D. punctata* are at least twice the size of those of *N. cinerea* (see Fig. 3 in Roth and Hahn, 1964), yet adults of the latter are considerably larger than field-collected adults of *D. punctata* (approximately 27 mm and 17 mm in length, respectively—Cochran, 1983a; WJB, unpubl. data). *Diploptera* females have three or four post-embryonic instars, compared with the usual seven to 13 in a sample of 11 other species of Blattaria (Willis et al., 1958). This suggests that *D. punctata* completes a substantial proportion of its juvenile development as an embryo, with a corresponding decrease in the duration of post-embryonic development. During embryogenesis,

Table 7.2. Changes in wet weight, water, and solids of cockroach eggs during embryogenesis (Roth and Willis, 1955a).

Species	Factors by which initial weights change, per egg		
	Wet weight	Water	Solids
Blatta orientalis	1.21	1.35	0.96
Blattella vaga	1.12	1.32	0.81
Blattella germanica	1.21	1.49	0.74
Nauphoeta cinerea	2.11	4.62	0.81
Diploptera punctata	73.47	85.80	49.28

closure of the dorsal body wall occurs at 19% of gestation, after which the embryos begin feeding on maternal secretions (Stay and Coop, 1973). Dorsal closure occurs at 46% of gestation time in *R. maderae* (Aiouaz, 1974), at 50% of gestation in *N. cinerea* (Imboden et al., 1978), and at 56% of gestation in *P. americana* (Lenoir-Rousseaux and Lender, 1970). Gestation of *D. punctata* embryos takes 63 days at 27°C (Stay and Coop, 1973); nymphs require just 43 to 52 days to become adults (Willis et al., 1958).

As might be expected of a group of embryos competing for food in a limited space, fewer eggs incubated by the mother results in larger nymphs. This was shown experimentally by Roth and Hahn (1964), who reduced the size of the litter in *D. punctata* by surgically removing one of the ovaries. Neonates in these broods were larger than those of control families, presumably because of the greater amount of nutritive material made available to the fewer developing embryos. In ovoviviparous *N. cinerea, R. maderae,* and *Eublaberus posticus,* however, the size of nymphs remains constant regardless of the number of incubated eggs (Roth and Hahn, 1964; Darlington, 1970). Nymphs within the same ootheca of *D. punctata* also can differ considerably in size depending on their position during development; embryos that have poor contact with the wall of the brood sac have less ready access to the nutritive secretion provided by the mother (Roth and Hahn, 1964). Neonate size, in turn, influences the number of stadia required to reach adulthood, the developmental response of individuals to their social environment, final adult size, and male sexual performance (Woodhead, 1984; Holbrook and Schal, 2004).

PARTHENOGENESIS

In a number of cockroach species, females are known to switch to an asexual mode of reproduction when isolated from males. The resultant offspring are always females, that is, these cockroaches display facultative thelytokous parthenogenesis. The phenomenon is known in *Blatta orientalis, B. germanica, Byr. fumigata, E. lapponicus, E. pallidus, N. cinerea, P. americana, P. fuliginosa, Polyphaga saussurei,* and *Su. longipalpa* (Roth and Willis, 1956; Barth, in Roth and Stay, 1962a; Brown, 1973a; Xian, 1998). Not all females of *N. cinerea* can reproduce by parthenogenesis; only those with a high level of heterozygosity are capable, and the ability tends to run in families (Corley et al., 2001). Parthenogenesis is rather common in *P. americana,* and can persist through two generations in the laboratory (Roth and Willis, 1956). Asexual reproduction, however, is clearly a fallback strategy that results in significantly reduced fitness in comparison to mated females. *Nauphoeta cinerea* virgins produce 10-fold fewer offspring than mated females, and nymphs are less viable, take longer to develop, have shorter adult life spans, and produce fewer offspring of their own when mated (Corley and Moore, 1999). Asexually produced oothecae, embryos, and hatched nymphs are often visibly deformed (Griffiths and Tauber, 1942a; Roth and Willis, 1956; Xian, 1998), and in *Ectobius,* few nymphs develop beyond the second instar (Brown, 1973a). Although the chromosome numbers of asexually produced embryos of *N. cinerea* ranged from 2n = 19 to 40, only those with the karyotype typical of the species (2n = 36) completed development to the hatching stage (Corley et al., 1999). Extreme variation in embryonic development within an ootheca can cause failure of the entire clutch. If few eggs develop, nymphs may be trapped in the oothecal casing, as hatch seems to require a group effort even in the thin, membranous oothecae of ovoviviparous cockroaches (Roth, 1974b).

Two cockroach species are known to be exclusively parthenogenetic. The best known is the cosmopolitan Surinam cockroach, *Pycnoscelus surinamensis.* This taxon is the asexual form of its sibling species *Pyc. indicus* (Roth, 1967b), and includes at least 21 diploid clones derived independently from sexual females and 11 triploid clones produced by backcrosses between clones and *Pyc. indicus.* There are more than 10 clones of *Pyc. surinamensis* in the southeastern United States alone (Roth and Cohen, 1968; Parker et al., 1977; Parker, 2002). In laboratory experiments females of *Pyc. surinamensis* tended to resist the overtures of male *Pyc. indicus,* but a few did mate and sperm transfer was successful. In these, the oocytes matured at the same rate as in virgins. Fertility was reduced, however, and all of the resultant offspring were female (Roth and Willis, 1961). In the bisexual *Pyc. indicus,* the oocytes of virgins develop slightly more slowly than those of mated females, but the proportion of oocytes that mature is the same. The oothecae, however, are almost always dropped without being retracted into the brood sac (Roth and Willis, 1961). Sperm in the spermathecae are required for normal oothecal retraction in this species (Stay and Gelperin, 1966), and if the ootheca is not quickly retracted, the enclosed eggs desiccate and die (Roth and Willis, 1955c). The evolution of parthenogenesis in *Pycnoscelus,* then, was dependent on overriding this dependence on sperm for oothecal retraction.

The number of eggs produced and matured by the obligately parthenogenetic *Pyc. surinamensis* is significantly less than that produced by sexual reproduction in its sister species (Roth, 1974b). Nonetheless, *Pyc. surinamensis* readily becomes established in a new location via a single nymph or adult, and has a widespread distribution (Roth, 1998b). It is found in tropical and subtropical habitats throughout the world, and in protected habitats, particularly greenhouses, in temperate climates (Roth, 1974b, 1998b). Its sexual sibling species *Pyc. indicus* is native to Indo-Malaysia and adjacent parts of Southeast Asia, and has colonized islands in the Pacific (Hawaii) and Indian (Mauritius) oceans. Both species may be found around human habitations, and both burrow in soil and are poor flyers. The widespread distribution of the asexual form is undoubtedly due to human transport, but the distribution pattern is also typical of geographic parthenogenesis (Niklasson and Parker, 1996), a condition in which a thelytokous race has a more extensive distribution than its sexual ancestor (Parker, 2002). *Pycnoscelus* has been used as a model to explore a variety of hypotheses on the subject (Gade and Parker, 1997; Niklasson and Parker, 1994; Parker, 2002; Parker and Niklasson, 1995).

Until recently, *Pyc. surinamensis* was the only case of obligatory parthenogenesis known in cockroaches. In 2003, a second case was reported in the Mediterranean blattellid species *Phyllodromica subaptera* by Knebelsberger and Bohn. The distribution of the sexual and asexual forms was studied by analyzing spermathecal contents and the sex of offspring. As in *Pycnoscelus,* the distribution of *Phy. subaptera* exhibits a pattern of geographic parthenogenesis: the asexual form is spread over most Mediterranean countries, while the bisexual forms are restricted to the Iberian peninsula. The parthenogenetic and sexual strains of *Phy. subaptera* cannot be distinguished by external morphology, suggesting that parthenogenesis is a relatively recent acquisition in the taxon.

FACTORS INFLUENCING REPRODUCTION

A variety of interacting factors are known to have an impact on the reproduction of female cockroaches, including food availability, body size, mating status, social contacts, and age (reviewed by Engelmann, 1970; Roth, 1970b). The presence of conspecifics accelerates reproduction in *B. germanica,* not only by influencing food intake but also via a more direct effect on juvenile hormone synthesis (Holbrook et al., 2000a). In *N. cinerea* maternal age is negatively correlated with fertility and lifetime fecundity. Old females take significantly longer than young ones to produce a first clutch. They also include fewer eggs per ootheca, and those eggs are slower to develop. Maternal age does not affect hatch rate, viability, nymphal development, or the reproductive potential of these nymphs when they became adults. While age does affect maternal fitness, then, it has no effect on the fitness of the offspring older females produce (Moore and Moore, 2001; Moore and Harris, 2003).

Species are differentially dependent on stored reserves for their first oviposition, varying from complete dependence (e.g., *R. maderae*—Roth, 1964b), to complete independence (e.g., *Pycnoscelus*—Roth and Stay, 1962a) (Table 7.3). Reproduction in relatively small blattellids can be closely tied to food availability. Females of *B. germanica* invest 34% of their pre-oviposition dry weight and 26% of their nitrogen into their first ootheca (Mullins et al., 1992). Female *Parc. fulvescens* typically store sufficient reserves to produce just one egg case, constituting 15–20% of her body weight (Cochran, 1986a; Lembke and Cochran, 1990). In larger species like *Periplaneta,* food intake is not necessary to mature the first batch of eggs, and females can produce up to five oothecae without feeding between successive ovipositions (Kunkel, 1966). Oothecae are just 7% of the weight of the unstarved female (Weaver and Pratt, 1981). Mating and feeding seem to have a synergistic effect in *N. cinerea* and *R. maderae,* since both stimuli are usually required for the

Table 7.3. Effect of starvation during the first preoviposition period in virgin and mated female cockroaches. See Roth (1970b) for citations of original work.

	Oocyte development[1]			
	Fed		Starved	
Species	Virgins	Mated	Virgins	Mated
Blattella germanica	+	+++	−	
Blattella vaga	+	+++	−	
Blaberus craniifer	+	+++	+	
Byrsotria fumigata	+	+++	+	
Eublaberus posticus	+−	+++	+−	+++
Nauphoeta cinerea	+	+++	+−	+++
Rhyparobia maderae	+−	+++	−	−
Pycnoscelus indicus	+++	+++	+++	+++
Pycnoscelus surinamensis	+++	+++	+++	+++
Diploptera punctata	−	+++	−	+++

[1](+++) develop and mature rapidly; (+) develop and mature; (+−) may or may not develop; (−) do not develop.

maximum rate of yolk deposition (Roth, 1964a, 1964b). Mating is necessary for initiation of yolk deposition in *D. punctata* (Engelmann, 1960; Roth and Stay, 1961), but has no effect on yolk deposition in *Byr. fumigata, Pyc. indicus,* or *B. germanica* (Roth and Stay, 1962a). Stimuli from feeding, drinking, mating, and social contact are required for the highest rates of yolk deposition in *P. americana.* A graded series of "sexually suppressed" females can be produced by withholding one or more of these stimuli (Weaver, 1984; Pipa, 1985).

EGG NUMBER AND SIZE

Comparisons of reproductive investment within a taxon require the resolution of differences attributable to body size. Although little information on the subject has been compiled for cockroaches, we do know that the body length of adults in the smallest species can be < 3% the length of the largest (Chapter 1), making them good candidates for investigations on the allometry of reproduction. At the species level there appears to be little relationship between the size of the mother and the packaging of the reproductive product. In the oviparous cockroaches, 18 mm long *Cartoblatta pulchra* females place about 95 eggs into an ootheca, more than any other species of Blattidae (Roth, 2003b). Ovoviviparous cockroaches average about 30 eggs per ootheca, but the relatively small *Panchlora* produces broods larger than a *Blaberus* 10 times its size and mass. *Panchlora nivea* is 2.5 cm long and internally incubates 60 or more eggs per clutch. The egg case is distorted into a semicircular or *J*-shape so that it may be internally accommodated (Roth and Willis, 1958b). The record, however, probably belongs to African *Gyna henrardi,* which somehow puts up to 243 eggs into a *z*-shaped ootheca that she stuffs into her brood sac (Grandcolas and Deleporte, 1998). Hatch must resemble the endless supply of clowns exiting a miniature car at the circus.

We know little regarding relative egg sizes among cockroaches. Two species with large post-ovulation investment are known to lay small eggs. In *C. punctulatus* eggs are only 44% of expected size for an oviparous cockroach of its dimensions (Nalepa, 1987). Most resources are channeled into an extensive period of post-hatch parental care and into the maintenance of the long-lived adults (Nalepa and Mullins, 1992). At hatch neonates in this species are tiny, blind, dependent, and fragile (Nalepa and Bell, 1997). Viviparous *D. punctata* also produces small eggs, with yolk insufficient to complete development (Roth, 1967d). As with all viviparous animals, supplying embryos with gestational nutrients places less reliance on producing large yolky oocytes. Neonates emerge at the precocial extreme of the developmental spectrum, with the largest relative size and shortest postembryonic development known among cockroaches.

EVOLUTION OF REPRODUCTIVE MODE

Of the two major divisions of the cockroaches, the superfamilies Blattoidea and Blaberoidea (McKittrick, 1964), most evolutionary drama with regard to reproductive mode is in the latter. It includes the Blattellidae, in which some species retain the egg case externally for the entire period of gestation, and where ovoviviparity arose independently in two different subfamilies. It also includes the Blaberidae, all of which incubate egg cases internally, suggesting that they have radiated since an ancestor acquired the trait. The sole viviparous genus, as well as the group that lost the oothecal covering, are in the Blaberidae. Of course, critical analysis of the pattern of reproductive evolution is dependent on the availability of robust phylogenies for the groups under study, but, as with most aspects of cockroach systematics, the relationships among several subgroups of the Blaberoidea are unsettled. In all phylogenetic hypotheses proposed so far, however, Blaberidae is most closely related to Blattellidae (Roth, 2003c), and some studies (Klass, 1997, 2001) suggest that blaberids are a subgroup of the Blattellidae.

The evolution of reproductive mode in cockroaches can be described with some confidence as a unidirectional trend from oviparity to viviparity, without character reversals. Reproduction is an extraordinarily complex process, with morphology, physiology, and behavior integrated and coordinated by neural and endocrine mechanisms. Transitions therefore tend to be irreversible due to genetic or physiological architecture, or because strong selection on offspring prevents them (Tinkle and Gibbons, 1977; Crespi and Semeniuk, 2004). An initial step in the evolution of ovoviviparity in cockroaches was likely to be facultative transport of the egg case, as in the oviparous type A species that retain oothecae until a suitable microhabitat is found. *Ectobius pallidus,* for example, typically deposits its egg case in one or two days, but has been reported to carry it 16 days or longer (Roth and Willis, 1958a). *Therea petiveriana* deposits the ootheca within a day of extrusion, but may retain it for as long as 90 hr if a suitably moist substrate is not available (Livingstone and Ramani, 1978). From this flexible starting point, the trend toward ovoviviparity would be exemplified by cockroaches that retain the egg case for the entire period of embryogenesis, but provide no materials additional to those originally in the egg case. Currently, there are no records of extant cockroaches that exhibit this pattern; the only oviparous type B species that has been studied, *B. germanica,* provides water and soluble materials to embryos. Obligate egg retention evolves when maternal tissues became responsive to the attached egg case; this recognition then induces further modifications of maternal function (Guillette, 1989).

Oothecal Rotation

The position of the ootheca while it is carried prior to deposition is taxonomically significant and important in understanding the evolution of reproductive mode in cockroaches (Roth, 1967a). All of the Blattoidea and some of the Blaberoidea carry the ootheca with the keel dorsally oriented. However, in some Blattellidae and in all of the Blaberidae, the female rotates the ootheca 90 degrees so that the keel faces laterad at the time it is either deposited on a substrate, carried externally for the entire period of embryogenesis, or retracted into the brood sac. Within the Blattellidae, rotation of the ootheca has been used as a taxonomic character to separate the non-rotators (Anaplectinae and Pseudophyllodromiinae) from the rotators (Blattellinae, Ectobiinae, and Nyctoborinae) (McKittrick, 1964). Most studies (McKittrick 1964; Roth, 1967a; Bohn, 1987; Klass, 2001) indicate that ootheca rotation evolved just once, and the recent phylogenetic tree of Klass and Meier (2006) (see Fig. P.1 in Preface) supports this view. One must be careful in determining oothecal rotation in museum specimens, as females may have been preserved while in the process of oothecal formation, prior to rotation. LMR found females with rotated oothecae from groups that do not normally exhibit this character; a museum worker had glued the oothecae to the females in an "incorrect" orientation. Some Polyphagidae exhibit a "primitive" or "false" type of rotation in which the ootheca is rotated and held by a "handle" or flange at the female's posterior end (Roth, 1967a). This type of rotation may have evolved as a way to prevent oothecae from being pulled off females as they move through sand (Fig. 2.6). The oothecae itself does not contact the female's vestibular tissues and ovoviviparity did not evolve in this group.

Transition to Live Bearing

Oothecal rotation is a key character when comparing the cockroach lineages that evolved ovoviviparity. Only one of the two subfamilies of Blattellidae exhibiting this reproductive mode rotates its egg case, but rotation occurs in all Blaberidae. Within the Blattellinae, the oviparous type B species, as exemplified by *B. germanica,* rotate the ootheca 90 degrees once it is formed and females carry it that way throughout gestation (Fig. 7.6). The ootheca is thus reoriented from its initial vertical position to one in which the long axes of the oocytes lay in the plane of the female's width. When first formed the egg cases are much

Fig. 7.6 *Blattella germanica* female carrying a fully formed ootheca (scale = mm). Photo courtesy of Donald Mullins.

taller than they are wide, like a package of frankfurters standing on end. Rotation likely evolved to prevent dislodgment of these egg cases as the morphologically flattened females scurried through crevices (Roth, 1968a, 1989a). Females of *B. germanica* that carry a rotated ootheca are able to crawl into spaces narrower than females carrying them in the vertical position (Wille, 1920). A gravid female one day before oviposition needs a space of 4.5 mm. A female with the ootheca carried in the vertical position requires 3.3 mm, and after the egg case is rotated the female can move into a space 2.9 mm high. Ovoviviparous cockroaches in the same subfamily as *Blattella* (e.g., *Stayella*) carry within their brood sac a rotated ootheca virtually identical to the externally carried, rotated egg case of *B. germanica* (Roth, 1984).

In the second blattellid subfamily with oviparous type B reproduction (Pseudophyllodromiinae), two species of *Lophoblatta* maintain the original vertical position of the ootheca while carrying it externally throughout gestation. These oothecae, however, are distinctly wider than high (Roth, 1968b). Ovoviviparous females in this subfamily (e.g., *Sliferia*) have similarly squat oothecae, and retract them while they are vertically oriented, without rotation. The two blattellid subfamilies, then, employ different but equivalent mechanisms for achieving the same end. An ootheca of dimensions appropriate for a crevice-

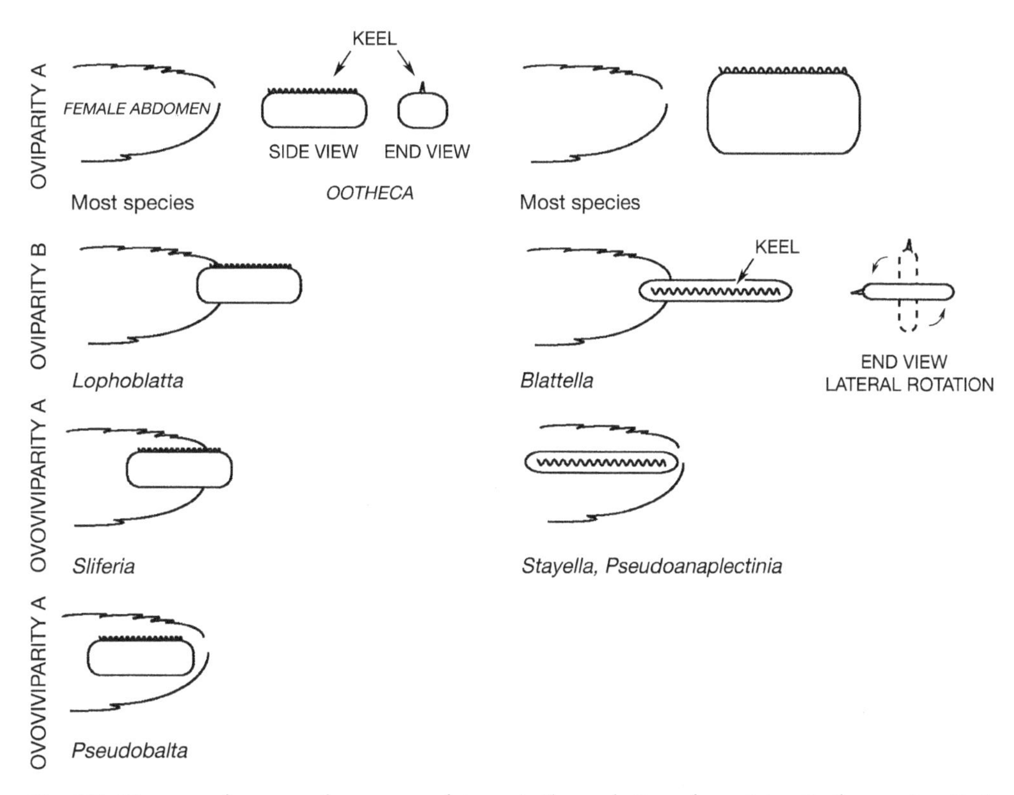

Fig. 7.7 Diagram of presumed sequence of stages in the evolution of ovoviviparity from oviparity in two subfamilies of Blattellidae. Note the difference in the orientation of the ootheca between the two subfamilies. Current evidence suggests that the oothecal rotation exhibited by the Blattellinae and by the ovoviviparous Blaberidae originated in a common ancestor.

dwelling insect to carry or internalize must be either squashed dorsoventrally or rotated so that it is as flat as the female (Fig. 7.7). Intermediate stages in parity mode are conspicuous in the Pseudophyllodromiinae. *Sliferia* is considered ovoviviparous; nonetheless the egg case is partially exposed while it is carried. Initially it was thought that these females were collected while still forming the ootheca. Now this condition is considered the norm, and points up the continuum of reproductive modes in this subfamily (Roth, 2003b).

All species in the ovoviviparous family Blaberidae carry a rotated egg case in their brood sac and are thought to have evolved from a *Blattella*-like ancestor (Roth and Willis, 1955c; Roth, 1967a; Mullins et al., 2002). Except for retraction of the egg case into the body, *B. germanica* exhibits all characteristics of an ovoviviparous cockroach (Roth and Willis, 1958a; Roth, 1970a). The oothecal case is thinner and less darkly colored than in other oviparous cockroaches, there is flow of water and other materials between mother and unhatched offspring, and oogenesis is suspended while females are carrying egg cases. The evolution of ovoviviparity would require only a minor transition from that starting point. Ovoviviparity evolved independently two or three times in cockroaches, but only in the blattellid/blaberid lineage (Roth, 1970a, 1989a): once in the Pseudophyllodromiinae, and once or twice in the clade that includes Blattellinae and Blaberidae. Viviparity evolved once, in *D. punctata* of the monogeneric subfamily Diplopterinae. Some authors also include *Calolampra* or *Phoetalia* in this subfamily (Roth, 2003c), so these genera may be logical targets for comparative study. Worldwide, Blattellidae is the largest cockroach family with about 1740 described species; there are approximately 1020 species of Blaberidae. The oviposition behavior is known in relatively few genera and species of these two families (Roth, 1982a).

Reduction and Loss of the Egg Case

In most oviparous type A cockroaches, the ootheca is a hard, dark, stiff structure completely covering the eggs. The dorsal keel is structurally complex, and the outer covering contains calcium oxalate crystals. These crystals comprise 8–15% of the dry weight of the ootheca in *P.*

americana, and are thought to have a structural and protective function (Stay et al., 1960; Rajulu and Renganathan, 1966), just as they do in plants that possess them (Hudgins et al., 2003). The oothecal casing is thinner and less rigid in species that externally carry the egg case (oviparous type B); calcium oxalate crystals are sparse in both *B. germanica* and *Loph. brevis* (Roth, 1968b). Ovoviviparous type A cockroaches typically produce a thin, soft, lightly colored ootheca that lacks a keel and which in some species only partially covers the eggs, particularly in later stages of gestation (Roth, 1968a) (Fig 7.5A); calcium oxalate is absent. This type of egg case is produced by Blaberidae and also *Sliferia,* one of few Blattellidae that retract their ootheca into a brood sac (Stay et al., 1960; Roth, 1968a). The nature of the ootheca, then, changes in parallel with stages of internalization of the egg case. It goes from having a rigid outer casing in those species that abandon the egg case, to a flexible, soft membrane in those that have internalized it. It has intermediate properties in those cockroaches that carry the ootheca externally during gestation, and has been completely lost in one derived lineage (Geoscapheini: ovoviviparous type B) (Roth and Willis, 1958a; Roth, 1968a, 1970a). Females exhibit a parallel regression of the morphological structures associated with oothecal production (reviewed by Nalepa and Lenz, 2000).

Oviparous cockroaches in protected environments, like social insect nests, also may exhibit reduction or loss of the egg case. The ootheca of *Attaphila fungicola,* for example, lacks a keel (Roth, 1971a), and several species of Nocticolidae have thin, transparent oothecal cases. *Nocticola termitophila* apparently lays its eggs singly, without any external covering (Roth, 1988). Termites, the "social cockroaches" (Chapter 9), exhibit a parallel loss of protective egg cases. The basal termite *Mastotermes darwiniensis* packages its eggs within a thin, flexible outer covering that lacks keel. The site and mode of production, associated morphological structures in the female, parallel arrangement of eggs, and discrete, tanned outer covering together indicate that the ootheca of *Mastotermes* is homologous with those of cockroaches (Nalepa and Lenz, 2000). All other termites lay their eggs singly, without a covering. Both the heart of a social insect colony and the brood sacs of live bearing cockroaches are moist, protected sites for incubating eggs, allowing for the reduction and eventual elimination of defensive structures in evolutionary time. The oothecal case is 86–95% protein (Table 4.5), so "it is no wild supposition that in the course of time the chitinous ootheca, being in these species a work of supererogation, will disappear" (Shelford, 1912b). Perhaps the main reason that the ootheca has not been completely eliminated in most ovoviviparous cockroaches is because it determines the orderly arrangement of eggs and therefore assures contact and exchange of water and other materials between each egg and the wall of the brood sac (Rugg and Rose, 1984b). A study of the Geoscapheini whose eggs are incubated in a disordered mass in the brood sac (Rugg and Rose, 1984c) (Fig. 7.5B) is the logical focal group for testing this hypothesis.

Selective Pressures

Most hypotheses offered to explain why live bearing has evolved in animals invoke agents affecting offspring viability as the selective pressure for an evolutionary shift in reproductive mode. Costs that accrue to mothers then either facilitate or constrain the transition. These may include reduced maternal mobility, with consequences for foraging efficiency and predator evasion, reduced fecundity, and the increased metabolic demands of carrying offspring throughout their development (Shine, 1985; Goodwin et al., 2002, among others). It is difficult, however, to use present-day characteristics of ovoviviparous or viviparous organisms as evidence for hypotheses on the evolution of these traits, as current habitats may be different from the habitats in which the reproductive modes first evolved (Shine, 1989). It is also important to note that each strategy has its benefits and liabilities in a given environment. Oviparity is not inherently inferior to ovoviviparity or viviparity just because it is the ancestral state. The problem of water balance in cockroaches, for example, is handled by each reproductive mode in different ways, each of which may be optimal in different habitats. Egg desiccation can be minimized if: (1) the ootheca is deposited in a moist environment, (2) the ootheca has a waterproofing layer, or (3) the female dynamically maintains water balance while the egg case is externally attached or housed in a brood sac (Roth, 1967d).

Increased Offspring Viability

McKittrick (1964) was of the opinion that the burial and concealment of oothecae by oviparous females is a response to pressure from parasitoids and cannibals. Although few studies directly address this question, some evidence suggests that concealing oothecae may attract rather than deter hymenopterous parasitoids. The mucopolysaccharides in the saliva used to attach egg cases to the substrate may act as kairomones, making oothecae more vulnerable to attack. Parasitic wasps may even expose buried oothecae by digging them out from their protective cover (Narasimham, 1984; Vinson and Piper, 1986; Benson and Huber, 1989). On the other hand, oothecae of *P. fuliginosa* that were glued to a substrate had a higher eclosion rate than those that were not glued, suggesting that salivary secretions may enhance egg viability in some unknown way (Gordon et al., 1994). Oothecae of

Fig 7.8 Parasitism of cockroach eggs. (A) *Anastatus floridanus* ovipositing into an ootheca carried by *Eurycotis floridana.* (B) Detail of oviposition by the parasitoid. Photos by L.M. Roth and E.R. Willis.

oviparous cockroaches are also prone to parasitism prior to deposition, while females are forming and carrying them. The window of vulnerability can be a wide one. Females of *Nyc. acaciana,* for example, can take 72 hr to form an ootheca (Deans and Roth, 2003). The parasitoid *Anastatus floridanus* (Eupelmidae) oviposits in egg cases attached to female *Eur. floridana* (Fig. 7.8) (Roth and Willis, 1954a). The cockroach can detect the presence of the wasp on the surface of the ootheca and tries to dislodge it with her hind legs (LMR, pers. obs.). *Blattella* spp. that carry egg cases externally until hatch are also vulnerable to egg parasitoids, and continue to carry the parasitized ootheca (Roth, 1985). External retention of egg cases, then, may be little better than concealment in conferring protection from parasitism.

The value of egg case burial lies primarily in protecting them from predation and cannibalism; concealment is almost 100% effective in saving oothecae from being devoured by other cockroaches (Rau, 1940). McKittrick et al. (1961) found that in *Eur. floridana,* burial of oothecae prevented cannibalism by conspecifics and predation by ants, carabids, rodents, and other predators. Conversely, exposed egg cases and those still attached to a female are subject to biting and cannibalism (Roth and Willis, 1954b; Willis et al., 1958; Gorton, 1979). These improprieties are countered with aggression on the part of the mother. Female *P. brunnea, P. americana,* and *Paratemnopteryx couloniana* drive other females away from exposed oothecae (Haber, 1920; Edmunds, 1957; Gorton, 1979). Two behavioral classes of female can be distinguished in *B. germanica;* females carrying oothecae are more aggressive than females that had not yet formed them (Breed et al., 1975). Aggressive behavior is favored despite its attendant risks, given that one nip taken from an ootheca can result in the death of the entire clutch from desiccation (Roth and Willis, 1955b).

Ovoviviparity is viewed as a solution to this constant battle against predators and parasites, and is thought to have appeared in the Mesozoic as an evolutionary response to cockroach enemies that first appeared during that time (Vishniakova, 1968). Parasitoids have not been detected in the oothecae of ovoviviparous blaberids (LMR, pers. obs.). The eggs are exposed to the environment for only the brief period of time between formation of the ootheca and its subsequent retraction into the body, allowing only a narrow time frame for parasitoid oviposition. Once in this enemy free space, the eggs are subject only to "the vicissitudes that beset the mother" (Roth and Willis, 1954b). Nonetheless, nymphs of ovoviviparous cockroaches are at risk from cannibalism at the time of hatch. Attempts by conspecifics to eat the hatchlings as the female ejects the ootheca have been noted and may include pulling the still attached egg case away from the mother (Willis et al., 1958). We note, however, that laboratory observations of cannibalism in cockroaches of any reproductive mode may be of little consequence in natural populations, with the exception of highly gregarious species like cave dwellers. Females of at least one species of the latter are known to be choosy about where they expel their neonates. Darlington (1970) reported that pregnant females of *Eub. posticus* preferred one chamber of the Tamana cave for giving birth, and migrated into that chamber from other parts of the cave. Defense against pathogens as agents of egg mortality is unstudied, despite the disease-conducive environments typical of cockroaches.

Parental Costs

Indirect reproductive costs of oviparity in cockroaches include the time, energy, and predation risks involved in concealing the ootheca in the environment and the metabolic expense of producing a protective oothecal case. The case consists primarily of quinone-tanned protein (Brunet and Kent, 1955) (Table 4.5), much of which can be recovered after hatch if the parent or neonates eat the embryonic membranes, unviable eggs, and the oothecal case after hatch (Roth and Willis, 1954b; Willis et al., 1958). In several species of cockroaches, oothecal predation by adults and the ingestion of oothecal cases after hatching by nymphs increases when other protein sources are lacking (WJB, unpubl. obs.).

Live bearing permits females to dispense with producing a thick, protective oothecal case, and allows them to channel the protein that would have been required for its manufacture into present or future offspring or into their own maintenance. Nonetheless, the burden of "wearing" the next generation may be metabolically expensive and impair mobility, with consequences for predator evasion and foraging efficiency. In *B. germanica,* however, Lee (1994) found no correlation between the physical load on the female and oxygen consumption, and in *N. cinerea* the mass-specific metabolic heat flux of pregnant females at rest was actually reduced in relation to non-pregnant females. This suggests that the energetic demands of gestation in these species do not translate into increased metabolic rates (Schultze-Motel and Greven, 1998). Still, most female cockroaches feed little, if at all, during gestation, even when offered food ad libitum in the laboratory (e.g., *Blattella*—Cochran, 1983b; Hamilton and Schal, 1988; *Rhyparobia*—Engelmann and Rau, 1965; *Trichoblatta*—Reuben, 1988). The most commonly offered explanation for fasting at this time is that the cumbersome bodies of pregnant females may increase their vulnerability to predation. This seems reasonable, given that, first, the mass of the reproductive product is 30% or more of female body weight in both *B. germanica* (Mullins et al., 1992; Lee, 1994) and *N. cinerea* (Schultze-Motel and Greven, 1998), and second, pregnant *N. cinerea* are demonstrably slower than virgin females of the same age (Meller and Greven, 1996a). Agility also may be affected. Ross (1929), however, opined that pregnant *B. germanica* "do not show any signs of being impeded by their burden" despite the clumsy ootheca dragging from their nether regions. Loss of agility may not be an issue in cockroaches that rely on crypsis or thanatosis to escape predators, but the larger body of gravid females requires a larger crevice in species that seek protective shelter (Koehler et al., 1994; Wille, 1920). It is unknown whether the physical burden of an egg clutch hinders flying in those species that depend on it for evasion. *Blattella karnyi* females can take to the air while carrying an impressive ootheca of up to 40 eggs (Roth, 1985).

In viviparous *D. punctata,* gravid females normally double their body weight during gestation but nonetheless forage; the nutrient secretion of the brood sac is derived from the maternal diet rather than stored nutrients, particularly in early pregnancy (Stay and Coop, 1974; WJB, unpubl. data). This species has hard, dome-shaped tegmina (common name = "beetle cockroach") and impressive defensive secretions (Eisner, 1958; Roth and Stay, 1958) that may permit some bravery when under attack by ants (Fig. 1.11A). Vertebrate predators, however, are threats, and lizards, toads, and birds have been observed eating them in the field (Roth and Stay, 1958; WJB, pers. obs.). It is possible that *D. punctata* females rely on readily accessible, predictable sources of high-quality food for supporting the explosive growth of their embryos. Their diet, however, appears little different from that of many other cockroaches.

Reduced Fecundity

One of the most significant costs exacted by carrying egg cases lies in terms of fecundity. Oviparous type A cockroaches have relatively high reproductive rates because the interval between successive oothecae is short, usually much shorter than the period of incubation. Females typically produce a second egg case long before the first laid hatches. Oviparous species with external egg retention as well as ovoviviparous females produce relatively few oothecae because oocytes do not mature in the ovaries while an ootheca is being carried. Viviparity is particularly expensive, in that female *D. punctata* have fewer eggs per oothecae, produce fewer oothecae per lifetime, and have a longer period of gestation than any other blaberid (Roth and Stay, 1961; Roth, 1967d). Consequently, the number of egg cases per lifetime decreases and the oviposition interval increases in the order oviparous, ovoviviparous, viviparous (Fig. 7.9) (Willis et al., 1958; Roth and Stay, 1959, 1962a; Breed, 1983).

Fecundity also appears reduced in cockroach species that exhibit parental care, particularly if the care involves feeding young dependents on bodily fluids. Such pabulum may be demanding in terms of the structures involved in its manufacture, the nutrients incorporated into the secretions, and the energy required to produce them.

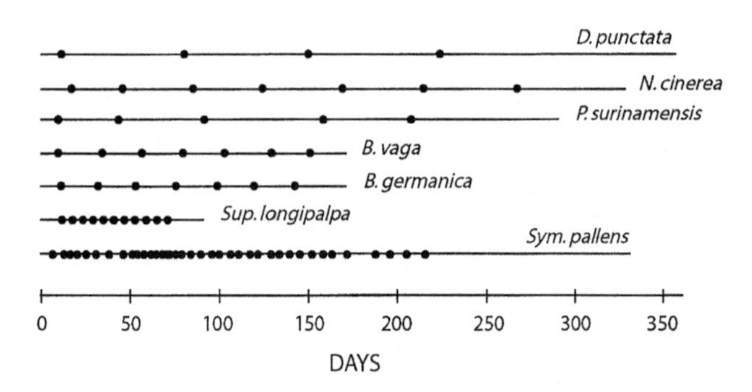

Fig. 7.9 Frequency of oviposition by individuals of different species of cockroach. Each dot represents the formation of an ootheca; the length of the line is the adult lifespan of the female. *Symploce pallens* (= *hospes*) and *Supella longipalpa* (Blattellidae) are oviparous and drop the ootheca shortly after it is formed. *Blattella germanica* and *B. vaga* (Blattellidae) carry their ootheca externally until the eggs hatch. The blaberids *Pycnoscelus surinamensis* (parthenogenetic) and *Nauphoeta cinerea* are ovoviviparous, and *Diploptera punctata* is viviparous. After Roth (1970a).

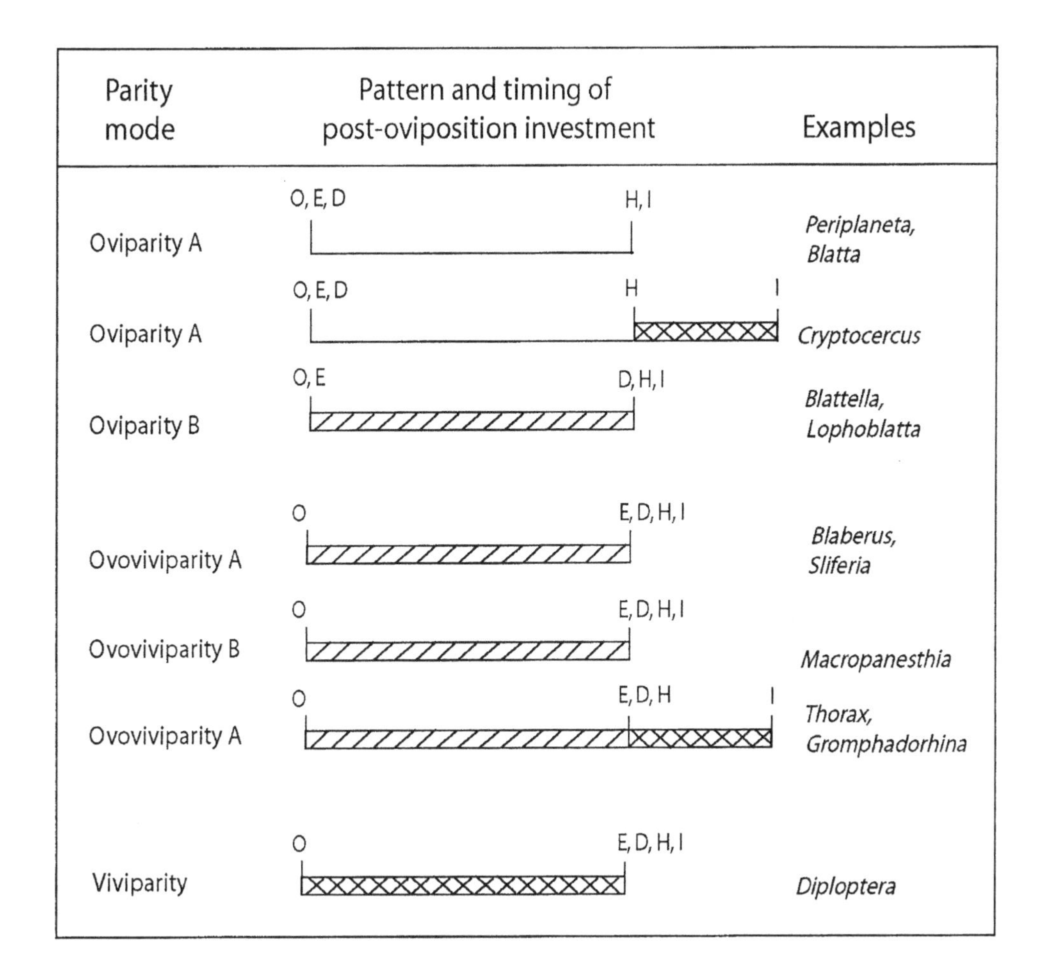

O = oviposition
E = extrusion of ootheca (eggs)
D = deposition of ootheca
H = hatch
I = independence of offspring

[hatched] = provisioning of water, small molecules
[cross-hatched] = provisioning of non-yolk nutrients

Fig 7.10 Post-oviposition provisioning in cockroaches. Oviposition refers to the release of eggs from the ovaries, while extrusion is the permanent expulsion of eggs from the body. Deposition is the dissociation of the egg case from the body. Independence is the ability of neonates to live apart from the parent(s). Modified from Nalepa and Bell (1997), with the permission of Cambridge University Press.

Perisphaerus sp. and *Thorax porcellana* both exhibit a reduction in the number of offspring per clutch as compared to other ovoviviparous species (Roth, 1981b).

PARENTAL INVESTMENT

In the majority of oviparous type A cockroaches females make their principal direct investment prior to fertilization, by supplying eggs with yolk nutrients. They then envelope the eggs in a protective covering and deposit them in a safe place for incubation. With the exception of *Cryptocercus*, there is no additional parental involvement. In species with external retention, like *Blattella*, embryos are dependent on yolk to fuel development but are also progressively supplied with water and some non-yolk nutrients during gestation (Fig. 7.10). This is likewise true of ovoviviparous cockroaches, but in several species neonates continue their dependence on maternally supplied nutrients for a period of time after hatch. These take the form of digestive fluids and glandular secretions; at least six types are known (Chapter 8). Cockroaches tend to have a very glandular integument, allowing for the repeated evolution of nutritive secretions from cuticular surfaces. Williford et al. (2004) recently demonstrated that proteins in the milk secreted by the brood sac of *Diploptera* are coded by genes from the same family (lipocalin) as those that code for a protein in the tergal gland secretion of *R. maderae* (Korchi et al., 1999).

One consequence of this variation in investment strategies is that it is not always easy to place cockroaches into distinct reproductive categories. There is a continuum between species that externally retain their egg cases and those that internalize them, obvious in Figs. 7.7 and 7.10. The location of the egg case during gestation differs in the *Lophoblatta-Sliferia-Pseudobalta* series, but the investment strategy is basically the same. Another example is a comparison between the viviparous *Diploptera* and the ovoviviparous *Gromphadorhina.* Both species apparently provision offspring on secretions that originate from the brood sac walls. *Diploptera* does so progressively, during gestation. *Gromphadorhina* and possibly other Blaberidae (*Byrsotria, Blaberus, Rhyparobia*) (Perry and Nalepa, 2003) expel it en masse for consumption by nymphs immediately after partition.

Termination of Investment

If a female cockroach has initiated a reproductive episode that is threatened for lack of food or other reasons, she has several options for converting reproductive investment back into somatic tissue, thereby maintaining and redirecting her resources (Elgar and Crespi, 1992). Termination of investment can occur at several points in the reproductive cycle. Prior to ovulation, starvation increases oocyte resorption in cockroaches (reviewed by Bell and Bohm, 1975). In *P. americana,* most starved females produce one, sometimes two, oothecae in addition to the one being produced when starvation is initiated (Bell, 1971). Large yolk-filled oocytes are retained in the ovaries of those females that do not deposit a second ootheca, and beginning on about the 10th day of starvation these oocytes are resorbed and the vitellogenins stored. When feeding resumes, these stored yolk proteins are rapidly incorporated into developing oocytes. In *Xestoblatta hamata,* both resorption of proximal oocytes and an extension of the interval between oothecae are common in the field and are the result of unsuccessful foraging (Schal and Bell, 1982; C. Schal, pers. comm. to WJB).

After ovulation, females have other mechanisms for terminating reproductive investment. Abortion can occur in laboratory cultures if gestating females are disturbed in *Pyc. surinamensis, Panchlora irrorata,* and *Blaberus craniifer* (Nutting, 1953b; Willis et al., 1958; Willis, 1966). It is unknown if and under what circumstances ovoviviparous and viviparous cockroaches jettison egg cases under natural circumstances; the possibility exists that they may relieve themselves of their oothecal burden if suddenly pursued by a predator in their natural habitat. This tactic may be more likely in those cockroaches that that use speed/agility to escape predators rather than crypsis or defensive sprays.

Post-partition, cannibalism can be a means of recovering and recycling a threatened reproductive investment. If disturbed when nymphs are freshly hatched, adults of *C. punctulatus* may cannibalize their entire brood (CAN, unpubl. obs.). Other cockroach species are known to eat their young (Roth and Willis, 1954b), and starved females are often more likely to do so (Roth and Willis, 1960; Rollo, 1984b; WJB, unpubl. obs.).

EIGHT

Social Behavior

The only useful outcome of my attempt to classify types of parental care into mutually exclusive sets was that it made clear that from many points of view by far the largest group of insects that exhibit parental care (is) the cockroaches.

—H.E. Hinton, *Biology of Insect Eggs*

It is difficult to conceive of any group of animals that are as universally and diversely social as cockroaches. Given the range of habitats they have mastered and their versatility in reproductive mode and feeding habits, it is unsurprising that they exhibit extraordinary variation in their social organization. Individual taxa are typically described as solitary, gregarious, or subsocial. We structure this chapter around those categories, treating each in turn, with the caveat that this simplistic pigeonholing masks the head-banging vexation we encountered in attempting to classify the social heterogeneity present. Cockroaches that live in family groups are a rather straightforward category, and domestic pests and a number of cave-dwelling species are without a doubt gregarious. For a variety of reasons many others elude straightforward classification. First, the majority of cockroaches are unstudied in the field, and the nature and frequency of social interactions have been specified in few species. With perhaps a score of exceptions, our concept of cockroach social organization is largely based on anecdotal evidence and brief observations noted during collection expeditions for museums. Second, cockroaches are often assigned social categories without specifying the employed criteria, and the terms describing their social tendencies have been used in a vague or inconsistent manner (discussed below). Third, evidence to date suggests that sociality in Blattaria is not as straightforward as it is in many insects. There is considerable spatial and temporal variation in social structure, influenced by, among other factors, the age and sex of the insects, environmental condition, physiological state, population density, and harborage characteristics. Fourth, many cockroaches are nocturnal and cryptic; consequently even those that live in laboratories can be full of surprises. Parental feeding behavior was only recently observed in *Gromphadorhina portentosa,* a species commonly kept in homes as pets, in laboratories for experiments, and in museums for educational purposes (Perry and Nalepa, 2003). Fifth, even closely related species can vary widely in social proclivities. The German cockroach *Blattella germanica* is strongly gregarious; it has been the test subject of the vast majority of studies on cockroach aggregation behavior. Its closely related congener *B. signata,* however, is apparently solitary (Tsai and Lee, 2001). Sixth, laboratory data can conflict with field descriptions. One example: studies on *Schultesia lampyridiformis* reared for 20 yr in the laboratory suggest that females use aggression to disperse nymphs after hatch (Van Baaren and Deleporte, 2001; Van Baaren et al., 2003). In the field (Brazil), however, Roth (1973a) found adults and nymphs living together in birds' nests. One nest contained 4 males, 8 females, and 29 nymphs, and other cockroach species were also present. Lastly, the division of species into group living and solitary categories is largely artificial in any case because most animal species are in an intermediate category, found in association with conspecifics at certain times of their lives, but not others (Krause and Ruxton, 2002).

These issues, and others, have bearing on phylogenetically based comparative analyses of cockroach social behavior. While these can be powerful tools for generating and testing ideas about the links between behavior and ecology, attempts to map social characteristics onto cladograms of cockroach taxa are premature. We are still early in the descriptive phase of cockroach social behavior, and unresolved phylogenies in many cases preclude meaningful comparative study. Some general trends are detectable and will be discussed below.

SOLITARY COCKROACHES

Currently, few cockroach species are convincingly classified as solitary, that is, leading separate lives except for a brief period of mating. One category of loners may be those cockroaches adapted to deep caves. Although they may cluster around food sources, troglobites are typically solitary animals, have wide home ranges, and meet only for mating (Langecker, 2000). The blattellid *Phyllodromica maculata* is considered solitary, as juveniles do not aggregate, nor are they attracted to filter paper contaminated by conspecifics (Gaim and Seelinger, 1984). *Paratemnopteryx couloniana* was called "relatively solitary" by Gorton (1979), but without statement of criteria. *Thanatophyllum akinetum* was described as solitary by Grandcolas (1993a). The insects spend much of their time motionless and flattened against dead leaves on the forest floor in French Guiana. Laboratory tests support the observation that individuals actively distance themselves from conspecifics (Van Baaren and Deleporte, 2001). A solitary, cryptic lifestyle is thought to allow them to escape detection by army ants (Grandcolas, 1998). Nonetheless, the female broods offspring for several hours following hatch, which is a subsocial interaction, albeit short term, between a mother and her offspring. *Lamproblatta albipalpus* was described as solitary by Gautier et al. (1988), but considered "weakly gregarious" by Gautier and Deleporte (1986). Males and females of this species are found together in resting sites, but their bodies are not in direct contact. Even strongly gregarious cockroaches, however, can be separated in space within a shelter under certain environmental conditions, for example, high relative humidity (Dambach and Goehlen, 1999).

AGGREGATIONS: WHAT CRITERIA?

A variety of nonexclusive criteria have been used to delineate cockroach aggregation behavior. These include their arrangement in space (are they in physical contact?), mechanisms that induce grouping (is a pheromone involved?), and the outcome of physical proximity (do group effects occur?). Aggregations have been described as mandatory, nonobligatory, strong, weak, and loose, without further detail. To most entomologists, mutual attraction is considered the primary criterion of aggregation behavior (Grassé, 1951; Sommer, 1974); group membership involves more than co-location, with individuals behaving in ways that maintain proximity to other group members. In practice, the distinction is not easily made, because in most cases both environmental and social influences play a role (Chopard, 1938). Many cockroaches predictably seek dark, humid, enclosed spaces as shelter, and live in close association with nutritional resources. The functional basis of a nonrandom distribution is especially vague for the vast majority of cockroaches regarded as crevice fauna: those found in small groups in small shelters, for example, under logs, in leaves, under stones, under loose bark. Eickwort (1981) suggested testing aggregation behavior by supplementing the resources of a group to see if it results in dispersion of the insects. Tsuji and Mizuno (1973) and Mizuno and Tsuji (1974) gave *Periplaneta americana, P. fuliginosa, P. japonica,* and *B. germanica* excess harborage and found that while adults and older nymphs shelter individually, young nymphs seek conspecifics. The results are difficult to interpret, because all these test species are commonly found in multigenerational aggregations.

What, then, are necessary and sufficient criteria for calling a cockroach gregarious? Are two nymphs found together considered a group? Do they have to be the same species? Are neonates that remain near a hatched ootheca for an hour before dispersing gregarious? What if they remain for 3 days? Do aggregation pheromones have to be involved? Do the insects have to be touching? The literature provides no easy answers. A broad range of variables influences the degree to which individuals are positive, neutral, or negative with regard to joining a group. These include genetics, physiology, informational state, geographic region, and the experimental protocol used to test them (Prokopy and Roitberg, 2001). Behavioral observations, distance measures, and association patterns in the field are all appropriate (Whitehead, 1999), but an explicit description of the criteria used in arriving at a social description is the logical first step.

Aggregations: Two Subdivisions

We divide cockroach aggregations into two categories, on the basis of the mechanism by which they are formed: cohort aggregations and affiliative aggregations. Cohort groups are formed by the non-dispersal of neonates after the hatch of an ootheca, and represent kin groups. Whether a cohesive sib group results in a cohort aggregation or is incorporated into an affiliative aggregation depends on the oviposition behavior of the female. The placement of an ootheca in an area remote from conspecifics by an oviparous female, or oviposition by a solitary ovoviviparous female will result in a group comprised solely of siblings. There are currently few reports of this kind of aggregation. In *Lanxoblatta emarginata,* group size is the mean brood size or slightly less, suggesting that in this case, aggregation of nymphs results from non-dispersal of a sib group (Grandcolas, 1993a). We suspect that some species of forest cockroaches whose nymphs live in the leaf litter form cohort aggregations. Affiliative aggregations are multigenerational groups that may include all developmental stages and both sexes. They are fluid societies formed by both the incorporation of cohorts of nymphs hatched into the group and by immigration. No genetic relationships are implied for affiliative aggregations, but they are not ruled out. Cockroaches that are urban pests form affiliative aggregations, and, along with cave cockroaches, are the best characterized in terms of gregarious behavior.

Relatedness within Groups

A key issue to address in the analysis of any social behavior is the degree of relatedness of group members; in cockroaches the variation is considerable. At one end of the spectrum, cockroach aggregations are not always species specific (Table 8.1). No overt agonistic encounters are observed in mixed-species groups, but, given the choice, individuals will usually associate with conspecifics (Brossut, 1975; Rust and Appel, 1985). *Blatta orientalis* and *B. germanica* mixed in the laboratory soon form segregated groups (Ledoux, 1945). Initially separated taxa, however, may eventually mingle if their habitat requirements coincide. Everaerts et al. (1997) placed two closely related Oxyhaloinae species, *Nauphoeta cinerea* and *Rhyparobia maderae,* together in laboratory culture. At first they stayed in monospecific groups, but the degree of mixing increased with time, and the taxa were randomly distributed by the fifth day. While intraspecific grouping in cockroaches should be considered the general rule, conditions of high density or scarcity of resources, such as suitable harborage or pockets of high humidity, may result in mixed groups. Mixed-species social groups also are reported from birds, hoofed mammals, primates, and fish, and these typically display gregarious behavior similar to that seen in single-species groups (Morse, 1980).

Although there are no available data on the relatedness of individuals in natural aggregations, populations of *B. germanica* within a building are more closely related than populations between buildings (C. Rivault, pers. comm. to CAN). There are also indications that aggregations are cohesive relative to other groups of the same species. In *B. germanica* almost no mixing of aggregations occurs, even if several are in close proximity (Metzger, 1995); mark-recapture studies show that only 15% of the animals left their initial site of capture (Rivault, 1990). In the cave cockroach *Eublaberus distanti,* 90% of individuals remained in the same group during a 30-day period (R. Brossut in Schal et al., 1984). Site constancy is also known in *P. americana* (Deleporte, 1976; Coler et al., 1987). It is

Table 8.1. Examples of mixed-species aggregations in cockroaches. Additional examples are given in Roth and Willis (1960).

Species	Harborage	Reference
Periplaneta americana, Eurycotis floridana	In stumps, under bark, in corded wood	Dozier (1920)
P. americana, Blatta orientalis, Blattella germanica	In cupboard of home	Adair (1923)
Schizopilia fissicollis, Lanxoblatta emarginata	Under bark	Grandcolas (1993a)
Schultesia lampyridiformis, Chorisoneura sp., *Dendroblatta onephia*	In bird's nest	Roth (1973a)
B. germanica, P. fuliginosa, P. americana	In cracked telephone pole	Appel and Tucker (1986)
Aglaopteryx diaphana, Nyctibora laevigata, Cariblatta insularis	In bromeliads, Jamaica	Hebard (1917)
Variety of combinations: *Blatta orientalis, P. americana, P. fuliginosa, Parcoblatta spp.*	In sewers	Eads et al. (1954, pers. comm. to LMR)

unclear, however, whether the insects are faithful to the group, to the physical location, or both.

Group Size and Composition

The size of a cockroach aggregation is ultimately controlled by its resource base. If food and water are adequate, the surface area of undisturbed dark harborage limits population size (Rierson, 1995). Favorable habitats can result in enormous populations. Roth and Willis (1957), for example, cite a case of 100,000 *B. germanica* in one four-room apartment. As with many other characteristics of urban and laboratory cockroaches, however, high population size and the tendency to form large aggregations are not typical of cockroaches in general. Although species that inhabit caves often live in large groups, individuals of most species are not at all crowded in nature. In Hawaii, aggregations of *Diploptera punctata* in dead dry leaves consisted of 2–8 adults, together with 5–8 nymphs (WJB and L. Kipp, unpubl. data). Researchers who study agonistic or mating behaviors of cockroaches in the laboratory are invariably amazed when they are unable to observe these activities in the field. Small groups of cockroaches are sometimes observed feeding and pairs may be seen copulating, but never in high numbers (Bell, 1990). In one 3-yr field study of cockroach behavior, only four instances of agonistic behavior were recorded, while in laboratory cages agonistic behavior occurred nearly continuously among males (WJB, unpubl. obs.).

Age- and sex-related variation in grouping tendencies are commonly reported in cockroaches (Gautier et al., 1988) and are no doubt related to the mating system and age-dependent fitness biases unique to a species or habitat. In most tested cockroaches the early instars have the strongest grouping tendencies, and in some they are the only stages that display gregarious behavior (e.g., Hafez and Afifi, 1956). All developmental stages are found in aggregations of *B. germanica* and *P. americana,* but young nymphs have the greatest tendency to remain in tight groups (Ledoux, 1945; Wharton et al., 1967; Bret et al., 1983; Ross and Tignor, 1986b). At hatch, neonates maintain a distance from each other, but aggregate as soon as the exoskeleton has hardened (Dambach et al., 1995). The gregarious behavior typical of young cockroaches is retained into later developmental stages in some species. Exceptions lie among the cave cockroaches, where older insects may show the strongest grouping tendencies; these differences appear related to habitat stratification. Adults and older nymphs are typically found aggregated on the walls of caves or hollow trees, utilizing crevices if present, and young nymphs burrow in guano or litter on the substrate (e.g., *Blaberus colloseus, Blab. craniifer, Blab. giganteus, Eublaberus posticus*) (Brossut, 1975; Farine et al., 1981; Gautier et al., 1988). Nonetheless, Darlington (1970) found that young nymphs of *Eub. posticus* aggregate strongly, but they do so independently of older stages, and aggregation pheromone is produced by all developmental stages of both *Eub. distanti* and *Blab. craniifer* (Brossut et al., 1974). Laboratory assays seldom take into account the habitat preferences of different stages, and we know nothing of the social tendencies of young cave cockroaches while under organic debris. Age-related distributional differences are known within the large affiliative aggregations typical of pest cockroaches. Young *B. germanica* typically cluster in the middle of the aggregation (Rivault, 1989). Fuchs and Sann (1981, in Metzger, 1995) found that first- and second-instar *B. germanica* create small independent aggregations and do not mingle with older conspecifics until the third instar.

There is a complex relationship between sex ratio, sexual status, and grouping behavior in affiliative aggregations. Ledoux (1945) noted that male nymphs of *B. germanica* showed significantly stronger aggregation tendencies than groups of females. Adult females of this species have the most influence on group composition, but these effects are moderated depending on the demographics of the group in question (Bret et al., 1983). The reproductive status of females was a factor, with gravid females promoting the strongest grouping behavior. The maturity of the egg cases carried by females was also influential. Adult males typically show little gregariousness and spend the least amount of time in shelters. The loss of gregarious behavior in males typically coincides with sexual maturity and the onset of competition for mates (Rocha, 1990).

An examination of group composition in the cockroaches listed by Roth and Willis (1960) indicates that aggregations of lesser known species in several cases do not contain adult males. The basic unit of some affiliative aggregations appears to be the uniparental family: groups of mothers together with their offspring. Species mentioned include females and young of *Ectobius albicinctus* found beneath stones (Blair, 1922), of *Polyphaga aegyptica* and *Polyp. saussurei* found in rodent burrows (Vlasov, 1933; Vlasov and Miram, 1937), and of *Arenivaga grata* collected from guano in bat caves (Ball et al., 1942). There are also occasional reports of cockroach aggregations consisting entirely of females, for example, *Arenivaga erratica* in burrows of kangaroo rats (Vorhies and Taylor, 1922), and aggregated females and dispersed or territorial males in *Apotrogia* sp. (= *Gyna maculipennis*) (Gautier, 1980).

Nothing is known about the immigration of unaffiliated cockroaches into established conspecific groups. Discrete aggregations collected in the field often mix together freely in the laboratory (e.g., *Panesthia cribrata*—O'Neill et al., 1987), but this is quite different from a solitary insect attempting to join an established group under natural conditions. When two isolated young nymphs of *P. americana* are placed in contact with each other, they undergo a "ritual of accommodation" which may become aggressive (Wharton et al., 1968). Behaviors include "sampling" each other's deposited saliva with palpi or antennae, stilting, tilting their bodies, bending their abdomens, antennal fencing, leg strikes, and biting. The decision to accept new members into the aggregation can be important when changing ecological conditions (e.g., food availability) alter the relationship between group size and fitness (Giraldeau and Caraco, 1993).

Choosing Shelter

Cockroaches use a variety of criteria in selecting harborage sites. In general, cockroaches orient to sheltered sites near food and water, and will remain true to a site as long as both are adequate (Ross et al., unpubl., in Bret et al., 1983; Rivault, 1990). Both the texture (Berthold, 1967) and orientation of surfaces (Bell et al., 1972) and the size of the harborage (Berthold and Wilson, 1967; Mizuno and Tsuji, 1974) are influential. Groups of cockroaches may segregate by body size, depending on the height of available space (reviewed by Roth and Willis, 1960). Small nymphs in the absence of older conspecifics prefer narrower crevices than do adults; however, they prefer larger harborages if other cockroaches are present, indicating that social stimuli supersede harborage height preferences (Tsuji and Mizuno, 1973; Koehler et al., 1994). Aggregation behavior of young nymphs is more pronounced in open areas than in shelters, suggesting that they may satisfy their thigmotactic tendencies with each other when the physical environment is devoid of tactile stimuli (Ledoux, 1945).

Pheromones

Pheromones rule the social world of cockroaches. The chemical repertoire includes both contact pheromones and volatiles, and these function as sex pheromones, attractants, arrestants, dispersants, alarm pheromones, trail pheromones, and mediators of kin recognition. Chemical stimuli help orchestrate cockroach aggregation behavior, and have been studied primarily for their potential in pest management.

Oviposition Pheromones

The location of first instars within their habitat is largely determined by the oviposition behavior of females, who tend to deposit their eggs near resources. Female *Periplaneta brunnea*, for example, generally glue their oothecae near a food supply (at least they do in 1 gal battery jars) (Edmunds, 1957). There is some evidence to suggest, however, that, like locusts (Lauga and Hatté, 1977; Loher, 1990), some cockroaches may employ oviposition pheromones. These serve to either convene gravid females in certain locations for egg laying, or attract them to sites where conspecifics have previously deposited oothecae. Edmunds (1952) found 184 oothecae of *Parcoblatta* sp. deposited in close proximity under tree bark. Similarly, oothecae of *Supella longipalpa* were found in clusters by Benson and Huber (1989). The authors observed ovipositing females deposit a drop of "genital fluid" on oothecae, and suggested that it contains a pheromone that attracts other females. Gravid females of *B. germanica* generally do not leave the harborage (Cochran, 1983b); consequently, first instars hatch into an aggregation (Rivault, 1989; Koehler et al., 1994). Stray females, however, may actively seek aggregations for oviposition. Escaped females of *B. germanica* in laboratory colonies laid their oothecae near a group of conspecific nymphs (Ledoux, 1945).

Aggregation Pheromones?

Enormous effort has been dedicated to localizing and characterizing the aggregation pheromone of pest cockroaches. The results, however, are still equivocal. Ledoux (1945) first proposed that aggregation in cockroaches was the result of mutual attraction of a chemical nature, and Ishii and Kuwahara (1967, 1968) identified fecal material as the source of the cue. Riding the wave of pheromone research during the 1960s, these authors dubbed the fecal chemical "aggregation pheromone." They suggested that it originates in the rectal pad cells and that it is applied to fecal pellets as they are being excreted. Cuticular waxes apparently absorbed the fecal pheromone also, as ether washings of the abdomen had higher activity than ether washings of other parts of the body. More recent work has identified more than 150 volatile and contact chemicals from German cockroach fecal pellets (Fuchs et al., 1985, in Metzger, 1995; Sakuma and Fukami, 1990). The attractiveness of individual components depends not only on the type of extraction used, but also the biological assay used to test them (reviewed by Dambach et al., 1995), and the stock or population of *B. germanica* used as test subjects. Mixtures of fecal compounds are generally more effective than single components (Scherkenbeck et al., 1999). Cuticular wax may be attractive independent of any chemicals absorbed from excretory material. Rivault et al. (1998) found that cuticular hydrocarbons alone, from any part of the body, can elicit aggregation behavior.

Fecal chemicals seem to function initially as short-

Table 8.2. Aggregation of cockroach nymphs on filter paper conditioned with the feces of other cockroach species. Six to eight trials were performed with each combination using 20 nymphs per run. Plus-signs represent significant aggregation to conditioned paper as compared to controls. From Bell et al. (1972).

	Species conditioning papers					
Nymph species	P. am.	B.o.	P.p.	E.p.	B.d.	B.f.
After 20 min						
P. americana	+	+	+	+	+	+
Blatta orientalis	+	+	–	+	–	–
Parc. pennsylvanica	–	–	–	+	–	–
Eub. posticus	–	–	–	+	+	–
Blab. discoidalis	–	+	–	–	+	–
Byr. fumigata	–	+	–	+	–	+
After 12 hr						
P. americana	+	+	+	+	+	+
Blatta orientalis	+	+	+	+	–	–
Parc. pennsylvanica	–	–	–	–	–	–
Eub. posticus	–	–	–	+	–	–
Blab. discoidalis	–	+	–	+	+	+
Byr. fumigata	–	–	+	+	–	+

range attractants (Ishii and Kuwahara, 1967; Bell et al., 1972; Roth and Cohen, 1973), then as arrestants (Burk and Bell, 1973). Nymphs halt their forward progress when they encounter a filter paper contaminated with feces; the response, however, is not strictly species specific (Bell et al., 1972; Roth and Cohen, 1973). Cockroaches prefer substrates contaminated by feces of their own species, but will aggregate on surfaces contaminated by distant relatives (Table 8.2). *Periplaneta americana* was attracted to paper contaminated by all species tested, and after 12 hr, *Parcoblatta pennsylvanica* was attracted to none, not even their own. Locomotor inhibition is enhanced by social interaction between assembled individuals; a nymph is more likely to stop on feces-contaminated filter paper if one or more nymphs are already in residence. Young nymphs are most responsive to the chemical cues, adults are intermediate, and middle instars the least (Bret and Ross, 1985; Runstrom and Bennett, 1990). Experience matters; nymphs that hatch in an aggregation are more likely to aggregate (Dambach et al., 1995).

The evidence suggests that the fecal substances that elicit aggregation behavior in cockroaches, then, are not pheromones in the classic sense, but a functional category of behavior-eliciting chemicals (Brossut, 1975). Their origin is unclear, they are poorly defined, and they lack specificity. Pheromones are, however, clearly implicated in two species, *Blab. craniifer* and *Eub. distanti*, where the origin of the intraspecific attractant has been traced to the mandibular glands (Brossut et al., 1974; Brossut, 1979). In these cockroaches the pheromone is secreted by all individuals at all times except during the molting period. The insects are unattractive from 72 hr before to 24 hr after ecdysis (Brossut et al., 1974; Brossut, 1975). This inactive period occurs because the mandibular gland is lined with cuticle (Noirot and Quennedy, 1974), which is shed along with the rest of the exoskeleton during molt.

Proximate Mechanisms: How Do They Aggregate?

If specific pheromones are not involved in many species, how do groups form? Aggregation in cockroaches is generally mediated by visual, acoustic, tactile, and/or olfactory stimuli (Grassé, 1951). The complication is that these often are not the only causes. Environmental factors, including light (Gunn, 1940), temperature (Gunn, 1935), and air movement (Cornwell, 1968) also play an important role. Humidity is a factor, although the degree to which it exerts an influence may be species specific (Roth and Willis, 1960). In some cockroaches, the lower the humidity, the stronger the tendency to aggregate (Sommer, 1974; Dambach and Goehlen, 1999). Response to these, as well as other environmental stimuli, results in the initial selection of a harborage, which is consequently marked with bodily secretions (Pettit, 1940); these then help mediate immigration into the group. In laboratory tests, 82% of *B. germanica* choose harborages previously inhabited by conspecifics (Berthold and Wilson, 1967). As the size of an aggregation increases, the collective signal of the mass should serve as an increasingly more powerful attractant to unassociated individuals. *Blattella germanica* will migrate from a less to a more colonized refuge; new refuges are colonized stepwise, with males (Denzer et al., 1988) or mid-size nymphs (Bret and Ross, 1985) as the first to arrive.

Kavanaugh (1977) suggested three mechanisms by which a group may assemble: (1) independent, individual responses to environmental gradients, leading to aggregation in an abiotically optimum location; (2) individual response to stimuli provided by other individuals, leading to group formation at a common location; (3) some combination of the two. Cockroaches, like many other animals, appear to employ the third mechanism, with the first and second involved sequentially. This approach was recently formalized by Deneubourg et al. (2002) and Jeanson et al. (2005). These authors conclude that cockroach aggregations are self-organized systems, resulting from interactions between individuals following simple rules based on local information. First, similar species-specific responses to the physical environment increase the probability that cockroaches converge in the same vicinity. Positive feedbacks and the modulation of individual behavior dependent on the proximity of conspecifics then result in group formation. Short-range volatiles, contact chemicals, physical contact, alterations in local microclimate, and perhaps sonic communication (Mistal et al., 2000) may all signal the presence of conspecifics and serve as cues for an individual to slow or stop locomotion. The response to these cues may be modulated by heterogeneities in the environment. Garnier et al. (2005) used a group of micro-robots modeled after cockroaches to demonstrate that the aggregation process is based on a simple set of behavioral rules. The robots were not only able to form aggregations, but could also make a collective choice when presented with two identical or different shelters. These broader approaches to cockroach aggregation behavior help account for much of the ambiguity in the literature on the subject, and aid in integrating cockroaches into the existing literature on grouping behavior in other animal systems.

Ultimate Causes: Why Do They Aggregate?

In cockroaches, gregarious behavior has a wide range of potential benefits, ranging from the simple advantage of safety in numbers, to group effects that have physiological and life history consequences. There are, however, no inherent advantages to group living, and the opposite is often true. Group members compete for food, shelter, and mates, and may burden each other with diseases and parasites (Alexander, 1974). It is reasonable to assume that aggregation in any animal involves both positive and negative components, and that observed social groups are the result of the balance of the two (Iwao, 1967; Vehrencamp, 1983). Fitness biases within a group will vary with species, habitat, resources, the age, sex, and reproductive status of individuals, and the demographics of the population.

Aggregations as Environmental Buffer

Although cockroaches are drawn to shelters with favorable temperature and humidity, to some extent cockroach aggregates are able to create their own microenvironment. Grouped cockroaches may better survive hostile dry conditions than loners in at least two species. Dambach and Goehlen (1999) found that as a result of respiration and diffusion, individuals of *B. germanica* are each surrounded by an envelope of water vapor. These individual diffusion fields overlap in aggregated insects, reducing net individual water loss. Aggregation behavior also reduces water loss in *G. portentosa;* Yoder and Grojean (1997) suggest that it is an adaptation for surviving the long tropical dry season of Madagascar. Documentation of seasonal changes in social behavior in the field would provide added support for this hypothesis.

Aggregations as Defense

Although cockroaches are known to have a variety of predators and a large number of weapons in their arsenal to defend against them, most available information relates to predation on individuals. Diurnal aggregations of inactive cockroaches, however, have properties that differ from active, nocturnal individuals and thus change the parameters of the predator-prey interaction. Cues that lead predators to prey are multiplied when prey aggregate (Hobson, 1978), and the rewards of finding such a concentrated source of food are greater. Since cockroaches typically assemble in inaccessible places (crevices, leaves, hollow logs, under bark, among roots), their apparency is presumably low to predators that rely primarily on visual cues. Conversely, cockroach aggregations may offer a more intense signal to olfactory hunters. At least one parasite is known to specialize on cockroach aggregations: eggs of the beetle *Ripidius pectinicornis* are laid in a cluster near cockroach aggregations, and early larval stages then locate their host (Barbier, 1947).

The greater number of available sensory receptors in an aggregation increases group capacity to sense potential predators. There is anecdotal evidence that vigilance behavior by peripheral insects may occur in aggregations of *P. americana*. Ehrlich's (1943) description depicts older

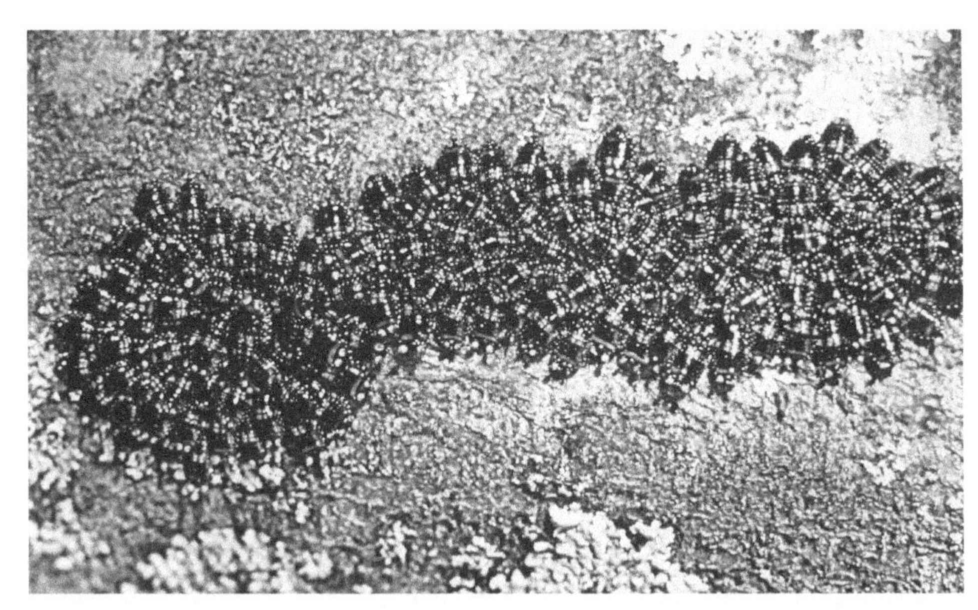

Fig. 8.1 Aggregation of nymphs of *Cartoblatta pulchra* on a tree trunk in Kenya. The nymphs are both aposematically colored and produce a sticky exudate on the terminal abdominal segments. Note that heads are oriented toward the center of the group (cycloalexy). Photo courtesy of Michel Boulard.

individuals serving as sentries on the periphery of the group; when danger approaches they warn the young with body movements. A more realistic interpretation, however, may be that members of the aggregation react to the evasive maneuvers of the first insect to detect a predator. Alarm pheromones have been described in *Eurycotis floridana* (Farine et al., 1997), *Therea petiveriana* (Farine et al., 2002), and cave-dwelling *Blaberus* spp. (Crawford and Cloudsley-Thompson, 1971; Gautier, 1974a; Brossut, 1983). The emission of these chemicals results in the rapid scattering of group members. Predators confronted by a confusing welter of moving targets presumably have trouble concentrating on individual prey. While defensive glands have been described in a large number of cockroaches (Roth and Alsop, 1978), whether the secretions of these glands function as weapons, signals, or both is in many cases untested. Certainly insects that exude or project defensive chemicals would benefit from an increase in point sources (Vulinec, 1990). One example of this type of defensive strategy is known among the Blattaria, although it may occur in others (e.g., *Dendroblatta sobrina*—Hebard, 1920a). Similar-sized nymphs of *Cartoblatta pulchra* (Blattinae) openly assemble on tree trunks in Tanganyika and Kenya (Fig. 8.1). One group, composed of 100–150 individuals, formed a rosette larger than a human hand. Individuals were polarized, with their heads facing the center of the group and their abdomens directed radially outward (cycloalexy). A brisk movement disperses the cockroaches, and they run into crevices in the tree trunk (Chopard, 1938). The insects are aposematically colored (black and orange), and each nymph displays a thick proteinaceous secretion on the terminal abdominal segments. This material originates from type 5 tergal glands (Fig. 5.11), is characteristic of many oviparous cockroaches (Fig. 4.7), and functions at least in part to protect them against ants (Roth and Alsop, 1978). Most known aposematic cockroach species are active during the day in relatively open areas and do not form conspicuous aggregations (e.g., *Platyzosteria ruficeps*—Waterhouse and Wallbank, 1967).

Aggregation and Nourishment

It has been suggested that one of the main functions of gregarious behavior in cockroaches is to signal to unassociated individuals the proximity of food and water (Wileyto et al., 1984). The addition of extra animals to a group, however, results in both added competition for food and higher travel costs (Chapman et al., 1995). Cockroaches in aggregations are central place foragers; they travel from a central location to forage elsewhere, then return to shelter. Short-range foraging is the rule in *B. germanica,* and food patches placed near shelters are depleted before patches placed farther away (Rivault and Cloarec, 1991; Rierson, 1995). When overcrowded, however, individuals are known to move more than 10 m (Owens and Bennett, 1983). Large, persistent aggregations no doubt depend on constant renewal of food resources in the vicinity of the harborage, such as dirty dishes left in the sink at every meal or the regular deposition of guano by bats.

In gregarious cockroaches, social facilitation in meeting nutritional requirements may occur within two contexts: (1) in locating and ingesting food away from the harborage, and (2) in the use of food originating from conspecifics within the harborage. Individuals of *B. germanica* forage individually but often converge on the same sites (Rivault and Cloarec, 1991), suggesting that there may be a social component to food finding. Trail pheromones (Chapter 9) may facilitate movement from the harborage to renewable food sources (a garbage can, for example). In habitats where food is unpredictable, ephemeral, or patchily distributed, a different form of social facilitation may occur. Cockroaches leave behind at feeding sites a variety of residues in the form of saliva, glandular deposits, and fecal pellets. Feeding sites that are "marked" by these residues may be more attractive than unmarked food patches because, whether or not foraging cockroaches are present, the food has been made "visible" by the traffic of conspecifics. If so, cockroaches exhibit the simplest form of food-related grouping behavior: local enhancement—the act of cueing on conspecifics for food information (Mock et al., 1988). Attraction to residues by cockroaches would be the chemical equivalent of the visual attraction of birds to feeding flocks, or the acoustic attraction of bats to the echolocation calls of conspecifics (Richner and Heeb, 1995). Cockroaches show a number of similarities to rats, which are nocturnal, omnivorous, central place foragers that leave chemical cues in the form of urine and fecal pellets on resources (food patches, nest sites) used by other rats. These residues provide a mechanism for social learning and are used in a variety of contexts (Galef, 1988; Laland and Plotkin, 1991).

The benefits of cueing on foraging conspecifics can be considerable for young nymphs, who do much better developmentally on the same food source if an adult is present. The adults seem to "condition" the food in some way, either by moistening it, breaking it into smaller pieces, or making initial excavations into a tough food item. Both *Blattella* and *Supella* have been observed depositing saliva on food (C. Schal, pers. comm. to WJB), and the development of *B. germanica* nymphs fed whole dog food pellets was slower by approximately 43% than nymphs that were fed the same food, but pulverized (Cooper and Schal, 1992).

Nutritional advantages of associating with conspecifics may also occur within the harborage. The exuvia, corpses, feces, exudates, oothecal cases, embryonic membranes, and unviable eggs produced by individuals in an aggregation as they progress through their lives are fed upon by other members of the group (reviewed by Nalepa, 1994) (Table 4.6). The presence of this proteinaceous food in the harborage may be of particular value to females and to young nymphs, as it is these stages that have the highest nitrogen requirements. Juveniles in particular may benefit from a ready source of high-quality food for several reasons. First, young insects have relatively small reserves, a high metabolism, and nutritional requirements that differ from those of adults (Slansky and Scriber, 1985; Rollo, 1986). Second, young cockroaches are inefficient in their foraging behavior, and typically do not forage far from shelter (Cloarec and Rivault, 1991; Chapter 4). Third, as noted above, young nymphs have difficulty processing physically hard food. High-quality, easily processed food that originates from conspecifics in their immediate vicinity may allow the young to pass more quickly through the stages during which they are most vulnerable.

Aggregation as a Source of Mates

In aggregation assays, *B. germanica* males displayed a stronger response to paper conditioned by virgin females than to paper conditioned by any other category, whereas the female response did not differ when presented with the residues of males, females, and juveniles (Wileyto et al., 1984). These authors postulated that males unassociated with an aggregation may be using the sexual information present in the residues to determine the composition of a group, and therefore to locate potential mates. They concluded that their results were consistent with the hypothesis that cockroaches aggregate for the purposes of mating.

Functional separation of aggregation pheromone and sexual pheromone is not always possible; sex ratios and reproductive status have a complex relationship with aggregation behavior in *B. germanica* (Sommer, 1974; Bret et al., 1983). Because females of this species produce a nonvolatile as well as a volatile sex pheromone (Nishida et al., 1974; Tokro et al., 1993), it is not surprising that males respond to their residues. Encounters between potential mates are increased by gregarious behavior; newly emerged virgin females occur in close proximity to males, and sexual communication over long distances is not required for mate finding (Metzger, 1995). A virgin, then, would not remain one for long in a group that already included adult males. The hypothesis of Wileyto et al. (1984) would be stronger if wandering males were attracted to groups that contained female nymphs in their penultimate instar, so that they were already present to compete for newly emerged virgins. The argument, however, has other flaws. Virgins leave residues regardless of whether they are isolated or in a group, and residues in a harborage are a mélange of all stages present. It is also unclear whether mating takes place within the aggregation in free populations. Rivault's (1989) work suggested that prior to the imaginal molt, *B. germanica* gather in high-density areas in the middle of the aggregate, looking for sexual partners. However, in a number of species, including *B. germanica* (Nojima et al., 2005), females produce volatile sex pheromones, and may move out of the group to release them. Females of *Blab. giganteus*, for example, have been observed calling on the outside of a tree that contained a large aggregation of conspecifics (C. Schal, pers. comm. to WJB). The age, sex, and kinship structure of a group will determine the optimal mating strategies open to an individual (Dunbar, 1979), and the disadvantages of mating in a group should not be ignored. Cockroaches typically require 30 min or longer to transfer a spermatophore (Roth and Willis, 1954b) and may be subject to harassment during that period of time (Chapter 6). The suggestion that cockroaches aggregate for the purposes of mating, then, may be true in some species or in some circumstances, but cannot be applied universally to gregarious species.

Aggregation and Group Effects

Group effects refer to morphological, physiological, or behavioral differences between animals that are grouped versus those of the same species that are bereft of social contact. The prolongation of the juvenile growth period in isolated nymphs is the best-studied group effect in cockroaches, occurs in a wide range of species (Table 8.3), and is discussed in Chapter 9 in relation to its evolutionary connection to caste control in termites. One benefit of accelerated development in grouped nymphs is that it moves them quickly through one of the riskiest stages of their lifecycle. The number of cockroach species examined for group effects is extremely limited relative to the number of species available for study; especially interesting would be a study of those in which nymphs seem to disperse shortly after hatch, like *Than. akinetum* (Grandcolas, 1993a). Altered juvenile growth rates, however, are not the only effect of social interaction. Like some other insects that aggregate (reviewed by Eickwort, 1981), molting in grouped cockroaches tends to be synchronized (Ishii and Kuwahara, 1967). This may be an evolutionary response to the threat of cannibalism, as all nymphs are vulnerable at the same time, and are incapable of feeding on each other until their mouthparts sclerotize.

Adult cockroaches also show group effects, which are manifested in physiology and behavior, can be species specific, and have a complex influence on reproductive success. In *B. germanica* the presence of another adult has an impact on how fast a female reproduces and how much she eats, but the former is at least partially independent of the latter (Gadot et al., 1989; Holbrook et al., 2000a). Komiyama and Ogata (1977) found that isolated females of this species deposit a greater number of oothecae than group-reared females, but the hatching success of those oothecae was considerably lower. In *Su. longipalpa*, group effects were primarily behavioral, and group composition rather than isolation was more influential on reproductive events. Neither oocyte growth nor calling behavior was affected by isolating virgin females, but the onset of calling and its diel periodicity were advanced in virgin females housed with other virgin females relative to females housed with either mated females or males that were unable to mate (Chon et al., 1990). Several studies have shown that isolated male cockroaches show a decreased reaction to female sex pheromone (Roth and Willis, 1952a; Wharton et al., 1954; Stürkow and Bodenstein, 1966); the social history of male *N. cinerea* is known to influence the amount of sex pheromone they produce (Moore et al., 1995).

Table 8.3. Cockroach species that exhibit group effects on development.

Blattidae: *Blatta orientalis, Eurycotis floridana, Periplaneta americana, P. australasiae, P. fuliginosa* (Willis et al., 1958)

Blattellidae: *Blattella germanica, B. vaga, Supella longipalpa* (Willis et al., 1958; Izutsu et al., 1970)

Blaberidae: *Diploptera punctata, Eurycotis floridana, Nauphoeta cinerea, Pycnoscelus surinamensis, Rhyparobia maderae* (Willis et al., 1958; Woodhead and Paulson, 1983)

A number of other behavioral effects can be induced by isolating cockroaches: the normal flight reaction to disturbance may be lost (Hocking, 1958), circadian rhythm may be altered (Metzger, 1995), the ability to learn may be affected (Gates and Allee, 1933), and activity increased (Hocking, 1958) or decreased (Cloudsley-Thompson, 1953). Aggressiveness was delayed in isolated male *N. cinerea* (Manning and Johnstone, 1970), but isolation increased aggressiveness in *Periplaneta* (Bell et al., 1973), *The. petiveriana* (Livingstone and Ramani, 1978), and several cave-dwelling Blaberidae (Gautier et al., 1988). Raisbeck (1976) found an aggression-stimulating substance produced by isolated *P. americana* that is masked or suppressed by "aggregation pheromone" when the insects live in groups.

Aggregations as Nurseries

Because the costs and benefits of grouping behavior vary with species, stage, sex, and environment, there is no simple answer to the question of why cockroaches aggregate. However, a persistent thread that runs through the previous sections relates to gregarious behavior in connection to benefits conferred on young nymphs. Regardless of the advantages other group members enjoy, affiliative aggregations may provide juveniles with all the necessities of early cockroach life. The benefits of aggregation behavior are often most pronounced in the young, which typically suffer the greatest mortality due to desiccation, starvation, predators, and cannibals (Eickwort, 1981). The more humid environment that surrounds an aggregate of cockroaches may be crucial for young nymphs, as their higher respiratory rate and smaller radius of action increases their dependence on local sources of moisture (Gunn, 1935). The company of conspecifics assures the rapid development of nymphs via group effects, and the presence of older developmental stages assures a supply of conspecific food and an inoculum of digestive microbiota (Chapter 5) within the harborage (Nalepa and Bell, 1997). Away from the harborage, it is possible that trail following and local enhancement allow young cockroaches access to better food sites than they would find by searching on their own. Young nymphs may also pick up adaptive patterns of behavior by living in social groups. Cockroaches can learn, retain, and recall information; this ability is a thorn in the side of urban entomologists attempting to develop effective baits for cockroach control (Rierson, 1995).

Costs of Aggregation

Two noteworthy potential costs of group living in gregarious cockroaches are the transmission of pathogens (Chapter 5) and the risk of cannibalism. Both the higher humidity and the intimate physical association typical of aggregations help promote infectious diseases. The cost may be direct, resulting in illness or death, or indirect, in the form of trade-offs ensuing from increased investment in the immune system. Cannibalism is usually a density-dependent behavior, in that high population levels may decrease the local food supply and lower attack thresholds. Injuries also may be more common in dense aggregations, resulting in scavenging of the crippled and dead. Vulnerable life stages such as oothecae and young or molting nymphs may be at risk regardless of group size (Dong and Polis, 1992; Elgar and Crespi, 1992). Young cockroaches typically suffer the highest mortality of any developmental stage (e.g., *B. germanica*—Sherron et al., 1982; *P. americana*—Wharton et al., 1967), in part because frequent ecdyses expose nymphs to injury and cannibalism. However, if the local food supply adequately meets the needs of the older group members, the advantages of living in a multigenerational group should outweigh the risks for young stages. Cannibalism is relatively unstudied in cockroaches (but see Gordon, 1959; Wharton et al., 1967), and the information we do have is sketchy. Young nymphs are described as the most cannibalistic in *P. americana* (Wharton et al., 1967, Roth, 1981a), but the behavior is rare in first to third instars of *B. germanica* (Pettit, 1940). While these findings may reflect species-specific differences, variation in cannibalistic behavior either within or among species may also be attributed to laboratory culture under different densities or feeding regimens.

There are additional costs to social behavior, particularly when groups become too large. These include increased competition for resources, decay in habitat quality, and increased attractiveness to predators (Parrish and Edelstein-Keshet, 1999). Overcrowded cockroaches may exhibit a breakdown in circadian rhythm, enhanced aggression, a prolonged nymphal period, supplementary juvenile stages, increased mortality, and decreased body size (Wharton et al., 1967). Optimal group size is no doubt variable and depends on both the taxon in question and available resources, but it has been calculated for one cockroach. Deleterious effects from crowding begin to occur in *B. germanica* when they exceed a level of 1.2 individuals/cm^2 in a harborage (Komiyama and Ogata, 1977). That the net gain of living in a group diminishes after the aggregate reaches a certain size is also reflected in cockroach chemical communication. The composition of the aggregation pheromone in *Eub. distanti* is known to vary with cockroach population density (Brossut, 1983), and dispersal pheromones have been found in the saliva of the German cockroach (Suto and Kumada, 1981; Ross and Tignor, 1986a). This pheromone counteracts fecal attractants and is most concentrated in the saliva of crowded, gravid females. It is thought to function as a space regulator within aggregations, force dispersal from crowded or otherwise unfavorable conditions, and deter cannibalism of young nymphs. Adult males react most strongly to the pheromone and are thought to be the main target group (Ross and Tignor, 1985; Faulde et al., 1990).

PARENTAL CARE

Most cockroaches show some form of parental care, in the broad sense: any form of parental behavior that promotes the survival, growth, and development of immatures, including the care of eggs or young inside or outside the parent's body, and the provisioning of young before or after birth (Tallamy and Wood, 1986; Clutton-Brock, 1991). Hinton (1981) considered cockroaches by far the largest group of insects that exhibit parental care, because he included ovoviviparity and viviparity in the category. Regardless of their reproductive mode, cockroaches characteristically care for their eggs in elaborate ways. In oviparous species, the care includes the production of oothecal cases, preparation of oothecal deposition sites, concealment of the oothecae, and defense of deposited oothecae. In ovoviviparous and viviparous females, the embryos are both protected and provisioned within the body of the female (Chapter 7). In this chapter the scope of parental care will be limited to enhancement of post-hatch offspring survival by one or both parents. The type of reproduction exhibited by a species

Fig. 8.2 Aposematically colored (dark brown with yellow-orange banding) female and nymphs of *Desmozosteria grossepunctata* found under a stone in mallee habitat, Western Australia. Photo courtesy of Edward S. Ross; identification by David Rentz.

does, however, influence parent-offspring interactions. The majority of cockroaches that exhibit any form of post-partition parental care are ovoviviparous; the internal retention of the egg case guarantees that the female is in the immediate vicinity of nymphs at hatch (Nalepa and Bell, 1997). Oviparous females that deposit the egg case shortly after its formation depart before neonates emerge and may produce several more egg cases before the first one deposited hatches. Thus, ovoviviparity results in a generational overlap in both time and space, providing ample opportunity for brooding behavior. The multiple origins of parental care among the ovoviviparous Blaberidae suggest that more elaborate forms of parent-offspring interactions then evolved from that starting point (Nalepa and Bell, 1997).

In 1983 Breed wrote that very little is known concerning post- hatching parent-offspring relationships in cockroaches. The situation has improved only slightly since that time. The majority of the cockroach species described as subsocial are known solely from brief notes taken during field collections, documenting females collected with offspring from harborages under bark, within logs, or under stones (Fig. 8.2). Examples include *Poeciloblatta* sp. (Scott, 1929), *Aptera fusca* (Skaife, 1954), and *Perisphaerus armadillo* (Karny, 1924). The variety of known subsocial interactions in cockroaches, however, is among the richest in the insects, and ranges from species in which females remain with neonates for a few hours, to biparental care that lasts several years and includes feeding the offspring on bodily fluids in a nest.

Brooding Behavior

The simplest form of parental care in cockroaches is brooding, defined as a short-term association of mother and neonates. In a number of ovoviviparous blaberids (e.g., *N. cinerea, Blab. craniifer*), young nymphs cluster under, around and sometimes on the female for varying periods of time after emergence. Most brooding associations last less than a day. Although observations of brooding behavior are based primarily on laboratory observations, Grandcolas (1993a) observed it in *Than. akinetum* in the field. The female was perched on a leaf when first instars emerged, and the nymphs aggregated beneath the mother's body for several hours prior to dispersing. In cockroaches known to brood, aggregation of the nymphs also occurs in the absence of the female; it is not solely predicated on all nymphs orienting to their mother as a common stimulus (Evans and Breed, 1984).

It is generally believed that brooding has a protective function; it takes several hours for the cuticle of neonates to harden, and soft, unpigmented nymphs are at risk from ants and cannibalism (Eickwort, 1981). The transfer of gut microbiota may also be a factor; short-term contact with the female may be necessary so that neonates secure at least one fecal meal (Nalepa and Bell, 1997). There are, however, no published observations or studies relating to the functional significance of brooding.

We place cockroaches that exhibit brooding behavior into a category separate from other subsocial species because short-term maternal presence alone defines the behavior. Although the female may stilt high on her legs to accommodate the nymphs beneath her (e.g., *Homalopteryx laminata*—Preston-Mafham and Preston-Mafham, 1993; *Nauphoeta cinerea*—Willis et al., 1958), there are currently no reports of active maternal feeding or defense in species placed in this group. More detailed study may indicate that at least some of these species are subsocial. We classified *G. portentosa* as exhibiting brooding behavior (Nalepa and Bell, 1997), when in fact it exhibits short-term, but elaborate parental care. After partition the female expels a sizable gelatinous mass that is eaten eagerly by neonates (Fig 8.3A) (Perry and Nalepa, 2003). Young nymphs then collect under the mother, who is aggressive to intruders and hisses at the slightest disturbance (Roth and Willis, 1960).

Subsocial Behavior

Parental care arose on a number of occasions within the ovoviviparous Blaberidae and elsewhere just once, in the oviparous Cryptocercidae. One extreme of the subsocial range is represented by *Byrsotria fumigata*. From what we currently know of parent-offspring interactions in this species, subsociality consists of no more than long-term brooding behavior. First instars are able to recognize their own mother and prefer to aggregate beneath her for the first 15 days after hatch (Liechti and Bell, 1975). More

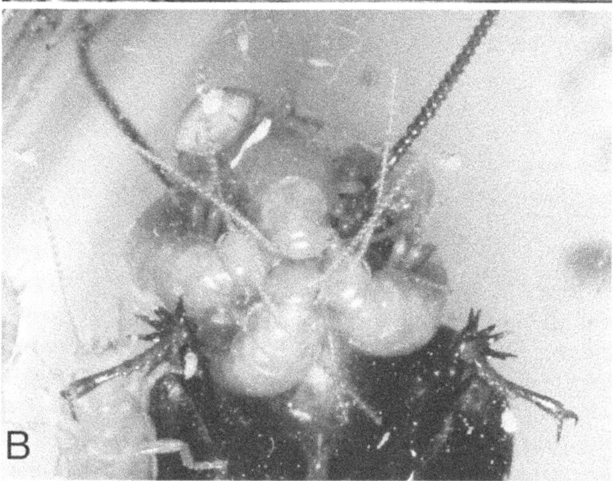

Fig. 8.3 (A) Newly hatched nymphs of *Gromphadorhina portentosa* feeding on secretory material expelled from the abdominal tip of the female (note left cercus). A new pulse of the material is just beginning to emerge. The oothecal case can be seen in the upper-right corner. Image captured from frame of videotape, courtesy of Jesse Perry. (B) Four young nymphs of *Salganea taiwanensis* feeding on the stomodeal fluids of the female, viewed through glass from below. Note antennae of the adult. Photo courtesy of Tadao Matsumoto.

elaborate forms of subsocial behavior include those species in which morphological modifications of the nymphs or the female facilitate parental care. Specializations of the juveniles include appendages that aid in clinging to the female, and adaptations of their mouthparts to facilitate unique feeding habits. Some females have evolved external brood chambers under their wing covers, and others have the ability to roll into a ball, pill bug-like (conglobulation), to protect ventrally clinging nymphs. Maternal care is the general rule, biparental care is recognized only in two taxa of wood-feeding cockroaches, and male uniparental care is unknown.

Parental Care on the Body

In several species of cockroach the protection and feeding of young nymphs occurs while the offspring are clinging to or attached to the body of the female. A simple form of this type of parental care is exhibited by *Blattella vaga,* an oviparous species that carries the ootheca until nymphs emerge. The female raises her wings, allowing freshly hatched nymphs to crawl under them. They appear to feed on material covering her abdomen, then scatter shortly afterward (Roth and Willis, 1954b, Fig. 65). More complex forms of this behavior are found among cockroaches in the Epilamprinae. Females in three genera (*Phlebonotus, Thorax,* and *Phoraspis*) (Roth, 2003a) have an external brood chamber, allowing them to serve as "armoured personnel carriers" (Preston-Mafham and Preston-Mafham, 1993). The tegmina are tough and dome-shaped, and cover a shallow trough-like depression in the dorsal surface of the abdomen, forming a space for protecting and transporting the young. The aquatic species *Phlebonotus pallens* carries about a dozen nymphs beneath its wing covers (Shelford, 1906b; Pruthi, 1933) (Fig. 8.4). In *Thorax porcellana* the maternal behavior lasts for about 7 weeks; 32–40 nymphs scramble into the brood chamber immediately after hatch and remain there during the first and second instars. Their legs are well adapted for clinging, with large pulvilli and claws. It is probable that nymphs feed on a pink material secreted from thin membranous areas on the dorso-lateral regions of the fourth, fifth, sixth, and seventh tergites of the mother. The mouthparts of first instars are modified with dense setae

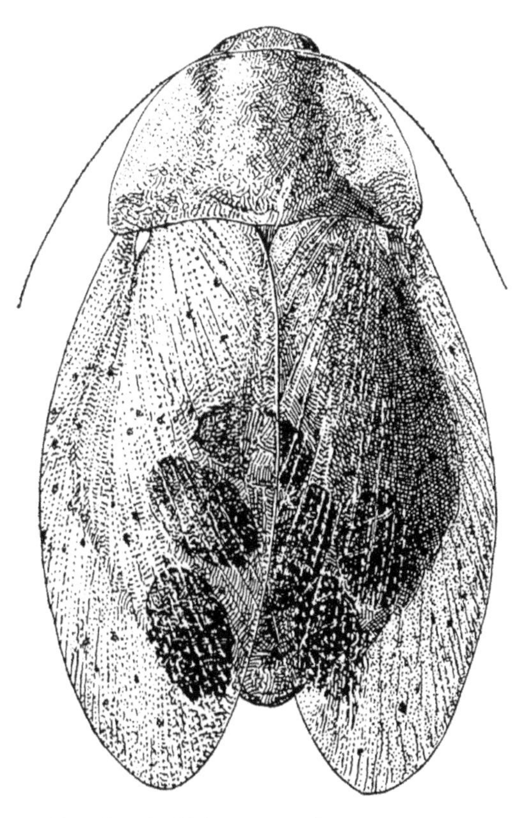

Fig. 8.4 Female of *Phlebonotus pallens* carrying nymphs beneath her tegmina. After Pruthi (1933).

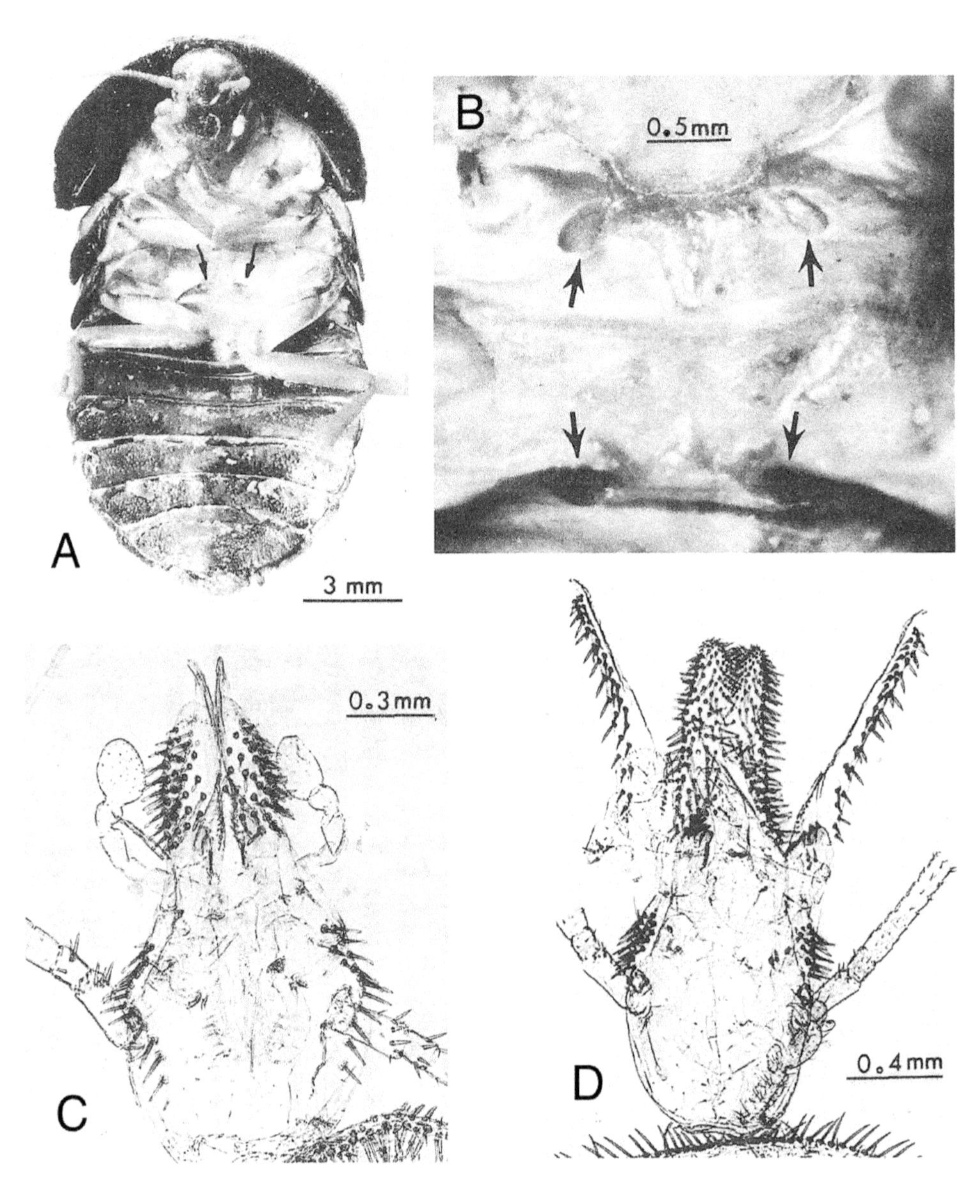

Fig. 8.5 *Perisphaerus* sp. from the Philippines. (A) Ventral view of adult female from Mt. Galintan; arrows indicate orifices between coxae. (B) Orifices (arrows) between coxae. (C) Head of probable first instar that was attached to an adult female. (D) Head of probable second instar that was attached to an adult female. From Roth (1981b); photos by L.M. Roth.

on the maxillae and labium, suggesting that they feed on a liquid diet. Midguts of young instars are filled with a pink material rather than the leaf chips they eat when older (Reuben, 1988). Jayakumar et al. (1994) and Bhoopathy (1998), however, suggest that young instars of this species may use a long, sharp mandibular tooth to pierce the tergites of the female and withdraw nourishment. First-instar nymphs removed from the mother do not live. Second-instar nymphs begin to make short forays from their maternal dome home to feed on dry leaves, and will survive if removed from their mother.

Among the Perisphaeriinae there are two recorded cases of nymphs clinging to the ventral surface of the mother for protection and nutrition. Nymphs of *Perisphaerus* cling to the female for at least two instars (Roth, 1981b). There are 17 species in this genus, but they are known almost exclusively from the study of museum specimens. First-instar nymphs are eyeless and have an elongate head and specialized galeae that suggest the intake of liquid food from the mother. There are four distinct orifices on the ventral surface of the female, with one pair occurring between the coxae of both the middle and hind legs (Fig. 8.5). Females have been collected with the mouthparts of a nymph inserted into one of these orifices; the "proboscis" of nymphs is 0.3 mm wide, about the same width as the intercoxal opening. The food of the nymphs may be glandular secretions or possibly hemolymph. The female can roll up into a ball with her clinging nymphs inside, rendering both the female and the nymphs she surrounds relatively impervious to attacks by ants (Fig. 1.11B). At least nine nymphs may be enclosed when the female assumes the defensive position. Other genera with the ability to conglobulate (e.g., *Pseudoglomeris*) may also exhibit this type of parental care. A similar defensive behavior occurs in species where the female "cups" her underside against a hard substrate (Fig. 8.6). In *Trichoblatta sericea,* well-developed pulvilli and claws of first-instar nymphs allow them to cling to the underside of the female for the first 2 to 3 days after hatching. The female secretes a milky fluid from her ventral side, which probably serves as food for the nymphs. Neonates isolated from their mother did not survive past the second instar (Reuben, 1988).

Parental Care in a Nest or Burrow

Nests and burrows typically reduce the biological hazards of the external environment and reinforce social behavior (Hansell, 1993). The structures offer protection from natural enemies and act as a buffer against temperature and moisture fluctuations. In subsocial cockroaches found in nests, one or both parents also actively defend the galleries against predators and conspecific intruders. Because these cockroaches nest in or near their food source (wood, leaf litter), parents can forage without leaving or carrying their offspring. Australian soil-burrowing cockroaches nest only where their food source is ample and forage close to the entrance (*Macropanesthia*), and so are absent from their family for only brief periods of time (Rugg and Rose, 1991; Matsumoto, 1992). Females with young are quite aggressive (D. Rugg, pers. comm. to CAN).

Fig. 8.6 Maternal care in an unidentified apterous cockroach collected in Namibia, ventral view. The female was clinging to a rock, with the elongated edges of the tergites serving to raise her venter above the substrate and form a brood covering "cup." The presence of ants (upper-right quadrant) in this field photo suggests that the behavior functions to defend young nymphs, although it is possible the female also supplies them with nutriment. Photo and information courtesy of Edward S. Ross.

Biparental care in a nest arose at least twice among wood-feeding cockroaches: in the ovoviviparous Panesthiinae and in the oviparous Cryptocercidae. These insects typically nest in damp, rotted logs, utilizing the wood itself as a food source; consequently, the young are never left untended. A wood-based diet may warrant the cooperation of both parents; wood-feeding has favored paternal investment not only in cryptocercids and some panesthiines, but also in passalid and scolytid beetles (Tallamy and Wood, 1986; Tallamy, 1994).

Cryptocercus is the only known oviparous cockroach with well-developed parental care, and is discussed in Chapter 9 in the context of its sister group relationship to termites. A recent study found that adult presence has a significant effect on offspring growth in families of *C. kyebangensis* (Park and Choe, 2003a), but the relative influence of parental care and group effects are yet to be determined. In gregarious *Periplaneta,* for example, single nymphs raised with adults grow and develop as rapidly as grouped nymphs (Wharton et al., 1968). All studied species in the wood-feeding blaberid genus *Salganea* live in biparental families (Matsumoto, 1987; Maekawa et al., 1999b, 2005). In *Sal. taiwanensis,* nymphs cling to the mouthparts of their parents and take liquids via stomodeal feeding (Fig. 8.3B). Removal of neonates from parental care results in high mortality; removed nymphs that live have a significantly longer duration of the first instar (T. Matsumoto and Y. Obata, pers. comm. to CAN).

Two different social structures have been reported for Australian wood-feeding panesthiines: both family groups and aggregations. Shaw (1925) reported that both *Panesthia australis* and *Pane. cribrata* (= *laevicollis*) live in family groups consisting of a pair of adults and nymphs in various stages of development. Matsumoto (1988) more recently studied *Pane. australis,* and found that of 29 social groups collected, the majority were families: 14 consisted of a female with nymphs, two were a male with nymphs, and two were an adult pair with nymphs. Groups never contained more than a single adult of either sex or an adult pair together with nymphs. The age of nymphs in the group ranged widely, however, so it is possible that the nymphs in these groups were aggregated individuals rather than a sibling group (T. Matsumoto, pers. comm. to CAN). The field studies of H. A. Rose (pers. comm. to CAN) indicate that neither *Pane. australis* nor any of the other wood-feeding Australian panesthiines are subsocial. Rugg and Rose (1984b) and O'Neill et al. (1987) found that while adult pairs with nymphs could be found in *Pane. cribrata* (12% of groups), the most commonly encountered groups (29%) were harems, consisting of a number of adult females, together with a single adult male and a number of nymphs. A possible reason for these discrepancies is that social structure in this genus may vary with habitat and population density. Harems seem to be common in areas of high population density, while family groups are generally found in marginal environments, or on the outer fringes of areas with high population density (D. Rugg, pers. comm. to CAN).

Parental Feeding of Offspring

Like other subsocial insects, the defense of offspring is a component of the behavioral repertoire of all cockroach species that exhibit parental care. Parents protect offspring in a nest, beneath the body, under wing covers, or directly attached to the body. A large number of cockroach species produce defensive secretions (Roth and Alsop, 1978) and females with young may be the most likely to employ them (e.g., *Thorax porcellana*—Reuben, 1988). More unique among subsocial insects is the variety of mechanisms by which cockroach parents are a direct source of food to their nymphs. Many species for which we have evidence of advanced parental care, as well as viviparous and possible ovoviviparous females, see to the nutritional needs of their offspring by feeding them on bodily fluids (Table 8.4). Parental food may be produced internally in a brood sac, expelled in a mass after hatch, secreted externally either dorsally or ventrally on the abdomen, or produced from either end of the digestive system. The materials transferred from parent to post-hatch offspring have not been analyzed in any cockroach species. The basis of the stomodeal feeding exhibited by *Salganea* (Fig. 8.3B) would be of particular interest, as *Periplaneta* is known to secrete at least two different types of saliva in response to stimulation from different neurotransmitters. One type of saliva has a dramatically higher proteinaceous component than the other (Just and Walz, 1994).

Maternal provisioning likely occurs in taxa additional to those listed in Table 8.4. Like *Gromphadorhina*, the blaberids *Byr. fumigata*, *Blaberus* sp., and *R. maderae* all have glandular cells in the brood sac that may secrete a post-hatch meal for neonates (references in Perry and Nalepa, 2003). The lateral abdominal tergites in most female Perisphaeriinae and in many Panesthiinae of both sexes have rows of glandular orifices of unknown function (Anisyutkin, 2003). The vast majority of ovoviviparous females have yet to be studied while alive. Even if a female does not provide bodily exudates, she may facilitate offspring feeding in other ways. There are two reports that young nymphs of *R. maderae* accompany their mother on nocturnal foraging trips (Séin, 1923; Wolcott, 1950).

If the standard diet of a species is one that can be handled more efficiently by adults than by juveniles (e.g., physically difficult food), then the most efficient way to convert it to a form usable by young nymphs may be via exudates from a parent. The young are offered a reliable,

Table 8.4. Parental care in cockroaches where post-hatch offspring are fed on the bodily secretions of adults (modified from Nalepa and Bell, 1997).

		Offspring	
Species	Subfamily	Location	Food source
Perisphaerus sp.	Perisphaeriinae	Cling ventrally	Hemolymph?[4]
Trichoblatta sericea	Perisphaeriinae	Cling ventrally	Sternal exudate[5]
Pseudophoraspis nebulosa	Epilamprinae	Cling ventrally	?[6]
Phlebonotus pallens	Epilamprinae	Under tegmina	?[6,7]
Thorax porcellana	Epilamprinae	Under tegmina	Tergal exudate[5]
Gromphadorhina portentosa	Oxyhaloinae	Abdominal tip of female	Secretion from brood sac?[8]
Salganea taiwanensis[1]	Panesthiinae	Mouthparts of adult	Stomodeal fluids[9]
Cryptocercus punctulatus, C. kyebangensis[1,2]	Cryptocercinae	Abdominal tip of adult	Hindgut fluids[10,11,12]
Blattella vaga[2,3]	Blattellinae	Under tegmina	Tergal exudate[13]

[1]Biparental families.
[2]Oviparous.
[3]Brief association.
[4]Roth (1981b).
[5]Reuben (1988).
[6]Shelford (1906a).
[7]Pruthi (1933).
[8]Perry and Nalepa (2003).
[9]T. Matsumoto and Y. Obata (pers. comm. to CAN).
[10]Seelinger and Seelinger (1983).
[11]Nalepa (1984).
[12]Park et al. (2002).
[13]Roth and Willis (1954).

easy-to-digest diet, thereby relieving them of the necessity of finding and processing their own food. Because the mother can meet at least part of the metabolic demands of "lactation" from her own bodily reserves, these cockroach juveniles are unaffected by temporary shortages of food items in the habitat during their phase of most rapid growth (Pond, 1983). The cockroach ability to store and mobilize nitrogenous materials via symbiotic fat body flavobacteria may be the basis for the variety of different food materials offered in parental provisioning (Nalepa and Bell, 1997, Chapter 5).

Altricial Development

After a parental lifestyle evolves, subsequent developmental adaptations often occur that reduce the cost of care and increase the dependency of offspring (Trumbo, 1996; Burley and Johnson, 2002). This is a universal trend, in that the developmental correlates of parental care are similar in both vertebrates and invertebrates. The pampered juveniles in these parental taxa are altricial, which in young cockroaches is evident in their blindness, delicate exoskeleton, and dependence on adults for food (Nalepa and Bell, 1997). Neonates of *Cryptocercus* are a good example of altricial development in cockroaches. First instars lack compound eyes; eye pigment begins developing in the second instar. The cuticle is pale and thin, with internal organs clearly visible through the surface of the abdomen. Gut symbionts are not established until the third instar, making young nymphs dependent on adults for food. First instars are small, averaging just 0.06% of their final adult dry weight. The small size of neonates is associated with the production of small eggs by the female. The length of the terminal oocyte is 5% of adult length, contrasting with 9–16% exhibited by six other species of oviparous cockroaches (Nalepa, 1996). Young nymphs of *Perisphaerus* also lack eyes; in one species at least the first two instars are blind (Roth, 1981b). We have little information on developmental trends in those cockroach species where females carry nymphs. It would be intriguing, however, to determine if, like marsupials, internal gestation in these species is truncated, with nymphs completing their early development in the female's external brood chamber.

Juvenile Mortality and Brood Reduction

Overall, insects that exhibit parental care may be expected to show low early mortality when compared to nonparental species (Itô, 1980). This pattern, however, does not seem to apply to the few species of subsocial cockroaches for which survivorship data are available. In *Macropanesthia,* mortality is about 35–40% by the time the nymphs disperse from the nest at the fifth to sixth instar (Rugg and Rose, 1991; Matsumoto, 1992). Both *Salganea esakii* and *Sal. taiwanensis* incubate an average of 15 eggs in the brood sac, but average only six nymphs (third instar) in young, field-collected families (T. Matsumoto and Y. Obata, pers. comm. to CAN). Family size of *Cryptocercus punctulatus* declines by about half during the initial stages; a mean of 73 eggs is laid, but families average only 36 nymphs prior to their first winter (Nalepa, 1988b, 1990). These data suggest that neonates may be subject to mortality factors such as disease or starvation despite the attendance of adults.

An alternative explanation for high neonate mortality in these species is that it represents an evolved strategy for adjusting parental investment after hatch (Nalepa and Bell, 1997). Unlike other oviparous cockroaches, in *Cryptocercus* the hatching of nymphs from the egg case is not simultaneous, but extended in time. Hatching asynchrony results in variation in competitive ability within a brood, a condition particularly conducive to the consumption of young offspring by older siblings (Polis, 1984). Nymphs of *C. punctulatus* 12 days old have been observed feeding on dead siblings, and attacks by nymphs on moribund siblings have also been noted. Age differentials within broods may allow older nymphs to monopolize available food, leading to the selective mortality of younger, weaker, or genetically inferior siblings. Necrophagy or cannibalism by adults or older juveniles may then recycle the somatic nitrogen of the lower-quality offspring back into the family (Nalepa and Bell, 1997). The production of expendable offspring to be eaten by siblings can be viewed as an alternative to producing fewer eggs, each containing more nutrients (Eickwort, 1981; Polis, 1981; Elgar and Crespi, 1992).

The behavioral mechanisms balancing supply (provisioning by parents) and demand (begging or solicitation by nymphs) are unstudied in subsocial cockroaches. In *Cryptocercus,* adults appear to offer hindgut fluids periodically, with juveniles competing for access to them. It is probable that, like piglets, nymphs that struggle the hardest to reach parental fluids will gain the biggest share. Competition for food may be a proximate mechanism for adjusting brood size and eliminating runts in other subsocial cockroaches as well. *Perisphaerus* sp. females possess just four intercoxal openings, but nine nymphs were associated with one of the museum specimens studied by Roth (1981b). Sibling rivalry for maternally produced food is also observable in *G. portentosa* and *Sal. taiwanensis* (Fig. 8.3). In *Cryptocercus,* there is some evidence of parent-offspring conflict in the amount of trophallactic food that an individual nymph receives. Adults can deny access to hindgut fluids by closing the terminal abdominal segments, like a clamshell. In the process of doing so the head of a feeding nymph is sometimes trapped, and the adult attempts to either fling it off with abdominal

Fig. 8.7 Nymphs of *Cryptocercus punctulatus* cooperatively feeding on a sliver of wood. Photo by C.A. Nalepa.

wagging, or to scrape it off by dragging it along the side of the gallery (CAN, unpubl. obs.).

It should be noted that in *Cryptocercus* there are cooperative as well as competitive behaviors among nymphs when procuring food. Wood is not only nutritionally poor and difficult to digest, but physically unyielding. Like young nymphs in aggregations, early developmental stages of *Cryptocercus* may need the presence of conspecifics to help acquire meals when they begin including wood in their diet. Nymphs have been observed feeding cooperatively on wood slivers pulled free by both siblings (Fig. 8.7) and adults (Nalepa, 1994; Park and Choe, 2003a).

Cost of Parental Care

Most cockroaches that exhibit parental care are subject to risks associated with brood defense and invest time in taking care of offspring. Other costs vary with the form and intensity of parental care. Brooding, for example, is a small investment on the part of the female in relation to potential returns (Eickwort, 1981). In females that carry offspring on their bodies, the burden may hinder locomotion and thus the ability to escape from predators. Energy expended on nest construction can detract from a parent's capacity for subsequent reproduction in those species where parental care occurs in excavated burrows. Insects that utilize nests may also invest time and energy in provisioning and hygienic activities (Tallamy and Wood, 1986). Feeding offspring on bodily secretions may drain stored reserves otherwise devoted to subsequent bouts of oogenesis. The metabolic expenditure may be particularly high in wood-feeding species, whose diet is typically low in nitrogenous materials. The high cost of parental care in *Cryptocercus* may account for their functional semelparity (Nalepa, 1988b), and has been proposed as a key precondition allowing for the evolution of eusociality in an ancestor they share with termites (Chapter 9). It is of interest then, that, another wood-feeding cockroach (*Salganea matsumotoi*) that lives in biparental groups and is thought to exhibit extensive parental care appears to have more than one reproductive episode (field data) (Maekawa et al., 2005).

In insects that do not nest in their food source, providing care to young may conflict with feeding opportunities, particularly in species whose diet consists of dispersed or ephemeral items that require foraging over substantial distances. One solution to is to carry one brood while gathering nutrients for subsequent brood development (Tallamy, 1994). To test this hypothesis, it is necessary to determine (1) if females feed while externally carrying nymphs, and (2) if females carrying nymphs are concurrently developing their next set of eggs, incubating eggs in the brood sac, or building reserves for the next brood. We found relevant information on two species. A *Pseudophoraspis nebulosa* female caught in the field with numerous neonates clinging to the undersurface of her abdomen was dissected, and her brood sac was empty (Shelford, 1906a). In *Tho. porcellana,* newly hatched nymphs remain in association with their mother for 45 days. After partition another ootheca is formed in 15 to 20 days, and gestation takes 45–52 days. There is therefore a period of time when the female is both internally incubating an ootheca in her brood sac and externally carrying nymphs on her back. However, these are sluggish insects that remain stationary in the leaves on which they feed (Reuben, 1988). At present, then, too little information is available for a fair evaluation of Tallamy's (1994) hypothesis.

SOCIAL INFLUENCES

Social behavior in cockroaches, as in other insects (Tallamy and Wood, 1986), is largely a function of the type, accessibility, abundance, persistence, predictability, and distribution of the food resources on which they depend. Large cockroach aggregations are found only where food is consistently renewed by vertebrates (bats, birds, humans). Biparental care is found only in wood-feeding cockroaches, whose diet is physically tough, low in nitrogen, and digested in cooperation with microorganisms. Young developmental stages in both aggregations and families rely at least in part on food originating from fellow cockroaches. Although predation pressure can alter social structure (Lott, 1991), and has been suggested as a selective pressure in cockroaches (Gautier et al., 1988), data with which we can evaluate its influence are scarce. Reproductive mode is unrelated to gregariousness; both oviparous and ovoviparous cockroaches aggregate. Subsocial cockroaches, however, are almost exclusively ovoviviparous. While the costs and benefits of social behavior for other developmental stages vary with a wide variety of factors, the benefactors in most cockroach social systems are young nymphs. Several uniquely blattarian characteristics influence cockroach social structure, such as the ability to mobilize stored nitrogenous reserves and the need for hatchlings to acquire an inoculum of gut microbes. Cockroaches also display similarities to not only other insect but also to vertebrate social systems (e.g., altricial development). They are thus potentially excellent models with which to test general hypotheses in social ecology.

NINE Termites as Social Cockroaches

Our ancestors were descended in early Cretaceous times from certain kind-hearted old cockroaches.

—W.M. Wheeler, "The Termitodoxa, of Biology and Society" (in the voice of a termite king)

It has long been known that termites (Isoptera), cockroaches (Blattaria), and mantids (Mantodea) are closely related (Wheeler, 1904; Walker, 1922; Marks and Lawson, 1962); they are commonly grouped as suborders of the order Dictyoptera (Kristensen, 1991). Although there is a general agreement on the monophyly of the order, during the past two decades the sister group relationships of these three taxa and the position of wood-feeding cockroaches in the family Cryptocercidae in relation to termites have been lively points of debate (see Nalepa and Bandi, 2000; Deitz et al., 2003; Lo, 2003 for further discussion). A variety of factors contribute to obscuring the relationships. First, fossil and molecular evidence indicate that these taxa radiated within a short span of time (Lo et al. 2000; Nalepa and Bandi, 2000). A rapid proliferation and divergence of the early forms would obscure branching events via short internal branches separating clades, instability of branching order, and low bootstrap values of the corresponding nodes (Philippe and Adoutte, 1996; Moore and Willmer, 1997). Second, heterochrony played a major role in the genesis and subsequent evolution of the termite lineage (Nalepa and Bandi, 2000). It is notoriously difficult to determine the phylogenetic relationships of organisms with a large number of paedomorphic characters (Kluge, 1985; Rieppel, 1990, 1993). Reductions and losses make for few morphological characters on which to base cladistic analysis, and parallel losses of characters by developmental truncation make it difficult to distinguish between paedomorphic and plesiomorphic traits (discussed in Chapter 2). Third, cockroaches in the particularly contentious family Cryptocercidae live and die within logs and have left no fossil record. Fourth, extant lineages of Dictyoptera represent the terminal branches of a once luxuriant tree, with many extinct taxa. Finally, several phylogenetic studies of the Dictyoptera have been problematic because of ambiguous character polarity, inadequate taxon sampling, and questionable reliability of the characters used for phylogenetic inference (for discussion, see Lo et al. 2000; Deitz et al., 2003; Klass and Meier, 2006).

The bulk of current evidence supports the classic view (Cleveland et al., 1934; Grassé

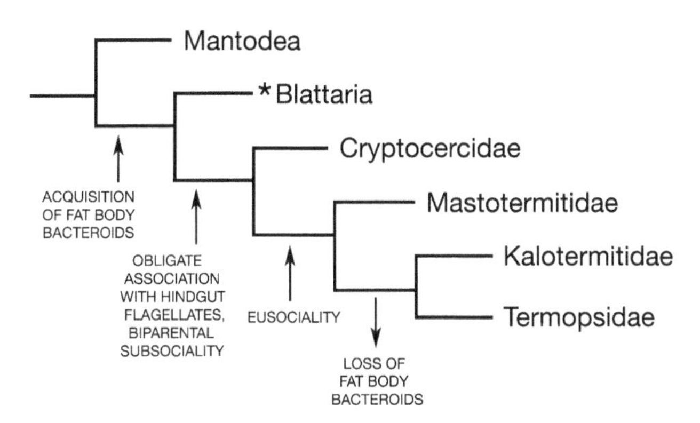

Fig. 9.1 Phylogenetic tree of Dictyoptera, after Deitz et al. (2003). Mantids branched first, Blattaria is paraphyletic with respect to the examined Isoptera (Mastotermitidae, Kalotermitidae, Termopsidae), and Cryptocercidae is the sister group to termites. The study was conducted utilizing the same morphological and biological data base used by Thorne and Carpenter (1992), however, polarity assumptions and uninformative characters were eliminated, characters, character states, and scorings were revised, and seven additional characters were added. The tree suggests a single acquisition of both symbiotic fat body bacteroids (*Blattabacterium*) and hindgut flagellates within the Dictyoptera. Bacteroids were subsequently lost in all termites but *Mastotermes;* oxymonadid and hypermastigid flagellates were lost in the "higher" termites (Termitidae—not included in tree). The sister group relationship of *Cryptocercus* and *Mastotermes* is supported by phylogenetic analysis of fat body endosymbionts (Fig. 5.7) and the cladistic analysis of Klass and Meier (Fig. P.1). *Blattaria denotes Blattaria except Cryptocercidae.

and Noirot, 1959) that Cryptocercidae is sister group to termites. It is not, however, a basal cockroach group as proposed by most early workers (e.g., McKittrick 1964, Fig. 1). Mantids branched first, with *Cryptocercus* + Isoptera forming a monophyletic group deeply nested within the paraphyletic cockroach clade (Fig. 9.1; see also Fig. P.1 in the Preface and Fig. 5.7). These relationships are supported by morphological analysis (Klass, 1995), by analysis of morphological and biological characters (Deitz et al., 2003; Klass and Meier, 2006), by Lo et al.'s (2000) analysis of three genes, and by Lo et al.'s (2003a) analysis of four genes in 17 taxa, the most comprehensive molecular study to date. The fossil record and the clock-like behavior of 16S rDNA of fat body endosymbionts in those lineages possessing them indicate that the radiation of mantids, termites, and modern cockroaches (i.e., without ovipositors) occurred during the late Jurassic–early Cretaceous (Vršanský, 2002; Lo et al., 2003a).

This phylogenetic hypothesis provides a parsimonious explanation for several key characters of Dictyoptera. An obligate relationship with Oxymonadida and Hypermastigida flagellates in the hindgut paunch first occurred in an ancestor common to *Cryptocercus* and termites, and was correlated with subsociality and proctodeal trophallaxis (Nalepa et al., 2001a). These gut flagellates were subsequently lost in the more derived Isoptera (Termitidae). Endosymbiotic bacteroids (*Blattabacterium*) in the fat body were acquired by a Blattarian ancestor, or acquired earlier in the dictyopteran lineage and subsequently lost in mantids. All termites but *Mastotermes* subsequently lost their *Blattabacterium* endosymbionts (Bandi and Sacchi, 2000, discussed below). The phylogenetic hypothesis depicted in Fig. 9.1, then, is consistent with a single acquisition and a single loss of each of the two categories of symbiotic associations. Eusociality evolved once, from a subsocial, *Cryptocercus*-like ancestor.

Lo (2003) offers two reasons for exercising some caution in the full acceptance of this phylogenetic hypothesis. First, for two of the genes that support the sister group relationship of *Cryptocercus* and termites, sequences are unavailable in mantids because they possess neither: 16S rDNA of bacteroids and those coding for endogenous cellulase. Second, because cockroach classification is in flux and taxon sampling is still relatively poor, additional data may alter tree topology. One possibility is that mantids may be the sister group of another lineage of cockroaches, which would render modern cockroaches polyphyletic with respect to both termites and mantids (Lo, 2003). Based on their examination of fossil evidence, Vršanský et al. (2002) suggested that contemporary cockroaches may be paraphyletic with respect to Mantodea as well as Isoptera.

The ancestor common to all three dictyopteran taxa was almost certainly cockroach-like (Nalepa and Bandi, 2000). Cockroaches are the most generalized of the orthopteroid insects (Tillyard, 1919), while Mantodea are distinguished by apomorphic characters associated with their specialized predatory existence. Both cockroaches and termites have predatory elements in them, although in termites it is probably limited to conspecifics (i.e., cannibalism). Mantids have short, straight alimentary canals (Ramsay, 1990), and like other predators (Moir, 1994), they neither have nor require gut symbionts. Elements of certain mantid behaviors are evident among extant cockroaches, such as the ability to grasp food with the forelegs (Fig. 9.2), and in some species, assumption of the "mantis posture" during intraspecific fights. A cockroach combatant may elevate the front portion of the body, raise the tegmina to 60 degrees or more above its back, fan the wings, and lash out with the mandibles and prothoracic legs (WJB, pers. obs.). Mantids, however, tend to lead open-air lives (Roy, 1999), and although some are known to guard egg cases, the suborder as a whole is solitary (Edmunds and Brunner, 1999). All extant termites, on the other hand, live in eusocial colonies, and have highly derived characters related to that lifestyle. There is little

Fig. 9.2 Similarity of feeding behavior in a cockroach and a mantid. (A) *Supella longipalpa* standing on four legs while grasping a food item with its spined forelegs. (B) Unidentified mantid feeding on a caterpillar, Zaire. Both photos courtesy of Edward S. Ross.

doubt that the evolution of eusociality was the event that rocketed the termite lineage into a new adaptive zone. A correlate of universal and complex social behavior among extant termites, however, is the difficulty in developing models of ancestral stages based on characters of living Isoptera. Because the best-supported phylogenetic hypotheses have termites nested within the Blattaria, we have license to turn to extant cockroaches, and in particular to *Cryptocercus*, in our search for a phylogenetic framework within which termite eusociality, and thus the lineage, evolved. It is a big topic, and one that can be explored from several points of view. Here we take a broad approach. We first examine how a variety of behaviors key to termite sociality and colony integration have their roots in behaviors displayed by living cockroach species. We then focus on cockroach development, its control, and how it can supply the raw material for the extraordinary developmental plasticity currently exhibited by the Isoptera. We address evolutionary shifts in developmental timing (heterochrony), and how these played crucial roles in the genesis and evolution of the termite lineage from Blattarian ancestors. We then turn to proximate causes of termite eusociality, first discussing how a wood diet and the symbionts involved in its digestion and assimilation provide a framework for the social transition. Finally, using young colonies of *Cryptocercus* as a model of the ancestral state, we show how a simple behavioral change, the assumption of brood care duties by the oldest offspring in the family, can account for all of the initial, defining characteristics of eusociality in termites.

THE BEHAVIORAL CONTINUUM

Striking ethological similarities in cockroaches and termites have been recognized since the early 1900s (Wheeler, 1904). These behavioral patterns probably arose in the stem group that gave rise to both taxa (Rau, 1941; Cornwell, 1968) and may therefore serve as points of departure when hypothesizing a behavioral profile of a termite ancestor. The most frequently cited behaviors shared by cockroaches and termites are those that regulate response to the physical environment. Both taxa are, in general, strongly thigmotactic (Fig. 3.7), adverse to light, and associated with warm temperatures and high humidity (Wheeler, 1904; Pettit, 1940; Ledoux, 1945). Additional shared behaviors include the use of conspecifics as food sources (Tables 4.6 and 8.4), the ability to transport food (Chapter 4), aggregation behavior, elaborate brood care (Chapter 8), hygienic behavior, allogrooming (Chapter 5), and antennal cropping, discussed below. The remaining behaviors common to Blattaria and Isoptera fall into one of two broad domains that we address in the following sections: those related to communication (vibrational alarm behavior, trail following, kin recognition) and those associated with nesting and building behavior (burrowing, substrate manipulation, behavior during excretion).

Communication: The Basis of Integrated Behavior

Complex communication is a hallmark of all social insects (Wilson, 1971). Most termites and cockroaches, however, differ from mantids and the majority of Hymenoptera in conducting all day-to-day activities, including foraging, in the dark. Both Blattaria and Isoptera rely heavily on non-visual mechanisms to orient to resources, to guide locomotion, and to communicate.

Vibrational Communication

Termites use vibratory signals in several functional contexts. Drywood termites, for example, assess the size of wood pieces by using the resonant frequency of the substrate (Evans et al., 2005). When alarmed, many termite species exhibit vertical (head banging) or horizontal oscillatory movements that catalyze increased activity throughout the colony (Howse, 1965; Stuart, 1969).

While cockroaches are known to produce a variety of acoustic stimuli in several functional contexts (Roth and Hartman, 1967), a recent review of vibrational communication included no examples of Blattaria (Virant-Doberlet and Cokl, 2004). It is known, however, that *Periplaneta americana* is capable of detecting substrate-borne vibration via receptors in the subgenual organ of the tibiae (Shaw, 1994b), and that male cockroaches use a variety of airborne and substrate-borne vibratory signals when courting females, including striking the abdomen on the substrate. Tropical cockroaches that perch on leaves during their active period may be able to detect predators or communicate with conspecifics via the substrate (Chapter 6). Adults and nymphs of *Cryptocercus* transmit alarm to family members via oscillatory movements nearly identical to those of termites (Cleveland et al., 1934; Seelinger and Seelinger, 1983).

Trail Following

In termites, trail following mediates recruitment and is a basic component of foraging behavior. In several species, the source of the trail pheromone is the sternal gland (Stuart, 1961, 1969; Peppuy et al., 2001). Cockroaches that aggregate are similar to eusocial insects in that there is a rhythmical dispersal of groups from, and return to, a fixed point in space (e.g., Seelinger, 1984), suggesting that cockroaches have navigational powers that allow them to either (1) resume a previously established membership in a group or (2) find their harborage. It is difficult to separate the two, and site constancy and homing ability may be a general characteristic of cockroaches regardless of their social patterns (Gautier and Deleporte, 1986). *Periplaneta americana* and *B. germanica* follow paths established by conspecifics as well as trails of fecal extracts (Bell et al., 1973; Kitamura et al., 1974; Miller and Koehler, 2000). Brousse-Gaury (1976) suggested that adult *P. americana* use the sternal gland to deposit a chemical trail during forays from the harborage. When the antennae of *P. americana* were crossed and glued into place, the cockroaches consistently turned in the opposite direction of a pheromonal trail in t-mazes, indicating that the mechanism employed is a comparison between the two antennae (Bell et al., 1973). There are indications of this kind of chemo-orientation in other species as well. The myrmecophile *Attaphila fungicola* follows foraging trails of its host ant (Moser, 1964), and female cockroaches that have recently buried oothecae may disturb the substrate in an attempt to obliterate odor trails from detection by cannibals (Rau, 1943).

Kin Recognition

Kin recognition is well developed in those cockroach species in which it has been sought. Juveniles of *B. germanica* are preferentially attracted to the odor of their own population or strain (Rivault and Cloarec, 1998). *Paratemnopteryx couloniana* females recognize their sisters (Gorton, 1979), first instars of *Byrsotria fumigata* recognize and orient to their own mother (Liechti and Bell, 1975), and juveniles of *Rhyparobia maderae* prefer to aggregate with siblings over non-siblings, a tendency most pronounced in first instars (Evans and Breed, 1984). Nymphs of *Salganea taiwanensis* up to the fifth instar are capable of distinguishing their parents from conspecific pairs (T. Matsumoto and Y. Obata, pers. comm. to CAN). Like termites (reviewed by Vauchot et al., 1998), non-volatile pheromones in the cuticular hydrocarbons can and do transfer among individuals via physical contact in cockroach aggregations (Roth and Willis, 1952a; Everaerts et al., 1997; discussed in Chapter 3).

Home Improvement: Digging, Burrowing, and Building

Among the social insects, termites are noted for the diversity and complexity of their nest architecture. Both fecal deposits and exogenous materials (soil, wood) transported by the mandibles are used as construction material, and the structure is made cohesive with a mortar of saliva and fecal fluid. Intricate systems of temperature regulation and ventilation are typically incorporated, resulting in a protected, climate-controlled environment for these vulnerable insects (Noirot and Darlington, 2000). Cockroaches exhibit rudimentary forms of these complex construction behaviors, providing support for the notion that termite construction skills are derivations of abilities already present in their blattarian ancestors (Rau, 1941, 1943).

A number of cockroach species tunnel in soil, leaf litter, guano, debris, rotten, and sometimes sound, wood (Chapters 2 and 3). Cockroaches also possess the morphological and behavioral requisites for more subtle excavation of substrates, as evidenced in oviparous cockroaches during the deposition and concealment of oothecae (Fig. 7.2) (McKittrick et al. 1961; McKittrick, 1964). On particulate substrates such as sand female Blattidae use a raking headstroke to dig a hole, but they gnaw crevices in more solid substances like wood. Blattellidae bite out mouthfuls of material on all substrate types. Legs may be used to help dig holes and to move debris away from the work site. *Euzosteria sordida* digs a hole using backstrokes of the head, followed by movement of each leg in turn to move sand away from the excavation site (Mackerras, 1965b). After the hole is the appropriate depth, the female has a "molding phase," during which she lines the bottom of the hole with a sticky layer of substrate particles mixed with saliva. The ootheca is then deposited in or near the hole, and adjusted into position with the mouthparts. A mixture of saliva and finely masticated substrate is applied to the surface of the egg case, and the remaining gaps are filled with dry material. The whole operation can last more than an hour (McKittrick et al. 1961; McKittrick, 1964). Females can be quite selective in their choice of building material. Rau (1943) noted that *Blatta orientalis* chooses large grains of sand and discards the small ones. In *P. americana* the egg case may be plastered with cockroach excrement dissolved in saliva (Rau, 1943). It should be noted in this regard that, like termites, cockroaches produce a heterogeneous mix of excretory products (Nalepa et al., 2001a). These may be distinguished in some species by the behavior of the excretor, the reaction of conspecifics in the vicinity, and the nature of the fecal material. Cockroaches that are domestic pests are well known for producing both solid fecal pellets and smears attached to the substrate. Both Lawson (1965) and Deleporte (1988) describe distinct and systematic defecation behaviors in *P. americana* that are reminiscent of termites during nest building. These include backing up prior to defecation, then dragging the tip of the abdomen on the substrate while depositing a fecal droplet.

Some cockroach species actively modify their living environment. *Arenivaga apacha* dwell in the burrows of kangaroo rats, within which they construct small living spaces lined with the nest material of their host (Chapter 3). The soil associated with these spaces is of unusually fine texture because the cockroaches work the soil with their mouthparts, reducing gravel-sized lumps to fine sand and silt-textured soil (Cohen and Cohen, 1976). *Eublaberus posticus* shapes the soft mass of malleable bat guano along the base of cave walls into irregular horizontal galleries (Fig. 9.3). These are subsequently consolidated by calcium carbonate from seepage water (Darlington, 1970). It is unclear whether the cockroaches actively build these structures or whether the hollows are epiphenomena, by-products of the insects' tendency to push themselves under edges and into small irregularities (Darlington, pers. comm. to CAN). The observation by Deleporte (1985) that various developmental stages of *P. americana* dig resting sites in clay walls suggests the former.

Cockroaches in the Cryptocercidae in many ways exhibit nest construction and maintenance behavior comparable to that of dampwood termites (Termopsidae). When initiating a nest, adult *Cryptocercus* actively excavate galleries; their tunnels are not merely the side effects of feeding activities. They eject frass from the nest, plug holes and gaps (Fig 9.4A), build pillars and walls to partition galleries, and erect barriers when their galleries approach those of families adjacent in the log (Nalepa, 1984,

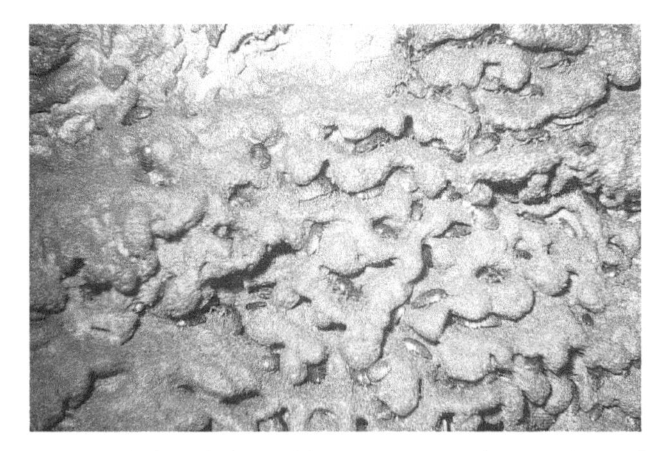

Fig. 9.3 Shelters fashioned from wet guano along the base of cave walls by *Eublaberus posticus*, Tamana main cave, Trinidad; note cockroaches in crevices. The insects may actively construct these structures, or they may result from the cockroach tendency to wedge into crevices. From Darlington (1970); photo and information courtesy of J.P.E.C. Darlington.

Fig. 9.4 Constructions of *Cryptocercus punctulatus*. (A) Detail of material used to plug holes and seal gaps; here it was sealing the interface between a gallery opening and the loose bark that covered it. Both fecal pellets (arrow) and small slivers of wood are present. (B) Sanitary behavior: fecal paste walling off the body of a dead adult (arrow) in a side chamber. An adult male was the only live insect present in the gallery system. Photos by C. A. Nalepa.

unpubl. obs.). Building activity is most common when the cockroaches nest in soft, well-rotted logs, and, like *Zootermopsis* and some other termites (Wood, 1976; Noirot and Darlington, 2000), excrement and masticated wood are the principal construction materials. If logs are damp, fecal pellets lose their discrete packaging and become a mass of mud-like frass.

Cryptocercus also exhibits a number of termite-like behaviors in maintaining a clean house. In addition to expelling frass from galleries, adults keep the nursery area (i.e., portion of the gallery with embedded oothecae) clear of fungal growth and the fecal mud that commonly lines the walls of galleries in the remainder of the nest (Nalepa, 1988a). They are known to eat dead nestmates, but, like termites (Weesner, 1953; Dhanarajan, 1978), *Cryptocercus* will bury unpalatable corpses in unused portions of the gallery (Fig. 9.4B).

DEVELOPMENTAL FOUNDATIONS

The influence of hemimetabolous development in the evolution of termite societies has long been recognized (Kennedy, 1947; Noirot and Pasteels, 1987). Unlike the holometabolous Hymenoptera, termite juveniles do not have to mature before they are capable of work. Hemimetabolous insects also tend to grow less between molts and molt more often over the course of development (Cole, 1980). This is due, at least in part, to differences in nutritional efficiency between the two groups. The conversion of digested food to body mass can be 50% greater in holometabolous insects, possibly because they do not need to produce and maintain a large mass of cuticle during the juvenile stage (Bernays, 1986).

Termite Development

In the Isoptera, day-to-day colony labor is the responsibility of juveniles—termites whose development has been truncated, either temporarily or permanently, relative to reproductives. Even terminal nonsexual stages (i.e., soldiers, and workers in some species) are considered immature, because they retain their prothoracic glands, which degenerate in all sexual forms. The only imagoes in the termitary are the king and queen (Noirot, 1985; Noirot and Pasteels, 1987; Noirot and Bordereau, 1989). The degree, permanence, timing, and reversal of developmental arrest, together with the organs subject to these changes, determine which developmental pathway is taken during the ontogeny of particular groups (Noirot and Pasteels, 1987; Roisin, 1990, 2000). This developmental flexibility is mediated by a combination of progressive, stationary, and reversionary molts, and is distinctive. Dedifferentiation of brachypterous nymphs in termites is the only known instance of a natural reversal of metamorphosis in insects (Nijhout and Wheeler, 1982). The extraordinary complexity and sophistication characteristic of termite development is nonetheless rooted in mechanisms of postembryonic development observed in non-eusocial insects (Bordereau, 1985). The developmental characteristics of cockroach ancestors, then, were the phylogenetic foundation on which termite polyphenisms were built.

Cockroach Development

Within a cockroach species, both the number and duration of instars that precede the metamorphic molt are variable, a trait unusual among hexapods (Heming, 2003). In *P. americana,* for example, the length of nymphal period can vary from 134 to 1031 days (Roth, 1981a)—nearly an order of magnitude. The number of molts in cockroaches varies from 5 or 6 to 12 or 13, and may or may not vary between the sexes. Within a species, variation in cockroach development occurs primarily in response to environmental conditions: low temperature, minor injuries, water or food deficits, or poor food quality (Tanaka, 1981; Mullins and Cochran, 1987). Even in laboratory cultures in which extrinsic influences have been minimized or controlled, however, the instar of metamorphosis remains variable, even in nymphs from the same ootheca (Kunkel, 1979; Woodhead and Paulson, 1983). There can be a lag of up to 9 mon between the appearance of the first and last adult among nymphs from the same sibling cohort of *Periplaneta australasiae* (Pope, 1953), and "runts"—nymphs stalled in the third or fourth instar when all others in the cohort have matured—have been noted in *P. americana* (Wharton et al., 1968). Kunkel (1979) describes the instar of metamorphosis in cockroaches as a polygenic trait with a great deal of environmental input involved in its expression. Significantly, there are records of both stationary and saltatory molts in cockroaches (Gier, 1947; Rugg and Rose, 1990). If the ancestor of the termites was like extant cockroaches, then it, too, possessed a tremendous amount of developmental plasticity prior to evolving eusociality.

Control of Development

An examination of conditions known to modify cockroach development may provide insight into the origins of termite polyphenism, the proximate causes of which are still little understood (Bordereau, 1985; Roisin, 2000). Here we focus on three extrinsic factors that may have influenced development as the termite lineage evolved: minor injuries, nourishment, and group effects. Each of these has a social component, in that each can be based on interactions with conspecifics rather than the external environment.

Injury and Development

There is a large body of literature indicating that minor wounds in cockroach juveniles delay development. Injuries to legs, cerci, and antennae result in an increased number of instars, in the prolonged duration of an instar, or both (Zabinski, 1936; Stock and O'Farrell, 1954; Willis et al., 1958; Tanaka et al., 1987). The developmental delay may be attributed to the allocation of limited resources, because energy and nutrients directed into wound repair and somatic regeneration are unavailable for progressive development (Kirkwood, 1981). This relationship between injury and development may be relevant to termites in two contexts. First, in a variety of lower termites, mutilation of the wing pads and occasionally other body parts is common (e.g., Myles, 1986). These injuries are hypothesized to result from the bites of nest mates, and they determine which individuals fly from the nest and which remain to contribute to colony labor. Injured individuals do not proceed to the alate stage, but instead undergo regressive or stationary molts (Roisin, 1994). The aggressive interactions that result in these injuries may be the expression of sibling manipulation if larvae, nymphs, or other colony members are doing the biting (Zimmerman, 1983; Myles, 1986), or they could indicate fighting among nymphs that are competing for alate status (Roisin, 1994).

A second, peculiar, termite behavior also may be linked to the physiological consequences of injury. After a dealate termite pair becomes established in its new nest, the male and female typically chew off several terminal segments of their own antennae, and/or those of their partner (e.g., *Archotermopsis*—Imms, 1919; *Cubitermes*—Williams, 1959; *Porotermes*—Mensa-Bonsu, 1976; *Zootermopsis*—Heath, 1903). This behavior is also recorded in several cockroach taxa. Nymphs of *B. germanica* self-prune their antennae (autotilly)—the ends are nipped off just prior to molting (Campbell and Ross, 1979). Although first and second instars of *Cryptocercus punctulatus* almost always have intact antennae, cropped antennae can be found in third instars and are common in fourth instars (Nalepa, 1990). Nymphs and adults of the myrmecophiles *Att. fungicola* and *Att. bergi* usually have mutilated antennae (Bolívar, 1901; Brossut, 1976), but Wheeler (1900) was of the opinion that it was the host ants that trimmed them for their guests. He likened it to the human habit of cropping the ears and tails of dogs. The developmental and/or behavioral consequences of antennal cropping are unknown for both termites and cockroaches.

Nutrition and Development

Cockroach development is closely attuned to nutritional status (Gordon, 1959; Mullins and Cochran, 1987). Poor food quality or deficient quantity results in a prolongation of juvenile development via additional molts and/or prolonged intermolts (Hafez and Afifi, 1956; Kunkel, 1966; Hintze-Podufal and Nierling, 1986; Cooper and Schal, 1992). Diets relatively high in protein produce the most rapid growth (Melampy and Maynard, 1937), and on diets lacking protein, nymphs survive for up to 8 mon, but eventually die without growing (Zabinski, 1929). The effect of nutrition on development is most apparent in early instars, corresponding to what is normally their period of maximum growth (Woodruff, 1938; Seamans and Woodruff, 1939). A nutrient deficiency in a juvenile cockroach results in a growth stasis, in which a semi-starved nymph "idles" until a more adequate diet is available. This plasticity in response to the nutritional environment is suggestive of the arrested development exhibited by workers (pseudergates) in lower termite colonies, and is hypothesized to be one of the key physiological responses underpinning the shift from subsocial to eusocial status in the termite lineage (Nalepa, 1994, discussed below).

Reproductive development is also closely regulated by the availability of food in cockroaches. Females stop or slow down reproduction until nutrients, particularly the amount and quality of ingested protein, is adequate (Weaver and Pratt, 1981; Durbin and Cochran, 1985; Pipa, 1985; Mullins and Cochran, 1987; Hamilton and Schal, 1988). In *P. americana* the initial response to lack of food is simply the slowing down of oocyte growth, but if starvation becomes chronic the corpora allata are turned off and reproduction effectively ceases. When food once again becomes available the endocrine system is rapidly reactivated and normal reproductive activity follows within a short time (Bell and Bohm, 1975). Kunkel (1966, 1975) used feeding as an extrinsically controllable cue for synchronizing both the molting of nymphs and the oviposition of females in *B. germanica* and *P. americana*. There is substantial evidence, then, that domestic cockroaches tightly modulate "high demand" metabolic processes such as reproduction and development in response to changes in food intake, and that both physiological processes can be controlled in individuals by manipulating their food source.

Group Effects and Development

Group effects (discussed in Chapter 8) can have a profound effect on the developmental trajectory of juvenile cockroaches and are known from at least three families of Blattaria (Table 8.3). Nymphs deprived of social contact typically have longer developmental periods, resulting from both decreased weight gain per stadium and increased stadium length (Griffiths and Tauber, 1942b; Willis et al., 1958; Wharton et al., 1968; Izutsu et al., 1970; Woodhead and Paulson, 1983). In *P. americana*, nymphs isolated at day 0 are one-half to one-third the size of grouped nymphs after 40 days (Wharton et al., 1968). The effect is cumulative, with no critical period. It occurs at any stage of development and is reversible at any stage (Wharton et al., 1967; Izutsu et al., 1970). Respiration of isolates may increase, and new proteins, expressed as electrophoretic bands, may appear in the hemolymph (Brossut, 1975; pers. comm. to CAN). The physiological consequences seem to be caused by a lack of physical contact (Pettit, 1940; Izutsu et al., 1970) and the presence of even one other individual can ameliorate the effects (Izutsu et al., 1970; Woodhead and Paulson, 1983). The means by which tactile stimuli orchestrate the physiological changes characteristic of the group effect in cockroaches is unknown. In termites, as in cockroaches, the physical proximity of conspecifics significantly increases the longevity and vigor of individuals, with just one nestmate as sufficient stimulus. This "reciprocal sensory intimacy" is thought to play a key, if unspecified, role in caste determination (Grassé, 1946; Grassé and Noirot, 1960).

Heterochrony: Evolutionary Shifts in Development

Termites are essentially the Peter Pans of the insect world—many individuals never grow up. Most colony members are juveniles whose progressive development has been suspended. Even mature adult termites exhibit numerous juvenile traits when compared to adult cockroaches, the phylogenetically appropriate reference group (Nalepa and Bandi, 2000). Termites therefore may be described as paedomorphic, a term denoting descendent species that resemble earlier ontogenetic stages of ancestral species (Reilly, 1994). The physical resemblance of termites and young cockroaches is indisputable, and is most obvious in the bodily proportions, the thin cuticle, and a short pronotum that leaves the head exposed. Cleveland et al. (1934) and Huber (1976) both noted the resemblance of early instars of *Cryptocercus* to larger termite species, with the major difference being the more rapid movement and longer antennae of *Cryptocercus* (Fig. 9.5). One advantage that termites gain by remaining suspended in this thin-skinned morphological state is the avoidance of a heavy nitrogenous (Table 4.5) investment in cuticle typical of older developmental stages of their cockroach relatives.

Cockroaches that are paedomorphic display a variety of termite-like characters such as thinning of the cuticle, eye reduction, and decrease in the size of the pronotal

Fig. 9.5 First instar of *Cryptocercus punctulatus*. Photo by C.A. Nalepa.

shield (e.g., *Nocticola australiensis*—Roth, 1988). These cockroaches are often wingless, but when wings are retained they can resemble those of termite alates. In *Nocticola babindaensis* and the genus *Alluaudellina* (= *Alluaudella*), the forewings and hindwings are nearly the same length, they considerably exceed the tip of the abdomen, both sets are membranous, and they have a reduced venation and anal lobe (Shelford, 1910a; Roth, 1988).

The expression of altered developmental timing in termites is not limited to morphological characters. It includes aspects of both behavior and physiology that are more characteristic of the juvenile rather than the adult stages of their non-eusocial relatives. Just as maturation of the body became truncated during paedomorphic evolution in the termite lineage, so did many features of behavioral and physiological development. Elsewhere in this chapter we noted several behaviors that are common to termites and cockroach taxa, including burrowing, building, substrate manipulation, trail following, and vibrational alarm behavior. There are additional behaviors crucial to termite social cohesion shared only with the *early developmental stages* of cockroaches (Nalepa and Bandi, 2000). In most cockroach species, young nymphs have the strongest grouping tendencies, and in some, early instars are the only stages that aggregate (Chapter 8). Early cockroach instars often display the most pronounced kin recognition (Evans and Breed, 1984), the most intense cannibalism (Wharton et al., 1967; Roth, 1981a), and the most frequent coprophagy (Nalepa and Bandi, 2000). Young *Periplaneta* nymphs affix fecal pellets to the substrate more often than do older stages (Deleporte, 1988). Antennal cropping is displayed in nymphs of two cockroach species, and it is only young developmental stages of *Cryptocercus* that allogroom (Seelinger and Seelinger, 1983). All of these behaviors are standard elements of the termite behavioral repertoire.

Many behaviors shared by termites and young cockroaches relate to food intake. Termites also resemble cockroach juveniles in aspects of digestive physiology and dietary requirements (Nalepa and Bandi, 2000). More so than older stages, early instars of cockroaches rely on conspecific food and ingested microbial protein to fuel growth, and are dependent on the metabolic contributions of microbial symbionts in both the gut and fat body for normal development. As termites evolved, they elaborated on this food-sharing, microbe-dependent mode instead of shifting to a more adult nutritional physiology during ontogenetic growth.

Caste control in termites also may be rooted in the developmental physiology of young cockroaches (Nalepa and Bandi, 2000). It is the early cockroach instars that are most susceptible to developmental perturbations related to nutrition, injury, and group effects (Woodruff, 1938; Seamans and Woodruff, 1939; Holbrook and Schal, 1998). Moreover, these stimuli are extrinsically controllable and may allow for manipulation of individual development by fellow colony members (Nalepa and Bandi, 2000).

In sum, a large number of the juvenile characters of their cockroach ancestors were co-opted by termites in the course of their evolution, and these were integral in the cascade of adaptations and co-adaptations that resulted in the highly derived, eusocial taxon it is today. Heterochrony is known to provide a basis for rapid divergence and speciation, because integrated character sets are typically under a system of hierarchical control (Gould, 1977; Futuyma, 1986). Simple changes in regulatory genes, then, can result in rapid, drastic phenotypic changes (Futuyma, 1986; Stanley, 1998).

WOOD DIET, TROPHALLAXIS, AND SYMBIONTS

That the character and direction of Isopteran evolution as a whole has been in the main determined by their peculiar food is obvious.

—Wheeler, *The Social Insects*

There are distinct advantages to living within your food source. Logs offer mechanical protection and refuge from a number of predators and parasites, with an interior temperature and humidity generally more moderate than that of the external environment. Abundant if low-quality food is always close at hand. One disadvantage is that when on this fixed diet, a wood-feeding dictyopteran would forfeit the opportunity to move within the habitat seeking specific nutrients and nitrogenous bonanzas (e.g., bird droppings) as its developmental and reproductive needs change. Reliance on slowly accumulated reserves and the use of food originating from conspecific sources, then, would become considerably more important, particularly in those stages with a high nitrogen demand—reproducing females and young nymphs (Nalepa, 1994).

Termites inherited from cockroaches a suite of interindividual behaviors that allow for nitrogen conservation at the colony level and provide a means of circulating it among individuals within the social group (Table 4.6). These include cannibalism, necrophagy, feeding on exuviae, and coprophagy. Two behaviors of particular note are allogrooming and trophallaxis, first, because they supply the organizational glue that keeps termite colonies cohesive and functional, and second, because among cockroaches these behaviors are only known from wood-feeding species. Allogrooming has been noted in *Panesthia* (M. Slaytor, pers. comm. to CAN) and *Cryptocercus*, and in the latter it occurs exactly as described in termites by Howse (1968). The groomer grazes on the body of a conspecific, and the insect being groomed responds by rotating its body or appendages into more accessible positions (Fig. 5.5B). As with termites, the nymph being tended may enter a trance-like state and afterward remain immobile for a short period of time before resuming activity (Nalepa and Bandi, 2000).

Trophallaxis is the circulatory system of a termite colony. It is the chief mechanism of disseminating water, nutrients, hormones, dead and live symbionts, and the metabolic products and by-products of the host and all its gut symbionts. Stomodeal trophallaxis (by mouth) occurs in all termite families, and proctodeal trophallaxis (by anus) occurs in all but the derived family Termitidae (McMahan, 1969; Breznak, 1975, 1982). Both types of trophallaxis occur in wood-feeding cockroaches, and in these taxa the behaviors occur in the context of parental care. *Salganea taiwanensis* feeds its young on oral secretions (T. Matsumoto, pers. comm. to CAN; Fig. 8.3B), and *Cryptocercus* adults feed young nymphs on hindgut fluids (Seelinger and Seelinger, 1983; Nalepa, 1984; Park et al., 2002).

Hindgut Protozoa

Digestion in *Cryptocercus* is comparable to that of lower termites in all respects. The hindgut is a fermentation chamber filled to capacity with a community of interacting symbionts, including flagellates, spirochetes, and bacteria that are free in the digestive tract, attached to the gut wall, and symbiotic with resident protozoans. Included are uricolytic bacteria, cellulolytic bacteria, methano-

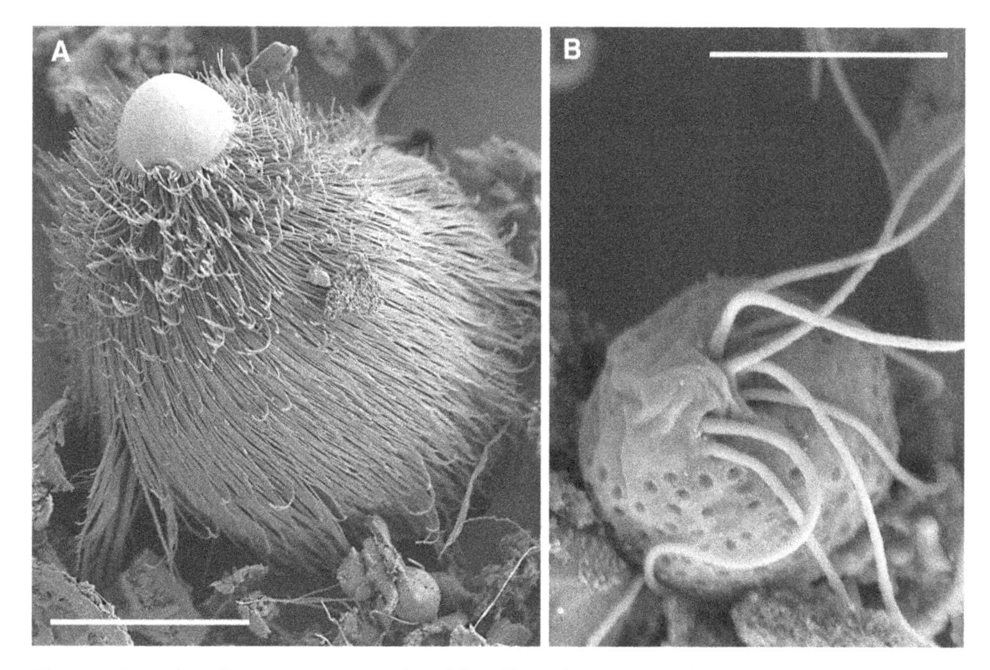

Fig. 9.6 Scanning electron micrographs of flagellates from the hindgut of *Cryptocercus punctulatus*. (A) The hypermastigote *Trichonympha* sp., scale bar = 25 μm. (B) The oxymonad *Saccinobaculus* sp., scale bar = 5 μm. Images courtesy of Kevin J. Carpenter and Patrick J. Keeling.

gens, and those capable of nitrogen fixation, as well as bacteria that participate in the biosynthesis of volatile fatty acids (Breznak et al., 1974; Breznak, 1982; Noirot, 1995).

The common possession of oxymonad and hypermastigid hindgut flagellates in *Cryptocercus* and lower termites (Fig. 9.6) is often a focal point in discussions of the evolutionary origins of termites. These protozoans are unusually large, making them good subjects for a variety of experimental investigations; some in the gut of *Cryptocercus* are 0.3 mm in length and visible to the unaided eye (Cleveland et al., 1934). They are unusually intricate, with singular morphological structures and a complex of bacterial symbionts of their own (e.g., Noda et al., 2006). They are unique; most are found nowhere in nature but the hindguts of these two groups (Honigberg, 1970). Finally, and of most interest for termite evolutionary biology, most are cellulolytic and interdependent with their hosts. For many years these flagellates were thought to be not only the sole mechanism by which dictyopteran wood feeders digested cellulose, but also the proximate cause of termite eusociality. Currently, however, neither of these hypotheses is fully supported, despite misconceptions that still abound in the literature.

Dependence on Flagellates for Cellulase?

All termites and all cockroaches examined to date produce their own cellulases, which are distinct from and unrelated to those produced by the hindgut flagellates (Watanabe et al., 1998; Lo et al., 2000; Slaytor, 2000; Tokuda et al., 2004). The common possession of a certain family of cellulase genes (GHF9) in termites, cockroaches, and crayfish suggest that these enzymes were established in the Dictyopteran lineage long before flagellates took up permanent residence in the hindguts of an ancestor of the termite-*Cryptocercus* clade (references in Lo et al., 2003b). At present, *Cryptocercus* and lower termites are considered to have a dual composting system (Nakashima et al., 2002; Ohkuma, 2003); cellulose is degraded by the combined enzymes of the host and the hindgut flagellates. Nonetheless, these hosts are dependent on the staggeringly complex communities of mutually interdependent co-evolved organisms from the Archaea, Eubacteria, and Eucarya in their digestive systems. The interactions of the microbes with each other and with their hosts are still poorly understood; however, exciting inroads are being made by the laboratories actively studying them, and the field is advancing quickly (e.g., Tokuda et al., 2004, 2005; Inoue et al., 2005; Watanabe et al., 2006). Products of cellulose degradation by gut protozoans may indirectly benefit the insect host by providing energy for anaerobic respiration and nitrogen fixation in gut bacteria (Bignell, 2000a; Slaytor, 2000). A comparison of gene expression profiles among castes of the termite *Reticulitermes flavipes* suggests that cellulases produced by the symbionts may be particularly important in incipient colonies (Scharf et al., 2005). This supports the idea that gut microbes may supply a metabolic boost at crucial points in host life history.

Flagellates Cause Eusociality?

Hindgut protozoans were crucial in the evolution of eusociality in their termite hosts, but not for the reasons usually cited. In termites, the hindgut flagellates die just prior to host ecdysis. A newly molted individual must reestablish its symbiosis by proctodeal trophallaxis from a donor nestmate, making group living mandatory. In the classic literature, this codependence of colony members was thought to be the main precondition for the evolution of eusociality in termites; the idea can be traced to the work of L.R. Cleveland (1934). While loss of flagellates at molt may enforce proximity, it provides no explanation for the defining characteristics of termite eusociality, namely, brood care, overlapping worker generations, and non-reproductive castes (Starr, 1979; Andersson, 1984). Moreover, the bulk of evidence suggests that protozoan loss at molt in termites did not precede eusociality. It is a secondary condition derived from eusociality of the hosts, and is associated with the physiology of developmental arrest and caste control (Nalepa, 1994).

Hindgut protozoans were crucial in the genesis of the termite lineage, because an obligate symbiotic relationship with them demands a reliable means of transmission between generations. The life history characteristics of a termite ancestor, as exemplified by *Cryptocercus*, combined with the physiology of encystment of these particular protozoans, mandate that this transmission could only occur via proctodeal trophallaxis (Nalepa, 1994). In an ancestor common to *Cryptocercus* and termites, flagellate cysts were presumably passed to hatchlings by intraspecific coprophagy in aggregations (Nalepa et al., 2001a). The physiology of encystment in these protists, however, does not allow for their transmission by adults. Their encystment is triggered by the molting cycle of the host; consequently they are passed in the feces only during the developmental stages of nymphs. Cysts are never found in the feces of adults or intermolts (Cleveland et al., 1934; Cleveland and Nutting, 1955; Cleveland et al., 1960). *Cryptocercus* is subsocial and semelparous. Most adults spend their entire lives nurturing one set of offspring. Consequently, older nymphs are not present in galleries when adults reproduce (Seelinger and Seelinger, 1983; Nalepa, 1984; Park et al., 2002). Coprophagy as a mechanism of intergenerational transmission is thus ruled out; adults do not excrete cysts, and older nymphs are absent from the social group. Cysts in the feces of molting *Cryptocercus* nymphs, as well as vestiges of the sexual/encystment process in termites (Grassé and Noirot, 1945; Cleveland, 1965; Messer and Lee, 1989), are a legacy of their distant gregarious past. In the ancestor *Cryptocercus* shared with termites, an obligate relationship with gut symbionts, intergenerational transmission via proctodeal trophallaxis, and subsociality were thus a co-evolved character set (Nalepa, 1991; Nalepa et al., 2001a). Proctodeal trophallaxis in young families of a *Cryptocercus*-like ancestor assured not only passage of cellulolytic flagellates between generations, but also passage of the entire complex of microorganisms present in the hindgut fluids. Trophallaxis thus conserved relationships between microbial taxa within consortia, allowing them to develop interdependent relationships by eliminating redundant pathways. The metabolic efficiency of these consortia consequently increased, shifting the cost-benefit ratio in favor of increased host reliance. The growing dependence of the host on gut microbes, in turn, reinforced selection for assured passage between generations via subsociality and trophallactic behavior. The switch from horizontal to vertical intergenerational transmission of gut fauna was thus one of the key influences in the transition from gregarious to *subsocial* behavior in the common ancestor of *Cryptocercus* and termites. It also set up one of the pivotal conditions allowing for the transition to eusociality by establishing the behavioral basis of trophallactic exchanges (Nalepa et al., 2001a).

The hypothesis that the loss of protozoan symbionts at molt was influential during the initial transition to eusociality, then, is not supported. The interdependence that the condition enforces on hosts nonetheless played a key role after the initial transition from subsociality to eusociality (detailed below). Subsequent hormonal changes related to developmental stasis and caste evolution, and the associated loss of protozoans at molt resulted in a "point of no return" (Hölldobbler and Wilson, 2005), when individuals became incapable of a solitary existence.

DOUBLE SYMBIOSIS: THE ROLE OF BACTEROIDS

A hindgut filled to capacity with a huge complex of interacting microbiota was not the only symbiotic association influential in the evolution of termite eusociality. Grassé and Noirot (1959) noted nearly a half-century ago that the two taxa bracketing the transition from cockroaches to termites share a unique double symbiosis: an association with cellulolytic flagellates in the hindgut, and endosymbiotic bacteria housed in the visceral fat body. *Cryptocercus* is the only cockroach that has the former symbiosis, which it shares with all lower termites, and

Fig. 9.7 Male and female dealate primary reproductives of *Mastotermes darwiniensis*. Photo by Kate Smith, CSIRO Division of Entomology.

Mastotermes (Fig. 9.7) is the only isopteran with the latter, which it shares with all examined Blattaria (Bandi et al., 1995; Lo et al., 2003a). *Mastotermes* has additional characters that ally the taxon with cockroaches, including a well-developed anal lobe in the hindwing and the packaging of eggs in an ootheca (Watson and Gay, 1991; Nalepa and Lenz, 2000; Deitz et al., 2003).

The bacteroid-uric acid circulation system was in place when termites evolved eusociality (Fig. 9.1), possibly allowing for the mobilization of urate-derived nitrogen from the fat body and its transfer among conspecifics via coprophagy and trophallaxis (Chapter 5). The endosymbiosis was subsequently lost in other termite lineages when these diverged from the Mastotermitidae (Bandi and Sacchi, 2000). Other termites sequester uric acid in the fat body, but without bacteroids, individuals lack the ability to mobilize it from storage. Stored reserves can only be used by colony members via cannibalism or necrophagy. Once ingested, the uric acid is broken down by uricolytic bacteria in the hindgut (Potrikus and Breznak, 1981; Slaytor and Chappell, 1994). Bacteroids were likely lost in most termites because two aspects of eusocial behavior made fat body endosymbionts redundant. The recycling of dead, moribund, and sometimes living nestmates, combined with the constant flow of hindgut fluids among nestmates via trophallaxis, allowed uricolytic gut bacteria to be a more cost-efficient option (Bandi and Sacchi, 2000). It is of note, then, that after eusociality evolved, the storage and circulation of uric acid and its breakdown products changed from one that occurs primarily at the level of individual physiology to one that occurs at the colony level. It is also of interest that proctodeal trophallaxis, a behavior linked to the presence of the hindgut symbionts, may have been influential in the loss of the fat body endosymbionts.

EVOLUTION OF EUSOCIALITY 1: BASELINE

A detailed examination of the biology of colony initiation in *Cryptocercus* lends itself to a logical, stepping-stone conceptual model of the evolution of the earliest stages of termite eusociality, with a clear directionality in the sequence of events. Female *C. punctulatus* lay a clutch of from one to four oothecae. Unlike other oviparous cockroaches (Fig. 7.1), nymphs do not hatch from the ootheca simultaneously. The majority of egg cases require 2–3 days for all neonates to exit (Nalepa 1988a). Laboratory studies further suggest that there is a lag of from 2–6 days between deposition of successive oothecae (Nalepa, 1988a, unpubl. data). Consequently, there can be an age differential of 2 or more weeks between the first and last hatched nymphs in large broods. These age differentials are corroborated by field studies. Families collected during autumn of their reproductive year can include second, third, and fourth instars (Nalepa, 1990), at which point development is suspended prior to the onset of their first winter.

Nymphs in these families hatch without the gut symbionts required to thrive on a wood diet; consequently, they rely on trophallactic food and fecal pellets (Fig. 5.4) from adults for nutrients. Parents apparently provide all of the dietary requirements of first-instar nymphs, and some degree of trophallactic feeding of offspring occurs until their hindgut symbioses are fully established. Individual nymphs probably have high nutritional requirements, since they gain considerable weight and go through a relatively quick series of molts after hatch. The young are potentially independent at the third or fourth instar (Nalepa, 1990, Table 2), but the family structure is generally maintained until parental death. Adults do not reproduce again. Because of their extraordinarily long developmental times (up to 8 yr, hatch to hatch, depending on the species—Chapter 3), adult *Cryptocercus* rarely, if ever, overlap with their adult offspring (CAN, unpubl.). In addition to providing food and microbes, parental care includes gallery excavation, defense of the family, and sanitation of the nest (Cleveland et al., 1934; Seelinger and Seelinger, 1983; Nalepa, 1984, 1990; Park et al., 2002). This degree of parental care exacts a cost. If eggs are removed from *Cryptocercus* pairs, 52% are able to reproduce during the following reproductive period. If parents

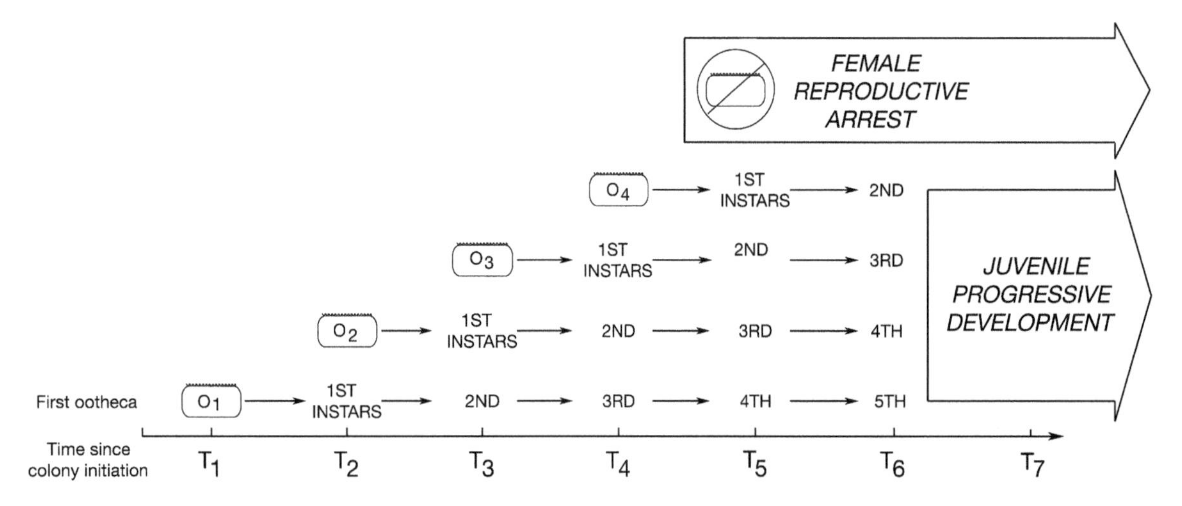

A. Colony initiation in a subsocial termite ancestor: adults feed dependents, oviposition is suspended

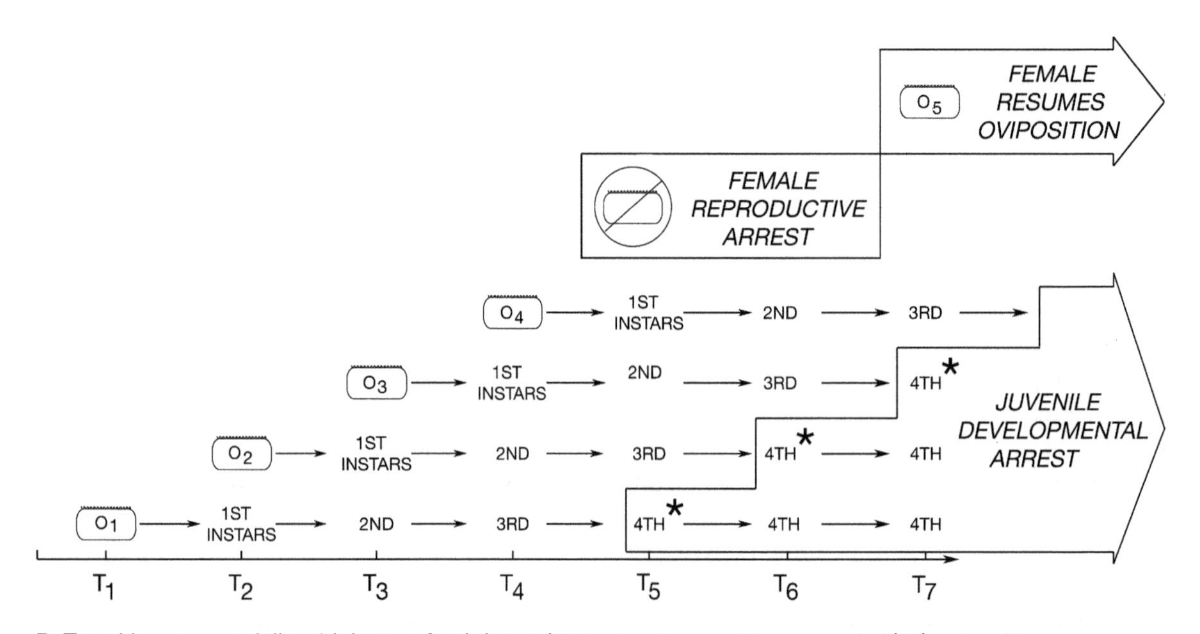

B. Transition to eusociality: 4th instars feed dependents, development is suspended (★), oviposition resumes

Fig. 9.8 Trophic shift model for transition from subsociality to initial stages of eusociality in a termite ancestor. (A) Baseline conditions. A series of egg cases are laid over a short period of time, resulting in age differentials within the brood. Adults feed all offspring; cost of parental care results in reproductive arrest. Juveniles develop slowly but progressively toward adulthood. (B) Transition to eusociality. Fourth instars begin feeding younger siblings; cost of alloparental care results in developmental arrest of juvenile caregivers. Female resumes oviposition. After Nalepa (1988b, 1994).

are allowed to take care of neonates for 3 mon prior to brood removal, however, only 12% oviposit the following summer. This suggests that parental care may deplete reserves that were accumulated over the course of their extended developmental period and are not easily replaced. Under the constraint of a wood diet, their apparent semelparity in the field can be attributed to the need for, and cost of, long-term parental care of the young (Nalepa, 1988b). The life history of a subsocial termite ancestor similar to that of *Cryptocercus* is depicted in Fig. 9.8A.

EVOLUTION OF EUSOCIALITY 2: TRANSITION

It is reasonable to assume that a termite ancestor packaged its eggs in oothecae, since the basal termite *Mastotermes* does so (Nalepa and Lenz, 2000). If the timing of oviposition in this ancestor was similar to that of *Cryptocercus*—a reproductive burst, with several oothecae laid within a relatively short time frame—nymphs in the family also exhibited age differentials. It is likely that reproduction was suspended as adults fed and otherwise cared for their dependent neonates, as reproductive stasis occurs in extant young termite families when adults are nurturing their first set of offspring (reviewed by Nalepa, 1994). This suggests that, as in *Cryptocercus*, parental care during colony initiation in the termite ancestor was costly.

The crucial step, and one that occurs during the ontogeny of extant termite colonies, is that older nymphs assume responsibility for feeding and maintaining younger siblings, relieving their parents of the cost of brood care and allowing them to invest in additional offspring (Fig. 9.8B). All defining components of eusociality (Michener, 1969; Wilson, 1971) follow. First, relieved of her provisioning duties, the female can redirect her reserves into oogenesis, and the result is a second cohort that overlaps with offspring produced during the first reproductive burst. Second, the assumption of responsibility for younger siblings by the oldest offspring in the family constitutes brood care. Third, by trophallactically feeding younger siblings, fourth instars are depleting reserves that could have been channeled into their own development, thus delaying their own maturation (Nalepa, 1988b, 1994). A single behavioral change, the switch from parental to alloparental care, thus represents the pathway for making a seamless transition between adaptive points, accounting with great parsimony for the defining components of the early stages of termite eusociality (Nalepa 1988b, 1994). A key life history characteristic in a *Cryptocercus*-like termite ancestor would be the extraordinarily extended developmental period the first workers face, even prior to assuming brood care duties. Tacking an addition developmental delay onto the half dozen or so years these nymphs already require to reach reproductive maturity may be a pittance when balanced against the additional eggs their already reproductively competent mother may be able to produce as a result of their alloparental behavior. A preliminary mathematical model indicates that when a key resource like nitrogen is scarce, the costs of delayed reproduction in these first workers are outweighed by the benefits accrued by their labor in the colony (Higashi et al., 2000).[4] A cockroach-like developmental plasticity supplied the physiological underpinnings for the social shift, as high-demand metabolic processes such as reproduction and development are tightly modulated in response to nutritional status in Blattaria. It is of particular interest, then, that in extant termites (*Reticulitermes*) two hexamerin genes may signal nutritional status and participate in the regulation of caste polyphenism (Zhou et al., 2006).

4. Masahiko Higashi was tragically killed in a boating accident in March 2000 (Bignell, 2000b) and never completed the study.

HETEROCHRONY REVISITED

The recognition that heterochronic processes play a fundamental role in social adaptations is increasingly recognized in birds and mammals (see references in Gariépy et al., 2001; Lawton and Lawton, 1986) but to date changes in developmental timing have not received the attention they deserve in studies of social insect evolution. Heterochrony is pervasive in termite evolution, and most aspects of isopteran biology can be examined within that framework (Nalepa and Bandi, 2000). The evolution of the initial stages of termite eusociality from subsocial ancestors described above is predicated on a behavioral heterochrony, an alteration in the timing of the expression of parental care (Nalepa, 1988b, 1994). Recently, behavioral heterochrony has been recognized as a key mechanism in hymenopteran social evolution as well (Linksvayer and Wade, 2005). Behavioral heterochronies often precede physiological changes, with the latter playing a subsequent supportive role (e.g., Gariépy et al., 2001); behavior changes first, developmental consequences follow. Development in the first termite workers was suspended as a result of the initial behavioral heterochrony in an ancestor, and selection was then free to shape a suite of interrelated juvenile characters, including allogrooming, kin recognition, coprophagy, and aggregation behavior. It has been noted that paedomorphic taxa frequently develop heightened social complexity, because the reduced aggression associated with juvenile appearance and demeanor enhances social interactions (e.g., Lawton and Lawton, 1986). After alloparental care became established in an ancestor, termite evolution escalated as the social environment, rather than the external environment, became the primary source of stimuli in shaping developmental trajectories (Nalepa and Bandi, 2000, Fig. 4). Major events were the rise of the soldier caste, the polyphyletic onset of an obligately sterile worker caste excluded from the imaginal pathway (Roisin, 1994, 2000), and the loss of gut flagellates at molt, making group living mandatory. The evolution of permanently sterile castes is outside the scope of this chapter. We do, however, note two conditions among extant young cockroaches that provide substructure for the genesis of polyphenism and division of labor. First, the potential for caste evolution would be stronger in an ancestor with a juvenile physiology, because young cockroaches are subject to the most powerful group effects. Social conditions during the early instars of *Diploptera punctata*, for example, can irreversibly fix future developmental trajectories (Holbrook and Schal, 1998). Second, evidence is increasing that the process of forming aggregations in cockroaches is a self-organized behavior (Deneubourg et al., 2002; Garnier et al., 2005; Jeanson et al., 2005). In eusocial insects, self-organization has been shaped by natural selection to produce task specialization, and plays a role in building behavior, decision making, synchronization of activities, and trail formation (Page and Mitchell, 1998; Camazine et al., 2001).

THE GROUND PLAN

Nature has set a very high bar for the attainment of eusociality, and only extraordinary environmental challenges and extraordinary circumstances in prior history can allow an organism to scale it (Hölldobbler and Wilson, 2005). In the termite ancestor, a nitrogen-deficient, physically difficult food source was undoubtedly the relevant environmental challenge, and costly brood care was an essential precedent. Nonetheless, the evolution of termite eusociality cannot be divorced from an entire suite of interrelated and influential morphological, behavioral, developmental, and life history characteristics. These include monogamy, altricial offspring, adult longevity, extended developmental periods, multiple relationships with microbial symbionts, proctodeal trophallaxis and other food-sharing behaviors, reproduction and development that closely track nutritional status, and semelparity with age differentials within the brood (Nalepa, 1984, 1994). So many conditions were interrelated, aligned, and influential in the transition that any attempt to reduce an explanation to a few basic elements is an oversimplification. It is important to note, however, that in integrated character sets such as these, selection on just one character can lead to changes in associated characters, and these changes can occur with a minimum of genetic change. It is in this manner that paedomorphic evolution often proceeds, with small tweaks in regulatory genes that result in maximum impact on an evolutionary trajectory (Gould, 1977; Futuyma, 1986; Stanley, 1998). It is also notable that all ground plan elements are found among extant cockroaches, and that the core process, as in other social insects (Hunt and Nalepa, 1994; Hunt and Amdam, 2005), is a shift in life history characters mediated by a nutrient-dependent switch.

TEN Ecological Impact

Is there nothing to be said about a cockroach which is nice?
It must have done a favor for somebody once or twice.
No one will speak up for it in friendly conversations.
Everyone cold-shoulders it except for its relations.
Whenever it is mentioned, people's faces turn to ice.
Is there nothing to be said about the cockroach which is nice?

—M.A. Hoberman, "Cockroach"

As a whole, cockroaches are considered garbage collectors in terrestrial ecosystems. They recycle dead plants, dead animals, and excrement, processes that are critical to a balanced environment. Here we describe some mechanisms by which cockroaches contribute to ecosystem functioning via the breakdown of organic matter and the release of nutrients. We also summarize their ecological impact on numerous floral, faunal, and microbial components of the habitats in which they live, on a variety of scales ranging from the strictly local to the global.

DETRITIVORY

Although they are rarely mentioned as such in soil science or ecology texts, the majority of cockroach species can be classified as soil fauna (Eisenbeis and Wichard, 1985). Many live in the upper litter horizon, some burrow into the mineral soil layer, and still others inhabit suspended soils. Cockroaches are also associated with decaying logs and stumps, rocks, living trees, and macrofungi, which are physically distinct from, but have biological links to, the soil (Wallwork, 1976). In the majority of these habitats, the core cockroach diet consists of dead plant material.

Because all species examined to date have endogenous cellulases (Scrivener and Slaytor, 1994b; Lo et al., 2000), cockroaches may act as primary consumers on at least some portion of ingested plant litter. There is no question, however, that the direct impact of any higher-level primary consumer does not rate mention when compared to soil microorganisms, which are universally responsible for breaking down complex carbohydrates and mineralizing nutrients in plant detritus in all ecosystems. As with other arthropod decomposers (Wardle, 2002), then, the most profound impact of cockroaches is indirect, and lies in their complex and multipartite interaction with soil microbes. The physical boundaries between cockroaches and microbial consortia in soil and plant litter, however, are not always obvious (Fig. 5.3), and the relationship is so complex as to make discrete classifications or discussion of individual roles arbitrary. Here we center on how cockroaches alleviate factors that constrain microbial decomposition, namely, the microbial lack of automotion and their dependence on water.

Although microbial communities account for most mineralization occurring in soil, they are dormant the majority of the time because of their inability to move toward fresh substrates once nutrients in their immediate surroundings are exhausted. Macroorganisms such as cockroaches remove this limitation on microbial activity via their feeding and locomotor activities, by fragmenting litter and thereby exposing new substrate to microbial attack, and by transporting microbes to fresh food (Lavelle et al., 1995; Lavelle, 2002). The physical acts of burrowing and channeling cause small-scale spatial and temporal variations in microbial processes (Meadows, 1991). These, in turn, effect major changes in the breakdown of woody debris (Ausmus, 1977) and leaf litter (Anderson, 1983), and may also influence ecological processes in other cockroach habitats such as soil, guano, abandoned termite nests, and the substrate under logs, bark, and stones. In addition to making substrate available for microbial colonization via physical disturbance and fragmentation, cockroaches transport soil microbes by carrying them in and on their bodies. This is particularly important in surface-foraging species that diurnally or seasonally take shelter under bark, in crevices, or in voids of rotting logs, where they inoculate, defecate, wet surface wood, affect nitrogen concentration, and contribute to bark sloughing (Wallwork, 1976; Ausmus, 1977).

A second factor that limits microbial decomposers is dependence on water (Lavelle et al., 1995). Cockroaches and other detritivores are able to mitigate this constraint, as the gut provides a moist environment for resident and ingested microbes. The hindgut also furnishes a stable temperature and pH, and a steady stream of fragmented, available substrate. In short, the detritivore gut provides an extremely favorable habitat if ingested microbes can elude the digestive mechanisms of the host. Fecal pellets, the end products of digestion, are similarly favorable habitats for microorganisms. Cockroaches on the floor of tropical forests consume huge quantities of leaf litter (Bell, 1990), thereby serving as mobile fermentation tanks that frequently and periodically dispense packets of microbial fast food. This alteration in the timing and spatial pattern of microbial decomposition may dramatically influence the efficient return of above-ground primary production to the soil. Fecal pellets also provide food for a legion of tiny microfauna, including Collembola, mites, protozoa, and nematodes. These feed on the bacteria and fungi growing on the pellets, as well as the fluids and metabolites resulting from excretory activity (Kevan, 1962).

Forests

In temperate climates, cockroaches are usually relegated to a minor role in soil biology because population densities can be low (e.g., *Ectobius* spp. in central Europe—Eisenbeis and Wichard, 1985). Similarly, in surveys of tropical forest litter, ants, mites, and springtails typically dominate in number, with cockroaches rating an incidental mention (e.g., Fittkau and Klinge, 1973). Cockroaches comprised just 3.0% of the arthropod biomass of the ground litter in a humid tropical forest in Mexico (Lavelle and Kohlmann, 1984), for example. On the other hand, cockroaches are very common in the leaf litter on the floor of the Pasoh Forest in West Malaysia, with 6.7 insects/m^2 (Saito, 1976). They are very well represented in several forest types in Borneo. Leakey (1987) cites a master's thesis by Vallack (1981) in which litter invertebrates were sampled in four forest types at Gunung Mulu in Sarawak. Cockroaches contributed an impressive 43% of the invertebrate biomass in alluvial forest, 33% in dipterocarp forest, 40% in heath forest, and 2% in a forest situated on limestone. A specific decomposer role has been quantitatively established for *Epilampra irmleri* in Central Amazonian inundation forests (Irmler and Furch, 1979). This species was estimated to be responsible for the consumption of nearly 6% of the annual leaf litter input. Given that seven additional cockroach species were noted in this habitat, the combined impact on decompositional processes may be considerable.

The ecological services of cockroaches are not limited to plant litter on the soil surface. Those species found in logs, treeholes, standing dead wood and branches, birds' nests, and plant debris trapped in epiphytes, lichens, mosses, and limb crotches in the forest canopy (i.e., suspended soils) are also members of the vertically stratified decomposer niche (Swift and Anderson, 1989). Cockroach species that feed on submerged leaf litter on stream bottoms and in tank bromeliads may have an impact in aquatic systems.

Wood Feeders

Wood-feeding cockroach species remove large quantities of wood from the surface but their contribution to soil fertility has yet to be explored. Both Panesthiinae and Cryptocercidae progressively degrade the logs they inhabit. They not only ingest wood, but also shred it without consumption when excavating tunnels. The abundant feces line galleries, pack side chambers, and are

Fig. 10.1 Decomposition of logs by *Cryptocercus punctulatus*, Mountain Lake Biological Station, Virginia. (A) Frass pile outside gallery entrance. (B) Small log hollowed and filled entirely with frass and fecal pellets. Photos by C.A. Nalepa.

pushed to the outside of the logs, no doubt influencing local populations of bacteria, fungi, and microfauna (Fig. 10.1). The typically substantial body size of these insects contributes to their impact; some species of *Panesthia* exceed 5 cm in length (Roth, 1979c). Although these two taxa are the best known, many cockroach species potentially influence log decomposition (Table 3.2).

Xeric Habitats

Cockroaches are known to participate in the breakdown of plant organic matter in deserts and other arid and semiarid landscapes, and have a direct and substantial impact on nutrient flow. *Anisogamia tamerlana* is the main consumer of plant litter in Turkmenistan deserts (Kaplin, 1995), and cockroaches in the genus *Heterogamia* are the most abundant detritivore in the Mediterranean coastal desert of Egypt. The latter dominate the arthropod fauna living beneath the canopy of desert shrubs, with up to 116,000 cockroaches/ha, comprising 82% of the arthropod biomass (Ghabbour et al., 1977; Ghabbour and Shakir, 1980). The daily food consumption of *An. tamerlana* is 17–18% of their dry body mass, with 57–69% assimilation. Females and juveniles consume 840–1008 g/ha dry plant debris and produce 259–320 g/ha of excrement (Kaplin, 1995). These cockroaches improve the status of desert soils via their abundant fecal pellets, the nitrogen content of which is 10 times that of their leaf litter food source (El-Ayouty et al., 1978).

Many of the ground-dwelling, wingless cockroaches of Australia are important in leaf litter breakdown. This is particularly true in stands of *Eucalyptus*, where litter production is high relative to other forest types, leaves decompose slowly, and more typical decomposers such as earthworms, isopods, and millipedes are uncommon (Matthews, 1976). The beautiful Striped Desert Cockroach *Desmozosteria cincta*, for example, lives among twigs and branches at the base of eucalypts (Rentz, 1996). In hummock grasslands and spinifex, genera such as *Anamesia* feed on the dead vegetation trapped between the densely packed stems (Park, 1990). The litter-feeding, soil-burrowing Geoscapheini are associated with a variety of Australian vegetation types ranging from dry sclerophyll to rainforest, and have perhaps the most potential ecological impact. First, they drag quantities of leaves, twigs, grass, and berries down into their burrows, thus moving surface litter to lower soil horizons. Second, they deposit excreta deep within the earth. Fecal pellets are abundant and large; those of *Macropanesthia rhinoceros* are roughly the size and shape of watermelon seeds. Third, burrowing by large-bodied insects such as these has profound physical and chemical effects on the soil. Burrows influence drainage and aeration, alter texture, structure, and porosity, mix soil horizons, and modify soil chemical profiles (Anderson, 1983; Wolters and Ekschmitt, 1997). The permanent underground lairs of *M. rhinoceros* have plastered walls and meander just beneath the soil surface before descending in a broad spiral (Fig. 10.2). The deepest burrows can be 6 m long, reach 1 m below the surface, and have a cross section of 4–15 cm. Burrows may be locally concentrated; the maximum density found was two burrows/m^2, with an average of 0.33/m^2 (Matsumoto, 1992; Rugg and Rose, 1991).

Cockroaches in arid landscapes nicely illustrate two subtleties of the ecological role of decomposers: first, an often mutualistic relationship with individual plants, and second, the key role of gut microbiota. In sparsely vegetated xeric habitats, the density of cockroaches generally varies as a function of plant distribution. In deserts, Polyphagidae are frequently concentrated under shrubs (Ghabbour et al., 1977), and the burrows of Australian Geoscapheini are often associated with trees. *Macropanesthia heppleorum* tunnels amid roots in *Callitris*-*Eucalyptus* forest, and *Geoscapheus woodwardi* burrows are located under overhanging branches of *Acacia* spp. in mixed open forest (Roach and Rentz, 1998). Not only are

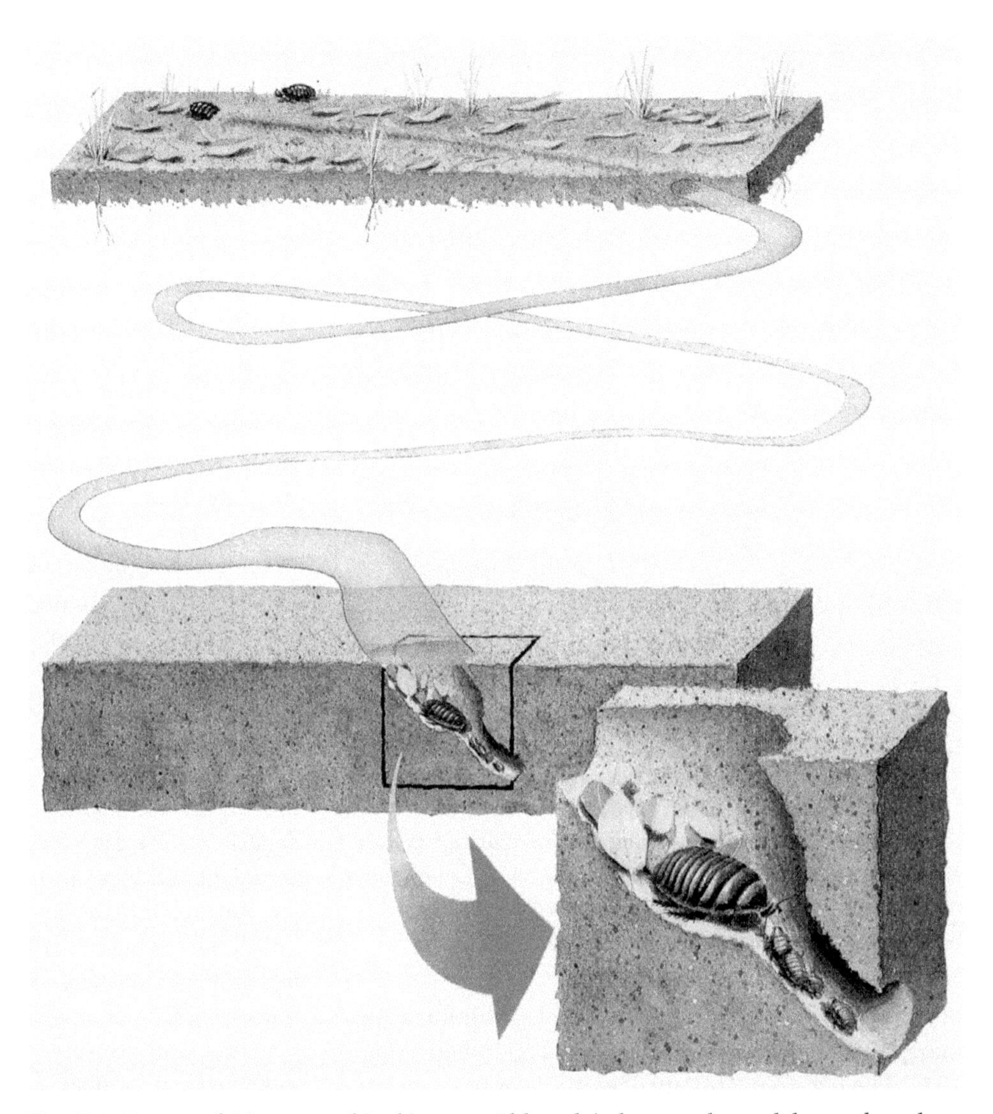

Fig. 10.2 Burrow of *Macropanesthia rhinoceros*. Although it does not descend deeper than about 1 m, the gently sloping spiral may be up to 6 m long. Near the bottom the tunnel widens to become a nesting chamber to rear young and to cache dried leaves. Drawing by John Gittoes, courtesy of Australian Geographic.

these cockroaches ideally located to collect plant litter, they are also positioned to take advantage of the shade, moisture retention, and root mycorrhizae provided by the plant. Reciprocally, the burrowing, feeding, and excretory activities of the cockroaches influence patterns of aeration, drainage, microbial performance, decomposition, and nutrient availability in the root zone of the plants (Anderson, 1983; Ettema and Wardle, 2002). This mutualistic relationship therefore may allow for peak performance by both parties in a harsh environment. It is a tightly coordinated positive feedback system in which decomposers improve the quantity and quality of their own resource (Scheu and Setälä, 2002).

Another alliance of ecological consequence occurs at a much smaller scale. Because the activity of soil microbes is dependent on water, decomposition in deserts occurs in pulses associated with precipitation. Ciliates, for example, occur in the soil in great numbers, but are active only in moisture films. As a consequence, microorganisms remain dormant most of the time and plant litter accumulates in deserts, restricting nutrient flow (Kevan, 1962; Taylor and Crawford, 1982). A significant resolution to this bottleneck lies in the digestive system of detritivores such as cockroaches. The gut environment allows for a relatively continuous rate of microbial activity, even during periods inimical to decomposition by free-living microbes in soil and litter. This relationship is present wherever cockroaches feed, but has a profound ecological significance in deserts and other extreme environments because it allows for decomposition during periods when it would not normally occur—in times of drought or excessive heat or cold (Ghabbour et al., 1977; Taylor and Crawford, 1982; Crawford and Taylor, 1984).

Significance of Cockroaches as Decomposers

The importance of plant litter decomposers to soil formation is unquestioned (Odum and Biever, 1984; Vitousek and Sanford, 1986; Whitford, 1986; Swift and Anderson, 1989; Meadows, 1991). Soils in turn provide an array of ecosystem services that are so fundamental to life that their total value could only be expressed as infinite (Daily et al., 1997). Detailing the contribution of cockroaches relative to other decomposers, however, is difficult. First, information is scarce. For any given ecosystem, it is the decomposers that receive the least detailed attention. Second, like most decomposers, cockroaches are so adaptable that they often do not have well defined ecological roles; functional redundancy among detritivores is high (Scheu and Setälä, 2002). Third, because of the intricate synergistic and antagonistic interactions among diverse bacteria, fungi, and invertebrates, decomposition is manifested in scales of space and time not easily observed or quantified. Decomposition occurs both internal and external to the gut, and at microscopic spatial scales. It operates via the creation of physical artifacts, like burrows and fecal pellets, which accumulate and continue to function in the absence of their creators. Effects can be localized and short term, or wide ranging and extended in time; wood decomposition in particular is a very long-term stabilizing force in forest ecosystems (Anderson et al., 1982; Anderson, 1983; Swift and Anderson, 1989; Wolters and Ekschmitt, 1997; Wardle, 2002).

Other problems in attempting to quantify the role of arthropods in decompositional processes are related to sampling bias; no one method works best for all groups and all soils (Wolters and Ekschmitt, 1997). The results of pitfall trapping, for example, can be difficult to interpret. No cockroaches were taken in unbaited pitfall traps in four habitats in Tennessee, but traps attracted quite a number of blattellids when bait (cornmeal, cantaloupe, fish) was added (Walker, 1957). Surface-collecting methodology such as soil and litter cores may not account for cockroach species that are only active after seasonal precipitation or those that shelter under bark, under stones, or in other concealed locations during the day. Sampling techniques for canopy arthropods also have methodological biases with regard to a given taxon, particularly those species in suspended soils and those that are seasonally present. Diurnal, seasonal, and spatial aggregation further complicate the proper estimation of abundance (Basset, 2001).

Members of the blattoid stem group undoubtedly played a major role in plant decomposition during the Paleozoic (Shear and Kukalová-Peck, 1990). The ecological significance of extant cockroaches, however, is usually assumed to be negligible (Kevan, 1993) because of their often low numbers during surveys (e.g., some Australian studies—Postle, 1985; Tanton et al., 1985; Greenslade and Greenslade, 1989). If considered in terms of biomass, however, their importance is magnified because of large individual body size relative to many other detritivores such as mites and Collembola. Basset (2001), in a review of studies conducted worldwide, concluded that cockroaches dominated in canopies, comprising an astonishing 24.3% of the invertebrate biomass (discussed in Chapter 3). The clumped distribution and social tendencies of many species also tends to increase their ecological impact. Cockroaches that aggregate in tree hollows, for example, directly benefit their host plant, as defecation steadily fertilizes the soil at the base of the tree (Janzen, 1976). Large, subsocial or gregarious wood-feeding cockroaches may be able to pulverize logs on a time scale comparable to, if not better than, termites. In this regard, several studies in montane environments report that cockroach population levels in plant litter are negatively correlated with the presence of termites, a group that strongly and predominantly influences the pattern of decomposition processes and whose ecological importance is clear. Surveys on Mt. Mulu in Sarawak, Borneo, indicate that the density of soil- and litter-dwelling termites declines with altitude (Collins, 1980). Cockroaches were present in low numbers at all altitudes, but individuals were larger and more numerous in upper montane forests, where they constituted 40% of the total macrofauna biomass. *Rhabdoblatta* was the most common genus at upper altitudes, found in all plots from 1130 m upward, but not below. The *Cryptocercus punctulatus* species complex dominates the saproxylic guild in the Southern Appalachian Mountains, and occupies the same niche as does the subterranean termite *Reticulitermes* at lower elevations (Nalepa et al., 2002). The same altitudinal trend was evident in soil and litter core samples taken on Volcán Barva in Costa Rica; the biomass of cockroaches fluctuated, but generally increased with altitude. Termites were not found above 1500 m, but cockroaches made up 61% of the biomass at that altitude (Atkin and Proctor, 1988). On Gunung Silam, a small mountain in Sabah, the altitudinal associations were reversed. At 280 m, cockroaches were 84% of the invertebrate biomass and termites were not found; at 870 m, termites were 25% of the biomass, while cockroaches were

< 1% (Leakey, 1987, Table 3). The reasons for these altitudinal changes in distribution were not causally related to measured changes in other site properties such as forest structure and soil organic matter in the Costa Rican study (Atkin and Proctor, 1988).

POLLINATION

Cockroaches are frequently observed on flowers and many readily feed on offered pollen and nectar (Roth and Willis, 1960). In temperate zones, Blattaria are only occasionally reported from blossoms. *Ectobius lapponicus* and *E. lividus* have been observed on flowers of the genera *Spirea, Filipendula,* and *Daucus* in Great Britain (Proctor and Yeo, 1972), and *Latiblattella lucifrons* feeds on pollen of *Yucca* sp. in southern Arizona (Ball et al., 1942). Nymphs of *Miriamrothschildia notulatus* and *Periplaneta japonica* and brachypterous adults of *Margattea satsumana* visit extrafloral nectaries at the base of fleshy, egg-like inflorescences of the low-growing root parasite *Balanophora* sp. on the floor of evergreen forests in Japan. Visits corresponded with cycles of evening nectar secretion, multiple plants were visited in succession, and pollen grains were observed attached to the tarsi and mouthparts of *Mar. satsumana.* All observed cockroaches, however, are flightless, suggesting to the authors that cross-pollination is unlikely to be effective (Kawakita and Kato, 2002). An association between cockroaches and flowering plants may be more widespread in the tropics. The strikingly colored *Paratropes bilunata* visits flowers of the Neotropical (Costa Rica) canopy species *Dendropanax arboreus* (Araliaceae). Cockroaches were observed flying during the day to successive inflorescences located 34 m above the ground, ignoring nearby flowers of a different species. The exposed condition of the anthers and stigma of *D. arboreus* and the observed floral fidelity of the cockroach suggest that *Parat. bilunata* is a likely pollinator (Perry, 1978; Roth, 1979a). Nagamitsu and Inoue (1997) offer more direct evidence that cockroaches can be the main pollinators of a plant species in the understory of a lowland mixed dipterocarp forest in Borneo. These authors observed blattellid cockroaches feeding on pollen and stigmatic exudate of *Uvaria elmeri* (Annonaceae) (Fig. 10.3). The visitation time of the cockroaches corresponded with nocturnal dehiscence of anthers, and pollen grains were observed in both the gut and on the undersurface of the head. Because few bees are typically found in canopy collections (Basset, 2001), cockroaches may be among those arthropods filling the pollinator niche in treetops. Of the known cases of cockroach pollination, the degree of floral specificity, distances between visited inflorescences, and consequent effect on gene flow in flowering plants have not been studied.

Fig. 10.3 Blattellid cockroach nymph feeding on pollen of *Uvaria elmeri* (Annonaceae) in lowland mixed dipterocarp forest in Borneo. From Nagamitsu and Inoue (1997). Photo courtesy of I. Nagamitsu, with permission of The American Journal of Botany.

FOOD CHAINS

Although cockroaches generally feed on dead plant and animal material, they are also well known as primary consumers. Many blattids in tropical forests are cryptic herbivores and some are overtly herbivorous, particular on young vegetation (Chapter 4). Roth and Willis (1960) were surprised that the role of cockroaches as plant pests is rarely discussed, and detailed the abundant records of the phenomenon in the literature. Most of the evidence comes from commercially grown crops, particularly in the tropics and in greenhouses. One field study, however, found that the frequency of herbivore damage on new leaves in rainforest canopy (Puerto Rico) was significantly correlated with the abundance of Blattaria (Dial and Roughgarden, 1995). It is therefore possible that cockroaches may have an undocumented but significant ecological and evolutionary impact on vascular tropical flora, as well as on nonvascular plants in the phylloplane.

At the next level of the food chain, cockroaches are prey for numerous taxa, including pitcher plants (*Sarracenia* and *Nepenthes* spp.) (Roth and Willis, 1960) and a variety of invertebrate and vertebrate predators (Fig. 10.4). The principal food of the grylloblattid *Galloisiana kurentzovi* in East Asia is juveniles of *Cryptocercus relictus* (Storozhenko, 1979), and small blattellid cockroaches climbing on low vertical twigs and grass blades constitute 92% of the prey of the Australian net-casting spider *Menneus unifasciatus* (Austin and Blest, 1979). In desert sand dunes of California, *Arenivaga investigata* makes up 23% of the prey biomass taken by the scorpion *Paruroctonus mesaensis* (Polis, 1979). Examination of the excrement of the South American frog *Phyllomedusa iheringii* indicates that cockroaches are a major part of its diet (Lagone, 1996). Blattellid cockroaches of the genus *Parcoblatta* are a high proportion of the menu of endangered red-cockaded woodpeckers (*Picoides borealis*) in the Coastal Plain of South Carolina (Horn and Hanula, 2002). Cockroaches were consistently taken by all observed birds, made up 50% of the overall diet, and were 69.4% of the prey fed to nestlings (Hanula and Franzreb, 1995; Hanula et al., 2000). *Pycnoscelus indicus* on Cousine Island in the Seychelles is the favored prey of the endangered magpie robin (*Copsychus sechellarum*) (S. Le Maitre, pers. comm. to LMR); the birds feed on American cockroaches as well. Attempts to control urban infestations of *Periplaneta americana* with toxic insecticides may have contributed to the decline of this species on Frégate Island. The birds feed close to human habitations and take advantage of dead and dying insecticide-treated cockroaches. Lethal doses accumulated in the birds, with subacute effects on their behavior. The current use of juvenile hormone analogs for cockroach control appears to result in good control of the pests while posing a negligible hazard to the birds (Edwards, 2004). These few examples (see Roth and Willis, 1960 for more) suffice to emphasize that in their role as prey, cockroaches may significantly influence the population structure of insectivores in terrestrial ecosystems. They may also be a link between terrestrial and aquatic food chains at river and stream edges, and in delicately balanced cave ecosystems. Cave-dwelling cockroaches accidentally introduced into water are one of the principal foods of some cavernicolous fishes; they are 26% of the diet of *Milyeringa veritas* (Humphreys and Feinberg, 1995). Cockroaches are considered the base of the food web in South African bat caves and support a large community of predators and parasites. Their feces are also an important food source for smaller invertebrates (Poulson and Lavoie, 2000). Hill (1981) noted that for most of the guano community in Tamana cave, Trinidad, the incoming supply of energy was in the form of cockroach, not bat, feces.

Fig. 10.4 Scorpion feeding on the ground-dwelling cockroach *Homalopteryx laminata*, Trinidad. Photo courtesy of Betty Faber.

At the top of the food chain, there are numerous reports of cockroaches preying on other insects (detailed by Roth and Willis, 1960). Most of these accounts are observations of opportunistic predation on a broad range of vulnerable taxa and life stages, particularly eggs and larvae. Instances of cockroaches controlling prey populations of crickets and bedbugs in urban settings are frequent in the historic literature but largely anecdotal and unverified. One ecological setting in which cockroaches do have potential for influencing population densities of prey is in caves (Chapter 4).

LARGE-SCALE EFFECTS

Cockroaches potentially influence biogeochemical cycles via two known pathways: nitrogen fixation and methane production. *Cryptocercus* is the only cockroach currently known to harbor gut microbes capable of fixing atmospheric nitrogen (Breznak et al., 1974), but spirochetes found in the hindgut of other species also may have the ability (Lilburn et al., 2001). Acetylene reduction assays indicate that adults and juveniles of *Cryptocercus* fix nitrogen at rates comparable to those of termites on a body weight basis (0.01–0.12 mg N day^{-1} g^{-1} wet weight) (Breznak et al., 1973; Breznak et al., 1974, 1975). The process provides a mechanism for nitrogen return to the ecosystem and may have a significant ecological impact (Nardi et al., 2002), particularly in the food chains of the montane mesic forests where *Cryptocercus* is the dominant macroarthropod feeding in rotting logs.

A more universal characteristic of cockroaches is an association with methanogenic bacteria in the hindgut and the consequent emission of methane. Almost all tropical cockroaches tested emit methane, regardless of the origin of specimens and their duration of laboratory captivity. Methane, carbon dioxide, and water are released synchronously in a resting cockroach, in slow periodic cycles that suggest the gases are respired (Bijnen et al., 1995, 1996). Among temperate species, North American *C. punctulatus* emits the gas (Breznak et al., 1974), but the European genus *Ectobius* does not (Hackstein and Strumm, 1994). Cockroaches (n = 34 species) produce an average of 39 nmol/g methane/h, with a maximum of

450 nmol/g/h (Hackstein, 1996). On a global scale, estimates of methane production by cockroaches vary widely and are debatable, given first, the paucity of data on which to base biomass estimates of field populations, and second, the finding that methane production varies with cockroach age and diet fiber content (Gijzen et al., 1991; Kane and Breznak, 1991). It has been suggested that cockroaches make a significant contribution to global methane, particularly in the tropics (Gijzen and Barugahare, 1992; Hackstein and Strumm, 1994). However, methane oxidation by bacteria in the soil may buffer the atmosphere from methane production by gut Archaea, and although cockroaches may be a gross source of methane, little to none of it may be escaping into the atmosphere. The sink capacity of the soil may exceed methane production by cockroaches, just as it does for termites (Eggleton et al., 1999; Sugimoto et al. 2000). Nonetheless, their typically large body size (relative to termites), and the tendency of many species to live in aggregations in enclosed spaces (e.g., treeholes, caves, logs) may engender atmospheric changes at a local level. Mamaev (1973), for example, collected more than 400 *C. relictus* from a single cedar log. On a per weight basis methane production by *C. punctulatus* is comparable to the termite *Reticulitermes flavipes* and may surpass levels emitted by ruminants (Breznak et al., 1974; Breznak, 1975).

OTHER ROLES

Cockroaches are part of the guild of arthropods that provide waste elimination services; they feed on the fecal material of animals in all trophic levels (Roth and Willis, 1957). While this behavior is most often noted in relation to disease transmission by pest species, it is likely that cockroaches also contribute to the rapid processing of excrement in natural settings (Fig. 5.2). Cockroaches habitually found in bird nests, mammal burrows, and the middens of social insects provide nest sanitation services for their hosts. MacDonald and Matthews (1983) suggest that nymphs of *Parcoblatta* help prolong the colony cycle of southern yellowjackets (*Vespula squamosa*) by scavenging colony debris and keeping fungal and protozoan populations suppressed. Cockroaches (probably *Periplaneta fuliginosa*) are frequently found in honeybee hives in North Carolina; their role in hive sanitation merits further investigation (D.I. Hopkins, pers. comm. to CAN).

In addition to acting as predators, prey, and regulators of microbial processes, cockroaches have ecological relationships with a variety of micro- and macrofauna. These include ecto- and endoparasites, parasitoids, and commensals (mites, for example). The burrows and tunnels of cockroaches that excavate solid substrates often serve as shelter for many additional tenants. The burrows of *M. rhinoceros* harbor a complex of other cockroaches (*Calolampra* spp., among others), beetles, silverfish, centipedes, frogs, and moths (Park, 1990; Rugg and Rose, 1991). One scarab (*Dasygnathus blattocomes*) has been collected nowhere else (Carne, 1978). Salamanders, centipedes, ground beetles, and springtails are frequently found in the galleries of *C. punctulatus* (Cleveland et al., 1934; CAN, unpubl.).

Within the human realm, cockroaches have both cultural and scientific significance. Several species are used as pets and pet food (McMonigle and Willis, 2000), and because they are robust under taxing conditions they make excellent fish bait. Urban pests serve as ideal subjects for a wide range of scientific studies. They are easily fed on commercially available pet chow, do not mind a dirty cage, withstand and even thrive under crowded conditions, and are prolific breeders. The relatively large size of some (e.g., *Periplaneta*) facilitates tissue and cell extraction, and their sizable organs are easily pierced with electrodes or cannulae. The cockroach nervous system is less cephalized than in many insects, making these insects excellent experimental models in neurobiology; two volumes have been written on the subject (Huber et al., 1990). Their overall lack of specialization makes them ideal for teaching students the basics of insect anatomy. They also readily lend themselves to laboratory experiments on the physiology of reproduction, nutrition, respiration, growth and metamorphosis, regeneration, chemical ecology, learning, locomotion, circadian rhythms, and social behavior (Bell, 1981, 1990). Therapeutic concoctions that include cockroaches are frequently cited in medical folklore, and their use as a diuretic has received some clinical support. Roth and Willis (1957) list 30 specific diseases and disorders where cockroaches have featured in treatment. When American jazz legend Louis Armstrong was a child, his mother fed him a broth made from boiled cockroaches whenever he was ill (Taylor, 1975). In southern China and in Chinatown in New York City, dried specimens of *Opisthoplatia orientalis* are still sold for medicinal purposes (Roth, 2003a), and *Blatta orientalis* is marketed on the Internet as a homeopathic medicine. Cockroaches produce a wide range of pheromones and defensive compounds, and may be rewarding subjects for pharmaceutical bioprospecting. Given the close association of cockroaches with rotting organic matter, a search for antimicrobials may be particularly fruitful (Roth and Eisner, 1961). The secretions used by some oviparous species to attach their oothecae to objects have been likened to superglue, as attempting to remove the egg cases either ruptures them or also pulls up the substrate (Edmunds, 1957; Deans and Roth, 2003). Cockroach guts, like termite guts (Ohkuma, 2003), may be a source of novel microorganisms with wide-ranging industrial applications.

CONSERVATION

Cockroaches are not generally considered a charismatic taxon; species that are threatened with extinction are unlikely to rally conservationists to action. They are nonetheless an integral part of a stable and productive ecosystem in tropical rainforest and other habitats. Cockroaches deserve our consideration and respect for the range of services they perform and for their membership in an intricate web of interdependent and interacting flora, fauna, and microbes. Many cockroach species live in habitats of conservation concern and are threatened by canopy removal, urbanization, and agricultural practices. Philopatric species with naturally small population sizes and specific habitat requirements are particularly vulnerable to perturbations (Pimm et al., 1995; Tscharntke et al., 2002; Boyer and Rivault, 2003). These taxa are frequently wingless, and their consequent low dispersal ability makes them vulnerable to habitat fragmentation and genetic bottlenecks. Several species of Australian burrowing cockroaches have restricted ranges and are affected by farming/forestry practices or by urbanization. The accompanying soil disturbance, soil compaction, and loss of their leaf litter food sources have devastated some populations of these unique insects (H.A. Rose, pers. comm. to CAN).

Caves are delicately balanced and vulnerable ecosystems whose resident cockroaches can be severely affected by guano compaction, guano collection, and other human disturbances (Braack, 1989). *Nocticola uenoi miyakoensis,* for example, became rare in the largest known limestone cave on Miyako-jima Island after it was opened to tourists (Asahina, 1974), and the invertebrate community of an Australian cave disappeared due to soil compaction by human visitors (Slaney and Weinstein, 1997a). According to Gordon (1996), the cave-dwelling species *Aspiduchus cavernicola* (Tuna Cave cockroach) living in a network of caves in southern Puerto Rico is officially classified as a "species at risk" by the U.S. Fish and Wildlife Service. Roth and Naskrecki (2003) recently described a new species of cave cockroach collected during a Conservation International survey of West African sites under threat from large-scale mining operations. The removal of cave cockroaches for scientific study also can have a significant impact on their populations (Slaney and Weinstein, 1997a).

Global warming and the resultant decrease in snow cover at high elevations may put cockroaches such as the New Zealand alpine species *Celatoblatta quinquemaculata* at risk (Sinclair, 2001). Although the species is physiologically protected against the cold, it relies on the thermal buffering effect of snow cover in particularly harsh winters. Reduced snow cover results in an increased number of freeze-thaw cycles and lower absolute minimum temperatures, making the "mild" winter more, rather than less, stressful to the insect.

Wood-feeding and other log-dependent cockroaches (Table 3.2) are sensitive to the ecological changes brought about by both modern forestry and human settlement and, like many saproxylic arthropods (Grove and Stork, 1999; Schiegg, 2000), may be used as habitat continuity indicators in ecological assessment. These insects rely on a resource whose removal from the ecosystem is the usual objective of forest management (Grove and Stork, 1999) and compete with lumber companies (Cleveland et al., 1934) and resident humans who prize coarse woody debris as fuel and building material. Wood-feeding cockroaches may survive canopy removal and subsequent desiccating conditions if logs of a size sufficient to provide a suitable microhabitat are left on the ground. *Cryptocercus primarius,* for example, has been collected from large-diameter logs in young re-growth forest in China (Fig. 10.5). More often, however, coarse woody debris left on the forest floor after logging operations is gathered and used as fuel (Nalepa et al., 2001b). Based on the work of Harley Rose (University of Sydney), the endemic Lord Howe Island wood-feeding cockroach *Panesthia lata* was recently listed by the New South Wales Scientific Committee as an endangered species (Adams, 2004). It has not been found on Lord Howe Island since the 1960s, probably because of rats introduced in 1918. Small numbers of the cockroach were recently discovered on Blackburn Island and Roach Island.

Litter-dwelling cockroaches can be sensitive habitat indicators. The Russian cockroach *Ectobius duskei,* normally found at levels of up to 10 individuals/m^2 in undisturbed steppe, disappears if these grasslands are plowed to grow wheat. If the fields are allowed to lie fallow, the cockroaches gradually become reestablished (Bei-Bienko, 1969, 1970). Although the species has been eliminated in intensely cultivated areas, a 1999 study found *E. duskei* well represented in the leaf litter of steppe meadows in the Samara district (Lyubechanskii and Smelyanskii, 1999).

The effect of disturbance on litter invertebrates depends not only on the type of disturbance, but also on site-specific factors. In the dry Mediterranean-type climate of western Australia cockroaches appear resilient to moderate disturbances. Cockroach numbers and species richness as measured by pitfall traps declined significantly after logging and fire, yet recovered within 48 mon.

Fig. 10.5 Li Li, Chinese Academy of Science, Kunming, and Wang De-Ming, Forest Bureau, Diqing Prefecture, opening a rotted log containing *Cryptocercus primarius* in a young regrowth spruce and fir forest at Napa Hai, Zhongdian Co., Yunnan Province, China. The cockroaches were found in large logs left on the forest floor after the forest was harvested; maximum regrowth was 10 cm in diameter. This site was immediately adjacent to a mature coniferous forest with logs also harboring the cockroach. Photo by C.A. Nalepa.

The insects showed no significant response to habitat fragmentation and livestock activity, but were most diverse where forest litter was thickest. The authors explain their results in terms of the fire ecology of the area. In seasonally dry habitats cockroaches appear to have a high degree of tolerance to recurrent disturbances and may aestivate in burrows or under bark during harsh conditions (Abenserg-Traun et al., 1996b; Abbott et al., 2003). There is a distinction, however, between cockroaches adapted to these habitats and those residing where the ecological equilibrium is much more precarious. Tropical rainforests, where the vast majority of cockroaches live, are under heavy assault (Wilson, 2003), and large numbers of described and undescribed species are being lost along with the natural greenhouses in which they dwell. Grandcolas, for example, estimated 181 cockroach species in a lowland tropical forest in French Guiana, with 67 species active in the understory during night surveys in one site (Grandcolas, 1991, 1994b). David Rentz (pers. comm. to CAN) has recorded 62 species of cockroaches, mostly blattellids, from his 0.65 ha of rainforest in Kuranda, Queensland (elev. 335 m asl). In one light trap study in Panama, 42% of 164 species captured were new to science (Wolda et al., 1983).

NEGATIVE IMPACT OF COCKROACHES

The negative impact of cockroaches introduced into non-native habitats is well documented. The handful of species that have invaded the man-made environment have had enormous economic significance as pests, as sources of allergens, as potential vectors of disease to humans and their animals, and as intermediate hosts for some parasites, such as chicken eye-worms. Exotic cockroaches have also been introduced into natural non-native ecosystems like caves (Samways, 1994) and islands, such as the Galapagos (Hebard, 1920b). In a survey of La Réunion and Mayotte in the Comoro Islands, 21 cockroach species were found, with introduced species more common than endemic species that use the same habitats. The abundant leaf litter and loose substrate typical of cultivated land was favorable habitat for the adventive species, particularly in irrigated plots (Boyer and Rivault, 2003). The Hawaiian Islands have no native cockroaches, but 19 introduced species (Nishida, 1992). *Periplaneta americana* has invaded a number of Hawaiian caves, and is thought to have contributed to the decline of the Kauai cave wolf spider (*Adelocosa anops*) by affecting its chief food source, cave amphipods. The cockroach opportunistically preys on immature stages of the amphipods, and competes with older stages at food sources (Clark, 1999). In Florida, laboratory studies indicate that the Asian cockroach *Blattella asahinai* may disrupt efforts to control pest aphids with parasitic wasps by feeding on parasitized aphid "mummies" (Persad and Hoy, 2004). Although this problem occurred primarily when the cockroaches were deprived of food for 24 hr, the high populations of Asian cockroaches that can occur in citrus orchards (up to 100,000/ha) (Brenner et al., 1988) guarantee that some are usually hungry.

OUTLOOK

The meager information we currently have on cockroach activities in natural habitats suggests that they may be key agents of nutrient recycling in at least some desert, cave, and forest habitats. They comprise the core diet for a variety of invertebrate and vertebrate taxa, and may play some role in pollination ecology, particularly in tropical canopies. Before we can begin to document and quantify their ecosystem services, however, more time, energy, and financial resources must be devoted to two specific areas of cockroach research.

The first and most obvious requirement is for basic information on the diversity, abundance, and biology of free-living species, as cockroaches remain a largely uninvestigated taxon. In 1960, Roth and Willis indicated that there were 3500 described species and estimated an additional 4000 unnamed species. Currently, most estimates are in the range of 4000 to 5000 living cockroaches, with at least that many yet to be described. Some of the most diverse families, such as Blattellidae, are strongly represented in tropical climes but very poorly studied (Rentz, 1996). Among described species, the observation by Hanitsch (1928) that "the life history of the insect begins in the net and ends in the bottle" still holds true for the vast majority. Core data on cockroach biology are derived nearly exclusively from insects that have been reared in culture and studied in the laboratory. How closely the results of these studies relate to Blattaria in natural habitats is in many cases questionable. Laboratory-reared cockroaches are domesticated animals typically kept in mixed sex, multiage groups within restricted, protected enclosures, and supplied with a steady, monotonous food source, ad lib water, and readily accessible mating partners. Most tropical species cultured in the United States are derived from just a few sources collected decades ago (LMR, pers. obs.), and are therefore apt to be lacking the variation expressed in free-living populations. The group dynamics (Chapter 8), locomotor ability (Akers and Robinson, 1983; Chapter 2), and fecundity (Wright, 1968) of laboratory cockroaches are known to differ from that of wild strains, and crowded rearing conditions and the inability to emigrate can result in artificially elevated levels of density-dependent behaviors such as aggression and cannibalism. Mira and Raubenheimer (2002) compared laboratory-reared *P. americana* to "feral" animals loose in their laboratory building and found that the free-range cockroaches had higher growth rates, additional nymphal stadia, greater resistance to starvation, and a higher numbers of endosymbiotic bacteria in the fat body. Field studies and experiments that incorporate a realistic simulation of field conditions are clearly desirable, incorporating as wide a range of taxa and habitat types as possible. A small army of eager young nocturnal scientists, and perhaps octogenarians, who cannot sleep anyway (LMR, pers. obs.), need to consider cockroaches as worthy subjects of observation and experimentation under natural conditions.

A second requisite for progress lies in bankrolling the training of a new generation of cockroach systematists, a need made especially acute with the passing of the second author of this volume (CAN, pers. obs.). Field studies will have little value if the subject of research efforts cannot be identified, or if collected vouchers languish undescribed in museum drawers. One of LMR's final publications sounded the call for "true systematists interested in studying the biology and classification of cockroaches," but recommended that "he or she marry a wealthy partner" (Roth 2003c).

Even if these two requirements are in some small measure met, progress in evaluating the ecological impact of cockroaches may be hindered unless we recognize the need for some attitudinal shifts in our approach to cockroach studies. First, evaluation of the role of cockroaches in the nutrient cycles of ecosystems demands a microbially informed perspective (Chapter 5). Relationships with microorganisms as food, on food, transient through the digestive tract, and resident in and on the body not only form the functional basis of cockroach performance on a plant litter diet, but also direct their impact on decompositional processes. Second, it might behoove us to keep the phylogenetic and ecological relationships of cockroaches and termites in mind when attempting to assess the role of Blattaria in ecosystems. Sampling and evaluation techniques employed in termite studies (e.g., Bignell and Eggleton, 2000) may also prove useful in studying their cryptic cockroach relatives. Scattered hints in the literature that the two taxa may be ecologically displacing each other in selected habitats would be well worth characterizing and quantifying. Third, and finally, as biologists we have a responsibility to help alter the lenses through which potential students as well as the general public characteristically regard the subjects of this book. A realistic image with which to begin public relations is that of inconspicuous workhorses, acting beneath the radar to move nutrients through the food web, maintain soil fertility, and support a variety of the complex and cascading processes that sustain healthy ecosystems.

Appendix

Assignation of the cockroach genera discussed in the text to superfamily, family, and subfamily; after Roth (2003c) unless otherwise indicated.

Blattoidea

Blattidae

Archiblattinae

Archiblatta

Blattinae

Blatta, Cartoblatta, Celatoblatta, Deropeltis, Eumethana, Hebardina, Neostylopyga, Pelmatosilpha, Periplaneta, Pseudoderopeltis

Lamproblattinae

Lamproblatta

Polyzosteriinae

Anamesia, Desmozosteria, Eurycotis, Euzosteria, Leptozosteria, Platyzosteria, Polyzosteria, Zonioploca

Tryonicinae

Angustonicus, Lauraesilpha, Methana, Pallidionicus, Pellucidonicus, Punctulonicus, Rothisilpha, Scabina, Tryonicus

Blaberoidea

Otherwise unplaced: *Neopolyphaga*[a]

Polyphagidae

Subfamily undetermined

Compsodes, Heterogamia,[b] *Homopteroidea, Leiopteroblatta, Myrmecoblatta, Oulopteryx, Tivia*

Polyphaginae

Anisogamia, Arenivaga, Austropolyphaga, Eremoblatta, Ergaula, Eucorydia, Eupolyphaga, Heterogamisca, Heterogamodes, Holocampsa, Homoeogamia, Hypercompsa, Polyphaga, Polyphagoides, Therea

Cryptocercidae

Cryptocercinae

Cryptocercus

Nocticolidae

Alluaudellina, Cardacopsis, Cardacus, Metanocticola, Nocticola, Spelaeoblatta, Typhloblatta

Blattellidae

Subfamily undetermined

Parellipsidion, Sphecophila

Anaplectinae

Anaplecta

Attaphilinae

Attaphila

Pseudophyllodromiinae

Aglaopteryx, Agmoblatta, Allacta, Amazonina, Balta, Cariblatta, Chorisoneura, Chorisoserrata, Dendroblatta, Ellipsidion, Euphyllodromia, Euthlastoblatta, Imblattella, Latiblattella, Lophoblatta, Macrophyllodromia, Margattea, Mediastinia, Nahublattella, Plecoptera, Prosoplecta, Pseudobalta, Riatia, Sliferia, Shelfordina, Sundablatta, Supella

Blattellinae

Beybienkoa, Blattella, Chorisia, Chromatonotus, Escala, Hemithyrsocera, Ischnoptera, Loboptera, Lobopterella, Miriamrothschildia, Nelipophygus, Neoloboptera, Neotemnopteryx, Neotrogloblattella, Nesomylacris, Nondewittea, Parasigmoidella, Paratemnopteryx, Parcoblatta, Pseudoanaplectinia, Pseudomops, Robshelfordia, Stayella, Symploce, Trogloblattella, Xestoblatta

Ectobiinae

Choristima, Ectobius, Phyllodromica

Nyctiborinae

Megaloblatta, Nyctibora, Paramuzoa, Paratropes

Blaberidae

A molecular phylogeny of blaberid subfamilies is given in Maekawa et al. (2003, Fig. 3)

Subfamily undetermined

Apotrogia, Compsolampra

Blaberinae

Archimandrita, Aspiduchus, Blaberus, Blaptica, Byrsotria, Eublaberus, Hyporichnoda, Lucihormetica, Monastria, Phoetalia

Panesthiinae

Ancaudellia, Caeparia, Geoscapheus,[c] *Macropanesthia,*[c] *Microdina, Miopanesthia, Neogeoscapheus,*[c] *Panesthia, Parapanesthia,*[c] *Salganea*

Zetoborinae

Capucina, Lanxoblatta, Parasphaeria, Phortioeca, Schizopilia, Schultesia, Thanatophyllum

Epilamprinae

Aptera, Calolampra, Colapteroblatta, Comptolampra,[d] *Dryadoblatta, Epilampra, Haanina, Homalopteryx, Litopeltis, Miroblatta, Molytria, Opisthoplatia, Phlebonotus, Phoraspis, Poeciloderrhis, Pseudophoraspis, Rhabdoblatta, Thorax, Ylangella*

Oxyhaloinae

Elliptorhina, Griffiniella, Gromphadorhina, Jagrehnia, Nauphoeta, Princisia, Rhyparobia, Simandoa

Pycnoscelinae

Pycnoscelus

Diplopterinae

Diploptera

Panchlorinae

Panchlora

Perisphaeriinae

Bantua, Compsagis, Cyrtotria, Derocalymma, Laxta, Neolaxta, Perisphaeria, Perisphaerus, Pilema, Poeciloblatta, Pseudoglomeris, Trichoblatta

Gyninae

Alloblatta, Gyna

a. According to Jayakumar et al. (2002).
b. According to Ghabbour et al. (1977).
c. Tribe Geoscapheini.
d. After Anisyutkin (1999).

Glossary

Accessory gland a secretory organ associated with the reproductive system.
Acrosome a cap-like structure at the anterior end of a sperm that produces enzymes aiding in egg penetration.
Aerobic growing or occurring in the presence of oxygen.
Alary pertaining to wings.
Alate the winged stage of a species.
Allogrooming grooming of one individual by another.
Alloparental care care of young dependents by individuals that are not their parents.
Anaerobic growing or occurring in the absence of oxygen.
Aphotic without sunlight of biologically significant intensity.
Aposematic possessing warning coloration.
Apterous without tegmina or wings.
Arolium (pl. arolia) an adhesive pad found at the tip of the tarsus, between the claws.
Autogrooming grooming your own body.
Batesian mimicry the resemblance of a palatable or harmless species (the mimic) to an unpalatable or venomous species (the model) in order to deceive a predator.
Bootstrap values a measure of the reliability of phylogenetic trees that are generated by cladistic methods.
Brachypterous having short or abbreviated tegmina and wings.
Brood sac an internal pouch where eggs are incubated in female cockroaches.
Brooding parental care where the females remain with newly hatched offspring for a short period of time, typically just until hardening of the neonate cuticle.
Bursa in the female, a sac-like cavity that receives the spermatophore during copulation.
Caudad toward the posterior, or tail end, of the body.
Cellulase an enzyme capable of degrading cellulose.
Cellulolytic causing the hydrolysis of cellulose.
Cellulose a complex carbohydrate that forms the main constituent of the cell wall in most plants.
Cephalic toward the anterior, or head end, of the body.
Cercus (pl. cerci) paired, usually multi-segmented, sensory appendages at the posterior end of the abdomen.

Chemotaxis the directed reaction of a motile organism toward (positive) or away from (negative) a chemical stimulus.
Chitin a polysaccharide constituent of arthropod cuticle.
Chitinase an enzyme capable of degrading chitin.
Circadian exhibiting 24-hr periodicity.
Clade a hypothesized monophyletic group of taxa sharing a closer common ancestry with one another than with members of any other clade.
Cladistic analysis a technique in which taxa are grouped based on the relative recency of common ancestry.
Clone the asexually derived offspring of a single parthenogenetic female.
Conglobulation the act of rolling up into a ball.
Consortium (pl. consortia) a group of different species of microorganisms that act together as a community.
Conspecific belonging to the same species.
Coprophagy the act of feeding on excrement.
Corpora allata a pair of small glandular structures, located immediately behind the brain, that produce juvenile hormone.
Coxa (pl. coxae) the basal segment of the leg.
Crepuscular active during twilight hours, dusk, and/or dawn.
Cryptic used of coloration and markings that allow an organism to blend with its surroundings.
Cuticle the non-cellular outer layer of the body wall of an arthropod.
Cycloalexy the formation of a rosette-shaped defensive aggregation.
Dealation wing removal.
Dehiscence the act of opening or splitting along a line of weakness.
Diapause a dormancy not immediately referable to adverse environmental conditions.
Dimorphism pertaining to a population or taxon having two, genetically determined, discontinuous morphological types. Sexual dimorphism: differing morphology between the males and females of a species.
Dipterocarp tree of the family Dipterocarpaceae.
Elytron (pl. elytra) a thickened, leathery, or horny front wing.
Embryogenesis the development of an embryo.
Emmet an ant (archaic).
Encapsulation the act of enclosing in a capsule.
Endemic native to, and restricted to, a particular geographic region.
Endophallus the inner eversible lining of the male intromittent organ.
Endosymbiont symbiosis in which one symbiont (the endosymbiont) lives within the body of the other.
Epigean living above the soil surface.
Epiphyll an epiphyte growing on a leaf.
Epiphyte an organism growing on the surface of a plant.
Euplanta(e) a swelling on a tarsal segment that facilitates adhesion to the substrate during locomotion.
Eusociality the condition where members of a social group are integrated and cooperate in taking care of the young, with non-reproductive individuals assisting those that produce offspring, and with an overlap of different generations contributing to colony labor.
Exuvium (pl. exuvia) the cast skin of an arthropod.
Fossorial adapted for or used in burrowing or digging.
Fungistatic referring to the inhibition of fungal growth.
Geophagy the act of feeding on soil.
Gestation the period of development of an embryo, from conception to hatch or birth.
Gonopore the external opening of a reproductive organ.
Gravid carrying eggs or young; pregnant.
Gregarious tending to assemble actively into groups or clusters.
Guild a group of species having similar ecological resource requirements and foraging strategies.
Gynandromorphs individuals of mixed sex, having some parts male and some parts female.
Hemimetabolous a pattern of development characterized by gradual changes, without distinct separation into larval, pupal, and adult stages.
Hemocyte a blood cell.
Heterochrony an evolutionary change in the onset or timing of the development of a feature relative to the appearance or rate of development of the same feature during the ontogeny of an ancestor.
Heteroploidy an organism or cell having a chromosome number that is not an even multiple of the haploid chromosome number for that species.
Heterotrophic used of organisms unable to synthesize organic compounds from inorganic substrates.
Heterozygosity the condition of having two different alleles at a given locus of a chromosome pair.
Holometabolous complete metamorphosis, having well-defined larval, pupal, and adult stages.
Homoplasy resemblance due to parallelism or convergent evolution rather than common ancestry.
Hyaline transparent, colorless.
Hypogean living underground.
Hypopharyngeal bladders a specialization of the mouthparts in some desert cockroaches that allows them to utilize atmospheric water.
Hypoxia oxygen deficiency.
Imago the adult stage of an insect.
Inquiline a species that lives within the burrow, nest, or domicile of another species.
Intercoxal referring to the area between the coxae, or basal portion of the legs.
Intromittent referring to something that allows, permits, or forces entry.
Iteroparous having repeated reproductive cycles.
Keel the raised crest running along the dorsal midline of an ootheca.
Macropterous tegmina and/or wings that are fully developed or only slightly shortened.
Mallee a thicket of dwarf, multi-stemmed Australian eucalypts.
Mechanoreceptor a sensory receptor that responds to mechanical pressure or distortion.
Metanotum the third dorsal division of the thorax.
Metathoracic referring to the third segment of the thorax.

Methanogens methane-producing bacteria.
Mimicry the close resemblance of one organism (the mimic) to another (the model) in order to deceive a third organism.
Monandrous (n. monandry) used of a female that mates with a single male.
Monophyletic referring to a group, including a common ancestor and all its descendents, derived from a single ancestral form.
Morphotype a collection of characteristics that determine the distinct physical appearance of an organism.
Mycetocyte a cell of the fat body specialized for housing bacterial symbionts.
Mycorrhiza(e) the symbiotic association of beneficial fungi with the small roots of some plants.
Myrmecophile an organism that spends part or all of its lifecycle inside of an ant nest.
Natal pertaining to birth.
Necrophagy feeding on corpses.
Neonates newborns.
Nuptial referring to the act or time of mating.
Ommatidium (pl. ommatidia) a single unit or visual section of a compound eye.
Omnivore (adj. omnivorous) feeding on a mixed diet of plant and animal material.
Ontogeny (adj. ontogenetic) the course of growth and development of an individual.
Oocyte a cell that produces eggs (ova) by meiotic division.
Oogenesis the formation, development, and maturation of female gametes.
Oviparity (adj. oviparous) producing an ootheca that is deposited in the external environment.
Ovoviviparity (adj. ovoviviparous) producing an ootheca that is withdrawn into the body and incubated in a brood sac; eggs have sufficient yolk to complete embryonic development. Typically, eggs hatch as the ootheca is expelled and active nymphs emerge.
Paedomorphosis retention of the juvenile characters of ancestral forms by the adults, or later ontogenetic stages, of their descendents.
Palp(s) a segmented, sensory appendage of the mouthparts.
Paraglossa(e) one of a pair of lobes at the tip of the "lower lip" (labium).
Paraphyletic a taxonomic group that does not include all the descendents of a common ancestor.
Paraproct(s) one of a pair of lobes bordering the anus.
Parthenogenesis the development of an individual from a female gamete that is not fertilized by a male gamete.
Phagocytosis the ingestion of solid particulate matter by a cell.
Phagostimulant anything that triggers feeding behavior.
Phallomere(s) sclerites of the male genitalia.
Phenology timing of the stages of the lifecycle, and its relation to weather and climate.
Phoresy (adj. phoretic) a symbiosis in which one organism is transported on the body of an individual of a different species.
Phylloplane the leaf surface, including the plants, algae, fungi, etc. associated with it.
Polyandrous (n. polyandry) used of a female that mates with more than one male.
Polyphenism the condition of having discontinuous phenotypes that lack genetic fixation.
Proctodeal referring to the hindgut.
Pronotum the first dorsal division of the thorax.
Protibiae the tibiae of the first set of legs.
Proventriculus the gizzard.
Pseudopenis an intromittent type male genital appendage that does not function to transfer sperm.
Pterothoracic referring to the wing-bearing segments of the thorax.
Quiescence a resting phase that occurs in direct response to deleterious physical conditions; it is terminated when conditions improve.
Rhizosphere the zone surrounding plant roots.
Sclerite a hardened plate of the exoskeleton bounded by sutures or membranous areas.
Sclerotized hardened.
Semelparous a life history where an organism reproduces just once in its lifetime.
Semi-voltine used of taxa that require 2 yr to develop to the adult stage of the lifecycle.
Seta(e) a bristle.
Spermatheca a receptacle for sperm storage in females.
Spermatophore a capsule containing sperm that is transferred from the male to the female during copulation.
Spiracle an external opening of the tracheal system; breathing pore.
Stadium the period between molts in a developing arthropod.
Sternal gland a gland on the ventral surface of the abdomen.
Stigmatic referring to the stigma, the upper end of the pistil in a flower.
Stomodeal referring to the foregut.
Subgenital plate a plate-like sclerite that underlies the genitalia.
Subsocial the condition in which one or both parents care for their own young.
Tarsus (pl. tarsi) the leg segment distally adjacent to the tibia; may be subdivided into segments (tarsomeres).
Taxon (pl. taxa) any group of organisms, populations, or taxonomic groups considered to be sufficiently distinct from other such groups as to be treated as a separate unit.
Tegmen (pl. tegmina) the thickened or leathery front wing of cockroaches and other orthopteroid insects.
Teneral a term applied to a recently molted, pale, soft-bodied arthropod.
Tergal glands glands on the dorsal surface of the abdomen; usually referring to those on males that entice females into position for copulatory engagement.
Tergite a sclerite of the dorsal surface of the abdomen.
Termitophile an organism that spends part or all of its lifecycle inside of a termite nest.
Thigmotaxis (adj. thigmotactic) a directed response of a motile organism to continuous contact with a solid surface.
Thorax the body region, located behind the head, which bears the legs and wings.

Tibia (pl. tibiae) the fourth segment of the leg, between the femur and the tarsus.

Trachea(e) a tube of the respiratory system.

Transovarial transmission the transmission of microorganisms between generations of hosts via the eggs.

Trichomes hair-like structures found on plant epidermis.

Troglomorphic having the distinct physical characteristics of an organism adapted to subterranean life.

Trophallaxis mutual or unilateral exchange of food between individuals.

Univoltine having one brood or generation per year.

Uric acid end product of nitrogen metabolism.

Uricolytic capable of breaking down uric acid.

Uricose glands male accessory glands that store and excrete uric acid.

Urocyte a cell in the fat body specialized for the storage of uric acid.

Vitellogenin yolk protein.

Viviparity (adj. viviparous) producing an ootheca that is withdrawn into the body and incubated in a brood sac. Eggs lack sufficient yolk to complete development, embryos rely on secretions from the brood sac walls for nourishment. Active nymphs emerge from the female.

Volant capable of flying.

References

Abbott, I., T. Burbidge, K. Strehlow, A. Mellican, and A. Wills. 2003. Logging and burning impacts on cockroaches, crickets and grasshoppers, and spiders in Jarrah forest, Western Australia. *Forest Ecology and Management.* 174:383–399.

Abbott, R.L. 1926. Contributions to the physiology of digestion in the Australian roach, *Periplaneta australasiae* Fab. *Journal of Experimental Zoology.* 44:219–253.

Abenserg-Traun, M., G.W. Arnold, D.E. Steven, G.T. Smith, L. Atkins, J.J. Viveen, and M. Gutter. 1996a. Biodiversity indicators in semi-arid, agricultural Western Australia. *Pacific Conservation Biology.* 2:375–389.

Abenserg-Traun, M., G.T. Smith, G.W. Arnold, and D.E. Steven. 1996b. The effects of habitat fragmentation and livestock grazing on animal communities in remnants of gimlet *Eucalyptus salubris* woodland in the Western Australian wheatbelt. I. Arthropods. *Journal of Applied Ecology.* 33:1281–1301.

Abrams, P.A., O. Leimar, S. Nylan, and C. Wiklund. 1996. The effect of flexible growth rates on optimal sizes and developmental times in a seasonal environment. *The American Naturalist.* 147:381–395.

Adair, E.W. 1923. Notes sur *Periplaneta americana* L. et *Blatta orientalis* L. (Orthop.). *Bulletin de la Societe Entomologique d'Égypte.* 7:18–38.

Adams, P. 2004. Lord Howe Island wood-feeding cockroach—endangered species listing. *New South Wales National Parks and Wildlife Service, Department of Environment & Conservation (NSW).* www.nationalparks.nsw.gov.au/npws.nsf/Content/Panesthia_lata_endangered_declaration. 4 February 2004.

Adiyodi, K.P., and R.G. Adiyodi. 1974. Control mechanisms in cockroach reproduction. *Journal of Science and Industrial Research.* 33:343–358.

Adrian, J. 1976. Gums and hydrocolloids in nutrition. *World Review of Nutrition and Dietetics.* 25:186–216.

Aiouaz, M. 1974. Chronologie du développement embryonnaire de *Leucophaea maderae* Fabr. (Insecte, Dictyoptère). *Archives de Zoologie Experimentale et Génerale.* 115:343–358.

Akers, R.C., and W.H. Robinson. 1983. Comparison of movement behavior in three strains of German cockroach, *Blattella germanica. Entomologia Experimentalis et Applicata.* 34:143–147.

Alexander, R.D. 1974. The evolution of social behavior. *Annual Review of Ecology and Systematics.* 5:325–383.

Alling, A., M. Nelson, and S. Silverstone. 1993. Life Under Glass: The Inside Story of Biosphere 2. The Biosphere Press, Oracle, AZ. 254 pp.

Alsop, D.W. 1970. Defensive glands of arthropods: comparative morphology of selected types. Ph.D. thesis, Cornell University, Ithaca, NY.

Ananthasubramanian, K.S., and T.N. Ananthakrishnan. 1959. The structure of the ootheca and egg laying habits of *Corydia petiveriana* L. *Indian Journal of Entomology.* 21:59–64.

Anderson, J. 1983. Life in the soil is a ferment of little rotters. *New Scientist.* 100:29–37.

Anderson, J.M., and D.E. Bignell. 1980. Bacteria in the food, gut contents and feces of the litter-feeding millipede, *Glomeris marginata. Soil Biology and Biochemistry.* 12:251–254.

Anderson, J.M., P. Ineson, and S.A. Huish. 1982. The effects of animal feeding activities on element release from deciduous forest litter and soil organic matter. *In* New Trends in Soil Biology. P. Lebrun, H.M. André, A. DeMedts, C. Grégoire-Wibo, and G. Wauthy, editors. International Colloquium of Soil Biology, Louvain la Neuve, Belgium. 87–100.

Anderson, J.M., and M.J. Swift. 1983. Decomposition in tropical forests. *In* Tropical Rain Forest: Ecology and Management. S.L. Sutton, T.C. Whitmore, and A.C. Chadwick, editors. Blackwell Scientific Publications, Oxford. 287–309.

Anderson, N.M. 1997. Phylogenetic tests of evolutionary scenarios: the evolution of flightlessness and wing polymorphism in insects. *Mémoires du Muséum National d'Histoire Naturelle.* 173:91–108.

Andersson, M. 1984. The evolution of eusociality. *Annual Review of Ecology and Systematics.* 15:165–189.

Anduaga, S., and G. Halffter. 1993. Nidification and feeding of *Liatonus rhinocerulus* (Bates) (Coleoptera: Scarabaeidae: Scarabaeinae). *Acta Zoologica Mexicana Nueva Serie.* 57:1–14.

Anisyutkin, L.N. 1999. Cockroaches of the subfamily Epilamprinae (Dictyoptera, Blaberidae) from the Indochina peninsula. *Entomological Review.* 79:434–454.

Anisyutkin, L.N. 2003. Contribution to knowledge of the cockroach subfamilies Paranauphoetinae (stat. n.), Perisphaeriinae and Panesthiinae (Dictyoptera: Blaberidae). *Zoosystematica Rossica.*12:55–77.

Annandale, N. 1906. Notes on the freshwater fauna of India. No. III. An Indian aquatic cockroach and beetle larva. *Journal and Proceedings of the Asiatic Society of Bengal, n.s.* 2:105–107.

Annandale, N. 1910. Cockroaches as predatory insects. *Records of the Indian Museum.* 3:201–202.

Appel, A.G., D.A. Rierson, and M.K. Rust. 1983. Comparative water relations and temperature sensitivity of cockroaches. *Comparative Biochemistry and Physiology.* 74A:357–361.

Appel, A.G., and M.K. Rust. 1986. Time activity budgets and spatial distribution patterns of the smokybrown cockroach *Periplaneta fuliginosa* (Dictyoptera Blattidae). *Annals of the Entomological Society of America.* 79:104–108.

Appel, A.G., and L.M. Smith. 2002. Biology and management of the smokybrown cockroach. *Annual Review of Entomology.* 47:33–55.

Appel, A.G., and J.B. Tucker. 1986. Occurrence of the German cockroach *Blattella germanica* (Dictyoptera: Blattellidae) outdoors in Alabama and Texas. *Florida Entomologist.* 69:422–423.

Archibold, O.W. 1995. Ecology of World Vegetation. Chapman and Hall, London. 510 pp.

Arnold, J.W. 1974. Adaptive features on the tarsi of cockroaches. *International Journal of Insect Morphology and Embryology.* 3:317–334.

Arnqvist, G., and T. Nilsson. 2000. The evolution of polyandry: multiple mating and female fitness in insects. *Animal Behaviour.* 60:145–164.

Asahina, S, 1960. Japanese cockroaches as household pest. *Japanese Journal of Sanitary Zoology.* 4:188–190.

Asahina, S. 1965. Taxonomic notes on Japanese Blattaria, III. On the species of the genus *Onychostylus* (in Japanese). *Japanese Journal of Sanitary Zoology.* 16:6–15.

Asahina, S. 1971. Notes on the cockroaches of the genus *Eucorydia* from the Ryukus, Taiwan, Thailand and Nepal. *Kontyû.* 39:256–262.

Asahina, S. 1974. The cavernicolous cockroaches of the Ryuku Islands. *Memoirs of the National Science Museum, Tokyo.* 7:145–157.

Atkin, L., and J. Proctor. 1988. Invertebrates in the litter and soil on Volcán Barva, Costa Rica. *Journal of Tropical Ecology.* 4:307–310.

Atkinson, T.H., P.G. Koehler, and R.S. Patterson. 1991. Catalogue and Atlas of the Cockroaches (Dictyoptera) of North America North of Mexico. *Miscellaneous Publications of the Entomological Society of America.* 78:1–86.

Atlas, R.M., and R. Bartha. 1998. Microbial Ecology: Fundamentals and Applications. Benjamin/Cummings Science Publishing, Menlo Park, CA. 694 pp.

Ausmus, B.S. 1977. Regulation of wood decomposition rates by arthropod and annelid populations. *Ecological Bulletin.* 25:180–192.

Austin, A.D., and A.D. Blest. 1979. The biology of two Australian species of dinopid spider. *Journal of Zoology, London.* 189:145–156.

Autrum, H., and W. Schneider. 1948. Vergleichende Untersuchungen über den Erschütterungssinn der Insekten. *Zeitschrift für Vergleichende Physiologie.* 31:77–88.

Ayyad, M.A., and I. Ghabbour. 1977. Systems analysis of Mediterranean desert ecosystems of northern Egypt. *Environmental Conservation.* 4:91–102.

Baccetti, B. 1987. Spermatozoa and phylogeny in orthopteroid insects. *In* Evolutionary Biology of Orthopteroid Insects. B.M. Baccetti, editor. John Wiley & Sons, New York. 12–112.

Back, E.A. 1937. The increasing importance of the cockroach, *Supella supellectilium* Serv., as a pest in the United States. *Proceedings of the Entomological Society of Washington.* 39:205–213.

Ball, E.D., E.R. Tinkham, R. Flock, and C.T. Vorhies. 1942.

The grasshoppers and other Orthoptera of Arizona. *University of Arizona College Agricultural Experiment Station Technical Bulletin.* 93:257–373.

Bandi, C., G. Damiani, L. Magrassi, A. Grigolo, R. Fani, and L. Sacchi. 1994. Flavobacteria as intracellular symbionts in cockroaches. *Proceedings of the Royal Society of London, Series B.* 257:43–48.

Bandi, C., and L. Sacchi. 2000. Intracellular symbiosis in termites. *In* Termites: Evolution, Sociality, Symbioses, Ecology. T. Abe, D.E. Bignell, and M. Higashi, editors. Kluwar Academic Publishers, Dordrecht. 261–273.

Bandi, C., M. Sironi, G. Damiani, L. Magrassi, C.A. Nalepa, U. Laudani, and L. Sacchi. 1995. The establishment of intracellular symbiosis in an ancestor of cockroaches and termites. *Proceedings of the Royal Society of London, Series B.* 259:293–299.

Bandi, C., M. Sironi, C.A. Nalepa, S. Corona, and L. Sacchi. 1997. Phylogenetically distant intracellular symbionts in termites. *Parassitologia.* 39:71–75.

Bao, N., and W.H. Robinson. 1990. Morphology and mating configuration of genitalia of the Oriental cockroach *Blatta orientalis* L. (Blattodea: Blattidae). *Proceedings of the Entomological Society of Washington.* 92:416–421.

Barbier, J. 1947. Observations sur la moeurs de *Rhipidius pectinicornis* Thunbg. et description de sa larve primaire (Col. Rhipiphoridae). *Entomologiste, Paris.* 3:163–180.

Barr, T.C.J. 1968. Cave ecology and the evolution of troglobites. *Evolutionary Biology.* 2:35–102.

Barr, T.C.J., and J.R. Holsinger. 1985. Speciation in cave faunas. *Annual Review of Ecology and Systematics.* 16:313–337.

Barrios, H. 2003. Insect herbivores feeding on conspecific seedlings and trees. *In* Arthropods of tropical forests. Spatio-temporal dynamics and resource use in the canopy. Y. Basset, V. Novotny, S.E. Miller, and R.L. Kitching, editors. Cambridge University Press, Cambridge. 282–290.

Barry, D. 2002. Armed roaches? Technology goes too far. *In* The Miami Herald. 3 November 2002.

Barth, F.G., H. Bleckmann, J. Bohnenberger, and E.A. Seyfarth. 1988. Spiders of the genus *Cupiennius* Simon 1891 (Araneae, Ctenidae). II. On the vibratory environment of a wandering spider. *Oecologia.* 77:194–201.

Barth, R.H. 1964. The mating behavior of *Byrsotria fumigata* (Guerin) (Blattidae, Blaberinae). *Behaviour.* 23:1–30.

Barth, R.H. 1968a. The comparative physiology of reproductive processes in cockroaches. Part I. Mating behaviour and its endocrine control. *Advances in Reproductive Physiology.* 3:167–207.

Barth, R.H. 1968b. The mating behavior of *Eurycotis floridana* (Walker) (Blattaria, Blattoidea, Polyzosteriinae). *Psyche.* 75:274–284.

Barth, R.H. 1968c. The mating behavior of *Gromphadorhina portentosa* (Schaum) (Blattaria, Blaberoidea, Blaberidae, Oxyhaloinae): an anomolous pattern for a cockroach. *Psyche.* 75:124–131.

Barton, H.A., M.R. Taylor, and N.R. Pace. 2004. Molecular phylogenetic analysis of a bacterial community in an oligotrophic cave environment. *Geomicrobiology Journal.* 21:11–20.

Basset, Y. 2001. Invertebrates in the canopy of tropical rain forests. How much do we really know? *Plant Ecology.* 153:87–107.

Basset, Y., H.-P. Aberlenc, H. Barrios, and G. Curletti. 2003a. Arthropod diel activity and stratification. *In* Arthropods of Tropical Forests: Spatio-Temporal Dynamics and Resource Use in the Canopy. Y. Basset, V. Novotny, S.E. Miller, and R.L. Kitching, editors. Cambridge University Press, Cambridge. 304–314.

Basset, Y., P.M. Hammond, H. Barrios, J.D. Holloway, and S.E. Miller. 2003b. Vertical stratification of arthropod assemblages. *In* Arthropods of Tropical Forests: Spatio-Temporal Dynamics and Resource Use in the Canopy. Y. Basset, V. Novotny, S.E. Miller, and R.L. Kitching, editors. Cambridge University Press, Cambridge. 17–27.

Basset, Y., N.D. Springate, H.P. Aberlenc, and G. Delvare. 1997. A review of methods for sampling arthropods in tree canopies. *In* Canopy Arthropods. N.E. Stork, J. Adis, and R.K. Didham, editors. Chapman and Hall, London. 27–52.

Baudoin, R. 1955. La physico-chimie des surfaces dans la vie des Arthropodes aériens, des miroirs d'eau, des rivages marins et lacustres et de la zone intercotidale. *Bulletin Biologique de la France et de la Belgique.* 89:16–164.

BBCNews. 23 December 2004. Giant cockroach among jungle find. http://news.bbc.co.uk/2/hi/asia-pacific/4121637.stm.

Beccaloni, G. 1989. Why not study cockroaches? *Amateur Entomologists' Society Bulletin.* 48:176–178.

Beebe, W. 1925. Jungle Days. G.P. Putnam's Sons, New York. 201 pp.

Beebe, W. 1951. Migration of insects (other than Lepidoptera) through Portachuelo Pass, Rancho Grande, north-central Venezuela. *Zoologica.* 36:255–266.

Beebe, W. 1953. Unseen Life of New York as a Naturalist Sees it. Duell, Sloan and Pearce, New York. 165 pp.

Bei-Bienko, G.Y. 1950. Blattodea. *In* Faune de l'URSS New Series 40. Akademiya Nauk SSSR, St. Petersburg. 332–336.

Bei-Bienko, G.Y. 1969. *Ectobius duskei* Adel. as a characteristic inhabitant of steppes in the USSR. *Memorie della Societa Entomologica Italiana.* 48:123–128.

Bei-Bienko, G.Y. 1970. New genera and species of cockroaches (Blattoptera) from tropical and subtropical Asia. *Entomological Review.* 48:528–548.

Bell, W.J. 1969. Continuous and rhythmic reproductive cycle observed in *Periplaneta americana. Biological Bulletin.* 137:239–249.

Bell, W.J. 1971. Starvation-induced oocyte resorption and yolk protein salvage in *Periplaneta americana. Journal of Insect Physiology.* 17:1099–1111.

Bell, W.J. 1981. The Laboratory Cockroach. Chapman and Hall, London. 161 pp.

Bell, W.J. 1990. Biology of the cockroach. *In* Cockroaches as Models for Neurobiology: Applications in Biomedical Research. Vol. 1. I. Huber, E.P. Masler, and B.R. Rao, editors. CRC Press, Boca Raton. 7–12.

Bell, W.J., and K.G. Adiyodi (eds.). 1982a. The American Cockroach. Chapman and Hall, London. 529 pp.

Bell, W.J., and K.G. Adiyodi. 1982b. Reproduction. *In* The

American Cockroach. W.J. Bell and K.G. Adiyodi, editors. Chapman and Hall, London. 343–370.

Bell, W.J., and M.K. Bohm. 1975. Oosorption in insects. *Biological Reviews.* 50:373–396.

Bell, W.J., T. Burk, and G.R. Sams. 1973. Cockroach aggregation pheromone: directional orientation. *Behavioral Biology.* 9:251–255.

Bell, W.J., C. Parsons, and E.A. Martinko. 1972. Cockroach aggregation pheromones: analysis of aggregation tendency and species specificity (Orthoptera: Blattidae). *Journal of the Kansas Entomological Society.* 45:414–421.

Bell, W.J., S.B. Vuturo, and M. Bennett. 1978. Endokinetic turning and programmed courtship acts of the male German cockroach. *Journal of Insect Physiology.* 24:369–374.

Belt, T. 1874. The Naturalist in Nicaragua. J.M. Dent & Sons, London. 306 pp.

Benson, E.P., and I. Huber. 1989. Oviposition behavior and site preference of the brownbanded cockroach *Supella longipalpa* (F.) (Dictyoptera: Blattellidae). *Journal of Entomological Science.* 24:84–91.

Benton, M.J., and G.W. Storrs. 1996. Diversity in the past: comparing cladistic phylogenies and stratigraphy. *In* Aspects of the Genesis and Maintenance of Biological Diversity. M.E. Hochberg, J. Clobert, and R. Barbault, editors. Oxford University Press, Oxford. 19–40.

Bernays, E.A. 1986. Evolutionary contrasts in insects: nutritional advantages of holometabolous development. *Physiological Entomology.* 11:377–382.

Bernays, E.A. 1991. Evolution of insect morphology in relation to plants. *Philosophical Transactions of the Royal Society of London, Series B.* 333:257–264.

Berrie, A.D. 1975. Detritus, micro-organisms and animals in fresh water. *In* The Role of Terrestrial and Aquatic Organisms in Decomposition Processes. J.M. Anderson and A. Macfadyen, editors. Blackwell Scientific Publications, Oxford. 323–338.

Berthold, R.J. 1967. Behavior of the German cockroach, *Blattella germanica* (L.), in response to surface textures. *Journal of the New York Entomological Society.* 75:148–153.

Berthold, R.J., and B.R. Wilson. 1967. Resting behavior of the German cockroach, *Blattella germanica. Annals of the Entomological Society of America.* 60:347–351.

Beutel, R.G., and S.N. Gorb. 2001. Ultrastructure of attachment specializations of hexapods (Arthropoda): evolutionary patterns inferred from a revised ordinal phylogeny. *Journal of Zoological Systematics and Evolutionary Research.* 39:177–207.

Bhoopathy, S. 1997. Microhabitat preferences among four species of cockroaches. *Journal of Nature Conservation.* 9:259–264.

Bhoopathy, S. 1998. Incidence of parental care in the cockroach *Thorax procellana* (Saravas) (Blaberidae: Blattaria). *Current Science.* 74:248–251.

Bidochka, M.J., R.J. St. Leger, and D.W. Roberts. 1997. Induction of novel proteins in *Manduca sexta* and *Blaberus giganteus* as a response to fungal challenge. *Journal of Invertebrate Pathology.* 70:184–189.

Bignell, D.E. 1976. Gnawing activity, dietary carbohydrate deficiency and oothecal production in the American cockroach (*Periplaneta americana*). *Experientia.* 32:1405–1406.

Bignell, D.E. 1977a. An experimental study of cellulose and hemicellulose degradation in the alimentary canal of the American cockroach. *Canadian Journal of Zoology.* 55:579–589.

Bignell, D.E. 1977b. Some observations on the distribution of gut flora in the American cockroach *Periplaneta americana. Journal of Invertebrate Pathology.* 29:338–343.

Bignell, D.E. 1978. Effects of cellulose in the diets of cockroaches. *Entomologia Experimentalis et Applicata.* 24:54–57.

Bignell, D.E. 1980. An ultrastructural study and stereological analysis of the colon wall in the cockroach, *Periplaneta americana. Tissue & Cell.* 12:153–164.

Bignell, D.E. 1981. Nutrition and digestion. *In* The American Cockroach. W.J. Bell and K.G. Adiyodi, editors. Chapman and Hall, New York. 57–86.

Bignell, D.E. 1984. Direct potentiometric determination of redox potentials of the gut contents in the termites, *Zootermopsis nevadensis* and *Cubitermes severus,* and 3 other arthropods. *Journal of Insect Physiology.* 30:169–174.

Bignell, D.E. 1989. Relative assimilations of 14 C-labelled microbial tissues and 14 C-plant fibre ingested with leaf litter by the millipede *Glomeris marginata* under experimental conditions. *Soil Biology and Biochemistry.* 21:819–828.

Bignell, D.E. 2000a. Introduction to symbiosis. *In* Termites: Evolution, Sociality, Symbioses, Ecology. T. Abe, D.E. Bignell, and M. Higashi, editors. Kluwer Academic Publishers, Dordrecht. 189–208.

Bignell, D.E. 2000b. Addition to the preface. *In* Termites: Evolution, Sociality, Symbioses, Ecology. T. Abe, D.E. Bignell, and M. Higashi, editors. Kluwer Academic Publishers, Dordrecht. xv–xvi.

Bignell, D.E., and P. Eggleton. 2000. Termites in ecosystems. *In* Termites: Evolution, Sociality, Symbioses, Ecology. T. Abe, D.E. Bignell, and M. Higashi, editors. Kluwar Academic, Dordrecht. 363–387.

Bijnen, F.G.C., J.H.P. Hackstein, P. Kestler, F.J.M. Harren, and J. Reuss. 1995. Fast laser photoacoustical detection of trace gases; respiration of arthropods. *Laser und Optoelektronik.* 27:68–72.

Bijnen, F.G.C., F.J.M. Harren, J.H.P. Hackstein, and J. Reuss. 1996. Intracavity CO laser photoacoustic trace gas detection: cyclic CH4, H2O and CO2 emission by cockroaches and scarab beetles. *Applied Optics.* 35:5357–5368.

Blackburn, D.G. 1999. Viviparity and oviparity: evolution and reproductive strategies. *In* Encyclopedia of Reproduction. Vol. 4. E. Knobil and J.D. Neill, editors. Academic Press, San Diego. 994–1003.

Blair, K.G. 1922. An entomological holiday in S. France. *Entomologist.* 55:147–151.

Bland, R.G., D.P. Slaney, and P. Weinstein. 1998a. Antennal sensilla on cave species of Australian *Paratemnopteryx* cockroaches (Blattaria: Blattellidae). *International Journal of Insect Morphology and Embryology.* 27:83–93.

Bland, R.G., D.P. Slaney, and P. Weinstein. 1998b. Mouthpart sensilla of cave species of Australian *Paratemnopteryx* cockroaches (Blattaria: Blattellidae). *International Journal of Insect Morphology and Embryology.* 27:291–3000.

Blatchley, W.S. 1920. Orthoptera of Northeastern America, with Special Reference to the Faunas of Indiana and

Florida. The Nature Publishing Company, Indianapolis. 784 pp.
Block, W. 1991. To freeze or not to freeze? Invertebrate survival of sub-zero temperatures. *Functional Ecology.* 5:284–290.
Blumenthal, H.J., and S. Roseman. 1957. Quantitative estimation of chitin in fungi. *Journal of Bacteriology.* 74:222–224.
Bodenstein, D. 1953. Studies on the humoral mechanisms in growth and metamorphosis of the cockroach *Periplaneta americana.* I. Transplantations of integumental structures and experimental parabioses. *Journal of Experimental Biology.* 123:189–232.
Bohn, H. 1987. Reversal of the left-right asymmetry in male genitalia of some Ectobiinae (Blattaria: Blattellidae) and its implications on sclerite homologization and classification. *Entomological Scandinavica* 18:293–303.
Bohn, H. 1991a. Revision of the *Loboptera* species of Morocco (Blattaria: Blattellidae: Blattellinae). *Entomologica Scandinavica.* 22:251–295.
Bohn, H. 1991b. Revision of the *Loboptera* species of Spain (Blattaria: Blattellidae). *Entomologica Scandinavica.* 21:369–403.
Bohn, H. 1993. Revision of the *panteli*-group of *Phyllodromica* in Spain and Morocco (Blattaria: Blattellidae: Ectobiinae). *Entomologica Scandinavica.* 24:49–72.
Bohn, H. 1999. Revision of the *carpetana*-group of *Phyllodromica* Fieber from Spain, Portugal and France (Insecta, Blattaria, Blattellidae, Ectobiinae). *Spixiana.* Supplement 25:1–102.
Bolívar, I. 1901. Un nuevo orthóptero mirmecófilo *Attaphila bergi. Comunicaciones del Museo Nacional de Buenos Aires.* 1:331–336.
Bordereau, C. 1985. The role of pheromones in caste differentiation. *In* Caste Differentiation in Social Insects. J.A.L. Watson, B.M. Okot-Kotber, and C. Noirot, editors. Pergamon Press, Oxford. 221–226.
Bowden, J., and J. Phipps. 1967. Cockroaches (*Periplaneta americana* (L.)) as predators. *Entomologist's Monthly Magazine.* 103:175–176.
Boyer, S., and C. Rivault. 2003. La Réunion and Mayotte cockroaches: impact of altitude and human activity. *Comptes Rendus Biologies.* 326:S210–S216.
Braack, L.E.O. 1989. Arthropod inhabitants of a tropical cave "island" environment populated by bats. *Biological Conservation.* 48:77–84.
Bracke, J.W., D.L. Cruden, and A.J. Markowetz. 1978. Effect of metronidazole on the intestinal microflora of the American cockroach, *Periplaneta americana* L. *Antimicrobial Agents and Chemotherapy.* 13:115–120.
Bracke, J.W., D.L. Cruden, and A.J. Markovetz. 1979. Intestinal microbial flora of the American cockroach, *Periplaneta americana* L. *Applied and Environmental Microbiology.* 38:945–955.
Bracke, J.W., and A.J. Markovetz. 1980. Transport of bacterial end products from the colon of *Periplaneta americana. Journal of Insect Physiology.* 26:85–89.
Breed, M.D. 1983. Cockroach mating systems. *In* Orthoptera Mating Systems. D.T. Gwynne and G.K. Morris, editors. Westview, Boulder. 268–284.
Breed, M.D., C.M. Hinkle, and W.J. Bell. 1975. Agonistic behavior in the German cockroach, *Blattella germanica. Zeitschrift für Tierpsychologie.* 39:24–32.
Breland, O.P., C.D. Eddleman, and J.J. Biesele. 1968. Studies of insect spermatozoa. I. *Entomological News.* 79:197–216.
Brenner, R.J., R.S. Patterson, and P.G. Koehler. 1988. Ecology, behavior and distribution of *Blattella asahinai* (Orthoptera: Blattellidae) in central Florida. *Annals of the Entomological Society of America.* 81:432–436.
Bret, B.L., and M.H. Ross. 1985. A laboratory study of German cockroach dispersal (Dictyoptera: Blattellidae). *Proceedings of the Entomological Society of Washington.* 87:448–455.
Bret, B.L., M.H. Ross, and G.I. Holtzman. 1983. Influence of adult females on within-shelter distribution patterns of *Blattella germanica* (Dictyoptera: Blattellidae). *Annals of the Entomological Society of America.* 76:847–852.
Breznak, J.A. 1975. Symbiotic relationships between termites and their intestinal microbiota. *Symposia of the Society for Experimental Biology.* 29:559–580, 6 plates.
Breznak, J.A. 1982. Biochemical aspects of symbiosis between termites and their intestinal microbiota. *In* Invertebrate-Microbial Interactions. J.M. Anderson, A.D.M. Rayner, and D.W.H. Walton, editors. Cambridge University Press, Cambridge. 173–203.
Breznak, J.A., W.J. Brill, J.W. Mertins, and H.C. Coppel. 1973. Nitrogen fixation in termites. *Nature.* 244:577–580.
Breznak, J.A., J.W. Mertins, and H.C. Coppel. 1974. Nitrogen fixation and methane production in a wood eating cockroach *Cryptocercus punctulatus* Scudder (Orthoptera: Blattidae). *University of Wisconsin Forestry Notes.* 184:1–2.
Bridwell, J.C., and O.H. Swezey. 1915. Entomological notes. *Proceedings of the Hawaiian Entomological Society.* 3:55–56.
Brodsky, A.K. 1994. The Evolution of Insect Flight. Oxford University Press, Oxford. 229 pp.
Bronstein, J.L. 1994. Conditional outcomes in mutualistic interactions. *Trends in Ecology and Evolution.* 9:214–217.
Bronstein, S.M., and W.E. Conner. 1984. Endotoxin-induced behavioral fever in the Madagascar cockroach, *Gromphadorhina portentosa. Journal of Insect Physiology.* 30:327–330.
Brooks, D.R. 1996. Explanations of homoplasy at different levels of biological organization. *In* Homoplasy: The Recurrence of Similarity in Evolution. M.J. Sanderson and L. Hufford, editors. Academic Press, San Diego. 3–36.
Brossut, R. 1970. L'interattraction chez *Blabera craniifer* Burm. (Insecta, Dictyoptera): Sécrétion d'une phéromone par les glandes mandibulaires. *Comptes rendus de l'Academie des Sciences, Paris.* 270:714–716.
Brossut, R. 1973. Evolution du système glandulaire exocrine céphalique des Blattaria et des Isoptera. *International Journal of Insect Morphology and Embryology.* 2:35–54.
Brossut, R. 1975. Pheromonal basis of gregarism and interattraction. *In* Pheromones and Defensive Secretions in Social Insects. C. Noirot, P.E. Howse, and G. LeMasne, editors. University of Dijon, Dijon. 67–85.
Brossut, R. 1976. Etude morphologique de la blatte myrmecophile *Attaphila fungicola* Wheeler. *Insectes Sociaux.* 23:167–174.

Brossut, R. 1979. Gregarism in cockroaches and in *Eublaberus* in particular. *In* Chemical Ecology: Odour Communication in Animals. F.J. Ritter, editor. Elsevier/North Holland Biochemical Press, Amsterdam. 237–246.

Brossut, R. 1983. Allomonal secretions in cockroaches. *Journal of Chemical Ecology.* 9:143–158.

Brossut, R., P. Dubois, and J. Rigaud. 1974. Le grégarisme chez *Blaberus craniifer:* isolement et identification de la phéromone. *Journal of Insect Physiology.* 20:529–543.

Brossut, R., P. Dubois, J. Rigaud, and L. Sreng. 1975. Etude biochemique de la sécrétion des glandes tergales des Blattaria. *Insect Biochemistry.* 5:719–732.

Brossut, R., and L.M. Roth. 1977. Tergal modifications associated with abdominal glandular cells in the Blattaria. *Journal of Morphology.* 151:259–298.

Brossut, R., and L. Sreng. 1985. L'univers chimique des blattes. *Bulletin Société Entomologique de France.* 90:1266–1280.

Brousse-Gaury, P. 1971a. Présence de mécanorécepteurs au niveau de la poche incubatrice de *Blabera fusca* Br. et *Leucophaea maderae* F., Dictyoptères Blaberidae. *Comptes Rendus Acad. Science Paris.* 272:2785–2787.

Brousse-Gaury, P. 1971b. Soies mécanoréceptrices et poche incubatrice de blattes. *Bulletin Biologique.* 105:337–343.

Brousse-Gaury, P. 1976. Glande sternale et balisage des pistes chez *Periplaneta americana* (L.). *Bulletin Biologique de la France et de la Belgique.* 110:395–420.

Brousse-Gaury, P. 1977. Starvation and reproduction in *Periplaneta americana* L.: Control of mating behaviour in the female. *In* Advances in Invertebrate Reproduction. Vol. 1. K.G. Adiyodi and R.G. Adiyodi, editors. Peralam-Kenoth, Kerala, India. 328–343.

Brousse-Gaury, P. 1981. Typologie et topographie des sensilles sur le tarse des mâles de *Periplaneta americana* L. (Dictyoptères, Blattidae). *Annales des Sciences Naturelles, Zoologie, Paris.* 3:69–94.

Brousse-Gaury, P., and F. Goudey-Perriere. 1983. Spermatophore et vitellogenèse chez *Blabera fusca* Br. (Dictyoptère, Blaberidae). *Comptes Rendus Acad. Science Paris.* 296:659–664.

Brown, E.B. 1952. Observations on the life history of the cockroach *Ectobius panzeri* Stephens (Orth., Blattidae). *Entomologist's Monthly Magazine.* 88:209–212.

Brown, V.K. 1973a. Aspects of the reproductive biology of three species of *Ectobius* (Dictyoptera: Blattidae). *Entomologia Experimentalis et Applicata.* 16:213–222.

Brown, V.K. 1973b. The overwintering stages of *Ectobius lapponicus* (L.) (Dictyoptera: Blattidae). *Journal of Entomology (A).* 48:11–24.

Brown, V.K. 1980. Developmental strategies in *Ectobius pallidus* (Dictyoptera: Blattidae). *International Journal of Invertebrate Reproduction.* 2:85–93.

Brown, V.K. 1983. Developmental strategies in British Dictyoptera: seasonal variation. *In* Diapause and Life Cycle Strategies. V.K. Brown and I. Hodel, editors. Dr W. Junk Publishers, The Hague. 111–125.

Brunet, P.C.J., and P.W. Kent. 1955. Mechanism of sclerotin formation: the participation of a Beta-glucoside. *Nature.* 175:819–820.

Buckland-Nicks, J. 1998. Prosobranch parasperm: sterile germ cells that promote paternity? *Micron.* 29:267–280.

Burk, T., and W.J. Bell. 1973. Cockroach aggregation pheromone: inhibition of locomotion (Orthoptera: Blattidae). *Journal of the Kansas Entomological Society.* 46:36–41.

Burley, N.T., and K. Johnson. 2002. The evolution of avian parental care. *Philosophical Transactions of the Royal Society of London B.* 357:241–250.

Camazine, S., J.-L. Deneubourg, N.R. Franks, J. Sneyd, G. Theraulaz, and E. Bonabeau. 2001. Self-Organization in Biological Systems. Princeton University Press, Princeton. 538 pp.

Camhi, J., and E.N. Johnson. 1999. High frequency steering maneuvers mediated by tactile cues: antennal wall-following in the cockroach. *Journal of Experimental Biology.* 202:631–643.

Campbell, F.L., and M.H. Ross. 1979. On the pruning of its flagella by the German cockroach during postembryonic development. *Annals of the Entomological Society of America.* 72:580–582.

Carne, P.B. 1978. *Dasygnathus blattocomes* sp. n. (Coleoptera: Scarabaeidae). *Journal of the Australian Entomological Society.* 17:91–93.

Carpenter, F.M. 1947. Early insect life. *Psyche.* 54:65–85.

Caudell, A.N. 1906. A new roach from the Philippines. *Canadian Entomologist.* 38:136.

Cazemier, A.E., J.H.P. Hackstein, H.J.M. Op den Camp, J. Rosenberg, and C. van der Drift. 1997a. Bacteria in the intestinal tract of different species of arthropods. *Microbial Ecology.* 33:189–197.

Cazemier, A.E., H.J.M. Op den Camp, J.H.P. Hackstein, and G.D. Vogels. 1997b. Fibre digestion in arthropods. *Comparative Biochemistry and Physiology.* 118A:101–109.

Chapman, C.A., R.W. Wrangham, and L.J. Chapman. 1995. Ecological constraints on group size: an analysis of spider monkey and chimpanzee subgroups. *Behavioral Ecology and Sociobiology.* 36:59–70.

Chapman, T., G. Arnqvist, J. Bangham, and L. Rowe. 2003. Sexual conflict. *Trends in Ecology & Evolution.* 18:41–47.

Chiang, A.S., A.P. Gupta, and S.S. Han. 1988. Arthropod immune system. I. Comparative light and electron microscopic accounts of immunocytes and other hemocytes of *Blattella germanica* (Dictyoptera: Blattellidae). *Journal of Morphology.* 198:257–268.

Chon, T.S., D. Liang, and C. Schal. 1990. Effects of mating and grouping on oocyte development and pheromone release activities in *Supella longipalpa* (Dictyoptera: Blattellidae). *Environmental Entomology.* 19:1716–1721.

Chopard, L. 1919. Zoological results of a tour in the Far East. The Orthoptères caverniculous de Birmanie et de la Peninsule Malaise. *Memoirs of the Asiatic Society of Bengal.* 6:341–396.

Chopard, L. 1925. La distribution géographique des Blattinae aptères ou subaptères. *Association Française pour l'avancement des Sciences, Congrès de Liège.* 1924:975–977.

Chopard, L. 1929. Orthoptera palearctica critica. VII. Lex polyphagièns de la faune paléarctique (Orth. Blatt.). *Eos.* 5:223–358.

Chopard, L. 1932. Un cas de micropthalmie liée a l'atrophie des ailes chez une blatte cavernicole. *In Livre du Centenaire, Société Entomologique de France.* 485–496.
Chopard, L. 1938. La Biologie des Orthoptères. Lechevalier, Paris. 564 pp.
Chopard, L. 1952. Description d'une Blatte xylicole du Mozambique (Dictyoptère). *Bulletin de la Société Entomologique de France.* 57:6–7.
Chopard, L. 1969. Description d'une intéressante Blatte du désert iranien (Dictyop. Polyphagidae). *Bulletin Société Entomologique de France.* 74:228–230.
Chown, S.L., and S.W. Nicolson. 2004. Insect Physiological Ecology: Mechanisms and Patterns. Oxford University Press, Oxford. 243 pp.
Christiansen, K. 1970. Survival of Collembola on clay substrates with and without food added. *Annales de Spéléologie.* 25:849–852.
Christy, J.H. 1995. Mimicry, mate choice, and the sensory trap hypothesis. *The American Naturalist.* 146:171–181.
Cisper, G., A.J. Zera, and D.W. Borst. 2000. Juvenile hormone titer and morph-specific reproduction in the wing polymorphic cricket, *Gryllus firmus. Journal of Insect Physiology.* 46:585–596.
Clark, A. 2003. Costs and consequences of evolutionary temperature adaptation. *Trends in Ecology & Evolution.* 18:573–581.
Clark, D.C., and A.J. Moore. 1995. Genetic aspects of communication during male-male competition in the Madagascar hissing cockroach: honest signalling of size. *Heredity.* 75:198–205.
Clark, J.R. 1999. Endangered and threatened wildlife and plants; final rule to list two cave animals from Kauai, Hawaii, as endangered. *Federal Register.* 65:2348–2357.
Cleveland, L.R. 1925. The effects of oxygenation and starvation on the symbiosis between the termite, *Termopsis,* and its intestinal flagellates. *Biological Bulletin.* 48:309–326.
Cleveland, L.R. 1965. Fertilization in *Trichonympha* from termites. *Archiv für Protistenkunde.* 108:1–5.
Cleveland, L.R., A.W.J. Burke, and P. Karlson. 1960. Ecdysone induced modifications in the sexual cycles of the protozoa of *Cryptocercus. Journal of Protozoology.* 7:229–239.
Cleveland, L.R., S.R. Hall, E.P. Sanders, and J. Collier. 1934. The wood-feeding roach *Cryptocercus,* its protozoa, and the symbiosis between protozoa and roach. *Memoirs of the American Academy of Arts and Sciences.* 17:185–342.
Cleveland, L.R., and W.L. Nutting. 1955. Suppression of sexual cycles and death of the protozoa of *Cryptocercus* resulting from change of hosts during molting period. *Journal of Experimental Zoology.* 130:485–513.
Cloarec, A., and C. Rivault. 1991. Age related changes in foraging in the German cockroach (Dictyoptera: Blattellidae). *Journal of Insect Behavior.* 4:661–673.
Cloudsley-Thompson, J.L. 1953. Studies in diurnal rhythms. III. Photoperiodism in the cockroach *Periplaneta americana* (L.). *Annals and Magazine of Natural History.* 6:705–712.
Cloudsley-Thompson, J.L. 1988. Evolution and Adaptation of Terrestrial Arthropods. Springer-Verlag, Berlin. 141 pp.
Clutton-Brock, T.H. 1991. The Evolution of Parental Care. Princeton University Press, Princeton. 352 pp.
Cocatre-Zilgein, J.H., and F. Delcomyn. 1990. Fast axon activity and the motor pattern in cockroach legs during swimming. *Physiological Entomology.* 15:385–392.
Cochran, D.G. 1979a. Comparative analysis of excreta and fat body from various cockroach species. *Comparative Biochemistry and Physiology.* 64A:1–4.
Cochran, D.G. 1979b. A genetic determination of insemination frequency and sperm precedence in the German cockroach. *Entomologia Experimentalis et Applicata.* 26:259–266.
Cochran, D.G. 1981. Comparative excreta analysis on various neotropical cockroaches and a leaf mantid. *Comparative Biochemistry and Physiology.* 70A:205–209.
Cochran, D.G. 1983a. Cockroaches—Biology and Control. World Health Organization, Geneva. 53 pp.
Cochran, D.G. 1983b. Food and water consumption during the reproductive cycle of female German cockroaches. *Entomologia Experimentalis et Applicata.* 34:51–57.
Cochran, D.G. 1985. Nitrogen excretion in cockroaches. *Annual Review of Entomology.* 30:29–49.
Cochran, D.G. 1986a. Biological parameters of reproduction in *Parcoblatta* cockroaches (Dictyoptera: Blattellidae). *Annals of the Entomological Society of America.* 79:861–864.
Cochran, D.G. 1986b. Feeding, drinking and urate excretory cycles in reproducing female *Parcoblatta* cockroaches. *Comparative Biochemistry and Physiology.* 84A:677–682.
Cochran, D.G., and D.E. Mullins. 1982. Physiological processes relating to nitrogen excretion in cockroaches. *Journal of Experimental Zoology.* 222:277–285.
Coelho, J.R., and A.J. Moore. 1989. Allometry of resting metabolic rate in cockroaches. *Comparative Biochemistry and Physiology.* 94A:587–590.
Cohen, A.C., and J.L. Cohen. 1976. Nest structure and micro-climate of the desert cockroach *Arenivaga apacha* (Polyphagidae, Dictyoptera). *Bulletin of the Southern California Academy of Sciences.* 75:273–277.
Cohen, A.C., and J.L. Cohen. 1981. Microclimate, temperature and water relations of two species of desert cockroaches. *Comparative Biochemistry and Physiology.* 69A:165–167.
Cohen, R.W. 2001. Diet balancing in the cockroach *Rhyparobia madera:* Does serotonin regulate this behavior? *Journal of Insect Behavior.* 14:99–111.
Cohen, R.W., S.L. Heydon, G.P. Waldbauer, and S. Friedman. 1987. Nutrient self-selection by the omnivorous cockroach *Supella longipalpa. Journal of Insect Physiology.* 33:77–82.
Cole, B. 1980. Growth rates in holometabolous and hemimetabolous insects. *Annals of the Entomological Society of America.* 73:489–491.
Coler, R.R., J.S. Elkinton, and R.G. Van Dreische. 1987. Density estimates and movement patterns of a population of *Periplaneta americana. Journal of the Kansas Entomological Society.* 60:389–396.
Coll, M., and M. Guershon. 2002. Omnivory in terrestrial arthropods: mixing plant and prey diets. *Annual Review of Entomology.* 47:267–297.
Collins, N.M. 1980. The distribution of soil macrofauna on the west ridge of Gunung (Mount) Mulu, Sarawak. *Oecologia.* 44:263–275.
Collins, N.M. 1989. Termites. *In* Tropical Rain Forest Ecosys-

tems. Vol. 14. Ecosystems of the World. B. H. Lieth and M.J.A. Werger, editors. Elsevier, Amsterdam. 455–471.

Conner, W.E., and M.N. Conner. 1992. Moths that go click in the night. *Wings.* 17:7–11.

Cooke, J.A.L. 1968. A further record of predation by cockroaches. *Entomologist's Monthly Magazine.* 104:72.

Cooper, R.A., and C. Schal. 1992. Differential development and reproduction of the German cockroach (Dictyoptera: Blattellidae) on three laboratory diets. *Journal of Economic Entomology.* 85:838–844.

Corbet, P.S. 1961. Entomological studies from a high tower in Mpanga forest, Uganda. XII. Observations on Ephemeroptera, Odonata, and some other orders. *Transactions of the Royal Entomological Society of London.* 113:356–368.

Cordero, C. 1995. Ejaculate substances that affect female reproductive physiology and behavior: honest or arbitrary traits? *Journal of Theoretical Biology.* 174:453–461.

Corley, L.S., J.R. Blankenship, and A.J. Moore. 2001. Genetic variation and asexual reproduction in the facultatively parthenogenetic cockroach *Nauphoeta cinerea:* implications for the evolution of sex. *Journal of Evolutionary Biology.* 14:68–74.

Corley, L.S., J.R. Blankenship, A.J. Moore, and P.J. Moore. 1999. Developmental constraints on the mode of reproduction in the facultatively parthenogenetic cockroach *Nauphoeta cinerea. Evolution & Development.* 1:90–99.

Corley, L.S., and A.J. Moore. 1999. Fitness of alternative modes of reproduction: developmental constraints and the evolutionary maintenance of sex. *Proceedings of the Royal Society of London B.* 266:471–476.

Cornwell, P.B. 1968. The Cockroach. Hutchinson and Co., Ltd., London. 391 pp.

Costerton, J.W. 1992. Pivotal role of biofilms in the focused attack of bacteria on insoluble substrates. *International Biodeterioration & Biodegradation.* 30:123–133.

Cott, H.B. 1940. Adaptive Coloration in Animals. Methuen, London. 508 pp.

Coxson, D.S., and N.M. Nadkarni. 1995. Ecological roles of epiphytes in nutrient cycles of forest ecosystems. *In* Forest Canopies. M.D. Lowman and N.M. Nadkarni, editors. Academic Press, San Diego. 495–543.

Crampton, G.C. 1932. A phylogenetic study of the head capsule in certain orthopteroid, psocoid, hemipteroid and holometabolous insects. *Bulletin of the Brooklyn Entomological Society.* 27:19–50.

Crawford, C.S., and J.L. Cloudsley-Thompson. 1971. Concealment behavior of nymphs of *Blaberus giganteus* L. (Dictyoptera: Blattaria) in relation to their ecology. *Revista Brasileira de Biologia.* 18:53–61.

Crawford, C.S., and E.C. Taylor. 1984. Decomposition in arid environments: role of the detritivore gut. *South African Journal of Science.* 80:170–176.

Creed, R.P.J., and J.R. Miller. 1990. Interpreting animal wall-following behavior. *Experientia.* 46:758–761.

Crespi, B.J., and C. Semeniuk. 2004. Parent-offspring conflict in the evolution of vertebrate reproductive mode. *The American Naturalist.* 163:635–653.

Crowell, H.H. 1946. Notes on an amphibious cockroach from the Republic of Panama. *Entomological News.* 57:171–172.

Cruden, D.L., and A.J. Markovetz. 1979. Carboxymethyl cellulose decomposition by intestinal bacteria of cockroaches. *Applied and Environmental Microbiology.* 38:369–372.

Cruden, D.L., and A.J. Markovetz. 1984. Microbial aspects of the cockroach hindgut. *Archives of Microbiology.* 138:131–139.

Cruden, D.L., and A.J. Markovetz. 1987. Microbial ecology of the cockroach gut. *Annual Review of Microbiology.* 41:617–643.

Culver, D.C. 1982. Cave Life, Evolution and Ecology. Harvard University Press, Cambridge, MA. 189 pp.

Culver, D.C., T.C. Kane, and D.W. Fong. 1995. Adaptation and natural selection in caves: the evolution of *Gammarus minus.* Harvard University Press, Cambridge, MA. 223 pp.

Cummins, K.W. 1974. Structure and function of stream ecosystems. *BioScience.* 24:631–641.

Curtis, C., I. Huber, and R.E. Calhoon. 2000. Fecundity of male *Blattella germanica* (Blattaria: Blattellidae) exposed to multiple virgin females. *Entomological News.* 11:371–374.

Cymorek, S. 1968. Adaptations in wood-boring insects: examples of morphological, anatomical, physiological and behavioral features. *In* Record of the 1968 Annual Convention of the British Wood Preserving Association, Cambridge. 161–180.

D.W. 1984. A certain death (photograph by R. A. Mendez). *Natural History.* 10:118–119.

Daan, S., and J.M. Tinbergen. 1997. Adaptation of life histories. *In* Behavioural Ecology, an Evolutionary Approach. J.R. Krebs and N.B. Davies, editors. Blackwell Science, Oxford. 311–333.

Daily, G.C., P.A. Matson, and P.M. Vitousek. 1997. Ecosystem services supplied by soil. *In* Nature's Services: Societal Dependence on Natural Ecosystems. G.C. Daily, editor. Island Press, Washington, DC. 113–132.

Dambach, M., and B. Goehlen. 1999. Aggregation density and longevity correlate with humidity in first instar nymphs of the cockroach (*Blattella germanica* L., Dictyoptera). *Journal of Insect Physiology.* 45:423–429.

Dambach, M., A. Stadler, and J. Heidelbach. 1995. Development of aggregation behavior in the German cockroach, *Blattella germanica* (Dictyoptera: Blattellidae). *Entomologia Generalis.* 19:129–141.

Danks, H.V. 1981. Arctic Arthropods: A Review of Systematics and Ecology with Particular Reference to the North American Fauna. Entomological Society of Canada, Ottawa. 608 pp.

Darlington, J.P.E.C. 1968. Biogenetics of *Eublaberus posticus.* Master's thesis, University of the West Indies, Trinidad.

Darlington, J.P.E.C. 1970. Studies on the ecology of the Tamana Caves with special reference to cave dwelling cockroaches. Ph.D. thesis, University of the West Indies, Trinidad. 224 pp.

Darlington, J.P.E.C. 1995a. Ecology and fauna of the Tamana Caves, Trinidad, West Indies. *Studies in Speleology.* 10:37–50.

Darlington, J.P.E.C. 1995b. A review of current knowledge about the Oropouche or Cumaca cave, Trinidad, West Indies. *Studies in Speleology.* 10:65–74.

Darlington, J.P.E.C. 1995–1996. Guanapo Cave. *Living World Journal of the Trinidad & Tobago Field Naturalists' Club.* 15–16.
Darlington, P.J.J. 1943. Carabidae of mountains and islands: data on the evolution of isolated faunas, and on atrophy of wings. *Ecological Monographs.* 13:37–61.
Darwin, C. 1859. The Origin of Species by Means of Natural Selection. John Murray, London. 502 pp.
Davey, K.G. 1960. A pharmacologically active agent in the reproductive system of insects. *Canadian Journal of Zoology.* 38:39–45.
Davey, K.G. 1965. Reproduction in the Insects. Oliver & Boyd, Edinburgh. 96 pp.
Davidson, D.W., S.C. Cook, R.R. Snelling, and T.H. Chua. 2003. Explaining the abundance of ants in lowland tropical forest canopies. *Science.* 300:969–972.
Day, M.F. 1950. The histology of a very large insect, *Macropanesthia rhinoceros* Sauss. (Blattidae). *Australian Journal of Scientific Research, Series B.* 3:61–75.
Dean, W.R.J., and J.B. Williams. 1999. Sunning behaviour and its possible influence on digestion in the whitebacked moosebird *Colius colius. The Ostrich.* 70:239–241.
Deans, A.R., and L.M. Roth. 2003. *Nyctibora acaciana* (Blattellidae: Nyctiborinae), a new species of cockroach from Central America that oviposits on ant-acacias. *Transactions of the American Entomological Society.* 129:267–283.
Deharveng, L., and A. Bedos. 2000. The cave fauna of Southeast Asia, origin, evolution and ecology. *In* Ecosystems of the World. Vol. 30: Subterranean Ecosystems. H. Wilkens, D.C. Culver, and W.F. Humphreys, editors. Elsevier, Amsterdam. 603–632.
Deitz, L.L., C.A. Nalepa, and K.D. Klass. 2003. Phylogeny of the Dictyoptera reexamined. *Entomologische Abhandlungen.* 61:69–91.
Dejean, A., and I. Olmsted. 1997. Ecological studies on *Aechmea bracteata* (Swartz) (Bromeliaceae). *Journal of Natural History.* 31:1313–1334.
Delcomyn, F. 1971. The locomotion of the cockroach *Periplaneta americana. Journal of Experimental Biology.* 54:443–452.
Deleporte, P. 1976. L'organization sociale chez *Periplaneta americana* (Dictyoptères). Aspects éco-éthologiques—Ontogenèse des relations inter-individuelles. Thése doctorat, 3e cycle, L'Université de Rennes.
Deleporte, P. 1985. Structure sociale et occupation de l'espace par la blatte *Periplaneta americana* (Dictyoptères). *Bulletin de la Societe Zoologique de France. Evolution et Zoologie.* 110:325–330.
Deleporte, P. 1988. Etude eco-ethologique et evolutive de *P. americana* et d'autres blattes sociales. Le Grade de Docteur d'Etat, U.F.R. Sciences de la Vie et de l'Environment. L'Universite de Rennes. 212 pp.
Deleporte, P., A. Dejean, P. Grandcolas, and R. Pellens. 2002. Relationships between the parthenogenic cockroach *Pycoscelus surinamensis* (Dictyoptera: Blaberidae) and ants (Hymenoptera: Formicidae). *Sociobiology.* 39:259–267.
Deleporte, P., D. Lebrun, and A. Lequet. 1988. Le gesier ou proventricule de *Cryptocercus punctulatus* Scudder et la phylogenie des Blattaria. *Actes Colloque Insectes Sociaux.* 4:353–358.
Demark, J.J., and L.P. Bennett. 1994. Diel activity cycles in nymphal stadia of the German cockroach (Dictyoptera: Blattellidae). *Journal of Economic Entomology.* 87:941–950.
Deneubourg, J.-L., A. Lioni, and C. Detrain. 2002. Dynamics of aggregation and emergence of cooperation. *Biological Bulletin.* 202:262–267.
Denic, N., D.W. Huyer, S.H. Sinal, P.E. Lantz, C.R. Smith, and M.M. Silver. 1997. Cockroach: the omnivorous scavenger. Potential misinterpretation of postmortem injuries. *American Journal of Forensic Medicine and Pathology.* 18:177–180.
Denno, R.F., C. Gratton, and G.A. Langellotto. 2001a. Significance of habitat persistence and dimensionality in the evolution of insect migration strategies. *In* Insect Movement: Mechanisms and Consequences. I. Woiwood, D.R. Reynolds, and C. Thomas, editors. CABI Publishing, London. 235–260.
Denno, R.F., D.J. Hawthorne, B.L. Thorne, and C. Gratton. 2001b. Reduced flight capability in British Virgin Island populations of a wing dimorphic insect: the role of habitat isolation, persistence and structure. *Ecological Entomology.* 26:25–36.
Denno, R.F., G.K. Roderick, K.L. Olmstead, and H.G. Dobel. 1991. Density-related migration in planthoppers (Homoptera: Delphacidae): the role of habitat persistence. *The American Naturalist.* 138:1513–1541.
Denzer, V.D.J., M.E.A. Fuchs, and G. Stein. 1988. Zum verhalten von *Blattella germanica* L.: aktionsradius und refugientreue. *Journal of Applied Entomology.* 105:330–343.
Deyrup, M., and F.W. Fisk. 1984. A myrmecophilous cockroach new to the United States (Blattaria: Polyphagidae). *Entomological News.* 95:183–185.
Dhanarajan, G. 1978. Cannibalism and necrophagy in a subterranean termite (*Reticulitermes lucifugus* var. *santonensis*). *The Malayan Nature Journal.* 31:237–251.
Dial, R., and J. Roughgarden. 1995. Experimental removal of insectivores from rain forest canopy: direct and indirect effects. *Ecology.* 76:1821–1834.
Diekman, L.J., and R.E. Ritzman. 1987. The effect of temperature on flight initiation in the cockroach *Periplaneta americana. Journal of Neurobiology.* 18:487–496.
Dillon, R.J., and A.K. Charnley. 1986. Invasion of the pathogenic fungus *Metarhizium anisopliae* through the guts of germfree desert locusts *Schistocerca gregaria. Mycopathlogia.* 96:59–66.
Dillon, R.J., and A.K. Charnley. 1995. Chemical barriers to gut infection in the desert locust: In vivo production of antimicrobial phenols associated with the bacterium *Pantoea agglomerans. Journal of Invertebrate Pathology.* 66:72–75.
Dillon, R.J., C.T. Vennard, and A.K. Charnley. 2000. Exploitation of gut bacteria in the locust. *Nature.* 403:851.
Dingle, H. 1996. Migration: The Biology of Life on the Move. Oxford University Press, New York. 474 pp.
Dix, N.J., and J. Webster. 1995. Fungal Ecology. Chapman and Hall, London. 549 pp.
Dong, Q., and G. Polis. 1992. The dynamics of cannibalistic populations: a foraging perspective. *In* Cannibalism: Ecology and Evolution Among Diverse Taxa. M.A. Elgar and

B.J. Crespi, editors. Oxford University Press, Oxford. 13–37.
Dow, J.A. 1986. Insect midgut function. *Advances in Insect Physiology.* 19:187–328.
Downer, R.G.H. 1982. Fat body and metabolism. *In* The American Cockroach. W.J. Bell and K.G. Adiyodi, editors. Chapman and Hall, London. 151–174.
Dozier, H.L. 1920. An ecological study of hammock and piney woods insects in Florida. *Annals of the Entomological Society of America.* 13:325–380.
Draser, B.S., and P.A. Barrow. 1985. Intestinal Microbiology. American Society of Microbiology, Washington, DC.
Dreisig, H. 1971. Diurnal activity in the dusky cockroach *Ectobius laponicus* L. (Blattodea). *Entomologia Scandinavica.* 2:132–138.
Dudek, D.M., and R.J. Full. 2000. Spring-like behavior of the legs of running insects. *American Zoologist.* 40:1002–1003.
Duffy, S.S. 1976. Arthropod allomones: chemical effronteries and antagonists. *In* Proceedings of the XV International Congress of Entomology. J.S. Packer and D. White, editors. Entomological Society of America, Washington, DC. 323–394.
Duman, J.G. 1979. Thermal-hysteresis factors in overwintering insects. *Journal of Insect Physiology.* 25:805–810.
Dunbar, R.I.M. 1979. Population demography, social organization, and mating strategies. *In* Primate Ecology and Human Origins: Ecological Influences on Social Organization. I.S. Bernstein and E.O. Smith, editors. Garland STPM Press, New York. 65–88.
Durbin, E.J., and D.G. Cochran. 1985. Food and water deprivation effects on reproduction in female *Blattella germanica. Entomologia Experimentalis et Applicata.* 37:77–82.
Durden, C.J. 1972. Systematics and morphology of Acadian Pennsylvanian blattoid insects (Dictyoptera: Palaeoblattina): a contribution to the classification and phylogeny of Palaeozoic insects. Ph.D. thesis, Yale University, New Haven.
Durden, C.J. 1988. Hamilton insect fauna. Kansas Geological Survey Guidebook Series 6:117–124.
Durier, V., and C. Rivault. 2001a. Effects of spatial knowledge and feeding experience on foraging choices in German cockroaches. *Animal Behaviour.* 62:681–688.
Durier, V., and C. Rivault. 2001b. Spatial knowledge of what type of food to find in a particular feeding site. *In* XXVII International Ethological Conference. Advances in Ethology. Vol. 36. R. Apfelbach, M. Fendt, S. Krämer, and B.M. Siemers, editors. Blackwell Wissenschafts-Verlag, Tübingen. 146.
Durier, V., and C. Rivault. 2003. Exploitation of home range and spatial distribution of resources in German cockroaches (Dictyoptera: Blattellidae). *Journal of Economic Entomology.* 96:1832–1837.
Duwel-Eby, L.E., L.M. Faulhaber, and R.D. Karp. 1991. Adaptive humoral immunity in the American cockroach. *In* Immunology of Insects and Other Arthropods. A.P. Gupta, editor. CRC Press, Boca Raton. 385–402.
Eads, R.B., F.J. Von Zuben, S.E. Bennett, and O.L. Walker. 1954. Studies on cockroaches in a municipal sewage system. *American Journal of Tropical Medicine and Hygiene.* 3:1092–1098.
Eaton, R.A., and M.D.C. Hale. 1993. Wood: Decay, Pests and Protection. Chapman and Hall, London. 546 pp.
Eberhard, W.G. 1985. Sexual Selection and Animal Genitalia. Harvard University Press, Cambridge, MA. 244 pp.
Eberhard, W.G. 1991. Copulatory courtship and cryptic female choice in insects. *Biological Reviews.* 66:1–31.
Eberhard, W.G. 1994. Evidence for widespread courtship during copulation in 131 insects and spiders, and implications for cryptic female choice. *Evolution.* 48:711–733.
Eberhard, W.G. 1996. Female Control: Sexual Selection by Crypic Female Choice. Princeton University Press, Princeton. 501 pp.
Eberhard, W.G. 2001. Multiple origins of a major novelty: movable abdominal lobes in male sepsid flies (Diptera: Sepsidae), and the question of developmental constraints. *Evolution & Development.* 3:206–222.
Edmunds, L.R. 1952. Some notes on the habits and parasites of native wood-roaches. *Entomological News.* 63:141–145.
Edmunds, L.R. 1957. Observations on the biology and life history of the brown cockroach *Periplaneta brunnea* Burmeister. *Proceedings of the Entomological Society of Washington.* 59:283–286.
Edmunds, M., and D. Brunner. 1999. Ethology of defenses against predators. *In* The Praying Mantids. F.R. Prete, H. Wells, P.H. Wells, and L.E. Hurd, editors. Johns Hopkins University Press, Baltimore. 276–299.
Edney, E.B. 1966. Absorption of water vapour from unsaturated air by *Arenivaga* sp. (Polyphagidae, Dictyoptera). *Comparative Biochemistry and Physiology.* 19:387–408.
Edney, E.B. 1967. Water balance in desert arthropods. *Science.* 156:1059–1066.
Edney, E.B. 1977. Water Balance in Land Arthropods. Springer-Verlag, Berlin. 282 pp.
Edney, E.B., P. Franco, and R. Wood. 1978. The responses of *Arenivaga investigata* (Dictyoptera) to gradients of temperature and humidity in sand studied by tagging with technetium 99m. *Physiological Zoology.* 51:241–255.
Edney, E.B., S. Haynes, and D. Gibo. 1974. Distribution and activity of the desert cockroach *Arenivaga investigata* (Polyphagidae) in relation to microclimate. *Ecology.* 55:420–427.
Edwards, J.P. 2004. Interactions between an endangered bird species, non-endemic insect pests, and insecticides: the deployment of insect growth regulators in the conservation of the Seychelles magpie-robin (*Copsychus sechellarum*). *In* Insect and Bird Interactions. H.F. Van Emden and M. Rothschild, editors. Intercept, Andover, Hampshire, UK. 121–145.
Eggleton, P., R. Homathevi, D.T. Jones, J.A. MacDonald, D. Jeeva, D.E. Bignell, R.G. Davies, and M. Maryati. 1999. Termite assemblages, forest disturbance and greenhouse gas fluxes in Sabah, East Malaysia. *Philosophical Transactions of the Royal Society of London B.* 354:1791–1802.
Ehrlich, H. 1943. Verhaltensstudien an der Schabe *Periplaneta americana* L. *Zeitschrift für Tierpsychologie.* 5:497–552.
Eickwort, G.C. 1981. Presocial insects. *In* Social Insects. Vol. 1. H.R. Hermann, editor. Academic Press, New York. 199–280.
Eisenbeis, G., and W. Wichard. 1985. Atlas on the Biology of Soil Arthropods. Springer-Verlag, Berlin. 437 pp.

Eisner, T. 1958. Spray mechanism of the cockroach *Diploptera punctata. Science.* 128:148–149.
El-Ayouty, E.Y., S.I. Ghabbour, and A.M. El-Sayyed. 1978. Role of litter and excreta of soil fauna in the nitrogen status of desert soils. *Journal of Arid Environments.* 1:145–155.
Elgar, M.A., and B.J. Crespi. 1992. Ecology and evolution of cannibalism. *In* Cannibalism: Ecology and Evolution Among Diverse Taxa. M.A. Elgar and B.J. Crespi, editors. Oxford University Press, Oxford. 1–12.
Endler, J.A. 1978. A predator's view of animal color patterns. *Evolutionary Biology.* 11:319–364.
Engelmann, F. 1957. Bau und Funktion des weiblichen Geschlechtsapparates bei der ovoviviparen Schabe *Leucophaea maderae* (Fabr.) und einige Beobachtungen über die Entwicklung. *Biologisches Zentralblatt.* 76:722–740.
Engelmann, F. 1959. The control of reproduction in *Diploptera punctata* (Blattaria). *Biological Bulletin.* 116:406–419.
Engelmann, F. 1960. Mechanisms controlling reproduction in two viviparous cockroaches. *Annals of the New York Academy of Sciences.* 89:516–536.
Engelmann, F. 1970. The Physiology of Insect Reproduction. Pergamon Press, New York. 307 pp.
Engelmann, F., and I. Rau. 1965. A correlation between the feeding and the sexual cycle in *Leucophaea maderae* (Blattaria). *Journal of Insect Physiology.* 11:53–64.
Ettema, C.H., and D.A. Wardle. 2002. Spatial soil ecology. *Trends in Ecology and Evolution.* 17:177–183.
Evans, H.E. 1968. Life on a Little Known Planet. Dutton, New York. 318 pp.
Evans, L.D., and M.D. Breed. 1984. Segregation of cockroach nymphs into sibling groups. *Annals of the Entomological Society of America.* 77:574–577.
Evans, M.E.G. 1990. Habits or habitats: do carabid locomotor adaptations reflect habitats or lifestyles? *In* The Role of Ground Beetles in Ecological and Environmental Studies. N.E. Stork, editor. Intercept Ltd, Andover, UK. 295–305.
Evans, M.E.G., and T.G. Forsythe. 1984. Comparison of adaptations to running, pushing and burrowing in some adult Coleoptera: especially Carabidae. *Journal of Zoology, London.* 202:513–534.
Evans, T.A., J.C.S. Lai, E. Toledano, L. McDowell, S. Rakotonarivo, and M. Lenz. 2005. Termites assess wood size by using vibration signals. *Proceedings of the National Academy of Sciences.* 102:3732–3737.
Everaerts, C., J.P. Farine, and R. Brossut. 1997. Changes of species specific cuticular hydrocarbon profiles in the cockroaches *Nauphoeta cinerea* and *Leucophaea maderae* reared in heterospecific groups. *Entomologia Experimentalis et Applicata.* 85:145–150.
Ewald, P.W. 1987. Transmission modes and evolution of the parasitism-mutualism continuum. *Endocytobiology III. Annals of the New York Academy of Sciences.* 503:295–306.
Ewing, L.S. 1967. Fighting and death from stress in a cockroach. *Science.* 155:1035–1036.
Ewing, L.S. 1972. Hierarchy and its relation to territory in the cockroach *Nauphoeta cinerea. Behaviour.* 42:152–174.
Failla, M.C., and A. Messina. 1987. Contribution of Blattaria to the biogeography of the Mediterranean area. *In* Evolutionary Biology of Orthopteroid Insects. B. Baccetti, editor. Ellis Horwood Limited, Chichester. 195–207.
Fairbairn, D.J. 1997. Allometry of sexual size dimorphism: pattern and process in the coevolution of body size in males and females. *Annual Review of Ecology and Systematics.* 28:659–687.
Farine, J.-P., R. Brossut, and C. A. Nalepa. 1989. Morphology of the male and female tergal glands of the woodroach *Cryptocercus punctulatus* (Insecta, Dictyoptera). *Zoomorphology.* 109:153–164.
Farine, J.-P., C. Everaerts, J.-L. LeQuere, E. Semon, R. Henry, and R. Brossut. 1997. The defensive secretion of *Eurycotis floridana* (Dictyoptera, Blattidae, Polyzosteriinae): chemical identification and evidence of an alarm function. *Insect Biochemistry and Molecular Biology.* 27:577–586.
Farine, J.-P., E. Semon, C. Everaerts, D. Abed, P. Grandcolas, and R. Brossut. 2002. Defensive secretions of *Therea petiveriana:* chemical identification and evidence of an alarm function. *Journal of Chemical Ecology.* 28:1629–1640.
Farine, J.-P., L. Sreng, and R. Brossut. 1981. L'interattraction chez *Nauphoeta cinerea* (Insecta, Dictyoptera): mise en évidence et étude préliminaire de la phéromone grégaire. *Comptes rendus de l'Academie des Sciences, Ser. III.* 292:781–784.
Farnsworth, E.G. 1972. Effects of ambient temperature, humidity, and age on wing-beat frequency of *Periplaneta* species. *Journal of Insect Physiology.* 18:827–839.
Faulde, M., M.E.A. Fuchs, and W. Nagl. 1990. Further characterization of a dispersion inducing contact pheromone in the saliva of the German cockroach *Blattella germanica* L. Blattodea Blattellidae. *Journal of Insect Physiology.* 36:353–360.
Faulhaber, L.M., and R.D. Karp. 1992. A diphasic immune response against bacteria in the American cockroach. *Immunology.* 75:378–381.
Fedorka, K.M., and T.A. Mousseau. 2002. Nuptial gifts and the evolution of male body size. *Evolution.* 56:590–596.
Feinberg, L., J. Jorgensen, A. Haselton, A. Pitt, R. Rudner, and L. Margulis. 1999. *Arthromitus* (*Bacillus cereus*) symbionts in the cockroach *Blaberus giganteus:* dietary influences on bacterial development and population density. *Symbiosis.* 27:109–123.
Fernando, W. 1957. New species of insects from Ceylon (1). *Ceylon Journal of Biological Science.* 1:7–18.
Fischer, K., and K. Fiedler. 2002. Reaction norms for age and size at maturity in response to temperature: a test of the compound interest hypothesis. *Evolutionary Ecology.* 16:333–349.
Fisk, F.W. 1977. Notes on cockroaches (Blattaria) from caves in Chiapas, Mexico and environs with descriptions of three new species. *Accademia Nazionale Dei Lincei.* 171:267–274.
Fisk, F.W. 1982. Key to the cockroaches of Central Panama. Part II. Flightless species. *Studies on Neotropical Fauna and Environment.* 17:123–127.
Fisk, F.W. 1983. Abundance and diversity of arboreal Blattaria in moist tropical forests of the Panama Canal area

and Costa Rica. *Transactions of the American Entomological Society.* 108:479–489.
Fisk, F.W., and C. Schal. 1981. Notes on new species of Epilamprine cockroaches from Costa Rica and Panama (Blattaria: Blaberidae). *Proceedings of the Entomological Society of Washington.* 83:694–706.
Fisk, F.W., M.V. Vargas, and F.B. Fallas. 1976. Notes on *Myrmecoblatta wheeleri* from Costa Rica (Blattaria: Polyphagidae). *Proceedings of the Entomological Society of Washington.* 78:317–322.
Fisk, F.W., and H. Wolda. 1979. Key to the cockroaches of Central Panama. Part I. Flying species. *Studies on Neotropical Fauna and Environment.* 14:177–201.
Fittkau, E.J., and H. Klinge. 1973. On biomass and trophic structure of the Central Amazonian rainforest ecosystem. *Biotropica.* 5:2–14.
Flock, R.A. 1941. The field roach *Blattella vaga. Journal of Economic Entomology.* 34:121.
Floren, A., and K.E. Linsenmair. 1997. Diversity and recolonization dynamics of selected arthropod species in a lowland forest in Sabah, Malaysia with special reference to Formicidae. *In* Canopy Arthropods. N.E. Stork, J. Adis, and R.K. Didham, editors. Chapman and Hall, London. 344–381.
Foerster, C.H. 2004. The circadian clock in the brain: a structural and functional comparison between mammals and insects. *Journal of Comparative Physiology A.* 190:601–613.
Fotedar, R., U.B. Shriniwas, and A. Verma. 1991. Cockroaches (*Blattella germanica*) as carriers of microorganisms of medical importance in hospitals. *Epidemiology and Infection.* 197:181–188.
Fraser, J., and M.C. Nelson. 1982. Frequency modulated courtship song in a cockroach. *Animal Behaviour.* 30:627–628.
Fraser, J., and M.C. Nelson. 1984. Communication in the courtship of a Madagascar hissing cockroach *Gromphadorhina portentosa.* 1. Normal courtship. *Animal Behaviour.* 32:194–203.
Friauf, J.J. 1953. An ecological study of the Dermaptera and Orthoptera of the Welaka area in northern Florida. *Ecological Monographs.* 23:79–126.
Friauf, J.J., and E.B. Edney. 1969. A new species of *Arenivaga* from desert sand dunes in southern California. *Proceedings of the Entomological Society of Washington.* 71:1–7.
Froggatt, W.W. 1906. Domestic insects: cockroaches (Blattidae). *Agricultural Gazette of New South Wales.* 2 May:440–447.
Full, R.J., K. Autumn, J.I. Chung, and A. Ahn. 1998. Rapid negotiation of rough terrain by the death-head cockroach. *American Zoologist.* 38(5):81A.
Full, R.J., and M.S. Tu. 1991. Mechanics of a rapid running insect: two-, four- and six-legged locomotion. *Journal of Experimental Biology.* 156:215–231.
Full, R.J., and A. Tullis. 1990. Capacity for sustained terrestrial locomotion in an insect: energetics, thermal dependence, and kinematics. *Journal of Comparative Physiology B.* 160:573–581.
Full, R.J., A. Yamauchi, and D.L. Jindrich. 1995. Maximum single leg force production: cockroaches righting on photoelastic gelatin. *Journal of Experimental Biology.* 198:2441–2452.
Futuyma, D.J. 1986. Evolutionary Biology. Sinauer Associates, Inc., Sunderland, MA. 600 pp.
Gadd, C.A., and D. Raubenheimer. 2000. Nutrient specific learning in an omnivorous insect: the American cockroach *Periplaneta americana* L. learns to associate dietary protein with the odors citral and carvone. *Journal of Insect Behavior.* 13:851–864.
Gade, B., and E.D.J. Parker. 1997. The effect of life cycle stage and genotype on desiccation tolerance in the colonizing parthogenetic cockroach *Pycnoscelus surinamensis* and its sexual ancestor *P. indicus. Journal of Evolutionary Biology.* 10:479–493.
Gadot, M., E. Burns, and C. Schal. 1989. Juvenile hormone biosynthesis and oocyte development in adult female *Blattella germanica:* effects of grouping and mating. *Archives of Insect Biochemistry and Physiology.* 11:189–200.
Gagné, W.C. 1979. Canopy-associated arthropods in *Acacia koa* and *Metrosideros* tree communities along an altitudinal transect on Hawaii island. *Pacific Insects.* 21:56–82.
Gaim, W., and G. Seelinger. 1984. Zu Oekologie und Verhalten der mitteleuropaeischen Schabe *Phyllodromica maculata* (Dictyoptera: Blattellidae). *Entomologia Generalis.* 9:135–142.
Galef, B.G.J. 1988. Imitation in animals: history, definition, and interpretation of data from the psychological laboratory. *In* Social Learning: Psychological and Biological Perspectives. T.R. Zentall and B.G.J. Galef, editors. Lawrence Erlbaum Associates, Hillsdale, NJ. 3–28.
Gariépy, J.-L., D.L. Bauer, and R.B. Cairns. 2001. Selective breeding for differential aggression in mice provides evidence for heterochrony in social behaviors. *Animal Behaviour.* 61:933–947.
Garnier, S., C. Jost, R. Jeanson, J. Gautrais, M. Asadpour, G. Caprari, and G. Theraulaz. 2005. Aggregation behavior as a source of collective decision in a group of cockroach-like robots. *In* Advances in Artificial Life, 8th European Conference on Artificial Life. S. Mathieu, S, Capcarrere, A.A. Freitas, P.J. Bentley, C.G. Johnson, and J. Timmis, editors. Springer, Berlin. 169–178.
Garthe, W.A., and M.W. Elliot. 1971. Role of intracellular symbionts in the fat body of cockroaches: influence on hemolymph proteins. *Experientia.* 27:593.
Gary, L. 1950. Controlling sewer insects and sewer odors. *Public Works.* 81:48–52.
Gates, M.F., and W.C. Allee. 1933. Conditioned behavior of isolated and grouped cockroaches on a simple maze. *Journal of Comparative Psychology.* 15:331–358.
Gautier, J.-Y. 1967. Immobilisation refléx liée à des excutations tactiles du pronotum chez les larves de *Blabera craniifer* (Burm.) normales ou recevant des implantations de corps allates. *Comptes rendus de l'Academie des Sciences, Paris.* 264:1319–1322.
Gautier, J.-Y. 1974a. Etude comparée de la distribution spatiale et temporelle des adultes de *Blaberus atropos* et *B. colosseus* (Dictyoptéres) dans cinq grottes de l'ile de Trinidad. *Revue du Comportement de Animale.* 9:237–258.
Gautier, J.-Y. 1974b. Processus de differenciation de l'organi-

zation sociale chez quelques especes de Blattes du genre *Blaberus:* aspects écologiques et éthologiques. Thèse de doctorat d'etat, L'Université de Rennes.
Gautier, J.-Y. 1980. Distribution spatiale et organisation sociale chez *Gyna maculipennis* (Insecte Dictyoptère) dans les cavernes et galeries de mines de la région de Belinga au Gabon. *Acta Oecologica Generalis* 1:347–358.
Gautier, J.-Y., and P. Deleporte. 1986. Behavioural ecology of a forest living cockroach, *Lamproblatta albipalpus* in French Guyana. *In* Behavioral Ecology and Population Biology. L.C. Drickamer, editor. Privat, I.E.C., Toulouse. 17–22.
Gautier, J.-Y., P. Deleporte, and C. Rivault. 1988. Relationships between ecology and social behavior in cockroaches. *In* The Ecology of Social Behavior. C.N. Slobodchikoff, editor. Academic Press, San Diego. 335–351.
Geissler, T.G., and C.D. Rollo. 1987. The influence of nutritional history on the response to novel food by the cockroach *Periplaneta americana. Animal Behaviour.* 35:1905–1907.
Gemeno, C., and C. Schal. 2004. Sex pheromones of cockroaches. *In* Advances in Insect Chemical Ecology. R.T. Cardé and J.G. Millar, editors. Cambridge University Press, New York. 179–247.
Gemeno, C., K. Snook, N. Benda, and C. Schal. 2003. Behavioral and electrophysiological evidence for volatile sex pheromones in *Parcoblatta* wood cockroaches. *Journal of Chemical Ecology.* 29:37–54.
Gentry, A.H., and C.H. Dodson. 1987. Diversity and biogeography of neotropical vascular epiphytes. *Annals of the Missouri Botanical Garden.* 74:205–233.
Ghabbour, S.I., W. Mikhaïl, and M.A. Rizk. 1977. Ecology of soil fauna of Mediterranean desert ecosystems in Egypt. I. Summer populations of soil mesofauna associated with major shrubs in the littoral sand dunes. *Revue d' Écologie et de Biologie du Sol.* 14:429–459.
Ghabbour, S.I., and W.Z.A. Mikhaïl. 1978. Ecology of soil fauna of Mediterranean desert ecosystems in Egypt. II. Soil mesofauna associated with *Thymelaea hirsuta. Revue d' Écologie et de Biologie du Sol.* 15:333–339.
Ghabbour, S.I., and S.H. Shakir. 1980. Ecology of soil fauna of Mediterranean desert ecosystems in Egypt. III. Analysis of *Thymelaea* mesofauna populations at the Mariut frontal plain. *Revue d' Écologie et de Biologie du Sol.* 17:327–352.
Gier, H.T. 1947. Growth rate in the cockroach *Periplaneta americana* (Linn.). *Annals of the Entomological Society of America.* 40:303–317.
Gijzen, H.J., and M. Barugahare. 1992. Contribution of anaerobic protozoa and methanogens to hindgut metabolic activities of the American cockroach, *Periplaneta americana. Applied and Environmental Microbiology.* 58:2565–2570.
Gijzen, H.J., C.A.M. Broers, M. Barughare, and C.K. Strumm. 1991. Methanogenic bacteria as endosymbionts of the ciliate *Nyctotherus ovalis* in the cockroach hindgut. *Applied and Environmental Microbiology.* 57:1630–1634.
Gijzen, H.J., C. van den Drift, M. Barugahare, and H.J.M. op den Camp. 1994. Effect of host diet and hindgut microbial composition on cellulolytic activity in the hindgut of the American cockroach, *Periplaneta americana. Applied and Environmental Microbiology.* 60:1822–1826.
Gilbert, J., and L. Deharveng. 2002. Subterranean ecosystems: a truncated functional biodiversity. *BioScience.* 52:473–481.
Gilbert, S.F., and J.A. Bolker. 2003. Ecological developmental biology: preface to the symposium. *Evolution & Development.* 5:3–8.
Gillott, C. 1983. 12. Arthropoda—Insecta. *In* Reproductive Biology of Invertebrates. III. Accessory Sex Glands. K.G. Adiyodi and R.G. Adiyodi, editors. John Wiley & Sons, Chichester. 319–471.
Gillott, C. 2003. Male accessory gland secretions: modulators of female reproductive physiology and behavior. *Annual Review of Entomology.* 48:163–184.
Giraldeau, L.A., and T. Caraco. 1993. Genetic relatedness and group size in an aggregation economy. *Evolutionary Ecology.* 7:429–438.
Gnaspini, P., and E. Trajano. 2000. Guano communities in tropical caves. *In* Ecosystems of the World. Vol. 30: Subterranean Ecosystems. H. Wilkens, D.C. Culver, and W.F. Humphreys, editors. Elsevier, Amsterdam. 251–268.
Goodman, S.M., and J.P. Benstead. 2003. The Natural History of Madagascar. University of Chicago Press, Chicago. 1709 pp.
Goodwin, N.B., N.K. Dulvey, and J.D. Reynolds. 2002. Life-history correlates of the evolution of live bearing in fishes. *Philosophical Transactions of the Royal Society of London B.* 357:259–267.
Gorb, S. 2001. Attachment devices of insect cuticle. Kluwer Academic Publishers, Dordrecht. 305 pp.
Gordner, P.F. 2001. Largest fossil cockroach found; site preserves incredible detail. http://researchnews.osu.edu/archive/bigroach.htm.
Gordon, D.G. 1996. The Compleat Cockroach. Ten Speed Press, Berkeley, CA. 178 pp.
Gordon, H.T. 1959. Minimal nutritional requirements of the German roach *Blattella germanica* L. *Annals of the New York Academy of Science.* 77:290–351.
Gordon, J.M., P.A. Zungoli, and L.W. Grimes. 1994. Population density effect on oviposition behavior in *Periplaneta fuliginosa* (Dictyoptera: Blattidae). *Annals of the Entomological Society of America.* 87:436–439.
Gorton, R.E.J. 1979. Agonism as a function of relationship in a cockroach *Shawella couloniana* (Dictyoptera: Blattellidae). *Journal of the Kansas Entomological Society.* 52:438–442.
Gorton, R.E.J. 1980. A comparative ecological study of the wood cockroaches in Northeastern Kansas. *The University of Kansas Science Bulletin.* 52:21–30.
Gorton, R.E.J., K.G. Colliander, and W.J. Bell. 1983. Social behavior as a function of context in a cockroach. *Animal Behaviour.* 31:152–159.
Goudey-Perriere, F., J.C. Baehr, and P. Brousse-Gaury. 1989. Relationship between haemolymphatic levels of juvenile hormone and the duration of the spermatophore in the bursa copulatrix of the cockroach *Blaberus craniifer* Burm. *Reproduction, Nutrition, Development.* 29:317–323.
Goudey-Perriere, F., H. Barreteau, C. Jacquot, P. Gayral, C. Perriere, and P. Brousse-Gaury. 1992. Influence of crowd-

ing on biogenic amine levels in the nervous system of the female cockroach *Blaberus craniifer* Burm. (Dictyoptera: Blaberidae). *Comparative Biochemistry and Physiology.* 103C:215–220.

Gould, G.E., and H.O. Deay. 1938. The biology of the American cockroach. *Annals of the Entomological Society of America.* 31:489–498.

Gould, S.J. 1977. Ontogeny and Phylogeny. Harvard University Press, Cambridge, MA. 501 pp.

Grandcolas, P. 1991. Les Blattes de Guyane Française: Structure du peuplement et étude éco-éthologique des Zetoborinae. Ph.D. thesis, L'Université de Rennes. 295 pp.

Grandcolas, P. 1993a. Habitats of solitary and gregarious species in the neotropical Zetoborinae (Insecta, Blattaria). *Studies on Neotropical Fauna and Environment.* 28:179–190.

Grandcolas, P. 1993b. Le genre *Paramuzoa* Roth, 1973: sa répartition et un cas de xylophagie chez les Nyctiborinae (Dictyoptera, Blattaria). *Bulletin de la Société Entomologique de France.* 98:131–138.

Grandcolas, P. 1994a. Evidence for hypopharynx protrusion and presumptive water vapour absorption in *Heterogamisca chopardi* Uvarov, 1936 (Dictyoptera: Blattaria: Polyphaginae). *Annales de la Societe Entomologique de France.* 30:361–362.

Grandcolas, P. 1994b. La richesse spécifique des communautés des blattes du sous-bois en forêt tropicale de Guyane Française. *Revue d' Ecologie (la Terre et la Vie).* 49:139–150.

Grandcolas, P. 1995a. Bionomics of a desert cockroach, *Heterogamisca chopardi* Uvarov, 1936 after the spring rainfalls in Saudi Arabia (Insecta, Blattaria, Polyphaginae). *Journal of Arid Environments.* 31:325–334.

Grandcolas, P. 1995b. Nouvelles données sur la genre *Alloblatta* Grandcolas, 1993 (Dictyoptera, Blattaria). *Bulletin de la Societe Entomologique de France.* 100:341–346.

Grandcolas, P. 1997a. *Gyna gloriosa,* a scavenger cockroach dependent on driver ants in Gabon. *African Journal of Ecology.* 35:168–171.

Grandcolas, P. 1997b. Habitat use and population structure of a polyphagine cockroach, *Ergaula capensis* (Saussure 1891) (Blattaria Polyphaginae) in Gabonese rainforest. *Tropical Ecology.* 10:215–222.

Grandcolas, P. 1997c. Systématique phylogénétique de la sousfamille des Tryonicinae (Dictyoptera, Blattaria, Blattidae). *In* Zoologia Neocaledonica, Memoir 171. J. Najt and L. Matile, editors. Museum of Natural History, Paris. 91–124.

Grandcolas, P. 1998. The evolutionary interplay of social behavior, resource use and antipredator behavior in Zetoborinae + Blaberinae + Gyninae + Diplopterinae cockroaches: a phylogenetic analysis. *Cladistics.* 14:117–127.

Grandcolas, P., and P. Deleporte. 1994. Escape from predation by army ants in *Lanxoblatta* cockroach larvae (Insecta, Blattaria, Zetoborinae). *Biotropica.* 26:469–472.

Grandcolas, P., and P. Deleporte. 1998. Incubation of zigzag-shaped oothecae in some ovoviviparous cockroaches *Gyna capucina* and *G. henrardi* (Blattaria: Blaberidae). *International Journal of Insect Morphology and Embryology.* 27:269–271.

Grassé, P.P. 1946. Societies animales et effet de groupe. *Experientia.* 15:365–408.

Grassé, P.P. 1951. Biocenotique et phenomene sociale. *Annals of Biology.* 27:153–160.

Grassé, P.P. 1952. Role des flagellés symbiotique chez les Blattes et les Termites. *Tijdschrift voor Entomologie.* 95:70–80.

Grassé, P.P., and C. Noirot. 1945. La transmission des flagelles symbiotiques et les aliments des termites. *Biological Bulletin of France and Belgium.* 79:273–297.

Grassé, P.P., and C. Noirot. 1959. L'évolution de la symbiose chez les Isopteres. *Experientia.* 15:365–408.

Grassé, P.P., and C. Noirot. 1960. L'isolement chez le termite a cou jaune (*Calotermes flavicollis* Fab.) et ses conséquences. *Insectes Sociaux.* 7:323–331.

Graves, P.N. 1969. Spermatophores of the Blattaria. *Annals of the Entomological Society of America.* 62:595–602.

Graves, R.C., J.S. Yoon, and E.J. Durbin. 1986. A gynandromorph in the Madagascar hissing cockroach *Gromphadorhina portentosa* (Blattodea: Blaberidae). *Annals of the Entomological Society of America.* 79:662–663.

Gray, J., and A.J. Boucot. 1993. Early Silurian nonmarine animal remains and the nature of the early continental ecosystem. *Acta Palaeontologica Polonica.* 38:303–328.

Greenberg, S., and B. Stay. 1974. Distribution and innervation of hairs in the brood sac of the cockroach, *Diploptera punctata* (Eschscholtz) (Dictyoptera: Blaberidae). *International Journal of Insect Morphology and Embryology.* 3:127–135.

Greenslade, P.J.M., and P. Greenslade. 1989. Ground layer invertebrate fauna. *In* Mediterranean Landscapes in Australia. J.C. Noble and R.A. Bradstock, editors. CSIRO, East Melbourne, Australia. 266–284.

Griffiths, J.T., and O.E. Tauber. 1942a. Fecundity, longevity, and parthenogenesis of the American roach, *Periplaneta americana* L. *Physiological Entomology.* 15:196–209.

Griffiths, J.T., and O.E. Tauber. 1942b. The nymphal development for the roach, *Periplaneta americana* L. *Journal of the New York Entomological Society.* 50:263–272.

Grillou, H. 1973. A study of sexual receptivity in *Blabera craniifer* Burm. (Blattaria). *Journal of Insect Physiology.* 19:173–193.

Grimaldi, D., and M.S. Engel. 2005. Evolution of the Insects. Cambridge University Press, New York. 755 pp.

Grove, S.J., and N.E. Stork. 1999. The conservation of saproxylic insects in tropical forests: a research agenda. *Journal of Insect Conservation.* 3:67–74.

Guillette, L.J., Jr. 1989. The evolution of vertebrate viviparity: morphological modifications and endocrine control. *In* Complex Organismal Functions: Integration and Evolution in Vertebrates. D.B. Wake and G. Roth, editors. John Wiley & Sons, Chichester, England. 219–233.

Gunn, D.L. 1935. The temperature and humidity relations of the cockroach. III. A comparison of temperature preference, and rates of dessication and respiration of *Periplaneta americana, Blatta orientalis* and *Blattella germanica. Journal of Experimental Biology.* 12:185–190.

Gunn, D.L. 1940. Daily activity rhythm of the cockroach. *Journal of Experimental Biology.* 17:267–277.

Gupta, A.P. 1947. On copulation and insemination in the

cockroach *Periplaneta americana* (Linn.). *Proceedings of the National Institute of Sciences, India.* 13:65–71.
Gupta, B.L., and D.S. Smith. 1969. Fine structural organization of the spermatheca in the cockroach *Periplaneta americana. Tissue & Cell.* 1:295–324.
Gurney, A.B. 1937. Studies in certain genera of American Blattidae (Orthoptera). *Proceedings of the Entomological Society of Washington.* 39:101–112.
Gurney, A.B. 1959. The largest cockroach. *Proceedings of the Entomological Society of Washington.* 61:133–134.
Gurney, A.B., and L.M. Roth. 1966. Two new genera of South American cockroaches superficially resembling *Loboptera,* with notes on bionomics (Dictyoptera, Blattaria, Blattellidae). *Psyche.* 73:196–207.
Guthrie, D.M., and A.R. Tindall. 1968. The Biology of the Cockroach. Edward Arnold Ltd., London. 408 pp.
Gwynne, D.T. 1984. Male mating effort, confidence of paternity, and insect sperm competition. *In* Sperm Competition and the Evolution of Animal Mating Systems. R.L. Smith, editor. Academic Press, London. 117–149.
Gwynne, D.T. 1998. Genitally does it. *Nature.* 393:734–735.
Haas, F., and J. Kukalova-Peck. 2001. Dermaptera hindwing structure and folding: new evidence for familial, ordinal and superordinal relationships within Neoptera. *European Journal of Entomology.* 98:445–509.
Haas, F., and R.J. Wootton. 1996. Two basic mechanisms in insect wing folding. *Proceedings of the Royal Society of London B.* 263:1651–1658.
Haber, V.R. 1920. Oviposition by a cockroach, *Periplaneta americana* Linn. (Orth.). *Entomological News.* 31:190–193.
Hackstein, J.H.P. 1996. Genetic and evolutionary constraints for the symbiosis between animals and methanogenic bacteria. *Environmental Monitoring and Assessment.* 42:39–56.
Hackstein, J.H.P., and C.K. Strumm. 1994. Methane production in terrestrial arthropods. *Proceedings of the National Academy of Sciences.* 91:5441–5445.
Hadley, N.F. 1994. Water relations of terrestrial arthropods. Academic Press, San Diego. 356 pp.
Hafez, M., and A.M. Afifi. 1956. Biological studies on the furniture cockroach *Supella supellectilium* Serv. in Egypt. *Bulletin de la Societe Entomologique d'Egypte.* 40:365–396.
Hagan, H.R. 1941. The general morphology of the female reproductive system of a viviparous roach, *Diploptera dytiscoides* (Serville). *Psyche.* 48:1–9.
Hales, R.A., and M.D. Breed. 1983. Female calling and reproductive behavior in the brown banded cockroach, *Supella longipalpa* (F.) (Orthoptera: Blattellidae). *Annals of the Entomological Society of America.* 76:239–241.
Hamilton, R.L., R.A. Cooper, and C. Schal. 1990. The influence of nymphal and adult dietary protein on food intake and reproduction in female brown-banded cockroaches. *Entomologia Experimentalis et Applicata.* 55:23–31.
Hamilton, R.L., D.E. Mullins, and D.M. Orcutt. 1985. Freezing tolerance in the woodroach *Cryptocercus punctulatus* (Scudder). *Experientia.* 41:1535–1536.
Hamilton, R.L., and C. Schal. 1988. Effects of dietary protein levels on reproduction and food consumption in the German cockroach (Dictyoptera: Blattellidae). *Annals of the Entomological Society of America.* 81:969–976.
Han, S.S., and A.P. Gupta. 1988. Arthropod immune system. V. Activated immunocytes (granulocytes) of the German cockroach, *Blattella germanica* L. (Dictyoptera: Blattidae) show increased number of microtubules and nuclear pores during immune reaction to foreign tissue. *Cell Structure and Function.* 13:333–343.
Hanitsch, R. 1923. On a collection of Blattidae from the Buitenzorg Museum. *Treubia.* 3:197–221.
Hanitsch, R. 1928. Spolia Metawiensia: Blattidae. *Bulletin of the Raffles Museum.* 1:1–44.
Hanitsch, R. 1933. XXI. The Blattidae of Mt. Kinabalu, British North Borneo. *Journal of the Federated Malay States Museum.* 17:297–337.
Hansell, M.H. 1993. The ecological impact of animal nests and burrows. *Functional Ecology.* 7:5–12.
Hanula, J.L., and K.E. Franzreb. 1995. Arthropod prey of nestling red-cockaded woodpeckers in the upper coastal plain of South Carolina. *Wilson Bulletin.* 107:485–495.
Hanula, J.L., D. Lipscomb, K.E. Franzreb, and S.C. Loeb. 2000. Diet of nestling red-cockaded woodpeckers at three locations. *Journal of Field Ornithology.* 71:126–134.
Harington, D. 1990. The Cockroaches of Stay More: A Novel. Vintage Books, New York. 337 pp.
Harris, W.E., and P.J. Moore. 2004. Sperm competition and male ejaculate investment in *Nauphoeta cinerea:* effects of social environment during development. *Journal of Evolutionary Biology.* 18:474–480.
Harris, W.E., and P.J. Moore. 2005. Female mate preference and sexual conflict: females prefer males that have had fewer consorts. *The American Naturalist.* 165:S64–S71.
Harrison, R.G. 1980. Dispersal polymorphisms in insects. *Annual Review of Ecology and Systematics.* 11:95–118.
Hartman, B., and L.M. Roth. 1967a. Stridulation by a cockroach during courtship behavior. *Nature.* 213:1243.
Hartman, B., and L.M. Roth. 1967b. Stridulation by the cockroach *Nauphoeta cinerea* (Olivier) during courtship behaviour. *Journal of Insect Physiology.* 13:579–586.
Hartman, H.B., L.P. Bennett, and B.A. Moulton. 1987. Anatomy of equilibrium receptors and cerci of the burrowing desert cockroach *Arenivaga* (Insecta: Blattodea). *Zoomorphology.* 107:81–87.
Hawke, S.D., and R.D. Farley. 1973. Ecology and behavior of the desert burrowing cockroach, *Arenivaga* sp. (Dictyoptera, Polyphagidae). *Oecologia.* 11:262–279.
Haydak, M.H. 1953. Influence of the protein level on the diet and longevity of cockroaches. *Annals of the Entomological Society of America.* 46:547–560.
Heath, H. 1903. The habits of California termites. *Biological Bulletin.* 4:47–63.
Hebard, M. 1916a. A new genus, *Cariblatta,* of the group Blattellites (Orthoptera, Blattidae). *Transactions of the American Entomological Society.* 42:147–186.
Hebard, M. 1916b. Studies in the group Ischnopterites (Orthoptera, Blattidae, Pseudomopinae). *Transactions of the American Entomological Society.* 42:337–383.
Hebard, M. 1917. The Blattidae of North America north of the Mexican boundary. *Memoirs of the American Entomological Society.* 2:255–258.
Hebard, M. 1920a. The Blattidae of Panama. *Memoirs of the American Entomological Society.* 4:1–154.
Hebard, M. 1920b. Expedition of the California Academy of

Sciences to the Galapagos Islands, 1905–1906. *Proceedings of the California Academy of Sciences,* 4th Ser. 2, Pt. 2:311–346.

Hebard, M. 1920 (1919). Studies in the Dermaptera, and Orthoptera of Colombia. *Transactions of the American Entomological Society.* 45:89–179.

Hebard, M. 1929. Studies in Malayan Blattidae (Orthoptera). *Proceedings of the Academy of Natural Sciences of Philadelphia.* 81:1–109.

Hebard, M. 1943. The Dermaptera and Orthopterous families Blattidae, Mantidae and Phasmidae of Texas. *Transactions of the American Entomological Society.* 68:239–311.

Hebard, M. 1945. The Orthoptera of the Appalachian Mountains in the vicinity of Hot Springs, Virginia, and notes on other Appalachian species and recent extensions of the known range of still other southeastern species. *Transactions of the American Entomological Society.* 71:77–97.

Heinrich, B. 2001. Racing the Antelope: What Animals Can Teach Us About Running and Life. HarperCollins, New York. 292 pp.

Helfer, J.R. 1953. How to Know the Grasshoppers, Cockroaches and their Allies. Wm. C. Brown Co., Dubuque, Iowa. 353 pp.

Hellriegel, B., and G. Bernasconi. 2000. Female-mediated differential sperm storage in a fly with complex spermathecae, *Scatophaga stercoraria. Animal Behaviour.* 59:311–317.

Hellriegel, B., and P.I. Ward. 1998. Complex female reproductive tract morphology: its possible use in postcopulatory female choice. *Journal of Theoretical Biology.* 190:179–186.

Heming, B.S. 2003. Insect Development and Evolution. Cornell University Press, Ithaca, NY. 494 pp.

Herreid, C.F.I., and R.J. Full. 1984. Cockroaches on a treadmill: aerobic running. *Journal of Insect Physiology.* 30:395–403.

Herreid, C.F.I., D.A. Prawel, and R.J. Full. 1981. Energetics of running cockroaches. *Science.* 212:331–333.

Higashi, M., N. Yamamura, and T. Abe. 2000. Theories on the sociality of termites. *In* Termites: Evolution, Sociality, Symbioses, Ecology. T. Abe, D.E. Bignell, and M. Higashi, editors. Kluwar Academic Publishers, Dordrecht. 169–187.

Hijii, N. 1983. Arboreal fauna in a forest. I. Preliminary observation on seasonal fluctuations in density, biomass, and faunal composition in a *Chamaecyparis obtusa* plantation. *Japanese Journal of Ecology.* 33:415–444.

Hill, S.B. 1981. Ecology of bat guano in Tamana Cave, Trinidad, W.I. *Proceedings of the 8th International Congress of Speleology.* 1:243–246.

Hinton, H.E. 1981. Biology of Insect Eggs. Vol. 1. Pergamon Press, Oxford.

Hintze-Podufal, C., and U. Nierling. 1986. Der Einfluss der nahrung auf entwicklung, wachstum und präreproduktionsphase von *Blaptica dubia* Stal. (Blaberoidea, Blaberidae). *Bulletin de la Société Entomologique Suisse.* 59:177–186.

Hoback, W.W., and D.W. Stanley. 2001. Insects in hypoxia. *Journal of Insect Physiology.* 47:533–542.

Hoberman, M.A. 1985. Cockroach. *In* The Oxford Book of Children's Verse in America. D. Hall, editor. Oxford University Press, Oxford. 274.

Hobson, E.S. 1978. Aggregating as a defense against predators in aquatic and terrestrial environments. *In* Contrasts in Behavior: Adaptations in the Aquatic and Terrestrial Environments. E.S. Reese and F.J. Lighter, editors. John Wiley & Sons, New York. 219–234.

Hocking, B. 1958. On the activity of *Blattella germanica* L. (Orthoptera: Blattidae). *Proceedings of the 10th International Congress of Entomology.* 2:201–204.

Hocking, B. 1970. Insect associations with the swollen thorn acacias. *Transactions of the Royal Entomological Society of London.* 122:211–255.

Hoffman, R.L., and J.A. Payne. 1969. Diplopods as carnivores. *Ecology.* 50:1096–1098.

Hohmann, R., F. Sinowatz, and E. Bamberg. 1978. Biochemical and histochemical examination of glycosidases in the genital tract of the cockroach *Blaberus craniifer. Zoologischer Anzeiger.* 200:379–385.

Holbrook, G.L., E. Armstrong, J.A.S. Bachmann, B.M. Deasy, and C. Schal. 2000a. Role of feeding in the reproductive 'group effect' in females of the German cockroach *Blattella germanica* (L.). *Journal of Insect Physiology.* 46:941–949.

Holbrook, G.L., J.A.S. Bachmann, and C. Schal. 2000b. Effects of ovariectomy and mating on the activity of the corpora allata in adult female *Blattella germanica* (L.) (Dictyoptera: Blattellidae). *Physiological Entomology.* 25:27–34.

Holbrook, G.L., and C. Schal. 1998. Social influences on nymphal development in the cockroach *Diploptera punctata. Physiological Entomology.* 23:121–130.

Holbrook, G.L., and C. Schal. 2004. Maternal investment affects offspring phenotypic plasticity in a viviparous cockroach. *Proceedings of the National Academy of Sciences.* 101:5595–5597.

Hölldobbler, B., and E.O. Wilson. 1990. The Ants. Belknap Press of Harvard University Press, Cambridge, MA. 732 pp.

Hölldobbler, B., and E.O. Wilson. 2005. Euociality: origin and consequences. *Proceedings of the National Academy of Sciences.* 102:13367–13371.

Holsinger, J.R. 2000. Ecological derivation, colonization, and speciation. *In* Ecosystems of the World. Vol. 30: Subterranean Ecosystems. H. Wilkens, D.C. Culver, and W.F. Humphreys, editors. Elsevier, Amsterdam. 399–415.

Honigberg, B.M. 1970. Protozoa associated with termites and their role in digestion. *In* Biology of Termites. Vol. 2. K. Krishna and F.M. Weesner, editors. Academic Press, New York. 1–36.

Hooper, R.G. 1996. Arthropod biomass in winter and the age of longleaf pines. *Forest Ecology and Management.* 82:115–131.

Horn, S., and J.L. Hanula. 2002. Life history and habitat associations of the broad wood cockroach *Parcoblatta lata* (Blattaria: Blattellidae) and other native cockroaches in the Coastal Plain of South Carolina. *Annals of the Entomological Society of America.* 95:665–671.

Howarth, F.G. 1983. Ecology of cave arthropods. *Annual Review of Entomology.* 28:365–389.

Howarth, F.G. 1988. Environmental ecology of North Queensland caves: or why are there so many troglobites in Australia. *In* 17th Biennial Conference of the Australian

Speleological Federation. L. Pearson, editor. Australian Speleological Federation, Lake Tinaroo, Far North Queensland. 76–84.

Howse, P.E. 1964. An investigation into the mode of action of the subgenual organ in the termite, *Zootermopsis angusticollis* Emerson and in the cockroach, *Periplaneta americana* L. *Journal of Insect Physiology.* 10:409–424.

Howse, P.E. 1965. On the significance of certain oscillatory movements in termites. *Insectes Sociaux.* 12:335–346.

Howse, P.E. 1968. On the division of labour in the primitive termite *Zootermopsis nevadensis* (Hagen). *Insectes Sociaux.* 15:45–50.

Hoyte, H.M.D. 1961a. The protozoa occurring in the hindgut of cockroaches. II. Morphology of *Nyctotherus ovalis. Parasitology.* 51:437–463.

Hoyte, H.M.D. 1961b. The protozoa occurring in the hindgut of cockroaches. III. Factors affecting the dispersal of *Nyctotherus ovalis. Parasitology.* 51:465–495.

Hubbell, T.H., and C.C. Goff. 1939. Florida pocket-gopher burrows and their arthropod inhabitants. *Proceedings of the Florida Academy of Sciences.* 4:127–166.

Huber, I. 1976. Evolutionary trends in *Cryptocercus punctulatus* (Blattaria: Cryptocercidae). *Journal of the New York Entomological Society.* 84:166–168.

Huber, I., E.P. Masler, and B.R. Rao. 1990. Cockroaches as models for neurobiology: applications in biomedical research. Vols. 1–2. CRC Press, Boca Raton.

Hudgins, J.W., T. Krekling, and V.R. Francheschi. 2003. Distribution of calcium oxalate crystals in the secondary phloem of conifers: a constitutive defense mechanism? *New Phytologist.* 159:677–690.

Hufford, L. 1996. Ontogenetic evolution, clade diversification, and homoplasy. *In* Homoplasy: The Recurrence of Similarity in Evolution. M.J. Sanderson and L. Hufford, editors. Academic Press, San Diego. 271–302.

Hughes, G.M. 1952. The co-ordination of insect movements. I. The walking movements of insects. *Journal of Experimental Biology.* 29:267–284.

Hughes, G.M., and P.J. Mill. 1974. Locomotion: terrestrial. *In* The Physiology of Insecta. Vol. 3. M. Rockstein, editor. Academic Press, New York. 335–379.

Hughes, M., and K.G. Davey. 1969. The activity of spermatozoa of *Periplaneta. Journal of Insect Physiology.* 15:1607–1616.

Humphrey, M., H.A. Rose, and D.J. Colgan. 1998. Electrophoretic studies of the cockroaches of the Australian endemic subfamily Geoscapheinae. *Zoological Journal of the Linnean Society.* 124:209–234.

Humphreys, W.F. 1993. Cave fauna in semi-arid tropical western Australia: a diverse relict wet-forest litter fauna. *Mémoires de Biospéologie.* 20:105–110.

Humphreys, W.F. 2000a. Background and glossary. *In* Ecosystems of the World. Vol. 30: Subterranean Ecosystems. H. Wilkens, D.C. Culver, and W.F. Humphreys, editors. Elsevier, Amsterdam. 3–14.

Humphreys, W.F. 2000b. The hypogean fauna of the Cape Range Peninsula and Barrow Island, Northwestern Australia. *In* Ecosystems of the World. Vol. 30: Subterranean Ecosystems. H. Wilkens, D.C. Culver, and W.F. Humphreys, editors. Elsevier, Amsterdam. 581–601.

Humphreys, W.F., and M.N. Feinberg. 1995. Food of the blind cave fishes of northwestern Australia. *Records of the Western Australian Museum.* 17:29–33.

Hunt, J.H. 2003. Cryptic herbivores of the rainforest canopy. *Science.* 300:916–917.

Hunt, J.H., and G.V. Amdam. 2005. Bivoltinism as an antecedent to eusociality in the paper wasp genus *Polistes. Science.* 308:264–267.

Hunt, J.H., and C.A. Nalepa. 1994. Nourishment, evolution and insect sociality. *In* Nourishment and Evolution in Insect Societies. J.H. Hunt and C.A. Nalepa, editors. Westview Press, Boulder. 1–19.

Hunter, F.M., and T.R. Birkhead. 2002. Sperm viability and sperm competition in insects. *Current Biology.* 12:121–123.

Hüppop, K. 2000. How do cave animals cope with the food scarcity in caves? *In* Ecosystems of the World. Vol. 30: Subterranean Ecosystems. H. Wilkens, D.C. Culver, and W.F. Humphreys, editors. Elsevier, Amsterdam. 159–188.

Ichinosé, T., and K. Zennyoji. 1980. Defensive behavior of the cockroaches *Periplaneta fuliginosa* Serville and *P. japonica* Karney (Orthoptera: Blattidae), in relation to their viscous secretion. *Applied Entomology and Zoology.* 14:400–408.

Imboden, H.B., J. Lanzrein, P. Delbecque, and M. Luscher. 1978. Ecdysteroids and juvenile hormone during embryogenesis in the ovoviviparous cockroach *Nauphoeta cinerea. General and Comparative Endocrinology.* 36:628–635.

Imms, A.D. 1919. II. On the structure and biology of *Archotermopsis,* together with descriptions of new species of intestinal protozoa, and general observations on the Isoptera. *Philosophical Transactions of the Royal Society of London.* 209:75–180.

Ingram, M.J., B. Stay, and G. Cain. 1977. Composition of milk from the viviparous cockroach, *Diploptera punctata. Insect Biochemistry.* 7:257–267.

Inoue, T., S. Moriya, M. Ohkuma, and T. Kudo. 2005. Molecular cloning and characterization of a cellulose gene from a symbiotic protist of the lower termite, *Coptotermes formosanus. Gene* 349:67–75.

Irmler, U., and K. Furch. 1979. Production, energy and nutrient turnover of the cockroach *Epilampra irmleri* Rocha e Silva & Aguiar, in Central-Amazonian inundation forest. *Amazonia.* 6:497–520.

Irving, P., L. Troxler, T.S. Heuer, B. M., C. Kopczynski, J.M. Reichhart, J.A. Hoffmann, and C. Hetru. 2001. A genome wide analysis of immune responses in Drosophila. *Proceedings of the National Academy of Sciences of the United States of America.* 98:15119–15124.

Ishii, S., and Y. Kuwahara. 1967. An aggregation pheromone of the German cockroach *Blattella germanica* L. (Orthoptera: Blattellidae). 1. Site of the pheromone production. *Applied Entomology and Zoology.* 2:203–217.

Ishii, S., and Y. Kuwahara. 1968. Aggregation of German cockroach (*Blattella germanica*) nymphs. *Experientia.* 24:88–89.

Itioka, T., M. Kato, H. Kaliang, M. Ben Merdeck, T. Nagamitsu, S. Sakai, S. Umah Mohamad, S. Yamane, A. Abdul Hamid, and T. Inoue. 2003. Insect responses to general flowering in Sarawak. *In* Arthropods of Tropical Forests: Spatio-Temporal Dynamics and Resource Use in the

Canopy. Y. Basset, V. Novotny, S.E. Miller, and R.L. Kitching, editors. Cambridge University Press, Cambridge. 126–134.

Itô, Y. 1980. Comparative Ecology. Cambridge University Press, Cambridge. 436 pp.

Iwao, S. 1967. Some effects of grouping on lepidopterous insects. *In* L'Effet de Groupe chez les Animaux. Editions du Centre National de la Recherche Scientifique, Paris. 185–212.

Izquierdo, I., and L. Medina. 1992. A new subterranean species of *Symploce* Hebard from Gran Canaria (Canary Islands) (Blattaria, Blattellidae). *Fragmenta Entomologica.* 24:39–44.

Izquierdo, I., and P. Oromi. 1992. Dictyoptera—Blattaria. *In* Encyclopaedia Biospeologica. Vol. 1. C. Juberthie and V. Decu, editors. Academie Roumaine, Bucharest. 295–300.

Izquierdo, I., P. Oromi, and X. Belles. 1990. Number of ovarioles and degree of dependence with respect to the underground environment in the Canarian species of the genus *Loboptera* Brunner (Blattaria, Blattellidae). *Memoirés de Biospélologie.* 17:107–111.

Izutsu, M., S. Veda, and S. Ishii. 1970. Aggregation effects on the growth of the German cockroach *Blattella germanica* (L.) (Blattaria: Blattellidae). *Applied Entomology and Zoology.* 5:159–171.

Jackson, L.L. 1983. Epicuticular lipid composition of the sand cockroach *Arenivaga investigata. Comparative Biochemistry and Physiology.* 74B:255–257.

Jaiswal, A.K., and M.B. Naidu. 1972. Studies on the reproductive system of the cockroach *Periplaneta americana* L. male reproductive system—Part I. *Journal of Animal Morphology and Physiology.* 19:1–7.

Jaiswal, A.K., and M.B. Naidu. 1976. Studies on the reproductive system of the cockroach *Periplaneta americana* L. male reproductive system—Part II. *Journal of Animal Morphology and Physiology.* 23:176–184.

Jamieson, B.G.M. 1987. The Ultrastructure and Phylogeny of Insect Spermatozoa. Cambridge University Press, Cambridge. 320 pp.

Jander, U. 1966. Untersuchungen zur Stammesgeschichte von Putzbewegungen von Tracheaten. *Zeitschrift für Tierpsychologie.* 23:799–844.

Janiszewski, J., and D. Wysocki. 1986. Body temperature of cockroaches: *Gromphadorhina brauneri* (Shelf.) and *Periplaneta americana* (L.) at high ambient temperatures. *Zoologica Poloniae.* 33:23–32.

Janzen, D.H. 1976. Why tropical trees have rotten cores. *Biotropica.* 8:110.

Janzen, D.H. 1977. Why fruits rot, seeds mold, and meat spoils. *The American Naturalist.* 111:691–713.

Jarvinen, O., and K. Vepsalainen. 1976. Wing dimorphism as an adaptive strategy in water striders (Gerris). *Hereditas.* 84:61–68.

Jayakumar, M., S.J. William, and K.S. Ananthasubramanian. 1994. Parental care in an Indian blaberid roach, *Thorax procellana. Geobios New Reports.* 13:159–163.

Jayakumar, M., S.J. William, N. Raja, K. Elumalai, and A. Jeyasankar. 2002. Mating behavior of a cockroach, *Neopolyphaga miniscula* (Dictyoptera: Blaberoidea). *Journal of Experimental Zoology, India.* 5:101–106.

Jeanson, R., C. Rivault, J.-L. Deneubourg, S. Blanco, R. Fournier, C. Jost, and G. Theraulaz. 2005. Self organized aggregation in cockroaches. *Animal Behaviour.* 69:169–180.

Jeyaprakash, A., and M.A. Hoy. 2000. Long PCR improves *Wolbachia* DNA amplification: wsp sequences found in 76% of sixty-three arthropod species. *Insect Molecular Biology.* 9:393–405.

Jindrich, D.L., and R.J. Full. 2002. Dynamic stabilization of rapid hexapodal locomotion. *Journal of Experimental Biology.* 205:2803–2823.

Johannes, R.E., and M. Satomi. 1966. Composition and nutritive value of fecal pellets of a marine crustacean. 11:191–197.

Johnson, C.G. 1976. Lability of the flight system: a context for functional adaptation. *Symposia of the Royal Entomological Society of London.* 7:217–234.

Jones, S.A., and D. Raubenheimer. 2001. Nutritional regulation in nymphs of the German cockroach, *Blattella germanica. Journal of Insect Physiology.* 47:1169–1180.

Joyner, K., and F. Gould. 1986. Conspecific tissues and secretions as sources of nutrition. *In* Nutritional Ecology of Insects, Mites, and Spiders. F.J. Slansky and J.G. Rodriguez, editors. John Wiley & Sons, New York. 697–719.

Juberthie, C. 2000a. Conservation of subterranean habitats and species. *In* Ecosystems of the World. Vol. 30: Subterranean Ecosystems. H. Wilkens, D.C. Culver, and W.F. Humphreys, editors. Elsevier, Amsterdam. 691–700.

Juberthie, C. 2000b. The diversity of the karstic and pseudokarstic hypogean habitats in the world. *In* Ecosystems of the World. Vol. 30: Subterranean Ecosystems. H. Wilkens, D.C. Culver, and W.F. Humphreys, editors. Elsevier, Amsterdam. 17–39.

Just, F., and B. Walz. 1994. Immunocytochemical localization of $Na+/K+$-ATPase and V-H+ ATPase in the salivary glands of the cockroach *Periplaneta americana. Cell & Tissue Research.* 278:161–170.

Kaakeh, W., B.L. Reid, and G.W. Bennett. 1996. Horizontal transmission of entomopathogenic fungus *Metarhizium anisopliae* (imperfect fungi: Hyphomycetes) and hydramethylnon among German cockroaches (Dictyoptera: Blattellidae). *Journal of Entomological Science.* 31:378–390.

Kaitala, A., and L. Hulden. 1990. Significance of spring migrations and flexibility in flight muscle histolysis in waterstriders Heteroptera Gerridae. *Ecological Entomology.* 15:409–418.

Kalmus, H. 1941. Physiology and ecology of cuticle color in insects. *Nature.* 148:428–431.

Kamimura, Y. 2000. Possible removal of rival sperm by the elongated genitalia of the earwig, *Euborellia plebeja. Zoological Science.* 17:667–672.

Kane, M.D., and J.A. Breznak. 1991. Effect of host diet on production of organic acids and methane by cockroach gut bacteria. *Applied and Environmental Microbiology.* 57:2628–2634.

Kaplin, V.G. 1995. Life history of the cockroach *Anisogamia tamerlana* Sauss. (Blattodea, Corydiidae) in the east Karakim. *Entomologicheskoe Obozrenie.* 74:287–298.

Kaplin, V.G. 1996. Daily activity, territorial and trophic associations of *Anisogamia tamerlana* Sauss. (Blattodea, Cory-

diidae) in Eastern Kara Kum. *Entomological Review.* 75:53–66.
Karlsson, B., and P.-O. Wickman. 1989. The cost of prolonged life: an experiment on a nymphal butterfly. *Functional Ecology.* 3:399–405.
Karny, H.H. 1924. Beiträge zur Malayischen Orthopterenfauna. V. Bemerkungen ueber einige Blattoiden. *Treubia.* 5:3–19.
Kavanaugh, D.H. 1977. An example of aggregation in *Scaphinotus* Subgenus *Brennus* Motschulsky (Coleoptera: Carabidae: Cychrini). *The Pan-Pacific Entomologist.* 53:27–31.
Kawakita, A., and M. Kato. 2002. Floral biology and unique pollination system of root holoparasites, *Balanophora kuroiwai* and *B. tobiracola* (Balanophoraceae). *American Journal of Botany.* 89:1164–1170.
Kayser, H. 1985. Pigments. *In* Comprehensive Insect Physiology, Biochemistry, and Pharmacology. Vol. 10. G.A. Kerkut and L.I. Gilbert, editors. Pergamon Press, New York. 368–415.
Kennedy, C.H. 1947. Child labor of the termite society versus adult labor of the ant society. *The Scientific Monthly.* 65:309–324.
Kevan, D.K.M. 1962. Soil Animals. H.F. & G. Witherby Ltd., London. 244 pp.
Kevan, D.K.M. 1993. Introducing orthopteroid insects (other than termites) and the soil. *Tropical Zoology.* Special Issue 1:61–83.
Khalifa, A. 1950. Spermatophore formation in *Blattella germanica. Proceedings of the Royal Society of London A.* 25:53–61.
Kidder, G.W. 1937. The intestinal protozoa of the wood-feeding roach *Panesthia. Parasitology.* 29:163–203, 10 plates.
King, L.E., J.E. Steele, and S.W. Bajura. 1986. The effect of flight on the composition of haemolymph in the cockroach *Periplaneta americana. Journal of Insect Physiology.* 32:649–655.
Kirkwood, T.B.L. 1981. Repair and its evolution: survival vs. reproduction. *In* Physiological Ecology: An Evolutionary Approach to Resource Use. C.R. Townsend and P. Calow, editors. Sinauer Associates, Inc., Sunderland, MA. 165–189.
Kistner, D. 1982. The Social Insects' Bestiary. *In* Social Insects. Vol. III. H.R. Hermann, editor. Academic Press, New York. 1–244.
Kitamura, C., H.S. Koh, and S. Ishii. 1974. Possible role of feces for directional orientation in the German cockroach *Blattella germanica* L. *Applied Entomology and Zoology.* 9:271–272.
Kitching, R.L., H. Mitchell, G. Morse, and C. Thebaud. 1997. Determinants of species richness in assemblages of canopy arthropods in rainforests. *In* Canopy Arthropods. N.E. Stork, J. Adis, and R.K. Didham, editors. Chapman and Hall, London. 131–150.
Klass, K.-D. 1995. Die Phylogeny der Dictyoptera. Ph.D. thesis, Fakultät für Biologie. Ludwig Maximilians Universität, München.
Klass, K.-D. 1997. The external male genitalia and the phylogeny of Blattaria and Mantodea. *Bonner Zoologische Monographien.* 42:1–341.
Klass, K.-D. 1998a. The ovipositor of Dictyoptera (Insecta): homology and ground plan of the main elements. *Zoologischer Anzeiger.* 236:69–101.
Klass, K.-D. 1998b. The proventriculus of Dicondylia, with comments on evolution and phylogeny in Dictyoptera and Odonata. *Zoologischer Anzeiger.* 237:15–42.
Klass, K.-D. 2001. Morphological evidence on blattarian phylogeny: "phylogenetic histories and stories" (Insecta, Dictyoptera). *Mitteilungen aus dem Museum für Naturkunde in Berlin, Deutsche Entomologische Zeitschrift.* 48:223–265.
Klass, K-D. 2003. Relationships among principal lineages of Dictyoptera inferred from morphological data. *Entomologische Abhandlungen.* 61:134–137.
Klass, K.-D., and R. Meier. 2006. A phylogenetic analysis of Dictyoptera (Insecta) based on morphological characters. *Entomologische Abhandlungen* 63:3–50.
Kluge, A.G. 1985. Ontogeny and phylogenetic systematics. *Cladistics.* 1:13–27.
Knebelsberger, T., and H. Bohn. 2003. Geographic parthenogenesis in the *subaptera*-group of *Phyllodromica* (Blattoptera, Blattellidae, Ectobiinae). *Insect Systematics & Evolution.* 34:427–452.
Koehler, P.G., and R.S. Patterson. 1987. The Asian roach invasion. *Natural History.* 11/87:29–35.
Koehler, P.G., C.A. Strong, and R.S. Patterson. 1994. Harborage width preferences of German cockroach (Dictyoptera: Blattellidae) adults and nymphs. *Journal of Economic Entomology.* 87:699–704.
Komiyama, M., and K. Ogata. 1977. Observations of density effects on the German cockroaches *Blattella germanica* (L.). *Japanese Journal of Sanitary Zoology.* 28:409–415.
Kopanic, R.J., G.L. Holbrook, V. Sevala, and C. Schal. 2001. An adaptive benefit of facultative coprophagy in the German cockroach *Blattella germanica. Ecological Entomology.* 26:154–162.
Korchi, A., R. Brossut, H. Bouhin, and J. Delachambre. 1999. cDNA cloning of an adult male putative lipocalin specific to tergal gland aphrodisiac secretion in an insect (*Leucophaea maderae*). *FEBS Letters.* 449:125–128.
Krajick, K. 2001. Cave biologists unearth buried treasure. *Science.* 293:2378–2381.
Kramer, K.J., A.M. Christensen, T.D. Morgan, J. Schaefer, T.H. Czapla, and T.L. Hopkins. 1991. Analysis of cockroach oothecae and exuviae by solid state ^{13}C-NMR spectroscopy. *Insect Biochemistry.* 21:149–156.
Kramer, S. 1956. Pigmentation in the thoracic musculature of cockroaches and related Orthoptera and the analysis of flight and stridulation. *Proceedings of the 10th International Congress of Entomology.* 1:569–579.
Krause, J., and G.D. Ruxton. 2002. Living in Groups. Oxford University Press, Oxford. 210 pp.
Kristensen, N.P. 1991. Phylogeny of extant hexapods. *In* The Insects of Australia. Vol. 1. CSIRO, editor. Melbourne University Press, Carleton, Victoria. 125–140.
Krivokhatskii, V.A. 1985. Experience with the monitoring of the burrow associations of the great gerbil *Rhombomys opimus* in Repetek Biosphere Preserve Turkmen—SSR

USSR (Abstract). *Izvestiya Akademii Nauk Turkmenskoi SSR Seriya Biologicheskikh Nauk.* 1985:27–32.

Kugimiya, S., R. Nishida, M. Sakuma, and Y. Kuwahara. 2003. Nutritional phagostimulants function as male courtship pheromone in the German cockroach *Blattella germanica. Chemoecology.* 13:169–175.

Kukor, J.J., and M.M. Martin. 1986. Nutritional ecology of fungus-feeding arthropods. *In* Nutritional Ecology of Insects, Mites, Spiders, and Related Invertebrates. F.J. Slansky and J.G. Rodriguez, editors. John Wiley & Sons, New York. 791–814.

Kulshrestha, V., and S.C. Pathak. 1997. Aspergillosis in German cockroach *Blattella germanica* (L.) (Blattoidea: Blattellidae). *Mycopathologia.* 139:75–78.

Kumar, R. 1975. A review of the cockroaches of West Africa and the Congo basin (Dictyoptera: Blattaria). *Bulletin de l'Institut fondamental d'Afrique noire, Serie A, Science naturelles.* 37:27–121.

Kunkel, J.G. 1966. Development and availability of food in the German cockroach *Blattella germanica* (L.). *Journal of Insect Physiology.* 12:227–235.

Kunkel, J.G. 1975. Cockroach molting. I. Temporal organization of events during molting cycle of *Blattella germanica* (L.). *Biological Bulletin.* 148:259–273.

Kunkel, J.G. 1979. A minimal model of metamorphosis: fat body competence to respond to juvenile hormone. *In* Current Topics in Insect Endocrinology and Nutrition. G. Bhaskaran, S. Friedman, and J.G. Rodriguez, editors. Plenum Press, New York. 107–129.

Labandeira, C.C. 1994. A compendium of fossil insect families. *Milwaukee Public Museum Contributions in Biology and Geology.* 88:1–71.

Labandeira, C.C., T.L. Phillips, and R.A. Norton. 1997. Oribatid mites and the decomposition of plant tissues in Paleozoic coal-swamp forests. *Palaios.* 12:319–353.

Lagone, J.A. 1996. Notes on *Phyllomedusa iheringii* Boulenger, 1885 (Amphibia, Anura, Hylidae). *Comunicaciones Zoologicas del Museo de Historia Natural de Montevideo.* 12:2–7.

Laird, T.B., P.W. Winston, and M. Braukman. 1972. Water storage in the cockroach *Leucophaea maderae* F. *Naturwissenschaften.* 59:515–516.

Laland, K.N., and H.C. Plotkin. 1991. Excretory deposits surrounding food sites facilitate social learning of food preferences in Norway rats. *Animal Behaviour.* 41:997–1005.

Lambiase, S., A. Grigolo, U. Laudani, L. Sacchi, and B. Baccetti. 1997. Pattern of bacteriocyte formation in *Periplaneta americana* (L.) (Blattaria: Blattidae). *International Journal of Insect Morphology and Embryology.* 26:9–19.

Langecker, T.G. 2000. The effects of continuous darkness on cave ecology and cavernicolous evolution. *In* Ecosystems of the World. Vol. 30: Subterranean Ecosystems. H. Wilkens, D.C. Culver, and W.F. Humphreys, editors. Elsevier, Amsterdam. 135–157.

Langellotto, G.A., R.F. Denno, and J.R. Ott. 2000. A trade-off between flight capability and reproduction in males of a wing-dimorphic insect. *Ecology.* 81:865–875.

Lauga, J., and M. Hatté. 1977. Acquisition de propriétés grégarisantes par la sable utilise a la ponte repétée des femelles grégaires de *Locusta migratoria* L. (Ins., Orthop). *Acridida.* 6:307–311.

Laurentiaux, D.M. 1963. Antiquite du dimorphism sexuel des Blattes. *Comptes Rendus Hebdomadaires des Seances de l'Academie des Sciences.* 257:3971–3974.

Lavelle, P. 2002. Functional domains in soils. *Ecological Research.* 17:441–450.

Lavelle, P., and B. Kohlmann. 1984. Étude quantitative de la macrofaune du sol dans une forêt tropicale humide du Mexique (Bonampak, Chiapas). *Pedobiologia.* 27:377–393.

Lavelle, P., C. Lattaud, D. Trigi, and I. Barois. 1995. Mutualism and biodiversity in soils. *Plant and Soil.* 170:23–33.

Lawless, L.S. 1999. Morphological comparisons between two species of *Blattella* (Dictyoptera: Blattellidae). *Annals of the Entomological Society of America.* 92:139–143.

Lawrence, R.F. 1953. The Biology of the Cryptic Fauna of Forests, with Special Reference to the Indigenous Forests of South Africa. A. A. Balkema, Cape Town. 408 pp.

Lawson, F.A. 1951. Structural features of the oothecae of certain species of cockroaches. *Annals of the Entomological Society of America.* 44:269–285.

Lawson, F.A. 1967. Ecological and collecting notes on eight species of *Parcoblatta* (Orthoptera: Blattidae) and certain other cockroaches. *Journal of the Kansas Entomological Society.* 40:267–269.

Lawson, F.A., and J.C. Thompson. 1970. Ultrastructual comparison of the spermathecae in *Periplaneta americana. Journal of the Kansas Entomological Society.* 43:418–434.

Lawson, J.W.H. 1965. The behaviour of *Periplaneta americana* in a critical situation and the variation with age. *Behaviour.* 24:210–228.

Lawton, M.F., and R.O. Lawton. 1986. Heterochrony, deferred breeding, and avian sociality. *Current Ornithology* 3:187–222.

le Patourel, G.N.J. 1993. Cold-tolerance of the oriental cockroach *Blatta orientalis. Entomologia Experimentalis et Applicata.* 68:257–263.

Leakey, R.J.G. 1987. Invertebrates in the litter and soil at a range of altitudes on Gunung Silam: a small ultrabasic mountain in Sabah. *Journal of Tropical Ecology.* 3:119–129.

Ledoux, A. 1945. Etude experimentale du gregarisme et de l'interattraction sociale chez les Blattidae. *Annales des Sciences Naturelles, Zoologie.* 7:76–103.

Lee, H.-J. 1994. Are pregnant females of the German cockroach too heavy to run? *Zoological Studies.* 33:200–204.

Lee, H.-J., and Y.-L. Wu. 1994. Mating effects on the feeding and locomotion of the German cockroach, *Blattella germanica. Physiological Entomology.* 19:39–45.

Lefeuvre, J.-C. 1971. Hormone juvénile et polymorphism alaire chez les Blattaria (Insecte, Dictyoptère). *Archives de Zoologie Experimentale et Génerale.* 112:653–666.

Leimar, O., B. Karlsson, and C. Wiklund. 1994. Unpredictable food and sexual size dimorphism in insects. *Proceedings of the Royal Society of London B.* 258:121–125.

Lembke, H.F., and D.G. Cochran. 1990. Diet selection by adult female *Parcoblatta fulvescens* cockroaches during the oothecal cycle. *Comparative Biochemistry and Physiology.* 95A:195–199.

Lenoir-Rousseaux, J.J., and T. Lender. 1970. Table de développement embryonnaire de *Periplaneta americana* (L.) Insecte, Dictyoptere. *Bulletin de la Societe Zoologique de France.* 95:737–751.

Lepschi, B.J. 1989. A preliminary note on the food of *Imblattella orchidae* Asahina (Blattellidae). *Australian Entomological Magazine.* 16:41–42.

Leuthold, R. 1966. Die Bewegungsaktivät der Weiblichen Schab *Leucophaea maderae* (F.) im Laufe des Fortpflazungszyklus und ihre experimentelle Beeinflussung. *Journal of Insect Physiology.* 12:1303–1333.

Lewis, S.M., and S.N. Austad. 1990. Sources of intraspecific variation in sperm precedence in red flour beetles. *The American Naturalist.* 135:351–359.

Liang, D., G.J. Blomquist, and J. Silverman. 2001. Hydrocarbon-released nestmate aggression in the Argentine ant, *Linepithema humile,* following encounters with insect prey. *Comparative Biochemistry and Physiology.* 129B:871–882.

Liang, D., and C. Schal. 1994. Neural and hormonal regulation of calling behavior in *Blattella germanica* females. *Journal of Insect Physiology.* 40:251–258.

Lieberstat, F., and J. Camhi. 1988. Control of sensory feedback by movement during flight in the cockroach. *Journal of Experimental Biology.* 136:483–488.

Liechti, P.M., and W.J. Bell. 1975. Brooding behavior of the Cuban burrowing cockroach *Byrsotria fumigata* (Blaberidae, Blattaria). *Insectes Sociaux.* 22:35–46.

Lilburn, T.G., K.S. Kim, N.E. Ostrom, K.R. Byzek, J.R. Leadbetter, and J.A. Breznak. 2001. Nitrogen fixation by symbiotic and free-living spirochetes. *Science.* 292:2495–2498.

Lin, T.M., and H.J. Lee. 1998. Parallel control mechanisms underlying locomotor activity and sexual receptivity of the female German cockroach, *Blattella germanica* (L.). *Journal of Insect Physiology.* 44:1039–1051.

Linksvayer, T.A., and M.J. Wade. 2005. The evolutionary origin and elaboration of sociality in the aculeate Hymenoptera: maternal effects, sib-social effects, and heterochrony. *Quarterly Review of Biology.* 80:317–336.

Livingstone, D., and R. Ramani. 1978. Studies on the reproductive biology. *Proceedings of the Indian Academy of Sciences B.* 87:229–247.

Lloyd, M. 1963. Numerical observations on the movements of animals between beech litter and fallen branches. *Journal of Animal Ecology.* 32:157–163.

Lo, N. 2003. Molecular phylogenetics of Dictyoptera: Insights into the evolution of termite eusociality and bacterial endosymbiosis in cockroaches. *Entomologische Abhandlungen.* 61:137–138.

Lo, N., C. Bandi, H. Watanabe, C.A. Nalepa, and T. Beninati. 2003a. Evidence for cocladogenesis between diverse dictyopteran lineages and their intracellular symbionts. *Molecular Biology and Evolution.* 20:907–913.

Lo, N. P. Luykx, R. Santoni, T. Beninati, C. Bandi, M. Casiraghi, W. Lu, E.V. Zakharov, and C.A. Nalepa. 2006. Molecular phylogeny of *Cryptocercus* wood-roaches based on mitochondrial COII and 16S sequences, and chromosome numbers in Palearctic representatives. *Zoological Science.* 23:393–398.

Lo, N., G. Tokuda, H. Wantanabe, H. Rose, M. Slaytor, K. Maekawa, C. Bandi, and H. Noda. 2000. Evidence from multiple gene sequences indicates that termites evolved from wood-feeding cockroaches. *Current Biology.* 10:801–804.

Lo, N., H. Watanabe, and M. Sugimura. 2003b. Evidence for the presence of a cellulase gene in the last common ancestor of bilateran animals. *Proceedings of the Royal Society of London B* (Supplement) 270:S69–S72.

Lockhart, A.B., P.H. Thrall, and J. Antonovics. 1996. Sexually transmitted diseases in animals: ecological and evolutionary implications. *Biological Reviews.* 71:415–471.

Lodha, B.C. 1974. Decomposition of digested litter. *In* Biology of Plant Litter Decomposition. C.H. Dickinson and G.J.P. Pugh, editors. Academic Press, London. 213–241.

Loher, W. 1990. Pheromones and phase transformation in locusts. *In* Biology of Grasshoppers. R.F. Chapman and A. Joern, editors. John Wiley & Sons, New York. 337–355.

Lopes, R.B., and S.B. Alves. 2005. Effect of *Gregarina* sp. parasitism on the susceptibility of *Blattella germanica* to some control agents. *Journal of Invertebrate Pathology.* 88:261–264.

Lott, D.F. 1991. Intraspecific Variation in the Social Systems of Wild Vertebrates. Cambridge University Press, Cambridge. 238 pp.

Lusis, O., T. Sandor, and J.-G. Lehoux. 1970. Histological and histochemical observations on the testes of *Byrsotria fumigata* Guer. and *Gromphadorhina portentosa. Canadian Journal of Zoology.* 48:25–30.

Lyubechanskii, I.I., and I.E. Smelyanskii. 1999. Structure of saprophagous invertebrate community on a catena in the steppe of trans-volga region. *Zoologicheskii Zhurnal.* 78:672–680.

MacDonald, J.F., and R.W. Matthews. 1983. Colony associates of the southern yellowjacket, *Vespula squamosa* (Drury). *Journal of the Georgia Entomological Society.* 18:555–559.

Machin, J., J.J.B. Smith, and G.J. Lampert. 1994. Evidence for hydration dependent closing of pore structures in the cuticle of *Periplaneta americana. Journal of Experimental Biology.* 192:83–94.

Mackerras, M.J. 1965a. Australian Blattidae (Blattodea). I. General remarks and revision of the genus *Polyzosteria* Burmeister. *Australian Journal of Zoology.* 13:841–882.

Mackerras, M.J. 1965b. Australian Blattidae (Blattodea). II. Revision of the genus *Euzosteria* Shelford. *Australian Journal of Zoology.* 13:883–902.

Mackerras, M.J. 1967a. Australian Blattidae (Blattodea). VI. Revision of the genus *Cosmoszoteria* Stal. *Australian Journal of Zoology.* 15:593–618.

Mackerras, M.J. 1967b. Australian Blattidae (Blattodea). VII. The *Platyzosteria* group, general remarks and revision of the subgenera *Platyzosteria* Brunner and *Leptozosteria* Tepper. *Australian Journal of Zoology.* 15:1207–1298.

Mackerras, M.J. 1967c. A blind cockroach from caves in the Nullarbor Plain (Blattodea: Blattellidae). *Journal of the Australian Entomological Society.* 6:39–44.

Mackerras, M.J. 1968a. Australian Blattidae (Blattodea). VIII. The *Platyzosteria* group; subgenus *Melanozosteria* Stal. *Australian Journal of Zoology.* 16:237–331.

Mackerras, M.J. 1968b. *Neolaxta monteithi*, gen. et sp. n. from eastern Australia (Blattodea: Blaberidae). *Journal of the Australian Entomological Society.* 7:143–146.

Mackerras, M.J. 1970. Blattodea (Cockroaches). *In* Insects of Australia. CSIRO, University of Melbourne Press, Melbourne. 262–274.

Maddrell, S.H.P., and B.O.C. Gardiner. 1980. The permeability of the cuticular lining of the insect alimentary canal. *Journal of Experimental Biology.* 85:227–237.

Maekawa, K., M. Kon, and K. Araya. 2005. New species of the genus *Salganea* (Blattaria, Blaberidae, Panesthiinae) from Myanmar, with molecular phylogenetic analyses and notes on social structure. *Entomological Science.* 8:121–129.

Maekawa, K., M. Kon, K. Araya, and T. Matsumoto. 2001. Phylogeny and biogeography of wood-feeding cockroaches, genus *Salganea* Stål (Blaberidae: Panesthiinae), in Southeast Asia based on mitochondrial DNA sequences. *Journal of Molecular Evolution.* 53:651–659.

Maekawa, K., N. Lo, O. Kitade, T. Miura, and T. Matsumoto. 1999a. Molecular phylogeny and geographic distribution of wood-feeding cockroaches in East Asian Islands. *Molecular Phylogenetics and Evolution.* 13:360–376.

Maekawa, K., N. Lo, H.A. Rose, and T. Matsumoto. 2003. The evolution of soil-burrowing cockroaches (Blattaria: Blaberidae) from wood-burrowing ancestors following an invasion of the latter from Asia into Australia. *Proceedings of the Royal Society of London B.* 270:1301–1307.

Maekawa, K., M. Terayama, M. Maryati, and T. Matsumoto. 1999b. The subsocial wood-feeding cockroach genus *Salganea* Stål from Borneo, with description of a new species (Blaberidae: Panesthiinae). *Oriental Insects.* 33:233–242.

Mamaev, B.M. 1973. Ecology of the relict cockroach (*Cryptocercus relictus*). *Ekologiya.* 4:70–73.

Mani, M.S. 1968. Ecology and Biogeography of High Altitude Insects. Dr W.S. Junk N.V. Publishers, Belinfante, NV. 527 pp.

Manning, A., and G. Johnstone. 1970. The effects of early adult experience on the development of aggressiveness in males of the cockroach, *Nauphoeta cinerea. Revue du Comportement Animal.* 4:12–16.

Manton, S.M. 1977. The Arthropoda: Habits, Functional Morphology, and Evolution. Clarendon Press, Oxford. 527 pp.

Markow, T.A. 1995. Evolutionary ecology and developmental stability. *Annual Review of Entomology.* 40:105–120.

Marks, E.P., and F.A. Lawson. 1962. A comparative study of the Dictyopteran ovipositor. *Journal of Morphology.* 111:139–172.

Marooka, S., and S. Tojo. 1992. Maintenance and selection of strains exhibiting specific wing form and body colour under high density conditions in the brown planthopper *Nilaparvata lugens* (Homoptera: Delphacidae). *Applied Entomology and Zoology.* 27:445–454.

Marquis, D. 1935a. archygrams. *In* the lives and times of archy and mehitabel. Doubleday Doran & Co., Garden City, NY. 257–260.

Marquis, D. 1935b. quarantined. *In* the lives and times of archy and mehitabel. Doubleday Doran & Co., Garden City, NY. 433.

Marquis, D. 1935c. a wail from little archy. *In* the lives and times of archy and mehitabel. Doubleday Doran & Co., Garden City, NY. 362–363.

Martin, J.L., and I. Izquierdo. 1987. Two new subterranean *Loboptera* Brun. and W. from El Hierro Island Canary Islands Spain Blattaria Blattellidae. *Fragmenta Entomologica.* 19:301–310.

Martin, J.L., and P. Oromi. 1987. Three new species of subterranean *Loboptera* Brun. and W. Blattaria Blattellidae and considerations on the subterranean environment of Tenerife Canary Islands Spain. *Annales de la Societe Entomologique de France.* 23:315–326.

Martin, M.M., and J.J. Kukor. 1984. Role of mycophagy and bacteriophagy in invertebrate nutrition. *In* Current Perspectives in Microbial Ecology. M.J. Klug and C.A. Reddy, editors. American Society for Microbiology, Washington, DC. 257–263.

Masaki, S., and T. Shimizu. 1995. Variability in wing form of crickets. *Research on Population Ecology.* 37:119–128.

Matsuda, R. 1979. Abnormal metamorphosis in arthropod evolution. *In* Arthropod Phylogeny. A.P. Gupta, editor. Van Nostrand Reinhold Co., New York. 137–256.

Matsuda, R. 1987. Animal Evolution in Changing Environments with Special Reference to Abnormal Metamorphosis. John Wiley & Sons, New York. 355 pp.

Matsumoto, T. 1987. Colony composition of the subsocial wood-feeding cockroaches *Salganea taiwanensis* Roth and *S. esakii* Roth (Blattaria: Panesthiinae). *In* Chemistry and Biology of Social Insects. J. Eder and H. Rembold, editors. Verlag J. Peperny, Munchen. 394.

Matsumoto, T. 1988. Colony composition of the wood-feeding cockroach, *Panesthia australis* Brunner (Blattaria, Blaberidae, Panesthiinae) in Australia. *Zoological Science.* 5:1145–1148.

Matsumoto, T. 1992. Familial association, nymphal development and population density in the Australian giant burrowing cockroach, *Macropanesthia rhinoceros* (Blattaria: Blaberidae). *Zoological Science.* 9:835–842.

Matsuura, K. 2001. Nestmate recognition mediated by intestinal bacteria in a termite, *Reticulitermes speratus. Oikos.* 92:20–26.

Matthews, E.G. 1976. Insect Ecology. University of Queensland Press, St. Lucia, Queensland. 226 pp.

McBrayer, J.F. 1973. Exploitation of deciduous leaf litter by *Apheloria montana* (Diplopoda: Eurydesmidae). *Pedobiologia.* 13:90–98.

McClure, H.E. 1965. Microcosms of Batu Caves. *Malayan Nature Journal.* 19:65–74.

McFall-Ngai, M.J. 2002. Unseen forces: the influence of bacteria on animal development. *Developmental Biology.* 242:1–14.

McFarlane, J.E., and I. Alli. 1985. Volatile fatty acids of frass of certain omnivorous insects. *Journal of Chemical Ecology.* 11:59–63.

McKeown, K.C. 1945. Australian Insects. Royal Zoological Society of New South Wales, Sydney. 303 pp.

McKinney, M.L. 1990. Trends in body size evolution. *In* Evolutionary Trends. K.J. McNamara, editor. University of Arizona Press, Tucson. 75–118.

McKittrick, F.A. 1964. Evolutionary Studies of Cockroaches.

Cornell University Agricultural Experiment Station Memoir. 197 pp.
McKittrick, F.A. 1965. A contribution to the understanding of cockroach-termite affinities. *Annals of the Entomological Society of America.* 58:18–22.
McKittrick, F.A., T. Eisner, and H.E. Evans. 1961. Mechanics of species survival. *Natural History.* 70:46–50.
McMahan, E.A. 1969. Feeding relationships and radioisotope techniques. *In* Biology of Termites. Vol. 1. K. Krishna and F.M. Weesner, editors. Academic Press, New York. 387–406.
McMonigle, O., and R. Willis. 2000. Allpet Roaches. Elytra and Antenna, Brunswick, OH. 40 pp.
Meadows, P.S. 1991. The environmental impact of burrows and burrowing animals—conclusions and a model. *Symposium of the Zoological Society of London.* 63:327–338.
Melampy, R.M., and L.A. Maynard. 1937. Nutrition studies with the cockroach, *Blattella germanica. Physiological Zoology.* 10:36–44.
Meller, P., and H. Greven. 1996a. Beobachtungen zur Laufgeschwindigkeit der viviparen Schabe *Nauphoeta cinerea* während des Fortpflanzungszyklus. *Acta Biologica Benrodis.* 8:19–31.
Meller, P., and H. Greven. 1996b. Locomotor activity patterns of the viviparous cockroach *Nauphoeta cinerea* and their relation to the reproductive cycle. *Zoologische Beitraege.* 37:217–245.
Mensa-Bonsu, A. 1976. The biology and development of *Porotermes adamsoni* (Froggatt) (Isoptera, Hodotermitidae). *Insectes Sociaux.* 23:155–156.
Messer, A.C., and M.J. Lee. 1989. Effect of chemical treatments on methane emission by the hindgut microbiota in the termite *Zootermopsis angusticollis. Microbial Ecology.* 18:275–284.
Metzger, R. 1995. Behavior. *In* Understanding and Controlling the German Cockroach. M.K. Rust, J.M. Owens, and D.A. Rierson, editors. Oxford University Press, New York. 49–76.
Michener, C.D. 1969. Comparative social behavior of bees. *Annual Review of Entomology.* 14:299–342.
Miller, D.M., and P.G. Koehler. 2000. Trail-following behavior in the German cockroach (Dictyoptera: Blattellidae). *Journal of Economic Entomology.* 93:1241–1246.
Miller, P.L. 1981. Respiration. *In* The American Cockroach. W.J. Bell and K.G. Adiyodi, editors. Chapman and Hall, London. 87–116.
Miller, P.L. 1990. Mechanisms of sperm removal and sperm transfer in *Orthetrum coerulescens* (Fabricius) (Odonata: Libellulidae). *Physiological Entomology.* 15:199–209.
Mira, A. 2000. Exuviae eating: a nitrogen meal? *Journal of Insect Physiology.* 46:605–610.
Mira, A., and D. Raubenheimer. 2002. Divergent nutrition-related adaptations in two cockroach populations inhabiting different environments. *Physiological Entomology.* 27:330–339.
Mistal, C., S. Takács, and G. Gries. 2000. Evidence for sonic communication in the German cockroach (Dictyoptera: Blattellidae). *The Canadian Entomologist.* 132:867–876.
Mizuno, T., and H. Tsuji. 1974. Harbouring behaviour of three species of cockroaches, *Periplaneta americana, P. japonica,* and *Blattella germanica. Japanese Journal of Sanitary Zoology.* 24:237–240.
Mock, D.W., T.C. Lamey, and D.B.A. Thompson. 1988. Falsifiability and the information center hypothesis. *Ornis Scandinavica.* 19:231–248.
Mohan, C.M., K.A. Lakshmi, and K.U. Devi. 1999. Laboratory evaluation of the pathogenicity of three isolates of the entomopathogenic fungus *Beauveria bassiana* (Bals.) vuillemin on the American cockroach (*Periplaneta americana*). *Biocontrol Science and Technology.* 9:29–33.
Moir, R.J. 1994. The 'carnivorous' herbivores. *In* The Digestive System in Mammals: Food, Form and Function. D.J. Chivers and P. Langer, editors. Cambridge University Press, Cambridge. 87–102.
Møller, A.P., S. Merino, C.R. Brown, and R.J. Robertson. 2001. Immune defense and host sociality: a comparative study of swallows and martins. *The American Naturalist.* 158:136–145.
Montrose, V.T., W.E. Harris, and P.J. Moore. 2004. Sexual conflict and cooperation under naturally occurring male enforced monogamy. *Journal of Evolutionary Biology.* 17:443–451.
Moore, A.J. 1990. Sexual selection and the genetics of pheromonally mediated social behavior in *Nauphoeta cinerea* (Dictyoptera: Blaberidae). *Entomologia Generalis.* 15:133–147.
Moore, A.J., and M.D. Breed. 1986. Mate assessment in a cockroach, *Nauphoeta cinerea. Animal Behaviour.* 34:1160–1165.
Moore, A.J., P.A. Gowaty, and P.J. Moore. 2003. Females avoid manipulative males and live longer. *Journal of Evolutionary Biology.* 16:523–530.
Moore, A.J., P.A. Gowaty, W. Wallin, and P.J. Moore. 2001. Fitness costs of sexual conflict and the evolution of female mate choice and male dominance. *Proceedings of the Royal Society of London B.* 268:517–523.
Moore, A.J., K.F. Haynes, R.F. Preziosi, and P.J. Moore. 2002. The evolution of interacting phenotypes: genetics and evolution of social dominance. *The American Naturalist.* 160:S143–S159.
Moore, A.J., N.L. Reagan, and K.F. Haynes. 1995. Conditional signalling strategies: effects of ontogeny, social experience and social status on the pheromonal signal of male cockroaches. *Animal Behaviour.* 50:191–202.
Moore, J., and P. Willmer. 1997. Convergent evolution in invertebrates. *Biological Reviews.* 72:1–60.
Moore, P.J., and W.E. Harris. 2003. Is a decline in offspring quality a necessary consequence of maternal age? *Proceedings of the Royal Society of London B.* 270:S192–S194.
Moore, P.J., and A.J. Moore. 2001. Reproductive ageing and mating: the ticking of the biological clock in female cockroaches. *Proceedings of the National Academy of Sciences.* 98:9171–9178.
Moore, T.E., S.B. Crary, D.E. Koditschek, and T.A. Conklin. 1998. Directed locomotion in cockroaches: "biobots." *Acta Entomologica Slovenica.* 6:71–78.
Moran, N.A. 2002. The ubiquitous and varied role of infection in the lives of animals and plants. *The American Naturalist.* 160:S1–S8.
Moret, Y., and P. Schmidt-Hempel. P. 2000. Survival for im-

munity: the price of immune system activation for bumblebee workers. *Science.* 290:1166–1168.
Morley, C. 1921. Nursery rhymes for the tender-hearted. *In* Hide and Seek. George H. Doran Co., New York. 120.
Morse, D.H. 1980. Behavioural Mechanisms in Ecology. Harvard University Press, Cambridge, MA. 383 pp.
Moser, J.C. 1964. Inquiline roach responds to trail-marking substance of leaf-cutting ants. *Science.* 143:148–149.
Mukha, D., B.M. Wiegmann, and C. Schal. 2002. Evolution and phylogenetic information content of the ribosomal DNA repeat in the Blattodea (Insecta). *Insect Biochemistry and Molecular Biology.* 32:951–960.
Mullins, D.E. 1982. Osmoregulation and excretion. *In* The American Cockroach. W.J. Bell and K.G. Adiyodi, editors. Chapman and Hall, London. 117–149.
Mullins, D.E., and D.G. Cochran. 1975a. Nitrogen metabolism in the American cockroach. I. An examination of positive nitrogen balance with respect to uric acid stores. *Comparative Biochemistry and Physiology.* 50A:489–500.
Mullins, D.E., and D.G. Cochran. 1975b. Nitrogen metabolism in the American cockroach. II. An examination of negative nitrogen balance with respect to mobilization of uric acid stores. *Comparative Biochemistry and Physiology.* 50A:501–510.
Mullins, D.E., and D.G. Cochran. 1987. Nutritional ecology of cockroaches. *In* Nutritional Ecology of Insects, Mites, Spiders, and Related Invertebrates. F.J. Slansky and J.G. Rodriguez, editors. John Wiley & Sons, New York. 885–902.
Mullins, D.E., and C.B. Keil. 1980. Paternal investment of urates in cockroaches. *Nature.* 283:567–569.
Mullins, D.E., C.B. Keil, and R.H. White. 1992. Maternal and paternal nitrogen investment in *Blattella germanica* (L.) (Dictyoptera; Blattellidae). *Journal of Experimental Biology.* 162:55–72.
Mullins, D.E., K.J. Mullins, and K.R. Tignor. 2002. The structural basis for water exchange between the female cockroach (*Blattella germanica*) and her ootheca. *Journal of Experimental Biology.* 205:2987–2996.
Murray, D.A.H., and R. Wicks. 1990. Injury levels for soil-dwelling insects in sunflower in the Central Highlands, Queensland. *Australian Journal of Experimental Agriculture.* 30:669–674.
Myles, T.G. 1986. Evidence of parental and/or sibling manipulation in three species of termites from Hawaii. *Proceedings of the Hawaiian Entomological Society.* 27:129–136.
Nadkarni, N.M., and J.T. Longino. 1990. Invertebrates in canopy and ground organic matter in a Neotropical montane forest, Costa Rica. *Biotropica.* 22:286–289.
Nagamitsu, T., and T. Inoue. 1997. Cockroach pollination and breeding system of *Uvaria elmeri* (Annonaceae) in a lowland mixed dipterocarp forest in Sarawak. *American Journal of Botany.* 84:208–213.
Nakashima, K., H. Watanabe, H. Saitoh, G. Tokuda, and J.I. Azuma. 2002. Dual cellulose-digesting system of the wood-feeding termite *Coptotermes formosanus. Insect Biochemistry and Molecular Biology.* 32:777–784.
Nalepa, C.A. 1984. Colony composition, protozoan transfer and some life history characteristics of the woodroach *Cryptocercus punctulatus* Scudder. *Behavioral Ecology and Sociobiology.* 14:273–279.
Nalepa, C.A. 1987. Life history studies of the woodroach *Cryptocercus punctulatus* Scudder (Dictyoptera: Cryptocercidae) and their implications for the evolution of termite eusociality. Ph.D. thesis, Entomology Department. North Carolina State University, Raleigh.
Nalepa, C.A. 1988a. Reproduction in the woodroach *Cryptocercus punctulatus* Scudder (Dictyoptera: Cryptocercidae): mating, oviposition and hatch. *Annals of the Entomological Society of America.* 81:637–641.
Nalepa, C.A. 1988b. Cost of parental care in *Cryptocercus punctulatus* Scudder (Dictyoptera: Cryptocercidae). *Behavioral Ecology and Sociobiology.* 23:135–140.
Nalepa, C.A. 1990. Early development of nymphs and establishment of hindgut symbiosis in *Cryptocercus punctulatus* (Dictyoptera: Cryptocercidae). *Annals of the Entomological Society of America.* 83:786–789.
Nalepa, C.A. 1991. Ancestral transfer of symbionts between cockroaches and termites: an unlikely scenario. *Proceedings of the Royal Society of London B.* 246:185–189.
Nalepa, C.A. 1994. Nourishment and the evolution of termite eusociality. *In* Nourishment and Evolution in Insect Societies. J.H. Hunt and C.A. Nalepa, editors. Westview Press, Boulder. 57–104.
Nalepa, C.A. 1996. Evolution of eusociality in termites: role of altricial offspring. *In* Proceedings of the 20th International Congress of Entomology, Florence, Italy. 396.
Nalepa, C.A. 2001. *Cryptocercus punctulatus* (Dictyoptera: Cryptocercidae) from monadnocks in the Piedmont of North Carolina. *Journal of Entomological Science.* 36:329–334.
Nalepa, C.A. 2003. Evolution in the genus *Cryptocercus* (Dictyoptera: Cryptocercidae): no evidence of differential adaptation to hosts or elevation. *Biological Journal of the Linnean Society.* 80:223–233.
Nalepa, C.A. 2005. *Cryptocercus punctulatus* (Dictyoptera, Cryptocercidae): dispersal events associated with rainfall. *Entomologist's Monthly Magazine.* 141:95–97.
Nalepa, C.A., and C. Bandi. 1999. Phylogenetic status, distribution, and biogeography of *Cryptocercus* (Dictyoptera: Cryptocercidae). *Annals of the Entomological Society of America.* 92:292–302.
Nalepa, C.A., and C. Bandi. 2000. Characterizing the ancestors: paedomorphosis and termite evolution. *In* Termites: Evolution, Sociality, Symbioses, Ecology. T. Abe, D.E. Bignell, and M. Higashi, editors. Kluwar Academic, Dordrecht. 53–75.
Nalepa, C.A., and W.J. Bell. 1997. Postovulation parental investment and parental care in cockroaches. *In* Social Behavior in Insects and Arachnids. J.C. Choe and B.J. Crespi, editors. Cambridge University Press, Cambridge. 26–51.
Nalepa, C.A., D.E. Bignell, and C. Bandi. 2001a. Detritivory, coprophagy, and the evolution of digestive mutualisms in Dictyoptera. *Insectes Sociaux.* 48:194–201.
Nalepa, C.A., G.W. Byers, C. Bandi, and M. Sironi. 1997. Description of *Cryptocercus clevelandi* (Dictyoptera: Cryptocercidae) from the northwestern United States, molecular analysis of bacterial symbionts in its fat body, and notes

on biology, distribution and biogeography. *Annals of the Entomological Society of America.* 90:416–424.

Nalepa, C.A., and S.C. Jones. 1991. Evolution of monogamy in termites. *Biological Reviews.* 66:83–97.

Nalepa, C.A., and M. Lenz. 2000. The ootheca of *Mastotermes darwiniensis* Froggatt (Isoptera: Mastotermitidae): homology with cockroaches. *Proceedings of the Royal Society of London B.* 267:1809–1813.

Nalepa, C.A., L. Li, W. Lu, and J. Lazell. 2001b. Rediscovery of the wood-eating cockroach *Cryptocercus primarius* (Dictyoptera: Cryptocercidae) in China, with notes on ecology and distribution. *Acta Zootaxonomica Sinica.* 26:184–190.

Nalepa, C.A., P. Luykx, K.-D. Klass, and L.L. Deitz. 2002. Distribution of karyotypes of the *Cryptocercus punctulatus* species complex (Dictyoptera: Cryptocercidae) in the Southern Appalachians: relation to habitat and history. *Annals of the Entomological Society of America.* 95:276–287.

Nalepa, C.A., and D.E. Mullins. 1992. Initial reproductive investment and parental body size in *Cryptocercus punctulatus* Scudder (Dictyoptera: Cryptocercidae). *Physiological Entomology.* 17:255–259.

Narasimham, A.U. 1984. Comparative studies on *Tetrastichus hagenowii* (Ratzeburg) and *T. asthenogmus* (Waterson), two primary parasites of cockroach oothecae, and on their hyperparasite *Tetrastichus* sp. (*T. miser* (Nees) group) (Hymenoptera: Eulophidae). *Bulletin of Entomological Research.* 74:175–189.

Nardi, J.B., R.I. Mackie, and J.O. Dawson. 2002. Could microbial symbionts of arthropod guts contribute significantly to nitrogen fixation in terrestrial ecosystems? *Journal of Insect Physiology.* 48:751–763.

Naskrecki, P. 2005. The Smaller Majority. Harvard University Press, Cambridge, MA. 278 pp.

Naylor, L.S. 1964. The structure and function of the posterior abdominal glands of the cockroach *Pseudoderopeltis bicolor* (Thunb.). *Journal of the Entomological Society of South Africa.* 27:62–66.

Nevo, E. 1999. Mosaic Evolution of Subterranean Mammals: Regression, Progression, and Global Convergence. Oxford University Press, Oxford. 413 pp.

Nigam, L.N. 1932. The life history of a common cockroach (*Periplaneta americana* Linneus). *Indian Journal of Agricultural Science.* 3:530–543.

Nijhout, H.F., and D.E. Wheeler. 1982. Juvenile hormones and the physiological basis of insect polymorphisms. *Quarterly Review of Biology.* 57:109–133.

Niklasson, M., and E.D.J. Parker. 1994. Fitness variation in an invading parthenogenetic cockroach. *Oikos.* 71:47–54.

Niklasson, M., and E.D.J. Parker. 1996. Human commensalism in relation to geographic parthenogenesis and colonizing/invading ability. *Journal of Evolutionary Biology.* 9:1027–1028.

Nishida, G.M. 1992. Hawaiian Terrestrial Arthropod Checklist. Bishop Museum, Honolulu, HI. 262 pp.

Nishida, R., H. Fukami, and S. Ishii. 1974. Sex pheromone of the German cockroach (*Blattella germanica* L.) responsible for male wing raising: 3,11-Dimethyl-2-nonacosanone. *Experientia.* 30:978–979.

Noda, S., T. Inoue, Y. Hongoh, M. Kawai, C.A. Nalepa, C. Vongkaluang, T. Kudo, and M. Ohkuma. 2006. Identification and characterization of ectosymbionts of distinct lineages in *Bacteroidales* attached to flagellated protists in the gut of termites and a wood-feeding cockroach. *Environmental Microbiology.* 8:11–20.

Noirot, C. 1985. Pathways of caste development in the lower termites. *In* Caste Differentiation in Social Insects. J.A.L. Watson, B.M. Okot-Kotber, and C. Noirot, editors. Pergamon Press, New York. 41–57.

Noirot, C. 1995. The gut of termites (Isoptera). Comparative anatomy, sytematics, phylogeny. I. Lower termites. *Annales de la Societe Entomologique de France, nouvelle série.* 31:197–226.

Noirot, C., and C. Bordereau. 1989. Termite polymorphism and morphogenetic hormones. *In* Morphogenetic Hormones of Arthropods: Roles in Histogenesis, Organogenesis, and Morphogenesis. A.P. Gupta, editor. Rutgers University Press, New Brunswick, NJ. 293–324.

Noirot, C., and J.P.E.C. Darlington. 2000. Termite nests: architecture, regulation and defence. *In* Termites: Evolution, Sociality, Symbioses, Ecology. T. Abe, D.E. Bignell, and M. Higashi, editors. Kluwar Academic Publishers, Dordrecht. 121–139.

Noirot, C., and J.M. Pasteels. 1987. Ontogenetic development and evolution of the worker caste in termites. *Experientia.* 43:851–860.

Noirot, C., and A. Quennedy. 1974. Fine structure of insect epidermal glands. *Annual Review of Entomology.* 19:61–80.

Nojima, S., G.M. Nishida, and Y. Kuwahara. 1999a. Nuptial feeding stimulants: A male courtship pheromone of the German cockroach, *Blattella germanica* (L.) (Dictyoptera: Blattellidae). *Naturwissenschaften.* 86:193–196.

Nojima, S., M. Sakuma, R. Nishida, and Y. Kuwahara. 1999b. A glandular gift in the German cockroach, *Blattella germanica* (L.) (Dictyoptera: Blattellidae): the courtship feeding of a female on secretions from male tergal glands. *Journal of Insect Behavior.* 12:627–640.

Nojima, S., C. Schal, F.X. Webster, R.G. Santangelo, and W.L. Roelofs. 2005. Identification of the sex pheromone of the German cockroach, *Blattella germanica. Science.* 307 (5712):1104–1106.

North, F.J. 1929. Insect life in the coal forests, with special reference to South Wales. *Report and Transactions of the Cardiff Naturalist's Society.* 62:16–44.

Nosil, P. 2001. Tarsal asymmetry, nutritional condition, and survival in water boatmen (*Callicorixa vulnerata*). *Evolution.* 55:712–720.

Novotny, V., Y. Basset, and R.L. Kitching. 2003. Herbivore assemblages and their food resources. *In* Arthropods of Tropical Forests: Spatio-Temporal Dynamics and Resource Use in the Canopy. Y. Basset, V. Novotny, S.E. Miller, and R.L. Kitching, editors. Cambridge University Press, Cambridge. 40–53.

Nutting, W.L. 1953a. Giant cockroaches of the genus *Blaberus* as laboratory animals. *Turtox News.* 31:134–136.

Nutting, W.L. 1953b. Observations on the reproduction of the giant cockroach *Blaberus craniifer. Psyche.* 60:6–14.

Nutting, W.L. 1969. Flight and colony foundation. *In* Biology of Termites. Vol. 1. K. Krishna and F.M. Weesner, editors. Academic Press, New York. 233–282.

O'Donnell, M.J. 1977a. Hypopharyngeal bladders and frontal

glands: novel structures involved in water vapour absorption in the desert cockroach, *Arenivaga investigata. American Zoologist.* 17:902.
O'Donnell, M.J. 1977b. Site of water vapour absorption in the desert cockroach *Arenivaga investigata. Proceedings of the National Academy of Sciences USA.* 74:1757–1760.
O'Donnell, M.J. 1981. Fluid movements during water vapour absorption by the desert burrowing cockroach *Arenivaga investigata. Journal of Insect Physiology.* 27:877–887.
O'Donnell, M.J. 1982. Hydrophilic cuticle—the basis for water vapour absorption by the desert burrowing cockroach, *Arenivaga investigata. Journal of Experimental Biology.* 99:43–60.
Odum, E.P., and L.J. Biever. 1984. Resource quality, mutualism, and energy partitioning in food chains. *The American Naturalist.* 124:360–376.
Ohkuma, M. 2003. Termite symbiotic systems: efficient biorecycling of lignocellulose. *Applied Microbiology and Biotechnology.* 61:1–9.
O'Neill, S.L., H.A. Rose, and D. Rugg. 1987. Social behaviour and its relationship to field distribution in *Panesthia cribrata* Saussure (Blattodea: Blaberidae). *Journal of the Australian Entomological Society.* 26:313–321.
Otronen, M. 1997. Sperm numbers, their storage and use in the fly *Dryomyza anilis. Proceedings of the Royal Society of London B.* 264:777–782.
Otronen, M., P. Reguera, and P.I. Ward. 1997. Sperm storage in the yellow dung fly *Scathophaga stercoraria:* identifying the sperm of competing males in separate female spermathecae. *Ethnology.* 103:844–854.
Owens, J.M., and G.W. Bennett. 1983. Comparative study of German cockroach population sampling techniques. *Environmental Entomology.* 12:1040–1046.
Pachamuthu, P., S.T. Kamble, T.L. Clark, and J.E. Foster. 2000. Differentiation of three phenotypically similar *Blattella* spp.: analysis with polymerase chain reaction-restriction fragment length polymorphism of DNA. *Annals of the Entomological Society of America.* 93:1138–1146.
Page, R.E., Jr., and S.D. Mitchell. 1998. Self-organization and the evolution of division of labor. *Apidologie.* 29:171–190.
Paoletti, M.G., R.A.J. Taylor, B.R. Stinner, and D.H. Stinner. 1991. Diversity of soil fauna in the canopy and forest floor of a Venezuelan cloud forest. *Journal of Tropical Ecology.* 7:373–383.
Park, A. 1990. Guess who's coming to tea. *Australian Geographic.* 18:30–45.
Park, Y.-C., and J.C. Choe. 2003a. Effect of parental care on offspring growth in the Korean wood-feeding cockroach *Cryptocercus kyebangensis. Journal of Ethology.* 21:71–77.
Park, Y.-C., and J.C. Choe. 2003b. Territorial behavior of the Korean wood-feeding cockroach, *Cryptocercus kyebangensis. Journal of Ethology.* 21:79–85.
Park, Y.-C., P. Grandcolas, and J.C. Choe. 2002. Colony composition, social behavior and some ecological characteristics of the Korean wood-feeding cockroach (*Cryptocercus kyebangensis*). *Zoological Science.* 19:1133–1139.
Park, Y.-C., K. Maekawa, T. Matsumoto, R. Santoni, and J.C. Choe. 2004. Molecular phylogeny and biogeography of the Korean woodroaches *Cryptocercus* spp. *Molecular Phylogenetics and Evolution.* 30:450–464.
Parker, E.D.J. 2002. Geographic parthenogenesis in terrestrial invertebrates: generalist or specialist clones? *In* Reproductive Biology of Invertebrates. Vol. XI. R.N. Hughes, editor. John Wiley & Sons, Chichester. 93–114.
Parker, E.D.J., and M. Niklasson. 1995. Desiccation resistance among clones in the invading parthenogenetic cockroach, *Pycnoscelus surinamensis:* a search for the general-purpose genotype. *Journal of Evolutionary Biology.* 8:331–337.
Parker, E.D.J., R.K. Selander, R.O. Hudson, and L.J. Lester. 1977. Genetic diversity in colonizing parthenogenetic cockroaches. *Evolution.* 31:836–842.
Parker, G.A. 1970. Sperm competition and its evolutionary consequences in the insects. *Biological Reviews.* 45:525–567.
Parker, G.G. 1995. Structure and microclimate of forest canopies. *In* Forest Canopies. M.D. Lowman and N.M. Nadkarni, editors. Academic Press, San Diego. 73–106.
Parrish, J.K., and L. Edelstein-Keshet. 1999. Complexity, pattern, and evolutionary trade-offs in animal aggregation. *Science.* 284:99–101.
Paulian, R. 1948. Observations sur la faune entomologique des nids de Ploceinae. *In* Proceedings of the 8th International Congress of Entomology, Stockholm. 454–456.
Payne, K. 1973. Some aspects of the ecology and behaviour of *Ectobius pallidus* (Olivier) (Dictyoptera). *Entomologist's Gazette.* 24:67–74.
Peck, S.B. 1990. Eyeless arthropods of the Galapagos Islands, Ecuador: composition and origin of the cryptozoic fauna of a young, tropical oceanic archipelago. *Biotropica.* 22:366–381.
Peck, S.B. 1998. A summary of diversity and distribution of the obligate cave-inhabiting faunas of the United States and Canada. *Journal of Cave and Karst Studies.* 60:18–26.
Peck, S.B., and L.M. Roth. 1992. Cockroaches of the Galapagos Islands, Ecuador, with descriptions of three new species (Insects: Blattodea). *Canadian Journal of Zoology.* 70:2217.
Pellens, R., and P. Grandcolas. 2003. Living in Atlantic forest fragments: life habits, behaviour, and colony structure of the cockroach *Monastria biguttata* (Dictyoptera, Blaberidae, Blaberinae) in Espiritu Santo, Brazil. *Canadian Journal of Zoology.* 81:1929–1937.
Pellens, R., P. Grandcolas, and D. da Silva-Netro. 2002. A new and independently evolved case of xylophagy and the presence of intestinal flagellates in the cockroach *Parasphaeria boleiriana* (Dictyoptera, Blaberidae, Zetoborinae) from the remnants of the Brazilian Atlantic Forest. *Canadian Journal of Zoology.* 80:350–359.
Peppuy, A., A. Robert, E. Semin, C. Ginies, M. Lettere, O. Bonnard, and C. Bordereau. 2001. (*Z*)-dodec-3-en-1-ol, a novel termite trail pheromone identified after solid phase microextraction from *Macrotermes annandalei. Journal of Insect Physiology.* 47:445–453.
Perriere, C., and F. Goudey-Perriere. 1988. Enzymatic activities in *Blaberus craniifer* Burm. (Dictyoptere, Blaberidae) spermatophore. *Bulletin de la Societe Zoologique de France.* 113:401–410.
Perry, D.R. 1978. *Paratropes bilunata* (Orthoptera: Blattidae): an outcrossing pollinator in a Neotropical wet forest

canopy. *Proceedings of the Entomological Society of Washington.* 80:656–657.
Perry, D.R. 1986. Life Above the Jungle Floor. Simon and Schuster, New York. 170 pp.
Perry, J., and C.A. Nalepa. 2003. A new mode of parental care in cockroaches. *Insectes Sociaux.* 50:245–247.
Persad, A.B., and M.A. Hoy. 2004. Predation by *Solenopsis invicta* and *Blattella asahinai* on *Toxoptera citicida* parasitized by *Lysiphlebus testaceipes* and *Lipolexis oregmae* on citrus in Florida. *Biological Control.* 30:531–537.
Pettit, L.C. 1940. The effect of isolation on growth in the cockroach *Blattella germanica* (L.) (Orthoptera: Blattidae). *Entomological News.* 51:293.
Philippe, H., and A. Adoutte. 1996. What can phylogenetic patterns tell us about the evolutionary processes generating biodiversity? *In* Aspects of the Genesis and Maintenance of Biological Diversity. M.E. Hochberg, J. Clobert, and R. Barbault, editors. Oxford University Press, Oxford. 41–59.
Pimm, S.L., G.J. Russell, and J.L. Gittleman. 1995. The future of biodiversity. *Science.* 269:347–350.
Pipa, R.L. 1985. Effects of starvation, copulation, and insemination on oocyte growth and oviposition by *Periplaneta americana* (Dictyoptera: Blattidae). *Annals of the Entomological Society of America.* 78:284–290.
Pitnick, S., T. Markow, and G.S. Spicer. 1999. Evolution of multiple kinds of female sperm-storage organs in *Drosophila. Evolution.* 53:1804–1822.
Plante, C.J., P.A. Jumars, and J.A. Baross. 1990. Digestive associations between marine detritivores and bacteria. *Annual Review of Ecology and Systematics.* 21:93–127.
Poinar, G. 1999. *Paleochordodes protus* n.g., n.sp. (Nematomorpha, Chordodidae), parasites of a fossil cockroach, with a critical examination of other fossil hairworms and helminths of extant cockroaches (Insecta: Blattaria). *Invertebrate Biology.* 118:109–115.
Polis, G. 1979. Prey and feeding phenology of the desert sand scorpion *Paruroctonus mesaensis* (Scorpionidae: Vaejovidae). *Journal of Zoology, London.* 188:333–346.
Polis, G. 1981. The evolution and dynamics of intraspecific predation. *Annual Review of Ecology and Systematics.* 12:225–251.
Polis, G. 1984. Intraspecific predation and "infant killing" among invertebrates. *In* Infanticide: Comparative and Evolutionary Perspectives. G. Hausfater and S.B. Hrdy, editors. Aldine, New York. 87–104.
Polis, G. 1991. Food webs in desert communities: complexity via diversity and omnivory. *In* The Ecology of Desert Communities. G. Polis, editor. University of Arizona Press, Tucson. 383–437.
Pond, C.M. 1983. Parental feeding as a determinant of ecological relationships in Mesozoic terrestrial ecosystems. *Palaeontologica.* 28:215–224.
Pope, P. 1953. Studies of the life histories of some Queensland Blattidae (Orthoptera). Part 1. The domestic species. *Proceedings of the Royal Society of Queensland.* 63:23–46.
Postle, A.C. 1985. Density and seasonality of soil and litter invertebrates at Dwellingup. *In* Soil and Litter Invertebrates of Australian Mediterranean-type Ecosystems. Vol. 12. P. Greenslade and J.D. Majer, editors. Western Australian Institute School of Biology Bulletin, Bentley, WA. 18–19.
Potrikus, C.J., and J.A. Breznak. 1981. Gut bacteria recycle uric acid nitrogen in termites: a strategy for nutrient conservation. *Proceedings of the National Academy of Sciences USA.* 78:4601–4605.
Poulson, T.L., and K.H. Lavoie. 2000. The trophic basis of subsurface ecosystems. *In* Ecosystems of the World. Vol. 30: Subterranean Ecosystems. H. Wilkens, D.C. Culver, and W.F. Humphreys, editors. Elsevier, Amsterdam. 231–249.
Poulson, T.L., and W.B. White. 1969. The cave environment. *Science.* 165:971–981.
Preston-Mafham, R., and K. Preston-Mafham. 1993. The Encyclopedia of Land Invertebrate Behaviour. The MIT Press, Cambridge. 320 pp.
Price, P.W. 2002. Resource-driven terrestrial interaction webs. *Ecological Research.* 17:241–247.
Princis, K., and D.K.M. Kevan. 1955. Cockroaches from Trinidad, B.W.I., with a few records from other parts of the Caribbean. *Opuscula Entomologia.* 20:149–169.
Proctor, M., and P. Yeo. 1972. The pollination of flowers. Taplinger Pub. Co., New York. 418 pp.
Prokopy, R.J. 1983. Visual detection of plants by herbivorous insects. *Annual Review of Entomology.* 28:337–364.
Prokopy, R.J., and B.D. Roitberg. 2001. Joining and avoidance behavior in non-social insects. *Annual Review of Entomology.* 46:631–665.
Pruthi, H.S. 1933. An interesting case of maternal care in an aquatic cockroach, *Phlebobotus pallens* Serv. (Epilamprinae). *Current Science (Bangalore).* 1:273.
Raisbeck, B. 1976. An aggression stimulating substance in the cockroach, *Periplaneta americana. Annals of the Entomological Society of America.* 69:793–796.
Rajulu, J.S., and K. Renganathan. 1966. On the stabilization of the ootheca of the cockroach *Periplaneta americana. Naturwissenschaften.* 53:136.
Ramsay, G.W. 1990. Mantodea (Insecta), with a review of aspects of functional morphology and biology. *Fauna of New Zealand.* 19:1–96.
Rau, P. 1940. The life history of the American cockroach, *Periplaneta americana* Linn. (Orthop.: Blattidae). *Entomological News.* 51:121–124, 151–155,186–189, 223–227, 273–278.
Rau, P. 1941. Cockroaches: the forerunners of termites (Orthoptera: Blattidae; Isoptera). *Entomological News.* 52:256–259.
Rau, P. 1943. How the cockroach deposits its egg-case; a study in insect behavior. *Annals of the Entomological Society of America.* 36:221–226.
Raubenheimer, D., and S.A. Jones. 2006. Nutritional imbalance in an extreme generalist omnivore: tolerance and recovery through complementary food selection. *Animal Behaviour.* 71:2153–1262.
Reddy, M.V. 1995. Litter arthropods. *In* Soil Organisms and Litter Decomposition in the Tropics. M.V. Reddy, editor. Westview Press, Boulder. 113–140.
Redheuil, M.E. 1973. Contribution a l'etude de la morphologie et du comportement de *Panesthia.* Thesis, Diplome d'Etudes Approfondies Ethologie, L'Université de Rennes.

Rehn, J.A.G. 1931. African and Malagasy Blattidae (Orthoptera), Part I. *Proceedings of the Academy of Natural Sciences of Philadelphia.* 83:305–387.

Rehn, J.A.G. 1932a. African and Malagasy Blattidae (Orthoptera), Part II. *Proceedings of the Academy of Natural Sciences of Philadelphia.* 84:405–511.

Rehn, J.A.G. 1932b. On apterism and subapterism in the Blattinae (Orthoptera: Blattidae). *Entomological News.* 43:201–206.

Rehn, J.A.G. 1945. Man's uninvited fellow-traveller—the cockroach. *Scientific Monthly.* 61:265–276.

Rehn, J.A.G. 1965. A new genus of symbiotic cockroach from southwest Africa (Orthoptera: Blattaria: Oxyhaloinae). *Notulae Naturae.* 374:1–8.

Rehn, J.W.H. 1951. Classification of the Blattaria as indicated by their wings (Orthoptera). *Memoirs of the American Entomological Society.* 14:1–134.

Reilly, S.M. 1994. The ecological morphology of metamorphosis: heterochrony and the evolution of feeding mechanisms in salamanders. *In* Ecological Morphology: Integrative Organismal Biology. P.C. Wainright and S.M. Reilly, editors. University of Chicago Press, Chicago. 319–338.

Rentz, D.C. 1987. *Imblattella orchidae* Asahina, an introduced cockroach associated with orchids in Australia (Blattodea: Blattellidae). *Australian Entomological Society News Bulletin.* May:44–45.

Rentz, D.C. 1996. Grasshopper Country: The Abundant Orthopteroid Insects of Australia. University of New South Wales Press, Sydney. 284 pp.

Reuben, L.V. 1988. Some aspects of the bionomics of *Trichoblatta sericea* (Saussure) and *Thorax porcellana* (Saravas = Saussure) (Blattaria). Ph.D. thesis, Department of Zoology. Loyola College, Madras, India. 211 pp.

Richards, A.G. 1963. The rate of sperm locomotion in the cockroach as a function of temperature. *Journal of Insect Physiology.* 9:545–549.

Richards, A.G., and M.A. Brooks. 1958. Internal symbiosis in insects. *Annual Review of Entomology.* 3:37–56.

Richards, A.M. 1971. An ecological study of the cavernicolous fauna of the Nullarbor Plain Southern Australia. *Journal of Zoology, London.* 164:1–60.

Richner, H., and P. Heeb. 1995. Is the information center hypothesis a flop? *Advances in the Study of Behavior.* 24:1–45.

Richter, K., and D. Barwolf. 1994. Behavioural changes are related to moult regulation in the cockroach, *Periplaneta americana. Physiological Entomology.* 19:133–138.

Ridgel, A.L., R.E. Ritzmann, and P.L. Schaefer. 2003. Effects of aging on behavior and leg kinematics during locomotion in two species of cockroaches. *Journal of Experimental Biology.* 206:4453–4465.

Ridley, M. 1988. Mating frequency and fecundity in insects. *Biological Reviews.* 63:509–549.

Ridley, M. 1989. The incidence of sperm displacement in insects: four conjectures, one corroboration. *Biological Journal of the Linnean Society.* 38:349–367.

Rieppel, O. 1990. Ontogeny—a way forward for systematics, a way backward for phylogeny. *Biological Journal of the Linnean Society.* 39:177–191.

Rieppel, O. 1993. The conceptual relationship of ontogeny, phylogeny, and classification: the taxic approach. *Evolutionary Biology.* 27:1–32.

Rierson, D.A. 1995. Baits for German cockroach control. *In* Understanding and Controlling the German Cockroach. M.K. Rust, J.M. Owens, and D.A. Rierson, editors. Oxford University Press, New York. 231–286.

Ritter, H.J. 1964. Defense of mate and mating chamber in a woodroach. *Science.* 143:1459–1460.

Rivault, C. 1983. Influence du groupement sur le developpement chez *Eublaberus distanti* (Dictyoptere, Ins.). *Insectes Sociaux.* 30:210–220.

Rivault, C. 1989. Spatial distribution of the cockroach, *Blattella germanica,* in a swimming bath facility. *Entomologia Experimentalis et Applicata.* 53:247–255.

Rivault, C. 1990. Distribution dynamics of *Blattella germanica* in a closed urban environment. *Entomologia Experimentalis et Applicata.* 57:85–91.

Rivault, C., and A. Cloarec. 1990. Food stealing in cockroaches. *Journal of Ethology.* 8:53–60.

Rivault, C., and A. Cloarec. 1991. Exploitation of food resources by the cockroach *Blattella germanica* in an urban habitat. *Entomologia Experimentalis et Applicata.* 61:149–158.

Rivault, C., and A. Cloarec. 1992a. Agonistic interactions and exploitation of limited food sources in *Blattella germanica* (L.). *Behavioural Processes.* 26:91–102.

Rivault, C., and A. Cloarec. 1992b. Agonistic interactions at a food source in the cockroach *Blattella germanica* L. *In* Biology and Evolution of Social Insects. J. Billen, editor. Leuven University Press, Leuven, Belgium. 295–300.

Rivault, C., and A. Cloarec. 1992c. Agonistic tactics and size asymmetries between opponents in *Blattella germanica* (L.) (Dictyoptera: Blattellidae). *Ethology.* 90:52–62.

Rivault, C., and A. Cloarec. 1998. Cockroach aggregation: discrimination between strain odors in *Blattella germanica. Animal Behaviour.* 55:177–184.

Rivault, C., A. Cloarec, and A. LeGuyader. 1993. Bacterial load of cockroaches in relation to the urban environment. *Epidemiology and Infection.* 110:317–325.

Rivault, C., A. Cloarec, and L. Sreng. 1998. Cuticular extracts inducing aggregation in the German cockroach *Blattella germanica* (L.). *Journal of Insect Physiology.* 44:909–918.

Roach, A.M.E., and D.C.F. Rentz. 1998. Blattodea. *In* Zoological Catalogue of Australia. Vol. 23. CSIRO, Australian Biological Resources Study. 21–162.

Roberts, S.K. 1960. Circadian activity rhythms in cockroaches. I. The free-running rhythms in steady state. *Journal of Cellular and Comparative Physiology.* 55:99–110.

Robertson, L.N., and G.B. Simpson. 1989. The use of germinating seed baits to detect soil insect pests before crop sowing. *Australian Journal of Experimental Agriculture.* 29:403–407.

Rocha e Silva Albuquerque, I., and S.M.R. Lopes. 1976. Blattaria de bromélia (Dictyoptera). *Revista Brasileira de Biologia.* 36:873–901.

Rocha e Silva Albuquerque, I., R. Tibana, J. Jurberg, and A.M.P. Rebordões. 1976. Contribuição para o conhecimento ecológico de *Poeciloderrhis cribosa* (Burmeister)

e *Poeciloderrhis verticalis* (Burmeister) comum estudo sobre a genitália externa (Dictyoptera: Blattariae). *Revue Suisse de Zoologie.* 36:239–250.
Rocha, I.R.D. 1990. Development of spacing patterns in *Nauphoeta cinerea* and *Henchoustedenia flexivitta* (Dictyoptera, Blattaria, Blaberidae). *Revista Brasileira de Entomologia.* 34:341–347.
Rodriguez, V., D. Windsor, and W.G. Eberhard. 2004. Tortoise beetle genitalia and demonstration of a sexually selected advantage for flagellum length in *Chelymorpha alternans* (Chrysomelidae, Cassidini, Stolaini). *In* New Developments in the Biology of Chrysomelidae. P. Jolivet, J.A. Santiago-Blay, and M. Schmitt, editors. SPB Academic Publishing, The Hague. 739–748.
Rodríguez-Gironés, M.A., and M. Enquist. 2001. The evolution of female sexuality. *Animal Behaviour.* 61:695–704.
Roesner, G. 1940. Zur Kenntnis der Lebensweise der Gewachshausschabe *Pycnoscelus surinamensis* L. *Die Gartenbauwissenschaft.* 15:184–225.
Roff, D.A. 1986. The evolution of wing dimorphism in insects. *Evolution.* 40:1009–1020.
Roff, D.A. 1990. The evolution of flightlessness in insects. *Ecological Monographs.* 60:389–421.
Roff, D.A. 1994. The evolution of flightlessness: is history important? *Evolutionary Ecology.* 8:639–657.
Roff, D.A., and D.J. Fairbairn. 1991. Wing dimorphisms and the evolution of migratory polymorphisms among the Insecta. *American Zoologist.* 31:243–251.
Roisin, Y. 1990. Reversibility of regressive molts in the termite *Neotermes papua. Naturwissenschaften.* 77:246–247.
Roisin, Y. 1994. Intragroup conflicts and the evolution of sterile castes in termites. *The American Naturalist.* 143:751–765.
Roisin, Y. 2000. Diversity and evolution of caste patterns. *In* Termites: Evolution, Sociality, Symbioses, Ecology. T. Abe, D.E. Bignell, and M. Higashi, editors. Kluwar Academic Publishers, Dordrecht. 95–119.
Rollo, C.D. 1984a. Resource allocation and time budgeting in adults of the cockroach *Periplaneta americana:* the interaction of behaviour and metabolic reserves. *Research on Population Ecology.* 26:150–187.
Rollo, C.D. 1984b. Variation among individuals and the effect of temperature on food consumption and reproduction in the cockroach *Periplaneta americana* Orthoptera Blattidae. *Canadian Entomologist.* 116:785–794.
Rollo, C.D. 1986. A test of the principle of allocation using two sympatric species of cockroaches. *Ecology.* 67:616–628.
Roonwal, M.L. 1970. Isoptera. *In* Taxonomist's Glossary of Genitalia of Insects. S.H. Tuxen, editor. Munksgaard, Copenhagen. 41–46.
Rosengaus, R.B., J.E. Moustakas, D.V. Calleri, and J.F.A. Traniello. 2003. Nesting ecology and cuticular microbial loads in dampwood (*Zootermopsis angusticollis*) and drywood termites (*Incisitermes minor, I. schwarzi, Cryptotermes cavifrons*). *Journal of Insect Science.* 3:31 (insectscience .org/3.31). 6 pp.
Rosengaus, R.B., J.F.A. Traniello, M.L. Lefebvre, and A.B. Maxmen. 2004. Fungistatic activity of the sternal gland secretion of the dampwood termite *Zootermopsis angusticollis. Insectes Sociaux.* 51:259–264.
Ross, H.H. 1929. The life history of the German cockroach. *Transactions of the Illinois State Academy of Sciences.* 1929:84–93.
Ross, M.H., and D.G. Cochran. 1967. A gynandromorph of the German cockroach, *Blattella germanica. Annals of the Entomological Society of America.* 60:859–860.
Ross, M.H., and D.E. Mullins. 1995. Biology. *In* Understanding and Controlling the German Cockroach. M.K. Rust, J.M. Owens, and D.A. Rierson, editors. Oxford University Press, New York. 21–47.
Ross, M.H., and K.R. Tignor. 1985. Response of German cockroaches *Blattella germanica* to a dispersant emitted by adult females. *Entomologia Experimentalis et Applicata.* 39:15–20.
Ross, M.H., and K.R. Tignor. 1986a. Response of German cockroaches to a dispersant and other substances secreted by crowded adults and nymphs (Blattodea: Blattellidae). *Proceedings of the Entomological Society of Washington.* 88:25–29.
Ross, M.H., and K.R. Tignor. 1986b. Response of German cockroaches to aggregation pheromones emitted by adult females. *Entomologia Experimentalis et Applicata.* 41:25–31.
Roth, L.M. 1962. Hypersexual activity induced in females of the cockroach *Nauphoeta cinerea. Science.* 138:1267–1269.
Roth, L.M. 1964a. Control of reproduction in female cockroaches with special reference to *Nauphoeta cinerea.* II. Gestation and postparturition. *Psyche.* 71:198–244.
Roth, L.M. 1964b. Control of reproduction in female cockroaches with special reference to *Nauphoeta cinerea.* I. First oviposition period. *Journal of Insect Physiology.* 10:915–945.
Roth, L.M. 1967a. The evolutionary significance of rotation of the oötheca in the Blattaria. *Psyche.* 74:85–103.
Roth, L.M. 1967b. Sexual isolation in the parthenogenetic cockroach *Pycnoscelus surinamensis* and application of the name *Pycnoscelus indicus* to its bisexual relative (Dictyoptera: Blattaria: Blaberidae: Pycnoscelinae). *Annals of the Entomological Society of America.* 60:774–779.
Roth, L.M. 1967c. Uricose glands in the accessory sex gland complex of male Blattaria. *Annals of the Entomological Society of America.* 60:1203–1211.
Roth, L.M. 1967d. Water changes in cockroach oothecae in relation to the evolution of ovoviviparity and viviparity. *Annals of the Entomological Society of America.* 60:928–946.
Roth, L.M. 1968a. Oothecae of Blattaria. *Annals of the Entomological Society of America.* 61:83–111.
Roth, L.M. 1968b. Oviposition behavior and water changes in the oothecae of *Lophoblatta brevis* Blatteria: Blattellidae: Plectopterinae. *Psyche.* 75:99–106.
Roth, L.M. 1968c. Reproduction in some poorly known species of Blattaria. *Annals of the Entomological Society of America.* 61:571–579.
Roth, L.M. 1969. The evolution of male tergal glands in the Blattaria. *Annals of the Entomological Society of America.* 62:176–208.

Roth, L.M. 1970a. Evolution and taxonomic significance of reproduction in the Blattaria. *Annual Review of Entomology.* 15:75–96.

Roth, L.M. 1970b. The stimuli regulating reproduction in cockroaches. *Centre National de la Recherche Scientifique, Paris.* 189:267–286.

Roth, L.M. 1971a. Additions to the oothecae, uricose glands, ovarioles, and tergal glands of Blattaria. *Annals of the Entomological Society of America.* 64:127–141.

Roth, L.M. 1971b. The male genitalia of Blattaria VIII. *Panchlora, Anchoblatta, Biolleya, Pelloblatta,* and *Achroblatta* (Blaberidae: Panchlorinae). *Psyche.* 78:296–305.

Roth, L.M. 1973a. Brazilian cockroaches found in birds nests, with descriptions of new genera and species. *Proceedings of the Entomological Society of Washington.* 75:1–27.

Roth, L.M. 1973b. Inhibition of oocyte development during pregnancy in the cockroach *Eublaberus posticus. Journal of Insect Physiology.* 19:455–469.

Roth, L.M. 1973c. The male genitalia of Blattaria. XI. Perisphaeriinae. *Psyche.* 80:305–347.

Roth, L.M. 1974a. Control of oötheca formation and oviposition in Blattaria. *Journal of Insect Physiology.* 20:821–844.

Roth, L.M. 1974b. Reproductive potential of bisexual *Pycnoscelus indicus* and clones of its parthenogenetic relative, *Pycnoscelus surinamensis. Annals of the Entomological Society of America.* 67:215–223.

Roth, L.M. 1977. A taxonomic revision of the Panesthiinae of the world. I. The Panesthiinae of Australia (Dictyoptera: Blattaria: Blaberidae). *Australian Journal of Zoology, Supplementary Series.* 48:1–112.

Roth, L.M. 1979a. Cockroaches and plants. *Horticulture.* August:12–13.

Roth, L.M. 1979b. A taxonomic revision of the Panesthiinae of the world. II. The genera *Salganea* Stål *Microdina* Kirby and *Caeparia* Stål (Dictyoptera: Blattaria: Blaberidae). *Australian Journal of Zoology, Supplementary Series.* 69:1–201.

Roth, L.M. 1979c. A taxonomic revision of the Panesthiinae of the world. III. The genera *Panesthia* Serville and *Miopanesthia* Serville (Dictyoptera: Blattaria: Blaberidae). *Australian Journal of Zoology, Supplementary Series.* 74:1–276.

Roth, L.M. 1980. Cave dwelling cockroaches from Sarawak, with one new species. *Systematic Entomology.* 5:97–104.

Roth, L.M. 1981a. Introduction. *In* The American Cockroach. W.J. Bell and K.G. Adiyodi, editors. Chapman and Hall, London. 1–14.

Roth, L.M. 1981b. The mother-offspring relationship of some blaberid cockroaches (Dictyoptera: Blattaria: Blaberidae). *Proceedings of the Entomological Society of Washington.* 83:390–398.

Roth, L.M. 1982a. Ovoviviparity in the blattellid cockroach, *Symploce bimaculata* (Gerstaecker) (Dictyoptera: Blattaria: Blattellidae). *Proceedings of the Entomological Society of Washington.* 84:277–280.

Roth, L.M. 1982b. A taxonomic revision of the Panesthiinae of the world. IV. The genus *Ancaudellia* Shaw, with additions to parts I–III, and a general discussion of distribution and relationships of the components of the subfamily (Dictyoptera: Blattaria: Blaberidae). *Australian Journal of Zoology, Supplementary Series.* 82:1–142.

Roth, L.M. 1984. *Stayella,* a new genus of ovoviviparous blattellid cockroaches from Africa (Dictyoptera: Blattaria: Blattellidae). *Entomologica Scandinavica.* 15:113–139.

Roth, L.M. 1985. A taxonomic revision of the genus *Blattella* Caudell (Dictyoptera, Blattaria: Blattellidae). *Entomologica Scandinavica Supplement.* 22:1–221.

Roth, L.M. 1987a. The genus *Neolaxta* Mackerras (Dictyoptera: Blattaria: Blaberidae). *Memoirs of the Queensland Museum.* 25:141–150.

Roth, L.M. 1987b. The genus *Tryonicus* Shaw from Australia and New Caledonia (Dictyoptera: Blattaria: Blattidae: Tryonicinae). *Memoirs of the Queensland Museum.* 25:151–167.

Roth, L.M. 1988. Some cavernicolous and epigean cockroaches with six new species and a discussion of the Nocticolidae (Dictyoptera: Blattaria). *Revue Suisse de Zoologie.* 95:297–321.

Roth, L.M. 1989a. *Sliferia,* a new ovoviviparous cockroach genus (Blattellidae) and the evolution of ovoviviparity in Blattaria (Dictyoptera). *Proceedings of the Entomological Society of Washington.* 91:441–451.

Roth, L.M. 1989b. Cockroach genera whose adult males lack styles. Part I. (Dictyoptera: Blattaria: Blattellidae). *Revue Suisse de Zoologie.* 96:747–770.

Roth, L.M. 1990a. Cockroaches from the Krakatau Islands (Dictyoptera: Blattaria). *Memoirs of the Museum of Victoria.* 50:357–378.

Roth, L.M. 1990b. A revision of the Australian Parcoblattini (Blattaria: Blattellidae: Blattellinae). *Memoirs of the Queensland Museum.* 28:531–596.

Roth, L.M. 1991a. Blattodea; Blattaria (Cockroaches). *In* The Insects of Australia. Vol. 1. CSIRO, I.D. Naumann, and others, editors. Cornell University Press, Ithaca, NY. 320–329.

Roth, L.M. 1991b. The cockroach genera *Beybienkoa,* gen. nov., *Escala* Shelford, *Eowilsonia,* gen. nov., *Hensaussurea* Princis, *Parasigmoidella* Hanitsch and *Robshelfordia* Princis (Dictyoptera: Blattaria: Blattellidae). *Invertebrate Taxonomy.* 5:553–716.

Roth, L.M. 1991c. A new cave dwelling cockroach from Western Australia (Blattaria: Nocticolidae). *Records of the Western Australian Museum.* 15:17–22.

Roth, L.M. 1991d. New combinations, synonymies, redescriptions, and new species of cockroaches, mostly Indo-Australian Blattellidae. *Invertebrate Taxonomy.* 5:953–1021.

Roth, L.M. 1992. The Australian cockroach genus *Laxta* Walker (Dictyoptera: Blattaria: Blaberidae). *Invertebrate Taxonomy.* 6:389–435.

Roth, L.M. 1994. The beetle-mimicking cockroach genera *Prosoplecta* and *Areolaria,* with a description of *Tomeisneria furthi* gen. n., sp. n. (Blattellidae: Pseudophyllodromiinae). *Entomologica Scandinavica.* 25:419–426.

Roth, L.M. 1995a. The cockroach genera *Hemithyrsocera* Saussure and *Symplocodes* Hebard (Dictyoptera: Blattellidae: Blattellinae). *Invertebrate Taxonomy.* 9:959–1003.

Roth, L.M. 1995b. New species and records of cockroaches

from western Australia. *Records of the Western Australian Museum.* 17:153–161.
Roth, L.M. 1995c. *Pseudoanaplectinia yumotoi,* a new ovoviviparous myrmecophilous cockroach genus and species from Sarawak (Blattaria: Blattellidae; Blattellinae). *Psyche.* 102:79–87.
Roth, L.M. 1995d. Revision of the cockroach genus *Homopteroidea* Shelford (Blattaria, Polyphagidae). *Tijdschrift voor Entomologie.* 138:103–116.
Roth, L.M. 1996. Cockroaches from the Seychelles Islands (Dictyoptera: Blattaria). *Journal of African Zoology.* 110:97–128.
Roth, L.M. 1998a. The cockroach genus *Colapteroblatta,* its synonyms *Poroblatta, Acroporoblatta,* and *Nauclidas,* and a new species of *Litopeltis* (Blattaria: Blaberidae; Epilamprinae). *Transactions of the American Entomological Society.* 124:167–202.
Roth, L.M. 1998b. The cockroach genus *Pycnoscelus* Scudder, with a description of *Pycnoscelus femapterus,* sp. nov. (Blattaria: Blaberidae: Pycnoscelinae). *Oriental Insects.* 32:93–130.
Roth, L.M. 1999a. Descriptions of new taxa, redescriptions, and records of cockroaches, mostly from Malaysia and Indonesia (Dictyoptera: Blattaria). *Oriental Insects.* 33:109–185.
Roth, L.M. 1999b. New cockroach species, redescriptions, and records, mostly from Australia, and a description of *Metanocticola christmasensis* gen. nov., sp. nov., from Christmas Island (Blattaria). *Records of the Western Australian Museum.* 19:327–364.
Roth, L.M. 2003a. Blattodea (Cockroaches). *In* Grzimek's Animal Life Encyclopedia. Vol. 3. M. Hutchins, D.A. Thoney, and M.C. McDade, editors. Gale, Detroit. 147–159.
Roth, L.M. 2003b. Some cockroaches from Africa and islands of the Indian Ocean, with descriptions of three new species (Blattaria). *Transactions of the American Entomological Society.* 129:163–182.
Roth, L.M. 2003c. Systematics and phylogeny of cockroaches (Dictyoptera: Blattaria). *Oriental Insects.* 37:1–186.
Roth, L.M., and D.W. Alsop. 1978. Toxins of Blattaria. *Handbook of Experimental Pharmacology.* 48:465–487.
Roth, L.M., and R.H. Barth. 1964. The control of sexual receptivity in female cockroaches. *Journal of Insect Physiology.* 10:965–975.
Roth, L.M., and R.H. Barth. 1967. The sense organs employed by cockroaches in mating behavior. *Behaviour.* 28:58–94.
Roth, L.M., and A.C. Cohen. 1973. Aggregation in Blattaria. *Annals of the Entomological Society of America.* 66:1315–1323.
Roth, L.M., and S.H. Cohen. 1968. Chromosomes of the *Pycnoscelus indicus* and *P. surinamensis* complex (Blattaria: Blaberidae: Pycnoscelinae). *Psyche.* 75:53–76.
Roth, L.M., and G.P. Dateo. 1964. Uric acid in the reproductive system of males of the cockroach *Blattella germanica. Science.* 146:782–784.
Roth, L.M., and G.P. Dateo. 1965. Uric acid storage and excretion by the accessory sex glands of male cockroaches. *Journal of Insect Physiology.* 11:1023–1029.
Roth, L.M., and G.P. Dateo. 1966. A sex pheromone produced by males of the cockroach *Nauphoeta cinerea. Journal of Insect Physiology.* 12:255–265.
Roth, L.M., and T. Eisner. 1961. Chemical defenses of arthropods. *Annual Review of Entomology.* 7:107–136.
Roth, L.M., and W. Hahn. 1964. Size of new-born larvae of cockroaches incubating eggs internally. *Journal of Insect Physiology.* 10:65–72.
Roth, L.M., and B. Hartman. 1967. Sound production and its evolutionary significance in the Blattaria. *Annals of the Entomological Society of America.* 60:740–742.
Roth, L.M., and G.C. McGavin. 1994. Two new species of Nocticolidae (Dictyoptera: Blattaria) and a rediagnosis of the cavernicolous genus *Spelaeoblatta* Bolívar. *Journal of Natural History.* 28:1319–1326.
Roth, L.M., and P. Naskrecki. 2001. Trophobiosis between a blattellid cockroach (*Macrophyllodromia* spp.) and fulgorids (*Enchophora* and *Copidocephala* spp.) in Costa Rica. *Journal of Orthoptera Research.* 10:189–194.
Roth, L.M., and P. Naskrecki. 2003. A new genus and species of cave cockroach (Blaberidae: Oxyhaloinae) from Guinea, West Africa. *Journal of Orthoptera Research.* 12:57–61.
Roth, L.M., and E.H. Slifer. 1973. Spheroid sense organs on the cerci of polphagid cockroaches (Blattaria: Polyphagidae). *International Journal of Insect Morphology and Embryology.* 2:13–24.
Roth, L.M., and W.H. Stahl. 1956. Tergal and cercal secretion of *Blatta orientalis* L. *Science.* 123:798–799.
Roth, L.M., and B. Stay. 1958. The occurrence of paraquinones in some arthropods, with emphasis on the quinone-secreting tracheal glands of *Diploptera punctata* (Blattaria). *Journal of Insect Physiology.* 1:305–308.
Roth, L.M., and B. Stay. 1959. Control of oocyte development in cockroaches. *Science.* 130:271–272.
Roth, L.M., and B. Stay. 1961. Oocyte development in *Diploptera punctata* (Eschscholtz) (Blattaria). *Journal of Insect Physiology.* 7:186–202.
Roth, L.M., and B. Stay. 1962a. A comparative study of oocyte development in false ovoviviparous cockroaches. *Psyche.* 69:165–208.
Roth, L.M., and B. Stay. 1962b. Oocyte development in *Blattella germanica* (L.) and *Blattella vaga* Heberd (Blattaria). *Annals of the Entomological Society of America.* 55:633–642.
Roth, L.M., and B. Stay. 1962c. Oocyte development in *Blattella germanica* and *Blattella vaga* (Blattaria). *Annals of the Entomological Society of America.* 55:633–642.
Roth, L.M., and E.R. Willis. 1952a. A study of cockroach behavior. *American Midland Naturalist.* 47:66–129.
Roth, L.M., and E.R. Willis. 1952b. Tarsal structure and climbing ability in cockroaches. *Journal of Experimental Zoology.* 119:483–518.
Roth, L.M., and E.R. Willis. 1954a. *Anastus floridanus* (Hymenoptera: Eupelmidae) a new parasite on the eggs of the cockroach *Eurycotis floridana. Transactions of the American Entomological Society.* 80:29–41.
Roth, L.M., and E.R. Willis. 1954b. The reproduction of cockroaches. *Smithsonian Miscellaneous Collections.* 122:1–49.

Roth, L.M., and E.R. Willis. 1955a. Intra-uterine nutrition of the "beetle-roach" *Diploptera dytiscoides* (Serv.) during embryogenesis, with notes on its biology in the laboratory (Blattaria: Diplopteridae). *Psyche.* 62:55–68.

Roth, L.M., and E.R. Willis. 1955b. Relation of water loss to the hatching of eggs from detached oothecae of *Blattella germanica* L. *Journal of Economic Entomology.* 48:57–60.

Roth, L.M., and E.R. Willis. 1955c. Water content of cockroach eggs during embryogenesis in relation to oviposition behavior. *Journal of Experimental Zoology.* 128:489–509.

Roth, L.M., and E.R. Willis. 1956. Parthenogenesis in cockroaches. *Annals of the Entomological Society of America.* 49:31–37.

Roth, L.M., and E.R. Willis. 1957. The medical and veterinary importance of cockroaches. *Smithsonian Miscellaneous Collections.* 134:147.

Roth, L.M., and E.R. Willis. 1958a. An analysis of oviparity and viviparity in the Blattaria. *Transactions of the American Entomological Society.* 83:221–238.

Roth, L.M., and E.R. Willis. 1958b. The Biology of *Panchlora nivea* with observations on the eggs of other Blattaria. *Transactions of the American Entomological Society.* 83:195–207.

Roth, L.M., and E.R. Willis. 1960. The biotic associations of cockroaches. *Smithsonian Miscellaneous Collections.* 141:1–470.

Roth, L.M., and E.R. Willis. 1961. A study of bisexual and parthenogenetic strains of *Pycnoscelus surinamensis* (Blattaria, Epilamprinae). *Annals of the Entomological Society of America.* 54:12–25.

Roy, R. 1999. Morphology and taxonomy. *In* The Praying Mantids. F.R. Prete, H. Wells, P.H. Wells, and L.E. Hurd, editors. Johns Hopkins University Press, Baltimore. 19–40.

Rugg, D. 1987. Aspects of the biology of an Australian wood-feeding cockroach, *Panesthia cribrata* (Blattodea: Blaberidae). Master's thesis, Department of Plant Pathology and Agricultural Entomology. University of Sydney, Sydney. 170 pp.

Rugg, D., and H.A. Rose. 1984a. Intraspecies association in *Panesthia cribrata* (Sauss.) (Blattodea: Blaberidae). *General and Applied Entomology.* 16:33–35.

Rugg, D., and H.A. Rose. 1984b. Reproductive biology of some Australian cockroaches (Blattodea: Blaberidae). *Journal of the Australian Entomological Society.* 23:113–117.

Rugg, D., and H.A. Rose. 1984c. The taxonomic significance of reproductive behavior in some Australian cockroaches. *Journal of the Australian Entomological Society.* 23:118.

Rugg, D., and H.A. Rose. 1989. Seasonal reproductive cycle in the Australian wood-feeding cockroach *Panesthia cribrata. Entomologia Generalis.* 14:189–195.

Rugg, D., and H.A. Rose. 1990. Nymphal development and adult longevity of the Australian wood-feeding cockroach *Panesthia cribrata* (Dictyoptera: Blaberidae). *Annals of the Entomological Society of America.* 83:766–775.

Rugg, D., and H.A. Rose. 1991. Biology of *Macropanesthia rhinoceros* (Dictyoptera: Blaberidae). *Annals of the Entomological Society of America.* 84:575–582.

Rundel, P.W., and A.C. Gibson. 1996. Adaptive strategies of growth form and physiological ecology in Neotropical lowland rain forest plants. *In* Neotropical Biodiversity and Conservation. A.C. Gibson, editor. University of California, Los Angeles. 33–71.

Runstrom, E.S., and G.W. Bennett. 1990. Distribution and movement patterns of German cockroaches (Dictyoptera: Blattellidae) within apartment buildings. *Journal of Medical Entomology.* 27:515–518.

Rust, M.K., and A.G. Appel. 1985. Intra- and interspecific aggregation in some nymphal blattellid cockroaches (Dictyoptera: Blattellidae). *Annals of the Entomological Society of America.* 78:107–110.

Rust, M.K., J.M. Owens, and D.A. Rierson. 1995. Understanding and Controlling the German Cockroach. Oxford University Press, New York. 430 pp.

Ruzicka, V. 1999. The first steps in subterranean evolution of spiders (Araneae) in Central Europe. *Journal of Natural History.* 33:255–265.

Sacchi, L., S. Corona, A. Grigolo, U. Laudani, M.G. Selmi, and E. Bigliardi. 1996. The fate of the endocytobionts of *Blattella germanica* (Blattaria: Blattellidae) and *Periplaneta americana* (Blattaria: Blattidae) during embryo development. *Italian Journal of Zoology.* 63:1–11.

Sacchi, L., A. Grigolo, M. Mazzini, E. Bigliardi, B. Baccetti, and U. Laudani. 1988. Symbionts in the oocytes of *Blattella germanica* (L.) (Dictyoptera: Blattellidae): their mode of transmission. *International Journal of Insect Morphology and Embryology.* 17:437–446.

Sacchi, L., C.A. Nalepa, E. Bigliardi, S. Corona, A. Grigolo, U. Laudani, and C. Bandi. 1998a. Ultrastructural studies of the fat body and bacterial endosymbionts of *Cryptocercus punctulatus* Scudder (Blattaria: Cryptocercidae). *Symbiosis.* 25:251–269.

Sacchi, L., C.A. Nalepa, E. Bigliardi, M. Lenz, C. Bandi, S. Corona, A. Grigolo, S. Lambiase, and U. Laudani. 1998b. Some aspects of intracellular symbiosis during embryo development of *Mastotermes darwiniensis* (Isoptera: Mastotermitidae). *Parassitologia.* 40:308–316.

Sacchi, L., C.A. Nalepa, M. Lenz, C. Bandi, S. Corona, A. Grigolo, and E. Bigliardi. 2000. Transovarial transmission of symbiotic bacteria in *Mastotermes darwiniensis* Froggatt (Isoptera: Mastotermitidae): ultrastructural aspects and phylogenetic implications. *Annals of the Entomological Society of America.* 93:1308–1313.

Saito, S. 1976. Studies on the productivity of soil animals in Pasoh Forest Reserve, West Malaysia. IV. Growth, respiration and food consumption of some cockroaches. *Japanese Journal of Ecology.* 26:37–42.

Sakaluk, S.K. 2000. Sensory exploitation as an evolutionary origin to nuptial food gifts in insects. *Proceedings of the Royal Society of London B.* 267:339–343.

Sakuma, M., and H. Fukami. 1990. The aggregation pheromone of the German cockroach *Blattella germanica* (L.) (Dictyoptera: Blattellidae): isolation and identification of the attractant components of the pheromone. *Applied Entomology and Zoology.* 25:355–368.

Samways, M.J. 1994. Insect Conservation Biology. Chapman and Hall, London. 358 pp.

Sanchez, C., F. Hernandez, P. Rivera, and O. Calderon. 1994.

Indigenous flora in cockroaches (Dictyoptera: Blattidae and Blattellidae): a bacteriological and ultrastructural analysis. *Revista de Biologia Tropical.* 42 (Suppl. 2):93–96.
Savage, D.C. 1977. Microbial ecology of the gastrointestinal tract. *Annual Review of Microbiology.* 31:107–133.
Schal, C. 1982. Intraspecific vertical stratification as a mate finding strategy in cockroaches. *Science.* 215:1405–1407.
Schal, C. 1983. Behavioral and physiological ecology and community structure of tropical cockroaches (Dictyoptera Blattaria). Ph.D. thesis, University of Kansas, Lawrence. 130 pp.
Schal, C., and W.J. Bell. 1982. Ecological correlates of paternal investment in a tropical cockroach. *Science.* 218:170–172.
Schal, C., and W.J. Bell. 1986. Vertical community structure and resource utilization in neotropical forest cockroaches. *Ecological Entomology.* 11:411–423.
Schal, C., J.-Y. Gautier, and W.J. Bell. 1984. Behavioural ecology of cockroaches. *Biological Reviews.* 59:209–254.
Schal, C., G.L. Holbrook, J.A.S. Bachmann, and V.L. Sevala. 1997. Reproductive biology of the German cockroach, *Blattella germanica:* juvenile hormone as pleiotropic master regulator. *Archives of Insect Biochemistry and Physiology.* 35:405–426.
Scharf, M.E., D. Wu-Scharf, X. Zhou, B.R. Pittendrigh, and G.W. Bennett. 2005. Gene expression profiles among immature and adult reproductive castes of the termite *Reticulitermes flavipes. Insect Molecular Biology.* 14:31–44.
Scharrer, B. 1946. The role of the corpora allata in the development of *Leucophaea maderae* (Orthoptera). *Endocrinology.* 38:35–45.
Scherkenbeck, J., G. Nentwig, K. Justus, J. Lenz, D. Gondol, G. Wendler, M. Dambach, F. Nischk, and C. Graef. 1999. Aggregation agents in German cockroach *Blattella germanica:* examination of efficacy. *Journal of Chemical Ecology.* 25:1105–1119.
Scheu, S., and H. Setälä. 2002. Multitrophic interactions in decomposer food webs. *In* Multitrophic Level Interactions. T. Tscharntke and B. Hawkins, editors. Cambridge University Press, Cambridge. 223–264.
Schiegg, K. 2000. Are there saproxylic beetles characteristic of high dead wood connectivity? *Ecography.* 23:579–587.
Schneider, J. 1977. Zür Variabilität der Flügel palaozoischer Blattodea (Insecta). Teil 1. *Freiberger Forschungshefte (C).* 326:87–105 (translation by A.U. Eben).
Schneider, J. 1978. Zur Taxonomie und Biostratigraphie der Blattodea (Insecta) des Karbon und Perm der DDR. *Freiberger Forschungshefte (C).* 340:1–152 (translation by A.U. Eben).
Schneider, J., and R. Werneburg. 1994. Neue Spiloblattinidae (Insecta, Blattodea) aus dem Oberkarbon und Unterperm von Mitteleuropa sowie die Biostratigraphie des Rotliegend. *Veröffentlichungen Naturhistorisches Museum Schloss Schleusingen.* 7/8:31–52.
Schoenly, K. 1983. Arthropods associated with bovine and equine dung in an ungrazed Chihuahuan desert ecosystem. *Annals of the Entomological Society of America.* 76:790–796.
Schowalter, T.D., and L.M. Ganio. 2003. Diel, seasonal and disturbance induced variation in invertebrate assemblages. *In* Arthropods of Tropical Forests: Spatio-Temporal Dynamics and Resource Use in the Canopy. Y. Basset, V. Novotny, S.E. Miller, and R.L. Kitching, editors. Cambridge University Press, Cambridge. 315–328.
Schultze-Motel, P., and H. Greven. 1998. Metabolic heat flux in pregnant females of the viviparous cockroach *Nauphoeta cinerea* (Blaberoidea: Blaberidae). *Entomologia Generalis.* 22:199–204.
Scott, H. 1929. On some cases of maternal care displayed by cockroaches and their significance. *Entomologist's Monthly Magazine.* 65:218–222.
Scriber, J.M., and F.J. Slansky. 1981. The nutritional ecology of immature insects. *Annual Review of Entomology.* 26:183–211.
Scrivener, A.M., and M. Slaytor. 1994a. Cellulose digestion in *Panesthia cribrata* Saussure: does fungal cellulase play a role? *Comparative Biochemistry and Physiology.* 107B:309–315.
Scrivener, A.M., and M. Slaytor. 1994b. Properties of the endogenous cellulase from *Panesthia cribrata* Saussure and purification of major endo-β-1,4-glucanase components. *Insect Biochemistry and Molecular Biology.* 24:223–231.
Scudder, S.H. 1886. The cockroach of the past. *In* The Cockroach. L.C. Miall and A. Denny, editors. Lovell Reeve & Co., London. 205–220.
Seamans, L., and L.C. Woodruff. 1939. Some factors influencing the number of molts of the German roach. *Journal of the Kansas Entomological Society.* 12:73–76.
Seastedt, T.R. 1984. The role of microarthropods in decomposition and mineralization processes. *Annual Review of Entomology.* 29:25–46.
Seelinger, G. 1984. Sex-specific activity patterns in *Periplaneta americana* and their relation to mate-finding. *Zeitschrift für Tierpsychologie.* 65:309–326.
Seelinger, G., and U. Seelinger. 1983. On the social organization, alarm and fighting in the primitive cockroach *Cryptocercus punctulatus. Zeitschrift für Tierpsychologie.* 61:315–333.
Seifert, R.P., and F.H. Seifert. 1976. Natural history of insects living in inflorescences of two species of *Heliconia. Journal of the New York Entomological Society.* 84:233–242.
Séin, F.J. 1923. Cucarachas. *Puerto Rico Insular Experiment Station Circular.* 64:1–12.
Shapiro, J.A. 1997. Multicellularity: the rule not the exception. *In* Bacteria as Multicellular Organisms. J.A. Shapiro and M. Dworkin, editors. Oxford University Press, New York. 14–49.
Shaw, E. 1918. Australian Blattidae, with descriptions of eleven new species. *Memoirs of the Queensland Museum, Brisbane.* 6:151–167.
Shaw, E. 1925. New genera and species (mostly Australasian) of Blattidae, with notes, and some remarks on Tepper's types. *Proceedings of the Linnean Society, New South Wales.* 1:171–213.
Shaw, S.R. 1994a. Detection of airborne sound by a cockroach 'vibration detector': a possible missing link in insect auditory evolution. *Journal of Experimental Biology.* 193:13–47.
Shaw, S.R. 1994b. Re-evaluation of the absolute threshold and response mode of the most sensitive known "vibra-

tion" detector, the cockroach's subgenual organ: a cochlea-like displacement threshold and a direct response to sound. *Journal of Neurobiology.* 25:1167–1185.

Shear, W.A., and J. Kukalová-Peck. 1990. The ecology of Paleozoic terrestrial arthropods: the fossil evidence. *Canadian Journal of Zoology.* 68:1807–1834.

Shelford, R. 1906a. Studies of the Blattidae. VI. Viviparity amongst the Blattidae. *Transactions of the Entomological Society of London.* 1906:509–514.

Shelford, R. 1906b. Studies of the Blattidae. VII. A new genus of symbiotic Blattidae. *Transactions of the Entomological Society of London.* 1906:515–519.

Shelford, R. 1906c. XIV. Studies of the Blattidae. III. Some new Blattidae from Sarawak, Borneo in the Hope Department, Oxford University Museum. *Transactions of the Entomological Society of London.* Part II. 1906:265–280.

Shelford, R. 1907. Aquatic cockroaches. *Zoologist,* Ser. 4. 11:221–226.

Shelford, R. 1908. XXVI. Some new genera and species of Blattidae, with notes on the form of the pronotum in the subfamily Perisphaeriinae. *Annals and Magazine of Natural History,* Ser. 8. 1:157–177.

Shelford, R. 1909. Notes on some amphibious cockroaches. *Records of the Indian Museum.* 3:125–127.

Shelford, R. 1910a. A new cavernicolous cockroach. *Annals and Magazine of Natural History,* Ser. 8. 6:114–116.

Shelford, R. 1910b. Orthoptera: Blattodea. *Sjostedt's Kilamanjaro-Meru Expedition.* 17:13–48.

Shelford, R. 1912a. Mimicry amongst the Blattidae; with a revision of the genus *Prosoplecta* Sauss., and the description of a new genus. *Proceedings of the Zoological Society of London.* 82:358–376.

Shelford, R. 1912b. The oothecae of Blattidae. *Entomologist's Record.* 24:283–287.

Shelford, R. 1916. A Naturalist in Borneo. T.F. Unwin, London. 331 pp.

Sherron, D.A., H.E.J. Wright, M.H. Ross, and M.H. Farrier. 1982. Density, fecundity, homogeneity, and embryonic development of German cockroach (*Blattella germanica* (L.)) populations in kitchens of varying degrees of sanitation (Dictyoptera: Blattellidae). *Proceedings of the Entomological Society of Washington.* 84:376–390.

Shimamura, H., S. Hori, H. Nagano, S.I. Matsunaga, and F. Urushizaki. 1994. Secondary kill effect of hydramethylnon bait against several species of cockroach. *Japanese Journal of Sanitary Zoology.* 45:97–100.

Shindo, J., and S. Masaki. 1995. Photoperiodic control of larval development in the semivoltine cockroach *Periplaneta japonica* (Blattidae: Dictyoptera). *Ecological Research.* 10:1–12.

Shine, R. 1985. The evolution of viviparity in reptiles: an ecological analysis. *In* The Biology of Reptilia. C. Gans and F. Billett, editors. John Wiley & Sons, New York. 606–694.

Shine, R. 1989. Ecological influences on the evolution of vertebrate viviparity. *In* Complex Organismal Functions: Integration and Evolution in Vertebrates. D.B. Wake and G. Roth, editors. John Wiley & Sons, New York. 263–278.

Sibley, R.M. 1981. Strategies of digestion and defecation. *In* Physiological Ecology: An Evolutionary Approach to Resource Use. C.R. Townsend and P. Calow, editors. Sinauer, Sunderland, MA. 109–139.

Silvestri, F. 1946. Prima nota su alcuni termitofili dell' Indocina. *Bollettino del Laboratorio di Entomologia Agraria Filippo Silvestri, Portici.* 6:313–330.

Simberloff, D.S., and E.O. Wilson. 1969. Experimental zoogeography of islands: the colonization of empty islands. *Ecology.* 50:278–286.

Simmons, L.W. 2001. Sperm Competition and its Evolutionary Consequences in Insects. Princeton University Press, Princeton. 434 pp.

Simon, D., and R.H. Barth. 1977a. Sexual behavior in the cockroach genera *Periplaneta* and *Blatta.* I. Descriptive aspects. *Zeitschrift für Tierpsychologie.* 44:80–107.

Simon, D., and R.H. Barth. 1977b. Sexual behavior in the cockroach genera *Periplaneta* and *Blatta.* III. Aggression and sexual behavior. *Zeitschrift für Tierpsychologie.* 44:306–322.

Simpson, B.S., R.E. Ritzmann, and A.J. Pollack. 1986. A comparison of escape behaviors of the cockroaches *Blaberus craniifer* and *Periplaneta americana. Journal of Neurobiology.* 17:405–419.

Sinclair, B.J. 1997. Seasonal variation in freezing tolerance of the New Zealand alpine cockroach *Celatoblatta quinquemaculata. Ecological Entomology.* 22:462–467.

Sinclair, B.J. 2000. Water relations of the freeze tolerant New Zealand alpine cockroach *Celatoblatta quinquemaculata* (Dictyoptera: Blattidae). *Journal of Insect Physiology.* 46:869–876.

Sinclair, B.J. 2001. Field ecology of freeze tolerance: interannual variation in cooling rates, freeze-thaw and thermal stress in the microhabitat of the alpine cockroach *Celatoblatta quinquemaculata. Oikos.* 93:286–293.

Sinclair, B.J., and S.L. Chown. 2005. Climatic variability and hemispheric differences in insect cold tolerance: support from southern Africa. *Functional Ecology.* 19:214–221.

Sinclair, B.J., J.M. Lord, and C.M. Thompson. 2001. Microhabitat selection and seasonality of alpine invertebrates. *Pedobiologia.* 45:107–120.

Singer, M.S., and E.A. Bernays. 2003. Understanding omnivory needs: a behavioral perspective. *Ecology.* 84:2532–2537.

Sirugue, D., O. Bonnard, J.-L. Le Quere, J.-P. Farine, and R. Brossut. 1992. 2-methylthiazolidine and 4-ethylguaiacol, male sex pheromone components of the cockroach *Nauphoeta cinerea* (Dictyoptera, Blaberidae): a reinvestigation. *Journal of Chemical Ecology.* 18:2261–2276.

Skaife, S.H. 1954. African Insect Life. Longmans, Green, New York. 387 pp.

Slaney, D.P. 2001. New species of Australian cockroaches in the genus *Paratemnopteryx* Saussure (Blattaria, Blattellidae, Blattellinae), and a discussion of some behavioural observations with respect to the evolution and ecology of cave life. *Journal of Natural History.* 35:1001–10012.

Slaney, D.P., and D. Blair. 2000. Molecules and morphology are concordant in discriminating among populations of cave cockroaches in the genus *Paratemnopteryx* Saussure (Blattodea: Blattellidae). *Annals of the Entomological Society of America.* 93:398–404.

Slaney, D.P., and P. Weinstein. 1996. Leaf litter traps for sampling orthopteroid insects in tropical caves. *Journal of Orthoptera Research.* 5:51–52.

Slaney, D.P., and P. Weinstein. 1997a. Conservation of cave fauna: more than just bats. *Memoirs of the Museum of Victoria.* 56:591–596.

Slaney, D.P., and P. Weinstein. 1997b. Geographical variation in the tropical cave cockroach *Paratemopteryx stonei* Roth (Blattellidae) in North Queensland, Australia. *International Journal of Speleology.* 25:1–14.

Slansky, F.J., and J.M. Scriber. 1985. Food consumption and utilization. *Comparative Insect Physiology, Biochemistry and Pharmacology.* 4:87–163.

Slaytor, M. 1992. Cellulose digestion in termites and cockroaches: what role do symbionts play? *Comparative Biochemistry and Physiology.* 103B:775–784.

Slaytor, M. 2000. Energy metabolism in the termite and its gut microbiota. *In* Termites: Evolution, Sociality, Symbioses, Ecology. T. Abe, D.E. Bignell, and M. Higashi, editors. Kluwer Academic Publishers, Dordrecht. 307–332.

Slaytor, M., and D.J. Chappell. 1994. Nitrogen metabolism in termites. *Comparative Biochemistry and Physiology.* 107B:1–10.

Smith, A.F., and C. Schal. 1990. The physiological basis for the termination of pheromone releasing behavior in the female brown-banded cockroach, *Supella longipalpa* (F.) (Dictyoptera: Blattellidae). *Journal of Insect Physiology.* 36:369–373.

Smith, D.C. 1992. The symbiotic condition. *Symbiosis.* 14:3–15.

Snart, J.O.H., M. Greenwood, R. Beck, and K.C. Highnam. 1984a. The functional morphology of the brood sac in two species of ovoviviparous cockroaches *Byrsotria fumigata* (Guerin) and *Gromphadorhina portentosa* (Schaum). 1. Scanning and light microscopy. *International Journal of Insect Morphology and Embryology.* 7:345–355.

Snart, J.O.H., M. Greenwood, R. Beck, and K.C. Highnam. 1984b. The functional morphology of the brood sac in two species of ovoviviparous cockroaches *Byrsotria fumigata* (Guerin) and *Gromphadorhina portentosa* (Schaum). 2. Transmission electron microscopy. *International Journal of Insect Morphology and Embryology.* 7:357–367.

Snodgrass, R.E. 1937. The male genitalia of orthopteroid insects. *Smithsonian Miscellaneous Collections.* 96:1–107.

Sommer, V.S.H. 1974. Aggregationsverhalten bei Schaben. *Angewandte Parasitologie.* 15:10–30.

Southwood, T.R.E. 1962. Migration of terrestrial arthropods in relation to habitat. *Biological Reviews.* 27:171–214.

Spirito, C.P., and D.L. Mushrush. 1979. Interlimb coordination during slow walking in the cockroach. I. Effects of substrate alterations. *Journal of Experimental Biology.* 78:233–243.

Sreng, L. 1979a. Phéromones et comportement sexuel chez *Nauphoeta cinerea* (Olivier) (Insecte, Dictyoptère). *Comptes rendus de l'Academie des Science, Paris.* 289:687–690.

Sreng, L. 1979b. Ultrastructure et chemie de la secretion des glandes tergales du male de *Blattella germanica* (Dictyoptera: Blattellidae). *International Journal of Insect Morphology and Embryology.* 8:213–227.

Sreng, L. 1984. Morphology of the sternal and tergal glands producing the sexual pheromones and the aphrodisiacs among cockroaches of the subfamily Oxyhaloinae. *Journal of Morphology.* 182:279–294.

Sreng, L. 1993. Cockroach mating behaviors, sex pheromones, and abdominal glands (Dictyoptera: Blaberidae). *Journal of Insect Behavior.* 6:715–735.

Stanley, S.M. 1998. Macroevolution, Pattern and Process. Johns Hopkins University Press, Baltimore. 332 pp.

Starr, C.K. 1979. Origin and evolution of insect eusociality: a review of modern theory. *In* Social Insects. Vol. 1. H.R. Hermann, editor. Academic Press, New York. 35–79.

Stay, B. 1962. The colleterial glands of cockroaches. *Annals of the Entomological Society of America.* 55:124–130.

Stay, B., and A.C. Coop. 1973. Developmental stages and chemical composition in embryos of the cockroach, *Diploptera punctata,* with observations on the effect of diet. *Journal of Insect Physiology.* 19:147–171.

Stay, B., and A.C. Coop. 1974. Milk secretion for embryogenesis in a viviparous cockroach. *Tissue and Cell.* 6:669–693.

Stay, B., and A. Gelperin. 1966. Physiological basis for ovipositional behaviour in the false ovoviviparous cockroach, *Pycnoscelus surinamensis* (L.). *Journal of Insect Physiology.* 12:1217–1226.

Stay, B., A. King, and L.M. Roth. 1960. Calcium oxalate in the oothecae of cockroaches. *Annals of the Entomological Society of America.* 53:79–86.

Stay, B., and L.M. Roth. 1958. The reproductive behavior of *Diploptera punctata* (Blattaria: Diplopteridae). *Proceedings of the 10th International Congress of Entomology.* 2:547–552 (1956).

Stein, W., and H. Haschemi. 1991. The influence of external factors on the migration behavior of the German cockroach *Blattella germanica* L. Blattodea, Blattellidae on a refuse tip. *Anzeiger fuer Schaedlingskunde Pflanzenschutz Umweltschutz.* 64:65–69.

Steinhaus, E.A. 1946. Insect Microbiology. Comstock Publishing Co., Inc., Ithaca, NY. 763 pp.

Stevenson, B.G., and D.L. Dindal. 1987. Functional ecology of coprophagous insects: a review. *Pedobiologia.* 30:285–298.

Stock, A., and A.F. O'Farrell. 1954. Regeneration and the moulting cycle in *Blattella germanica* L. *Australian Journal of Biological Science.* 7:302–307.

Stokes, D.R., J.G. Malamud, and D.A. Schreihofer. 1994. Gender specific developmental transformation of a cockroach bifunctional muscle. *Journal of Experimental Zoology.* 268:364–376.

Stone, F.D. 1988. The cockroaches of North Queensland caves and the evolution of tropical troglobites. *In* Australian Speleological Federation Tropicon Conference. L. Pearson, editor. Australian Speleological Federation, Lake Tinaroo, Far North Queensland. 88–93.

Stork, N.E. 1991. The composition of the arthropod fauna of Bornean lowland rain forest trees. *Journal of Tropical Ecology.* 7:161–180.

Storozhenko, S.Y. 1979. Behavioral and habitation conditions of the grylloblattid *Galloisiana kurentzovi* in the southern primorski krai—SFSR USSR (in Russian; English abstract). *Biologicheskie Nauki Moscow.* 2:18–21.

Stout, J.D. 1974. Protozoa. *In* Biology of Plant Litter Decomposition. Vol. 2. C.H. Dickenson and G.J.F. Pough, editors. Academic Press, London. 385–420.

Strohecker, H.F. 1937. An ecological study of some Orthoptera of the Chicago area. *Ecology.* 18:231–250.

Stuart, A.M. 1961. Mechanism of trial laying in two species of termites. *Nature.* 189:419.

Stuart, A.M. 1969. Social behavior and communication. *In* Biology of Termites. Vol. 1. K. Krishna and F.M. Weesner, editors. Academic Press, New York. 193–232.

Stürkow, B., and W.G. Bodenstein. 1966. Location of the sex pheromone in the American cockroach *Periplaneta americana* (L.). *Experientia.* 22:851–853.

Sueuer, J., and T. Aubin. 2006. When males whistle at females: complex FM acoustic signals in cockroaches. *Naturwissenschaften.* 93:500–505.

Sugimoto, A., D.E. Bignell, and J.A. MacDonald. 2000. Global impact of termites on the carbon cycle and atmospheric trace gases. *In* Termites: Evolution, Sociality, Symbioses, Ecology. T. Abe, D.E. Bignell, and M. Higashi, editors. Kluwer Academic Publishers, Dordrecht. 409–435.

Suto, C., and N. Kumada. 1981. Secretion of dispersion-inducing substance by the German cockroach, *Blattella germanica* L. (Othoptera: Blattellidae). *Applied Entomology and Zoology.* 16:113–120.

Swallow, J.G., and G.S. Wilkinson. 2002. The long and short of sperm polymorphisms in insects. *Biological Reviews.* 77:153–182.

Swarbeck, E. 1946. Notes on insect life on Mt. Buffalo. *Victorian Naturalist.* 63:19–23.

Swift, M.J., and J.M. Anderson. 1989. Decomposition. *In* Ecosystems of the World. Vol. 14B: Tropical Rain Forest Ecosystems. H. Lieth and M.J.A. Werger, editors. Elsevier, Amsterdam. 547–569.

Swift, M.J., O.W. Heal, and J.M. Anderson. 1979. Decomposition in Terrestrial Ecosystems. University of California Press, Berkeley. 372 pp.

Takagi, M. 1978. Ecological studies on the smoky brown cockroach, *Periplaneta fuliginosa.* II. A rearing experiment of the nymphal development outdoors in Tsu, Mie prefecture. *Mie Medical Journal.* 27:85–92.

Takahashi, R. 1926. Observations on the aquatic cockroach *Opisthoplatia maculata* (in Japanese). *Dôbuts Zasshi, Tokyo.* 38:89–92.

Takahashi, S., and C. Kitamura. 1972. Occurrence of phenols in the ventral glands of the American cockroach, *Periplaneta americana* (L.) (Orthoptera: Blattidae). *Applied Entomology and Zoology.* 4:199–206.

Tallamy, D.W. 1994. Nourishment and the evolution of paternal investment in subsocial arthropods. *In* Nourishment and Evolution in Insect Societies. J.H. Hunt and C.A. Nalepa, editors. Westview Press, Boulder. 21–55.

Tallamy, D.W., and T.K. Wood. 1986. Convergence patterns in social insects. *Annual Review of Entomology.* 31:369–390.

Tanaka, K. 1981. Regulation of body size during larval development in the German cockroach *Blattella germanica. Journal of Insect Physiology.* 27:587–592.

Tanaka, K., M. Ohtake-Hashiguchi, and E. Ogawa. 1987. Repeated regeneration of the German cockroach legs. *Growth.* 51:282–300.

Tanaka, K., and S. Tanaka. 1997. Winter survival and freeze tolerance in a northern cockroach, *Periplaneta japonica* (Blattidae: Dictyoptera). *Zoological Science.* 14:849–853.

Tanaka, S. 1994. Evolution and physiological consequences of de-alation in crickets. *Researches on Population Ecology.* 36:137–143.

Tanaka, S. 2002. Temperature acclimation in overwintering nymphs of a cockroach, *Periplaneta japonica:* walking on ice. *Journal of Insect Physiology.* 48:571–583.

Tanaka, S., and D.H. Zhu. 2003. Presence of three diapauses in a subtropical cockroach: control mechanisms and adaptive significance. *Physiological Entomology.* 28:323–330.

Tanton, M.T., A.J. Campbell, and H.M.G. Thomas. 1985. Invertebrates from litter under selected eucalypt and pine forests in the Australian Capital Territory. *In* Soil and Litter Invertebrates of Some Australian Mediterranean-type Ecosystems. Vol. 12. P. Greenslade and J.D. Majer, editors. Western Australian School of Biology Bulletin, Bentley, WA. 91–93.

Taubes, G. 2000. Biologists and engineers create a new generation of robots that imitate life. *Science.* 288:80–83.

Taylor, E.C., and C.S. Crawford. 1982. Microbial gut symbionts and desert detritivores. *Scientific Reviews on Arid Zone Research.* 1:37–52.

Taylor, R.L. 1975. Butterflies in My Stomach: or, Insects in Human Nutrition. Woodbridge Press Publishing Company, Santa Barbara, CA. 224 pp.

Teder, T., and T. Tammaru. 2005. Sexual size dimorphism within species increases with body size in insects. *Oikos.* 108:321–334.

Tepper, J.G.O. 1893. The Blattidae of Australia and Polynesia. *Transactions of the Royal Society of South Australia.* 17:25–126.

Tepper, J.G.O. 1894. The Blattariae of Australia and Polynesia: supplementary and additional descriptions and notes. *Transactions of the Royal Society of South Australia.* 18:165–189.

Thorne, B.L., and J.M. Carpenter. 1992. Phylogeny of the Dictyoptera. *Systematic Entomology.* 17:253–268.

Thornhill, R. 1983. Cryptic female choice and its implications in the scorpionfly *Harpovittacus nigriceps. The American Naturalist.* 122:765–788.

Thornhill, R., and J. Alcock. 1983. The Evolution of Insect Mating Systems. Harvard University Press, Cambridge, MA. 547 pp.

Thornton, I.W.B., T.R. New, R.A. Zann, and P.A. Rawlinson. 1990. Colonization of the Krakatau Islands by animals: a perspective from the 1980s. *Philosophical Transactions of the Royal Society of London.* 328:131–165.

Thrall, P.H., J. Antonovics, and J.D. Bever. 1997. Sexual transmission of disease and host mating systems. *The American Naturalist.* 149:485–506.

Tillyard, R.J. 1919. Mesozoic insects of Queensland, No. 6, Blattoidea. *Proceedings of the Linnean Society of New South Wales.* 44:358–382.

Tillyard, R.J. 1926. The Insects of Australia and New Zealand. Angus & Robertson, Ltd., Sydney. 560 pp.

Tinkham, E.R. 1948. Faunistic and ecological studies on the Orthoptera of the Big Bend Region of Trans-Pecos Texas, with especial reference to the Orthopteran zones and faunae of midwestern North America. *American Midland Naturalist.* 40:521–563.

Tinkle, D.W., and J.W. Gibbons. 1977. The distribution and evolution of viviparity in reptiles. *Miscellaneous Publications of the University of Michigan.* 154:1–55.

Tokro, P.G., R. Brossut, and L. Sreng. 1993. Determination of sex pheromone in females of *Blattella germanica* L. *Insect Science and its Application.* 14:115–126.

Tokuda, G., N. Lo, H. Watanabe, G. Arakawa, T. Matsumoto, and H. Noda. 2004. Major alteration of the expression site of endogenous cellulases in members of an apical termite lineage. *Molecular Ecology.* 13:3219–3228.

Tokuda, G., N. Lo, and H. Watanabe. 2005. Marked variations in patterns of cellulase activity against crystalline- vs. carboxymethylcellulose in the digestive systems of diverse, wood feeding termites. *Physiological Entomology.* 30:372–380.

Tracy, R.L., and G.E. Walsberg. 2002. Kangaroo rats revisited: re-evaluating a classic case of desert survival. *Oecologia.* 133:449–457.

Travis, J. 1994. Evaluating the adaptive role of morphological plasticity. *In* Ecological Morphology: Integrative Organismal Biology. P.C. Wainright and S.M. Reilly, editors. University of Chicago Press, Chicago. 99–122.

Trewick, S.A. 2000. Molecular evidence for dispersal rather than vicariance as the origin of flightless insect species on the Chatham Islands, New Zealand. *Journal of Biogeography.* 27:1189–1200.

Troyer, K. 1984. Microbes, herbivory and the evolution of social behavior. *Journal of Theoretical Biology.* 106:157–169.

Trumbo, S.T. 1996. Parental care in invertebrates. *Advances in the Study of Behavior.* 25:3–51.

Tsai, C.-W., and H.-J. Lee. 2000. Circadian locomotor rhythm masked by the female reproduction cycle in cockroaches. *Physiological Entomology.* 25:63–73.

Tsai, C.-W., and H.-J. Lee. 2001. Analysis of specific adaptation to a domicile habitat: A comparative study of two closely related cockroach species. *Journal of Medical Entomology.* 38:245–252.

Tscharntke, T., I. Steffan-Dewenter, A. Kruess, and C. Thies. 2002. Characteristics of insect populations on habitat fragments: a mini-review. *Ecological Research.* 17:229–239.

Tsuji, H., and T. Mizuno. 1973. Behavioural interaction between two harbouring individuals of the smoky brown cockroach, *Periplaneta fuliginosa* S. *Japanese Journal of Sanitary Zoology.* 24:65–72.

Ullrich, B., M. Vollmer, W. Stoecker, and V. Storch. 1992. Hemolymph protein patterns and coprophagous behavior in *Oniscus asellus* L. (Crustacea, Isopoda). *Invertebrate Reproduction and Development.* 21:193–200.

Vahed, K. 1998. The function of nuptial feeding in insects: a review of empirical studies. *Biological Reviews.* 73:43–78.

Vallack, H.W. 1981. Ecological studies in a tropical rainforest on limestone in Gunung Mulu National Park, Sarawak. M.Sc. thesis, University of Stirling, United Kingdom.

Van Baaren, J., A.-S. Bonhomme, P. Deleporte, and J.S. Pierre. 2003. Behaviors promoting grouping or dispersal of mothers and neonates in ovoviviparous cockroaches. *Insectes Sociaux.* 50:45–53.

Van Baaren, J., and P. Deleporte. 2001. Comparison of gregariousness in larvae and adults of four species of zetoborine cockroaches. *Entomologia Experimentalis et Applicata.* 99:113–119.

Van Baaren, J., P. Deleporte, and P. Grandcolas. 2002. Cockroaches of French Guiana Icteridae birds nests. *Amazonia.* 17:243–248.

Van Herrewege, C. 1973. Contribution a l'étude des Blattaria de la faune Malgache. II. Description de huit espèces nouvelles appartenant aux genres *Gromphadorhina* Brunner v.W. et *Elliptorhina* gen. nov. *Bulletin de la Société Linnéenne de Lyon, 42 année, spécial du 150 anniversaire:*75–103.

van Hoek, A.H.A.M., T.A. van Alen, V.S.I. Sprakel, J.H.P. Hackstein, and G.D. Vogels. 1998. Evolution of anaerobic ciliates from the gastrointestinal tract: phylogenetic analysis of the ribosomal repeat from *Nyctotherus ovalis* and its relatives. *Molecular Biology and Evolution.* 15:1195–1206.

van Lear, D.H. 1996. Dynamics of coarse woody debris in southern forest ecosystems. *In* Biodiversity and coarse woody debris in southern forests. Vol. SE-94. J.W. McMinn and D.A.J. Crossley, editors. USDA Forest Service Technical Report. 10–17.

van Soest, P.J. 1994. Nutritional Ecology of the Ruminant. Cornell University Press, Ithaca, NY. 476 pp.

van Wyk, L.E. 1952. The morphology and histology of the genital organs of *Leucophaea maderae* (Fab.) (Blattidae, Orthoptera). *Journal of the Entomological Society of South Africa.* 15:3–62.

Vandel, A. 1965. Biospeleology: The Biology of Cavernicolous Animals. Pergamon Press, Oxford. 524 pp.

Vannier, G., and S.I. Ghabbour. 1983. Effect of rising ambient temperature on transpiration in the cockroach *Heterogamia syriaca* Sauss. from the Mediterranean coastal desert of Egypt. *In* New Trends in Soil Biology. P. Lebrun, H.M. André, A. De Medts, C. Grégoire-Wibo, and G. Wauthy, editors. Dieu-Brichart, Louvain la Neuve. 441–453.

Vauchot, B., E. Pruvost, A.-G. Bagneres, G. Riviere, M. Roux, and J.-L. Clement. 1998. Differential adsorption of allospecific hydrocarbons by the cuticles of two termite species, *Reticulitermes santonensis* and *R. lucifugus grassei*, living in a mixed colony. *Journal of Insect Physiology.* 44:59–66.

Vehrencamp, S.L. 1983. A model for the evolution of despotic versus egalitarian societies. *Animal Behaviour.* 31:667–682.

Verrett, J.M., K.B. Green, L.M. Gamble, and F.C. Crochen. 1987. A hemocoelic *Candida* parasite of the American cockroach Dictyoptera Blattidae. *Journal of Economic Entomology.* 80:1205–1212.

Vidlička, L. 1993. Seasonal dynamics of vertical migration and distribution of cockroach *Ectobius sylvestris* (Blattaria: Blattellidae: Ectobiinae). *Biologia, Bratislava.* 48:163–166.

Vidlička, L. 2002. The new cockroach species from the genus *Chorisoserrata* from Laos (Blattaria: Blattellidae: Pseudophyllodromiinae). *Entomological Problems.* 32:145–147.

Vidlička, L., and A. Huckova. 1993. Mating of the cockroach *Nauphoeta cinerea* (Blattodea: Blaberidae). I. Copulatory behavior. *Entomological Problems.* 24:69–73.
Vidlička, L., P. Vršanský, and D.E. Shcherbakov. 2003. Two new troglobitic cockroach species of the genus *Spelaeoblatta* (Blaberidae: Nocticolidae) from North Thailand. *Journal of Natural History.* 37:107–114.
Vijayalekshmi, V.R., and K.G. Adiyodi. 1973. Accessory sex glands of male *Periplaneta americana* (L.). Part I. Quantitative analysis of some non-enzymatic components. *Indian Journal of Experimental Biology.* 11:512–514.
Vinson, S.B., and G.L. Piper. 1986. Source and characterization of host recognition kairomones of *Tetrastichus hagenowii,* a parasitoid of cockroach eggs. *Physiological Entomology.* 11:459–468.
Virant-Doberlet, M., and A. Čokl. 2004. Vibrational communication in insects. *Neotropical Entomology.* 33:121–134.
Vishniakova, V.N. 1968. Mesozoic blattids with external ovipositors and peculiarities of their reproduction. *In* Jurassic Insects of Karatau. B.B. Rohdendorf, editor. Akademiya Nauk SSSR, Ordelenie Obschej Biologii, Moscow. 55–86 (In Russian, translation by M. E. Barbercheck).
Vitousek, P.M., and R.L.J. Sanford. 1986. Nutrient cycling in moist tropical forest. *Annual Review of Ecology and Systematics.* 17:137–167.
Vlasov, P. 1933. Die Fauna der Wohnhohlen von *Rhombomys opimus* Licht. und *Spermophilopsis leptodactylus* Licht. in der Umgebung von Aschhabad. *Zoologischer Anzeiger.* 101:143–158.
Vlasov, P., and E.F. Miram. 1937. Cockroaches and Orthoptera from the burrows around Ashkhabad. *Trudy Soveta po Izucheniyu Proizvoditel'nykh.Sil, Akademiya Nauk, S.S.S.R., Seriya Turkmenskaia.* 9:259–262.
Vorhies, C.T., and W.P. Taylor. 1922. Life history of the kangaroo rat, *Dipodomys spectabilis spectabilis* Merriam. *U.S. Department of Agriculture Bulletin.* 1091:1–40.
Vršanský, P. 1997. *Piniblattella* gen. nov.—the most ancient genus of the family Blattellidae (Blattodea) from the Lower Cretaceous of Siberia. *Entomological Problems.* 28:67–79.
Vršanský, P. 2002. Origin and early evolution of mantises. *AMBA Projecty.* 6:3–16.
Vršanský, P., V.N. Vishniakova, and A.P. Rasnitsyn 2002. Order Blattida Latreille, 1810. The cockroaches (= Blattodea Brunner von Wattenvill, 1882). *In* History of Insects. A.P. Rasnitsyn and D.L.J. Quicke, editors. Kluwar Academic Publishers, Dordrecht. 263–270.
Vršanský, P. 2003. Umenocoleoidea—an amazing lineage of aberrant insects (Insecta Blattaria). *AMBA Projecty.* 7:1–32.
Vulinec, K. 1990. Collective security: aggregation by insects as a defense. *In* Adaptive Mechanisms and Strategies of Prey and Predators. D.L. Evans and J.O. Schmidt, editors. SUNY Press, Albany. 251–288.
Wagner, D.L., and J.K. Liebherr. 1992. Flightlessness in insects. *Trends in Ecology and Evolution.* 7:216–220.
Wake, D.B. 1991. Homoplasy: the result of natural selection, or evidence of design limitations. *The American Naturalist.* 138:543–567.
Walker, E.M. 1919. The terminal abdominal structures of Orthopteroid insects: a phylogenetic study. *Annals of the Entomological Society of America.* 12:267–316.
Walker, E.M. 1922. The terminal structures of orthopteroid insects: a phylogenetic study. II. The terminal abdominal structures of the male. *Annals of the Entomological Society of America.* 15:1–87.
Walker, J.A., and H.A. Rose. 1998. Oöthecal structure and male genitalia of the Geoscapheinae and some Australian *Panesthia* Serville (Blattodea: Blaberidae). *Australian Journal of Entomology.* 37:23–26.
Walker, J.A., D. Rugg, and H.A. Rose. 1994. Nine new species of Geoscapheinae (Blattodea: Blaberidae) from Australia. *Memoirs of the Queensland Museum, Brisbane.* 35:263–284.
Walker, P.A. 1965. The structure of the fat body in normal and starved cockroaches as seen through the electron microscope. *Journal of Insect Physiology.* 11:1625–1631.
Walker, T.J.J. 1957. Ecological studies of the arthropods associated with certain decaying materials in four habitats. *Ecology.* 38:262–276.
Walker, W.F. 1980. Sperm utilization strategies in non-social insects. *The American Naturalist.* 115:780–799.
Waller, D.A., and J.C. Moser. 1990. Invertebrate enemies and nest associates of the leaf-cutting ant *Atta texana* (Buckley) (Formicidae, Attini). *In* Applied Myrmecology: A World Perspective. R.K. Vandermeer, K. Jaffe, and A. Cedeno, editors. Westview Press, Boulder. 256–273.
Wallwork, J.A. 1976. The Distribution and Diversity of Soil Fauna. Academic Press, London. 355 pp.
Waloff, N. 1983. Absence of wing polymorphism in the arboreal, phytophagous species of some taxa of temperate Hemiptera: an hypothesis. *Ecological Entomology.* 8:229–232.
Walter, D.E., and D.J. O'Dowd. 1995. Life on the forest phylloplane: hairs, little houses, and myriad mites. *In* Forest Canopies. M.D. Lowman and N.M. Nadkarni, editors. Academic Press, San Diego. 325–351.
Walthall, W.W., and H.B. Hartman. 1981. Receptors and giant interneurons signalling gravity orientation information in the cockroach *Arenivaga. Journal of Comparative Physiology A.* 142:359–370.
Wang, C.H., H.T. Yang, and Y.S. Chow. 1995. The controlling effects of abamectin and hydramethylnon for the Australian cockroach *Periplaneta australasiae* (F.) (Orthoptera: Blattellidae), in Taiwan. *Journal of Entomological Science.* 30:154–163.
Ward, P.I. 1993. Females influence sperm storage and use in the yellow dung fly *Scatophaga stercoraria* (L.). *Behavioral Ecology and Sociobiology.* 32:313–319.
Wardle, D.A. 2002. Communities and Ecosystems: Linking the Aboveground and Belowground Components. Princeton University Press, Princeton. 392 pp.
Warnecke, U., and C. Hintze-Podufal. 1990. Sexualdimorphism der flugel bei der ovoviviparen schaben-art *Blaptica dubia* (Blattoidea: Blaberoidea: Blaberidae). *Entomologia Generalis.* 20:185–194.
Watanabe, H. 1983. Effects of repeated aerial applications of insecticides for pine-wilt disease on arboreal arthropods in a pine stand. *Journal of the Japanese Forestry Society.* 65:282–287.

Watanabe, H., Y. Kobayashi, M. Sakura, Y. Matsumoto, and M. Mizunami. 2003. Classical olfactory conditioning in the cockroach *Periplaneta americana*. *Zoological Science.* 20:1447–1454.

Watanabe, H., S. Noda, Tokuda, and N. Lo. 1998. A cellulase gene of termite origin. *Nature.* 394:330–331.

Watanabe, H., and S. Ruaysoongnern. 1989. Estimation of arboreal arthropod density in a dry evergreen forest in Northeastern Thailand. *Journal of Tropical Ecology.* 5:151–158.

Watanabe, H., A. Takase, G. Tokuda, A. Yamada, and N. Lo. 2006. Symbiotic "Archaezoa" of the primitive termite *Mastotermes darwiniensis* still play a role in cellulase production. *Eukaryotic Cell.* 5:1571–1576.

Waterhouse, D.F., and J.W. McKellar. 1961. The distribution of chitinase activity in the body of the American cockroach. *Journal of Insect Physiology.* 6:185–195.

Waterhouse, D.F., and B.E. Wallbank. 1967. 2-methylene butanal and related compounds in the defensive scent of *Platyzosteria* cockroaches (Blattidae: Polyzosteriinae). *Journal of Insect Physiology.* 13:1657–1669.

Watson, J.A.L., and F.J. Gay. 1991. Isoptera (termites). *In* The Insects of Australia. Vol. 1. CSIRO, Melbourne University Press, Carlton, Victoria. 330–347.

Watson, J.T., R.E. Ritzmann, S.N. Zill, and A.J. Polack. 2002. Control of obstacle climbing in the cockroach, *Blaberus discoidalis*. I. Kinematics. *Journal of Comparative Physiology A.* 188:39–53.

Weaver, R.J. 1984. Effects of food and water availability and of NCA-1 section upon juvenile hormone biosynthesis and oocyte development in adult female *Periplaneta americana*. *Journal of Insect Physiology.* 30:831–838.

Weaver, R.J., and G.E. Pratt. 1981. Effects of starvation and feeding upon corpus allatum activity and oocyte growth in adult female *Periplaneta americana*. *Journal of Insect Physiology.* 27:75–83.

Webb, D.P. 1976. Regulation of deciduous forest litter decomposition by soil arthropod feces. *In* The Role of Arthropods in Forest Ecosystems. W.J. Mattson, editor. Springer-Verlag, New York. 57–69.

Wedell, N., M.J.G. Gage, and G.A. Parker. 2002. Sperm competition, male prudence and sperm limited females. *Trends in Ecology & Evolution.* 17:313–320.

Weesner, F.M. 1953. Biology of *Tenuirostritermes tenuirostris* (Desneux) with emphasis on caste development. *University of California (Berkeley) Publications in Zoology.* 57:251–302.

Weidner, H. 1969. *Rhabdoblatta stipata* (Walker, 1868), eine im Wasser lebende Schabe (translation by Christof Stumpf). *Entomologische Zeitschrift.* 79:101–106.

Weinstein, P. 1994. Behavioral ecology of tropical cave cockroaches: preliminary field studies with evolutionary implications. *Journal of the Australian Entomological Society.* 33:367–370.

Weinstein, P., and D.P. Slaney. 1995. Invertebrate faunal survey of Rope Ladder Cave, Northern Queensland: a comparative study of sampling methods. *Journal of the Australian Entomological Society.* 34:233–236.

Weis-Fogh, T. 1967. Respiration and ventilation in locusts and other flying insects. *Journal of Experimental Biology.* 47:561–587.

Wendelken, P.W., and R.H. Barth. 1971. The mating behavior of *Parcoblatta fulvescens* (Saussure and Zehntner) (Blattaria, Blaberoidea, Blattellidae, Blattellinae). *Psyche.* 78:319–329.

Wendelken, P.W., and R.H. Barth. 1985. On the significance of pseudofemale behavior in the Neotropical cockroach genera *Blaberus, Archimandrita* and *Byrsotria*. *Psyche.* 92:493–503.

Wendelken, P.W., and R.H. Barth. 1987. The evolution of courtship phenomena in Neotropical cockroaches of the genus *Blaberus* and related genera. *Advances in Ethology.* 27:1–98.

Werren, J.H., D. Windsor, and L. Guo. 1995. Distribution of *Wolbachia* among Neotropical arthropods. *Proceedings of the Royal Society of London B.* 262:97–204.

Wharton, D.R.A., J.E. Lola, and M.L. Wharton. 1967. Population density, survival, growth, and development of the American cockroach. *Journal of Insect Physiology.* 13:699–716.

Wharton, D.R.A., J.E. Lola, and M.L. Wharton. 1968. Growth factors and population density in the American cockroach, *Periplaneta americana*. *Journal of Insect Physiology.* 14:637–653.

Wharton, D.R.A., G.L. Miller, and M.L. Wharton. 1954. The odorous attractant of the American cockroach *Periplaneta americana* (L.). 1. Quantitative aspects of the attractant. *Journal of General Physiology.* 37:461–469.

Wharton, D.R.A., and M.L. Wharton. 1965. The cellulase content of various species of cockroaches. *Journal of Insect Physiology.* 11:1401–1405.

Wharton, D.R.A., M.L. Wharton, and J.E. Lola. 1965. Cellulase in the cockroach, with special reference to *Periplaneta americana* (L.). *Journal of Insect Physiology.* 11:947–959.

Wharton, M.L., and D.R.A. Wharton. 1957. The production of sex attractant substance and of oothecae by the normal and irradiated American cockroach, *Periplaneta americana*. *Journal of Insect Physiology.* 1:229–239.

Wheeler, W.M. 1900. A new myrmecophile from the mushroom gardens of the Texan leaf-cutting ant. *The American Naturalist.* 34:851–862.

Wheeler, W.M. 1904. The phylogeny of termites. *Biological Bulletin.* 8:29–37.

Wheeler, W.M. 1910. Ants: Their Structure, Development and Behavior. Columbia University Press, New York. 663 pp.

Wheeler, W.M. 1911. A desert cockroach. *Journal of the New York Entomological Society.* 19:262–263.

Wheeler, W.M. 1920. The termitodoxa, or biology and society. *The Scientific Monthly.* February:113–124.

Wheeler, W.M. 1928. The Social Insects: Their Origin and Evolution. Harcourt, Brace and Company, New York. 378 pp.

White, T.C.R. 1985. When is an herbivore not an herbivore? *Oecologia.* 67:596–597.

White, T.C.R. 1993. The Inadequate Environment: Nitrogen and the Abundance of Animals. Springer-Verlag, Berlin. 425 pp.

Whitehead, H. 1999. Testing association patterns of social animals. *Animal Behaviour.* 57:F26–F29.
Whitford, W.G. 1986. Decomposition and nutrient cycling in deserts. *In* Pattern and Process in Desert Ecosystems. W.G. Whitford, editor. University of New Mexico Press, Albuquerque. 93–117.
Whiting, M.F., S. Bradler, and T. Maxwell. 2003. Loss and recovery of wings in stick insects. *Nature.* 421:264–267.
Wickler, W. 1968. Mimicry in Plants and Animals. McGraw-Hill, New York. 253 pp.
Wileyto, E.P., G.M. Boush, and L.M. Gawin. 1984. Function of cockroach (Orthoptera: Blattidae) aggregation behavior. *Environmental Entomology.* 13:1557–1560.
Wille, J. 1920. Biologie und Bekämpfung der deutschen Schabe (*Phyllodromia germanica* L.). *Monographien zur angewandten Entomologie.* 5:1–140.
Williams, R.M.C. 1959. Flight and colony foundation in two *Cubitermes* species (Isoptera: Termitidae). *Insectes Sociaux.* 6:205–218.
Williford, A., B. Stay, and D. Bhattacharya. 2004. Evolution of a novel function: nutritive milk in the viviparous cockroach, *Diploptera punctata. Evolution & Development.* 6:67–77.
Willis, E.R. 1966. Biology and behavior of *Panchlora irrorata. Annals of the Entomological Society of America.* 59:514–516.
Willis, E.R. 1969. Bionomics of three cockroaches (*Latiblattella*) from Honduras. *Biotropica.* 1:41–46.
Willis, E.R. 1970. Mating behavior of three cockroaches (*Latiblattella*) from Honduras. *Biotropica.* 2:120–128.
Willis, E.R., and N. Lewis. 1957. The longevity of starved cockroaches. *Journal of Economic Entomology.* 50:438–440.
Willis, E.R., G.R. Riser, and L.M. Roth. 1958. Observations on reproduction and development in cockroaches. *Annals of the Entomological Society of America.* 51:53–69.
Willis, E.R., and L.M. Roth. 1959. Gynandromorphs of *Byrsotria fumigata* (Guérin) (Blattaria: Blaberinae). *Annals of the Entomological Society of America.* 52:420–429.
Wilson, E.O. 1971. The Insect Societies. Harvard University Press, Cambridge, MA. 548 pp.
Wilson, E.O. 2003. The Future of Life. Alfred A. Knopf, New York. 229 pp.
Winchester, N.N., and V. Behan-Pelletier. 2003. Fauna of suspended soils in an *Ongokea gore* tree in Gabon. *In* Arthropods of Tropical Forests: Spatio-Temporal Dynamics and Resource Use in the Canopy. Y. Basset, V. Novotny, S.E. Miller, and R.L. Kitching, editors. Cambridge University Press, Cambridge. 102–109.
Wobus, U. 1966. Der Einfluss der Lichtintensität auf die circadiene laufaktivität der Schabe *Blaberus craniifer* Burm. (Insecta: Blattariae). *Biologisches Zentralblatt.* 85:305–323.
Wolcott, G.N. 1950. The insects of Puerto Rico. *Journal of the Agricultural University of Puerto Rico (1948).* 32:1–224.
Wolda, H., and F.W. Fisk. 1981. Seasonality of tropical insects. II. Blattaria in Panama. *Journal of Animal Ecology.* 50:827–838.
Wolda, H., F.W. Fisk, and M. Estribi. 1983. Faunistics of Panamanian cockroaches (Blattaria). *Uttar Pradesh Journal of Zoology.* 3:1–9.
Wolda, H., and S.J. Wright. 1992. Artificial dry season rain and its effects on tropical insect abundance and seasonality. *Proceedings of the Koninklijke Nederlandse Akademie van Wetenschappen.* 95:535–548.
Wolters, V., and K. Ekschmitt. 1997. Gastropods, Isopods, Diplopods, and Chilopods: Neglected groups of the decomposer food web. *In* Fauna in Soil Ecosystems: Regulating Processes, Nutrient Fluxes, and Agricultural Production. G. Benckiser, editor. Marcel Dekker, Inc., New York. 265–306.
Wood, T.G. 1976. The role of termites (Isoptera) in decomposition processes. *In* The Role of Terrestrial and Aquatic Organisms in Decomposition Processes. J.M. Anderson, editor. Blackwell Scientific Publications, Oxford. 145–168.
Woodhead, A.P. 1984. Effect of duration of larval development on sexual competence in young male *Diploptera punctata. Journal of Insect Physiology.* 9:473–477.
Woodhead, A.P. 1985. Sperm mixing in the cockroach *Diploptera punctata. Evolution.* 39:159–164.
Woodhead, A.P., and C.R. Paulson. 1983. Larval development of *Diploptera punctata* reared alone and in groups. *Journal of Insect Physiology.* 29:665–668.
Woodruff, L.C. 1938. The normal growth rate of *Blattella germanica* L. *Journal of Experimental Zoology.* 79:145–167.
Worland, M.R., B.J. Sinclair, and D.A. Wharton. 1997. Ice nucleator activity in a New Zealand alpine cockroach *Celatoblatta quinquemaculata* (Dictyoptera: Blattidae). *Cryo-Letters.* 18:327–334.
Worland, M.R., D.A. Wharton, and S.G. Byars. 2004. Intracellular freezing and survival in the freeze tolerant alpine cockroach *Celatoblatta quinquemaculata. Journal of Insect Physiology.* 50:225–232.
Wren, H.N., J.L. Johnson, and D.G. Cochran. 1989. Evolutionary inferences from a comparison of cockroach nuclear DNA and DNA from their fat body and egg endosymbionts. *Evolution.* 43:276–281.
Wright, C.G. 1968. Comparative life histories of chlordane-resistant and nonresistant German cockroaches. *Journal of Economic Entomology* 61:1317–1320.
Wyttenbach, R., and T. Eisner. 2001. Use of defensive glands during mating in a cockroach (*Diploptera punctata*). *Chemoecology.* 11:25–28.
Xian, X. 1998. Effects of mating on oviposition, and possibility of parthogenesis of three domestic cockroach species, the American cockroach, *Periplaneta americana;* the Smoky brown cockroach, *Periplaneta fuliginosa;* and the German cockroach, *Blattella germanica. Medical Entomology and Zoology.* 49:27–32.
Yoder, J.A., and N.C. Grojean. 1997. Group influence on water conservation in the giant Madagascar hissing-cockroach, *Gromphadorhina portentosa* (Dictyoptera: Blaberidae). *Physiological Entomology.* 22:79–82.
Yokoi, N. 1990. The sperm removal behavior of the yellow spotted longicorn beetle *Psacothea hilaris* (Coleoptera: Cerambycidae). *Applied Entomology and Zoology.* 25:383–388.
Young, A.M. 1983. Patterns of distribution and abundance in small samples of litter-inhabiting Orthoptera in some Costa-Rican Cacao plantations. *Journal of the New York Entomological Society.* 91:312–327.
Zabinski, J. 1929. The growth of blackbeetles and of cock-

roaches on artificial and incomplete diets. Part 1. *Journal of Experimental Biology.* 6:360–386.

Zabinski, J. 1936. Inconstancy of the number of moults during the post-embryonal development of certain Blattidae. *Annales Musei Zoologici Polonici.* 11:237–240.

Zera, A.J., and R.F. Denno. 1997. Physiology and ecology of dispersal polymorphism in insects. *Annual Review of Entomology.* 42:207–231.

Zervos, S. 1987. Notes on the size distribution of a New Zealand cockroach, *Celatoblatta vulgaris. New Zealand Journal of Zoology.* 14:295–297.

Zhang, J., A.M. Scrivener, M. Slaytor, and H.A. Rose. 1993. Diet and carbohydrase activities in three cockroaches, *Calolampra elegans* Roth and Princis, *Geoscapheus dilatatus* Saussure and *Panesthia cribrata* Saussure. *Comparative Biochemistry and Physiology.* 104A:155–161.

Zhang, R., L.S. Chen, and J.T. Chang. 1990. Induction and isolation of an antibacterial peptide in *Periplaneta americana. Acta Entomologica Sinica.* 33:7–13.

Zhou, X., F.M. Oi, and M.E. Scharf. 2006. Social exploitation of hexamerin: RNAi reveals a major caste-regulatory factor in termites. *Proceedings of the National Academy of Sciences.* 103:4499–4504.

Zhu, D.H., and S. Tanaka. 2004a. Photoperiod and temperature affect the life cycle of a subtropical cockroach, *Opisoplatia* (sic) *orientalis:* seasonal pattern shaped by winter mortality. *Physiological Entomology.* 29:16–25.

Zhu, D.H., and S. Tanaka. 2004b. Summer diapause and nymphal growth in a subtropical cockroach: response to changing photoperiod. *Physiological Entomology.* 29:78–83.

Zimmerman, R.B. 1983. Sibling manipulation and indirect fitness in termites. *Behavioral Ecology and Sociobiology.* 12:143–145.

Zompro, O., and I. Fritzsche. 1999. *Lucihormetica fenestrata* n. gen., n. sp., the first record of luminescence in an orthopteroid insect (Dictyoptera: Blaberidae: Blaberinae: Brachycolini). *Amazonia.* 15:211–219.

Zuk, M., and A.M. Stoehr. 2002. Immune defense and host life history. *The American Naturalist.* 160:S9–S22.

Zunino, M. 1991. Food relocation behavior: a multivalent strategy of Coleoptera. *In* Advances in Coleopterology. M. Zunino, X. Belles, and M. Blas, editors. European Association of Coleopterology, Barcelona. 297–314.

Zurek, L. 1997. The biotic associations of cockroaches—an aspect of cockroach success. *Symbiosis News.* 1:5.

Zurek, L., and B.A. Keddie. 1996. Contribution of the colon and colonic bacterial flora to metabolism and development of the American cockroach *Periplaneta americana* L. *Journal of Insect Physiology.* 42:743–748.

Zurek, L., and B.A. Keddie. 1998. Significance of methanogenic symbionts for development of the American cockroach, *Periplaneta americana. Journal of Insect Physiology.* 44:645–651.

Index

www.ingramcontent.com/pod-product-compliance
Ingram Content Group UK Ltd.
Pitfield, Milton Keynes, MK11 3LW, UK
UKHW010808151224
452485UK00002B/4